582.28/WEI

-162559

WEIDENBORNER, M

Mycotoxins in feeds

Mycotoxins in Feedstuffs

Martin Weidenbörner

Mycotoxins in Feedstuffs

 Springer

Dr. Martin Weidenbörner
Pützhecke 5
Bonn 53229
Germany

Library of Congress Control Number: 2007923600

ISBN-13: 978-0-387-46411-4   e-ISBN-13: 978-0-387-46412-1

Printed on acid-free paper.

9 8 7 6 5 4 3 2 1

springer.com

My family

Acknowledgements

I wish to thank Dr. Ursula Monnerjahn-Karbach for valuable advice. I would also like to thank Renate Frohn, Ursula Kleinheyer-Thomas, Verena Bormann and Mr. Krämer from Bran Library for Medicine, Science and Agricultural Science, Bonn University, for providing the literature to prepare this book. I appreciate the great patience of my family, especially of my wife, which was essential for writing this book. My father is thanked for the very careful review of the total manuscript.

Abbreviations

cg	conventionally grown
e	estimated
ec	exact composition
eg	ecologically grown
Fdk	*Fusarium*-damaged kernels
hlk	healthy looking kernels
is	indifferent samples
na	not analyzed
nc	no comment
ncac	no comment about consumption
ncaps	no comment about positive samples
nd	not detected
og	organically grown
tr	traces
s	selected
sa	samples
wp	whole plant

Introduction

Feedstuff is a common standard for each kind of food for animals, which are in the charge of man and serve as food. Feed for livestock is of special interest. The quality of feed is responsible for the health of animals and indirectly for the quality of human nutrition. Agriculturally used plants, such as numerous grains, oil seeds and nuts, root crops, and to a smaller extent, many forage crops are susceptible to mycotoxin contamination. Fungal and in the end mycotoxin contaminated feed may be involved in modern livestock production practice (confined rearing on high-density diets) because plant feedstuff especially from multiple sources may be used for feeding. The mixing of mycotoxin contaminated pecan, walnut, or other nut meats into feedstuff is one example. The nuts are pressed to recover the oil while most of the toxin is concentrated in the residual meats. The press cake usually is diverted into animal feed channels. The amount of these (protein supplement) ingredients, while small, could cause problems in the health of animal and human.

Available data suggest that the mold and mycotoxin problem is largely one of the worldwide feed management. Especially individual farm silos and feed troughs are major sites of toxin production in mold-contaminated feeds. Guidelines for the investigation and amelioration of feedstuff quality in different countries have been prepared.

Mycotoxin contamination of feeds occurs as a result of crop invasion by field fungi. Here drought of plants, close planting, competition from weeds, reduced fertilization and other factors are causing stress to plants. Another reason is the growth of storage fungi in improperly stored crops and/or after processing the plants into food products or animal feed. A lot of (microscopic) fungi like *Fusarium* spp. and *Alternaria* spp. (field fungi) as well as *Aspergillus* spp. and *Penicillium* spp. (storage fungi) start growth if suitable conditions prevail. It is well documented that mycotoxin contamination is more regularly associated with low-grade cereals in undeveloped countries. Normally, in developed countries these cereals do not enter the human food chain. Storage fungi also grow during mixing and delivery of grain and animal feed. In the worst case these fungi produce secondary metabolites, the so-called "mycotoxins". They can be very harmful for animals (and man). In 1960 the Turkey "X" disease was caused by a toxin (aflatoxin) produced by molds which belong to the *Aspergillus flavus* group. It resulted in the death of 100,000 turkeys in UK.

The presence of mycotoxins in feedstuffs (maize and cottonseed are particularly affected) could cause serious mycotoxicosis syndromes in animals. At high levels they cause even death shortly after exposure. At lower levels, they can cause disorders in various organs. Decreased feed efficiencies, low weight gain, impaired immunity and other symptoms of acute mycotoxicosis may arise high costs to livestock producers. Especially in equivocal cases of illness of the animal and/or unspecific syndromes of decline in performance feedstuff should be investigated for mycotoxins. Furthermore, these fungal metabolites can end up in meat, milk, or eggs, which are important components of the human food chain. Nowadays consumers pay more attention to their health and therefore a higher quality of food is necessary.

The present book gives an overview of mycotoxins in feedstuffs and ingredients. It lists the degree of contamination, concentration and country of origin/detection for each case of mycotoxin contamination. Besides these information, the book shows whether a feedstuff/ingredient is predisposed for a mycotoxin contamination (number of mycotoxins as well as number of citations concerning one feedstuff).

"Mycotoxins in Feedstuffs" will therefore contribute to more information and transparency in the human food chain. It is especially suitable for responsibles in the feed industry (e.g. supervisors of feed, feed producers, feeding traders) just as ministries, offices and departments of farming and environmental control on the national and international level, offices, associations, agricultural chambers, meat industry, veterinaries, farmers (especially livestock breeders), mycologists, mycotoxicologists, biologists, chemists, supervisors in food and feed quality control, lawyers and experts in law of food and feed, students of respective fields and further groups of interests.

The feeds and/or ingredients are listed in alphabetical order. Terms with a comma are following, than terms with definitions in brackets. This was done as far as possible. In some cases, the country of detection is not necessarily the country of origin, but information was lacking concerning the country of origin of such imports in the original literature. If only "imported" occurs after the country of investigation no more data were available in the original literature. Sometimes the original literature neither contained the producing country nor the addition "imported". In these cases also no indications were given.

Authors are invited to send **accepted** manuscripts or already **published** articles (in English) in **recommended** journals to the author of the book. After critical review relevant results may be published in a later edition.

Alfalfa may contain the following mycotoxins:

Aflatoxin B$_1$
incidence: 6/55, conc. range: 13–65 µg/kg, Ø conc.: 34.7 µg/kg, country: India[247]
see also *Ambadi* cake, animal feedstuffs (dairy cake), bagasse, barley, bengalgram husk, bird food, bird food, wild, biri testa, blackgram, blackgram husk, bran, broiler mixed feed, calf fattening mixed feed, calf fattening mixed feed (containing 4–20% peanut products), *Carthamus* cake, castor cake, cereals, cereal products, chick pea, coconut cake, cocos, concentrate, mixed, concentrates, cotton cake, cottonseed, cottonseed (dehulled), cottonseed cake, cottonseed extract, cottonseed meal, cottonseed meal (ammoniated), cottonseed meal (decorticated), cottonseed meats, cottonseed products, crumbles, crumbles, grower, cycad meal, dairy cattle feed, dairy cattle feed (containing 2–5% peanut products), dairy cattle feed (containing 6–10% peanut products), dairy cattle feed (containing 6–12% peanut products), dairy cattle feed (containing more than 20% peanut products), diets, mixed, dog food, egg production mixed feed, feed, feed and ingredients, feed, compound, feed, layer, feed, mixed, feed (beef), feed (broilers), feed (calf), feed (cat), feed (cattle), feed (chicken), feed (dairy), feed (dog), feed (dug), feed (fish), feed (gluten), feed (horse), feed (miscellaneous), feed (pig), feed (poultry), feed (poultry, pig), feed (rabbit), feed (sheep), feed (50–60% maize), fish meal, grain by-products, grains (no specification), greengram, hay/silage, horsegram, husk, *Jagni* cake, legume mixture, linseed, linseed cake, livol, *Mahua* cake, maize, maize germ, maize gluten, maize grits, maize husk, maize meal, maize oil cake, maize powder, maize screenings, maize, ground, maize, hybrid, maize, preharvest, maize, yellow, maize (dark grains), *Makhana* (*Euryale ferox* Salisb) puffs, manioc, milk production mixed feed, mung testa, murkool, mustard cake, neem cake, niger cake, oats, palm kernel expeller cake, palm kernels, palm products, peanut cake, peanut cake (deoiled), peanut expeller, peanut hay, peanut meal, peanuts, peanut, kernels, peanut, shells, pellets, finisher, pig meal and pellets, pigeon pea, poultry feeds (peanut containing), rapeseed cake, redgram husk, rice, rice bran, rice bran (deoiled), rice chaff, rice crack, rice germ, rice germ cake, rice meal, rice straw, rice (damaged), rice (polish), safflower cake, sal seed cake, sesame, sesame cake, sorghum, soybean meal, soybeans, sunflower, sunflower cake, sunflower flour, tapioca, wheat, wheat bran, wheat bran and chana testa

Diacetoxyscirpenol
incidence: 1/7345, conc.: nc, country: Hungary[209]
see also barley, feed, feed components, feed (dog), feed (fish), feed (mink), feed (pig), feed (poultry), feed (reindeer), feeds, grain, feedstuff, forage grass, grain, mixed feed, grains (no specification), maize, maize gluten, oat and barley (hammer-milled), oats, peanuts, soybeans, wheat

Ochratoxin A
incidence: 4/7345, conc. range: nc, country: Hungary[209]
see also barley, barley, oats, barley (high moisture), barley-soybean diet, bird food, domestic, bird food, wild, broilers feed, cereal grains, citrus pulp, coconut, expeller, corn cob mix silage, diet (dairy cow), diet (poultry), diet (starter), dog food, eat, egg production mixed feed, feed, feed wheat, oat and barley, feed, commercial mix, feed, mixed, feed, mixed (pelleted), feed (broilers), feed (cat), feed (cattle), feed (cereals), feed (pig), feed ec (pig), feed (poultry), feed ec (poultry), feed (poultry, pig), feed ec (rabbit), feed (trout), grain, mixed feed, grains, mixed, grains (heated), hay, horse bean, maize, maize feed, milo, maize gluten, maize meal, maize, white, *Makhana* (*Euryale ferox* Salisb) puffs, milk production mixed feed, millet, oat and barley (hammer-milled), oats, palm products, peanut cake, peas, peas and beans, pet food, pig feedstuffs, pig grower diet, pig meal, piglet diet, poultry feedstuffs, rice bran, rice

germ, rice germ cake, rye, sorghum, soybean groats, sunflower, sunflower seeds, extracted, tapioca, triticale, *Vicia faba*, wheat, wheat and barley, wheat bran, wheat hay, wheat, oats

T-2 TOXIN
incidence: 2/7345, conc. range: nc, country: Hungary[209]
see also barley, bran, diet (grower), diet (poultry), feed, feed components, feed, layer, feed, mixed, feed (dog), feed (fish), feed (mink), feed (pig), feed (poultry), feedstuff, forage grass, grain, mixed feed, grains (no specification), hay, maize, maize germ/bran, maize gluten, maize meal, maize screenings, maize stalks (pith), oat and barley (hammer-milled), oats, peanuts, pig grower diet, piglet diet, rye, sorghum, triticale, wheat

ZEARALENONE
incidence: 1/7345, conc.: nc, country: Hungary[209]
see also barley, barley and feed, barley, husked, barley, unhusked (naked), bone meal, bran, broilers feed, *Carthamus* cake, chick pea, concentrate, mixed, corn cob mix silage, cotton cake, cottonseed, cottonseed cake, diet (dairy cow), diet (poultry), diets (mixed), feed, feed components, feed, mixed, feed, mixed (primarily maize also maize, oats, wheat), feed (bran), feed (broiler chicken), feed (cattle), feed (chicken), feed (dairy), feed (developing pig), feed (mill run, from wheat), feed (miscellaneous), feed (pig), feed (poultry), feed (poultry, pig), feed (starter chicken), feedstuff, fish meal, forage grass, grain, bruised, grain, mixed feed, grains (no specification), hay, maize, maize ears, maize flakes, maize germ, maize germ/bran, maize gluten, maize kernels, maize meal, maize oil cake, maize screenings, maize stalks (pith), maize, "Baby", maize, hybrid, maize, shelled, maize, unshelled, maize, white, maize grain, artificially dried, maize grain, crib dried, maize grain, ensiled, milk production mixed feed, oats, *Paspalum palidosum*, straw, peanut hulls/skins, rice bran, rice germ, rice germ cake, rye, silage, sorghum, soybeans, soybeans, extracted, sunflower cake, tapioca, triticale,

wheat, wheat bran, wheat bran and chana testa, wheat soya meal

***Ambadi* cake**   may contain the following mycotoxins:

AFLATOXIN B$_1$
incidence: 1/2, conc. range: 59.4 µg/kg, country: India[253]
see also alfalfa, animal feedstuffs (dairy cake), bagasse, barley, bengalgram husk, bird food, bird food, wild, biri testa, blackgram, blackgram husk, bran, broiler mixed feed, calf fattening mixed feed, calf fattening mixed feed (containing 4–20 % peanut products), *Carthamus* cake, castor cake, cereals, cereal products, chick pea, coconut cake, cocos, concentrate, mixed, concentrates, cotton cake, cottonseed, cottonseed (dehulled), cottonseed cake, cottonseed extract, cotton-seed meal, cottonseed meal (ammoniated), cottonseed meal (decorticated), cottonseed meats, cottonseed products, crumbles, crumbles, grower, cycad meal, dairy cattle feed, dairy cattle feed (containing 2–5 % peanut products), dairy cattle feed (containing 6–10 % peanut products), dairy cattle feed (containing 6–12 % peanut products), dairy cattle feed (containing more than 20 % peanut products), diets, mixed, dog food, egg production mixed feed, feed, feed and ingredients, feed, compound, feed, layer, feed, mixed, feed (beef), feed (broilers), feed (calf), feed (cat), feed (cattle), feed (chicken), feed (dairy), feed (dog), feed (dug), feed (fish), feed (gluten), feed (horse), feed (miscellaneous), feed (pig), feed (poultry), feed (poultry, pig), feed (rabbit), feed (sheep), feed (50–60 % maize), fish meal, grain by-products, grains (no specification), greengram, hay/silage, horsegram, husk, *Jagni* cake, legume mixture, linseed, linseed cake, livol, *Mahua* cake, maize, maize germ, maize gluten, maize grits, maize husk, maize meal, maize oil cake, maize powder, maize screenings, maize, ground, maize, hybrid, maize, preharvest, maize, yellow, maize (dark grains), *Makhana* (*Euryale ferox* Salisb) puffs, manioc, milk

production mixed feed, mung testa, murkool, mustard cake, neem cake, niger cake, oats, palm kernel expeller cake, palm kernels, palm products, peanut cake, peanut cake (deoiled), peanut expeller, peanut hay, peanut meal, peanuts, peanut, kernels, peanut, shells, pellets, finisher, pig meal and pellets, pigeon pea, poultry feeds (peanut containing), rapeseed cake, redgram husk, rice, rice bran, rice bran (deoiled), rice chaff, rice crack, rice germ, rice germ cake, rice meal, rice straw, rice (damaged), rice (polish), safflower cake, sal seed cake, sesame, sesame cake, sorghum, soybean meal, soybeans, sunflower, sunflower cake, sunflower flour, tapioca, wheat, wheat bran, wheat bran and chana testa

Animal feed (maize)   may contain the following mycotoxins:

Aflatoxins
incidence: 33/70, conc. range: $\leq$3872 μg/kg, Ø conc.: 369 μg/kg, country: USA[280]
incidence: 2/74, conc. range: $\leq$133 μg/kg, Ø conc.: 69 μg/kg, country: USA[280]
incidence: 29/39, conc. range: $\leq$2839 μg/kg, Ø conc.: 406 μg/kg, country: USA[280]
incidence: 1/3, conc.: 2 μg/kg, country: USA[280]
see also animal feed (mixed), barley, copra meal, cottonseed, cottonseed fines, cottonseed meal, cottonseed meats, feed, feed ingredients (miscellaneous), feed, mixed, feed (excluding peanuts, suspect), feed (goat), feed (pig), feed (poultry), feedstuff, grain, grain, mixed feed, maize, maize, shelled, millet, peanut cake, peanut meal, peanut meal and by-products, peanuts, protein concentrates, rice, rice bran, sorghum, soybean meal, sunflower

Animal feed (mixed)   may contain the following mycotoxins:

Aflatoxins
incidence: 2/40, conc. range: $\leq$105 μg/kg, Ø conc.: 94 μg/kg, country: USA[280]
incidence: 5/6*, conc. range: $\leq$67 μg/kg, Ø conc.: 36 μg/kg, country: USA[280], *imported

see also animal feed (maize), barley, copra meal, cottonseed, cottonseed fines, cottonseed meal, cottonseed meats, feed, feed ingredients (miscellaneous), feed, mixed, feed (excluding peanuts, suspect), feed (goat), feed (pig), feed (poultry), feedstuff, grain, grain, mixed feed, maize, maize, shelled, millet, peanut cake, peanut meal, peanut meal and by-products, peanuts, protein concentrates, rice, rice bran, sorghum, soybean meal, sunflower

Animal feedstuffs (dairy cake)   may contain the following mycotoxins:

Aflatoxin B$_1$
incidence: 1/1, conc.: 116 μg/kg, country: Saudi Arabia[288]
see also alfalfa, *Ambadi* cake, bagasse, barley, bengalgram husk, bird food, bird food, wild, biri testa, blackgram, blackgram husk, bran, broiler mixed feed, calf fattening mixed feed, calf fattening mixed feed (containing 4–20 % peanut products), *Carthamus* cake, castor cake, cereals, cereal products, chick pea, coconut cake, cocos, concentrate, mixed, concentrates, cotton cake, cottonseed, cottonseed (dehulled), cottonseed cake, cottonseed extract, cottonseed meal, cottonseed meal (ammoniated), cottonseed meal (decorticated), cottonseed meats, cottonseed products, crumbles, crumbles, grower, cycad meal, dairy cattle feed, dairy cattle feed (containing 2–5 % peanut products), dairy cattle feed (containing 6–10 % peanut products), dairy cattle feed (containing 6–12 % peanut products), dairy cattle feed (containing more than 20 % peanut products), diets, mixed, dog food, egg production mixed feed, feed and ingredients, feed, compound, feed, layer, feed, mixed, feed (beef), feed (broilers), feed (calf), feed (cat), feed (cattle), feed (chicken), feed (dairy), feed (dog), feed (dug), feed (fish), feed (gluten), feed (horse), feed (miscellaneous), feed (pig), feed (poultry), feed (poultry, pig), feed (rabbit), feed (sheep), feed (50–60 % maize), fish meal, greengram, grain by-products, grains (no specification), greengram, hay/silage, horsegram, husk, *Jagni*

cake, legume mixture, linseed, linseed cake, livol, *Mahua* cake, maize, maize germ, maize gluten, maize grits, maize husk, maize meal, maize oil cake, maize powder, maize screenings, maize, ground, maize, hybrid, maize, preharvest, maize, yellow, maize (dark grains), *Makhana* (*Euryale ferox* Salisb) puffs, manioc, milk production mixed feed, mung testa, murkool, mustard cake, neem cake, niger cake, oats, palm kernel expeller cake, palm kernels, palm products, peanut cake, peanut cake (deoiled), peanut expeller, peanut hay, peanut meal, peanuts, peanut, kernels, peanut, shells, pellets, finisher, pig meal and pellets, pigeon pea, poultry feeds (peanut containing), rapeseed cake, redgram husk, rice, rice bran, rice bran (deoiled), rice chaff, rice crack, rice germ, rice germ cake, rice meal, rice straw, rice (damaged), rice (polish), safflower cake, sal seed cake, sesame, sesame cake, sorghum, soybean meal, soybeans, sunflower, sunflower cake, sunflower flour, tapioca, wheat, wheat bran, wheat bran and chana testa

AFLATOXIN B_2
incidence: 1/1, conc.: 10 μg/kg, country: Saudi Arabia[288]
see also bird food, bird food, wild, biri testa, blackgram, blackgram husk, cottonseed, cottonseed cake, cottonseed extract, cotton-seed meal, cottonseed meal (ammoniated), cottonseed meats, dog food, egg production mixed feed, feed, compound, feed (cat), feed (cattle), feed (dog), feed (pig), feed (poultry), feed (rabbit), feed (sheep), fish meal, horsegram, maize, maize gluten, maize husk, maize, ground, maize, preharvest, mung testa, mustard cake, niger cake, peanut cake, peanut cake (deoiled), peanut expeller, peanut hay, peanut meal, peanuts, peanut, kernels, redgram husk, rice bran, rice bran (deoiled), rice chaff, rice meal, rice (polish), sal seed cake, sesame cake, sorghum, soybean meal, soybeans, wheat, wheat bran

AFLATOXIN G_1
incidence: 1/1, conc.: 10 μg/kg, country: Saudi Arabia[288]

see also bird food, bird food, wild, concentrate, mixed, cottonseed, cottonseed cake, feed (cat), feed (chicken), feed (dog), maize, maize, ground, maize, preharvest, meat meal, milk production mixed feed, murkool, peanut cake, peanut expeller, peanut hay, peanut meal, peanuts, rice bran, rice germ, sorghum, soybeans, wheat, wheat bran, wheat bran and chana testa

AFLATOXIN G_2
incidence: 1/1, conc.: 1 μg/kg, country: Saudi Arabia[288]
see also bird food, feed (cat), feed (dog), maize, maize, preharvest, peanut expeller, peanut hay, peanut meal, peanuts, soybeans

Bagasse may contain the following mycotoxins:

AFLATOXIN B_1
incidence: 1/2, conc.: 9.3 μg/kg, country: India[247]
see also alfalfa, *Ambadi* cake, animal feed-stuffs (dairy cake), barley, bengalgram husk, bird food, bird food, wild, biri testa, blackgram, blackgram husk, bran, broiler mixed feed, calf fattening mixed feed, calf fattening mixed feed (containing 4–20 % peanut products), *Carthamus* cake, castor cake, cereals, cereal products, chick pea, coconut cake, cocos, concentrate, mixed, concentrates, cotton cake, cottonseed, cottonseed (dehulled), cottonseed cake, cottonseed extract, cottonseed meal, cottonseed meal (ammoniated), cottonseed meal (decorticated), cottonseed meats, cottonseed products, crumbles, crumbles, grower, cycad meal, dairy cattle feed, dairy cattle feed (containing 2–5 % peanut products), dairy cattle feed (containing 6–10 % peanut products), dairy cattle feed (containing 6–12 % peanut products), dairy cattle feed (containing more than 20 % peanut products), diets, mixed, dog food, egg production mixed feed, feed, feed and ingredients, feed, compound, feed, layer, feed, mixed, feed (beef), feed (broilers), feed (calf), feed (cat), feed (cattle), feed (chicken),

feed (dairy), feed (dog), feed (dug), feed (fish), feed (gluten), feed (horse), feed (miscellaneous), feed (pig), feed (poultry), feed (poultry, pig), feed (rabbit), feed (sheep), feed (50–60 % maize), fish meal, grain by-products, grains (no specification), greengram, hay/silage, horsegram, husk, *Jagni* cake, legume mixture, linseed, linseed cake, livol, *Mahua* cake, maize, maize germ, maize gluten, maize grits, maize husk, maize meal, maize oil cake, maize powder, maize screenings, maize, ground, maize, hybrid, maize, preharvest, maize, yellow, maize (dark grains), *Makhana* (*Euryale ferox* Salisb) puffs, manioc, milk production mixed feed, mung testa, murkool, mustard cake, neem cake, niger cake, oats, palm kernel expeller cake, palm kernels, palm products, peanut cake, peanut cake (deoiled), peanut expeller, peanut hay, peanut meal, peanut, kernels, peanut, shells, peanuts, pellets, finisher, pig meal and pellets, pigeon pea, poultry feeds (peanut containing), rapeseed cake, redgram husk, rice, rice bran, rice bran (deoiled), rice chaff, rice crack, rice germ, rice germ cake, rice meal, rice straw, rice (damaged), rice (polish), safflower cake, sal seed cake, sesame, sesame cake, sorghum, soybean meal, soybeans, sunflower, sunflower cake, sunflower flour, tapioca, wheat, wheat bran, wheat bran and chana testa

Barley Barley for feed may contain the following mycotoxins:

3-ACETYLDEOXYNIVALENOL
incidence: 4/15*, conc. range: 20–96 µg/kg, Ø conc.: 58 µg/kg, country: Finland[9], *imported?
incidence: 54/240, conc. range: 2–224 µg/kg, Ø conc.: 35.9 µg/kg, country: Germany[123]
incidence: 21/44, conc. range: 3–17 µg/kg, Ø conc.: 5 µg/kg, country: Germany[130]
incidence: 5/7*, conc. range: 1300–5300 µg/kg, Ø conc.: 2780 µg/kg, country: Canada[202], *ncac
incidence: 3/23, conc. range: 230–380 µg/kg, country: Canada[410]

see also feed components, feed (poultry), feeds, grain, feeds, industrial, feedstuffs (rapeseed, turnip, fish meal, concentrates), maize, maize ears, maize gluten, maize kernels, maize meal, maize screenings, oats, wheat

15-ACETYLDEOXYNIVALENOL
incidence: 14/153, conc. range: 2–18 µg/kg, Ø conc.: 6.2 µg/kg, country: Germany[123]
incidence: 5/7*, conc.range: 400–2400 µg/kg, Ø conc.: 1280 µg/kg, country: Canada[202], *ncac
incidence: 1/94, conc.: 5 µg/kg, country: Canada[240]
incidence: 2/23, conc. range: 110–160 µg/kg, country: Canada[410]
see also feed components, feed, commercial mix, feed (poultry), maize, maize ears, maize germ, maize germ/bran, maize gluten, maize kernels, maize meal, maize screenings, maize, "Baby", oats, silage, wheat

3,15-ACETYLDEOXYNIVALENOL
incidence: 6/7*, conc. range: tr–400 µg/kg, country: Canada[202], *ncac

15-*O*-ACETYL-4-DEOXYNIVALENOL
incidence: 6/1494*, conc. range: 90–190 µg/kg, Ø conc.: 120 µg/kg, country: Canada[323]; *ncac

AFLATOXIN B$_1$
incidence: 3/3, conc. range: 10–200 µg/kg, Ø conc.: 103.3 µg/kg, country: Australia[121]
incidence: 2/2, conc. range: 100–200 µg/kg, Ø conc.: 150 µg/kg, country: Australia[181]
see also alfalfa, *Ambadi* cake, animal feedstuffs (dairy cake), bagasse, bengalgram husk, bird food, bird food, wild, biri testa, blackgram, blackgram husk, bran, broiler mixed feed, calf fattening mixed feed, calf fattening mixed feed (containing 4–20 % peanut products), *Carthamus* cake, castor cake, cereals, cereal products, chick pea, coconut cake, cocos, concentrate, mixed, concentrates, cotton cake, cottonseed, cottonseed (dehulled), cottonseed cake, cottonseed extract, cottonseed meal, cottonseed meal (ammoniated), cottonseed

meal (decorticated), cottonseed meats, cottonseed products, crumbles, crumbles, grower, cycad meal, dairy cattle feed, dairy cattle feed (containing 2–5 % peanut products), dairy cattle feed (containing 6–10 % peanut products), dairy cattle feed (containing 6–12 % peanut products), dairy cattle feed (containing more than 20 % peanut products), diets, mixed, dog food, egg production mixed feed, feed, feed and ingredients, feed, compound, feed, layer, feed, mixed, feed (beef), feed (broilers), feed (calf), feed (cat), feed (cattle), feed (chicken), feed (dairy), feed (dog), feed (dug), feed (fish), feed (gluten), feed (horse), feed (miscellaneous), feed (pig), feed (poultry), feed (poultry, pig), feed (rabbit), feed (sheep), feed (50–60 % maize), fish meal, grain by-products, grains (no specification), greengram, hay/silage, horsegram, husk, *Jagni* cake, legume mixture, linseed, linseed cake, livol, *Mahua* cake, maize, maize germ, maize gluten, maize grits, maize husk, maize meal, maize oil cake, maize powder, maize screenings, maize, ground, maize, hybrid, maize, preharvest, maize, yellow, maize (dark grains), *Makhana* (*Euryale ferox* Salisb) puffs, manioc, milk production mixed feed, mung testa, murkool, mustard cake, neem cake, niger cake, oats, palm kernel expeller cake, palm kernels, palm products, peanut cake, peanut cake (deoiled), peanut expeller, peanut hay, peanut meal, peanut, kernels, peanut, shells, peanuts, pellets, finisher, pig meal and pellets, pigeon pea, poultry feeds (peanut containing), rapeseed cake, redgram husk, rice, rice bran, rice bran (deoiled), rice chaff, rice crack, rice germ, rice germ cake, rice meal, rice straw, rice (damaged), rice (polish), safflower cake, sal seed cake, sesame, sesame cake, sorghum, soybean meal, soybeans, sunflower, sunflower cake, sunflower flour, tapioca, wheat, wheat bran, wheat bran and chana testa

AFLATOXIN B
incidence: 13/14, conc. range:
≥ 25–>100 μg/kg, country: France[46]

see also cocoa (oil cake), cottonseed cake, feed, feed (cattle), feed (maize, gluten), feed (pig), feed (poultry), flour (wheat), lucern (dried), maize, maize gluten, maize grains, oats, peanut (oil cake), rice, broken, sorghum, soybean (oil cake), sunflower (oil cake), wheat

AFLATOXINS
incidence: 1/23, conc. range: 51–500 μg/kg, country: Australia[21]
see also animal feed (maize), animal feed (mixed), copra meal, cottonseed, cottonseed fines, cottonseed meal, cottonseed meats, feed, feed ingredients (miscellaneous), feed, mixed, feed (excluding peanuts, suspect), feed (goat), feed (pig), feed (poultry), feedstuff, grain, grain, mixed feed, maize, maize, shelled, millet, peanut cake, peanut meal, peanut meal and by-products, peanuts, protein concentrates, rice, rice bran, sorghum, soybean meal, sunflower

ALTERNARIOL METHYL ETHER
incidence: 1/10, conc.: 16 μg/kg, country: Germany[294]
see also barley, oats, cereals, mixture, oats, sunflower, wheat, wheat, oats

BEAUVERICIN
incidence: 14/14*, conc. range: tr–19 μg/kg, country: Finland[205], *ncac
incidence: 8/8*, conc. range: tr–13 μg/kg, country: Finland[205], *ncac
see also maize, oats, wheat, wheat, summer, wheat, winter

CITRININ
incidence: 2/14, conc. range: 160–1000 μg/kg, Ø conc.: 580 μg/kg, country: Denmark[24]
incidence: 1/1, conc.: 800 μg/kg, country: Denmark[82]
incidence: 7/45*, conc. range: 1–8 μg/kg, country: UK[94], *imported?
incidence: 1/94, conc.: 10 μg/kg, country: Canada[240]
incidence: 4/269*, conc. range: 30–480 μg/kg, Ø conc.: 182.5 μg/kg, country: Sweden[261], *ncac

incidence: 1/1, conc.: 71 µg/kg, country:
UK?[288]
see also barley, oats, barley-soybean diet,
cottonseed cake, feed, feed, mixed, feed
(cattle), feed (pig), fish meal, hay, maize,
maize, white, *Makhana* (*Euryale ferox* Salisb)
puffs, oats, palm products, peas and beans,
rice bran, rice germ, wheat, wheat and other
grains (moldy), wheat bran

DEOXYNIVALENOL
incidence: 161/222, conc. range:
50–1200 µg/kg, Ø conc.: 268 µg/kg, country:
Hungary[6]
incidence: 15/15*, conc. range: 8–465 µg/kg,
Ø conc.: 77 µg/kg, country: Finland[9],
*imported?
incidence: 6/86, conc. range: 10–100 µg/kg,
country: Germany[68]
incidence: 6/7, conc. range: 5–50 µg/kg,
country: Finland[112]
incidence: 194/240, conc. range:
2–4764 µg/kg, Ø conc.: 134.6 µg/kg, country:
Germany[123]
incidence: 43/44, conc. range: 4–4764 µg/kg,
Ø conc.: 399 µg/kg, country: Germany[130]
incidence: 13/25*, conc. range: 8–223 µg/kg,
Ø conc.: 45.5 µg/kg, country: Norway[132];
*probably feed
incidence: 21/25*, conc. range: 6–2139 µg/kg,
Ø conc.: 243.4 µg/kg, country: Norway[132],
*probably feed
incidence: 35/35*, conc. range:
≤10–10,000 µg/kg, country: Germany[192],
*ncac
incidence: 2/14, conc. range: 80–160 µg/kg,
Ø conc.: 120 µg/kg, country: Sweden[197]
incidence: 1/6, conc.: 50 µg/kg, country:
Sweden[197]
incidence: 4/32, conc. range: 60–150 µg/kg,
Ø conc.: 90 µg/kg, country: Sweden[197]
incidence: 7/7*, conc. range:
260–24,000 µg/kg, Ø conc.: 13,794 µg/kg,
country: Canada[202], *ncac
incidence: 48/68, conc. range: 20–99 µg/kg
(47 sa), 126 µg/kg (1 sa), country: UK[203]
incidence: 1/6*, conc.: 390 µg/kg, country:
Poland[214], *ncac

incidence: 799/1095*, conc. range:
30–13,330 µg/kg, country: Norway[224],*ncac
incidence: 2/2*, conc. range: 310–350 µg/kg,
Ø conc.: 330 µg/kg, country: Japan[230],*ncac
incidence: 19/25*, conc. range:
tr–40,400 µg/kg, Ø conc.: 4400 µg/kg,
country: Japan[231], *ncac
incidence: 2/4*, conc. range: tr–22 µg/kg,
country: France[232],*ncac
incidence: 3/5*, conc. range: tr–30 µg/kg,
country: France[232],*ncac
incidence: 3/5*, conc. range: tr µg/kg,
country: Denmark[232],*ncac
incidence: 3/6*, conc. range: tr µg/kg,
country: unknown[232],*ncac
incidence: 4/4*, conc. range: 140–4500 µg/kg,
Ø conc.: 1520 µg/kg, country: Norway[235],
*ncac
incidence: 7/7*, conc. range: <30–93 µg/kg,
country: Norway[235],*ncac
incidence: 4/94, conc. range: 13–70 µg/kg,
Ø conc.: 30.6 µg/kg, country: Canada[240]
incidence: 71/94*, conc. range: 5–3780 µg/kg,
Ø conc.: 270 µg/kg, country: Japan[265] *ncac
incidence: 17/1494*, conc. range:
50–2250 µg/kg, Ø conc.: 740.6 µg/kg,
country: Canada[323]; *ncac
incidence: 50/116?, Ø conc.: 1370 µg/kg,
country: Canada[346]
incidence: 12/29*, conc. range: ≤163 µg/kg,
Ø conc.: 49 µg/kg, country: Lithuania[399],
*ncac
incidence: 84/116, conc. range: ≤9110 µg/kg,
Ø conc.: 1370 µg/kg, country: Canada[410]
incidence: 28/40*, conc. range: 5–857 µg/kg,
Ø conc.: 79.3 µg/kg, country: Sweden[418],
*ncac
see also barley, husked, barley, unhusked
(naked), barley (pressed), bone meal, bran,
broilers feed, calf fattening mixed feed,
coconut, expeller, corn cob mix silage,
cottonseed, cottonseed cake, dairy cattle feed,
egg production mixed feed, feed, feed
components, feed, commercial mix, feed,
mixed, feed, mixed (primarily maize), feed
(barley), feed (cattle), feed (chicken), feed
(dog), feed (fish), feed (mill run, from
wheat), feed (mink), feed (pig), feed

(poultry), feed (reindeer), feeds, grain, feeds, industrial, feedstuff, feedstuffs (rapeseed, turnip, fish meal, concentrates), fish meal, grain, mixed feed, grains, mixed, grains (no specification), maize, maize ears, maize fibre, maize germ, maize germ/bran, maize germ meal, maize gluten, maize kernels, maize meal, maize powder, maize screenings, maize stalks (pith), maize, "Baby", maize, hybrid, maize, white, oats, rice bran, rice germ cake, rye, silage, sorghum, soybeans, triticale, wheat, wheat and barley, wheat, red hard winter, wheat, soft white winter, wheat, spring, wheat, winter

DIACETOXYSCIRPENOL
incidence: 5/45, conc. range:
300–17,000 µg/kg, country: Germany[68]
incidence: 11/89, conc. range: 200–700 µg/kg, country: Germany[68]
see also alfalfa, feed, feed components, feed (dog), feed (fish), feed (mink), feed (pig), feed (poultry), feed (reindeer), feeds, grain, feedstuff, forage grass, grain, mixed feed, grains (no specification), maize, maize gluten, oat and barley (hammer-milled), oats, peanuts, soybeans, wheat

ENNIATIN A
incidence: 14/14*, conc. range: 2–950 µg/kg, Ø conc.: 87.93 µg/kg, country: Finland[205], *ncac
incidence: 7/8*, conc. range: tr–59 µg/kg, country: Finland[205], *ncac
see also wheat, summer

ENNIATIN A₁
incidence: 14/14*, conc. range:
18–2000 µg/kg, Ø conc.: 310.3 µg/kg, country: Finland[205], *ncac
incidence: 8/8*, conc. range: tr–570 µg/kg, country: Finland[205], *ncac
see also oats, rye, wheat, wheat, summer, wheat, winter

ENNIATIN B
incidence: 14/14*, conc. range:
150–9760 µg/kg, Ø conc.: 1828.6 µg/kg, country: Finland[205], *ncac

incidence: 8/8*, conc. range: 44–3980 µg/kg, Ø conc.: 1260 µg/kg, country: Finland[205], *ncac
see also oats, rye, wheat, wheat, summer, wheat, winter

ENNIATIN B₁
incidence: 14/14*, conc. range:
89–5720 µg/kg, Ø conc.: 860.1 µg/kg, country: Finland[205], *ncac
incidence: 8/8*, conc. range: tr–3240 µg/kg, country: Finland[205], *ncac
see also oats, rye, wheat, wheat, summer, wheat, winter

FUMONISIN B₁
incidence: 21/29, conc. range:
200–11,600 µg/kg, Ø conc.: 1900 µg/kg, country: Spain[155]
see also bird food, wild, dog food, feed, feed, complete ration, feed, general, feed, layer, feed, maize-based, feed, mixed, feed, pelleted ration, feed, screenings, feed, sweet, feed (broilers), feed (cat), feed (chicken), feed (dog), feed (gluten), feed (horse), feed (maize), feed (pig), feed (poultry), feed (rat), feed (rodent), forage grass, maize, maize and maize screenings, maize bran, maize ears, maize fine fractions, maize flakes, maize germ, maize germ/bran, maize germ meal, maize gluten, maize grits, maize kernels, maize meal, maize powder, maize screenings, maize, "Baby", maize, ground, maize, preharvest, maize, sweet feed, maize/oats mix, rat chow, silage, sorghum, soybeans, wheat

FUMONISIN B₂
incidence: 1/29, conc.: 500 µg/kg, country: Spain[155]
see also bird food, wild, dog food, feed, feed, maize-based, feed, mixed, feed (cat), feed (dog), feed (gluten), feed (horse), feed (maize), feed (poultry), feed (rodent), maize, maize bran, maize fine fractions, maize flakes, maize germ, maize germ/bran, maize germ meal, maize gluten, maize grits, maize kernels, maize meal, maize powder, maize screenings, maize, "Baby", maize, ground, maize, preharvest, rat chow, wheat

Fusarenone-X
incidence: 4/1095*, conc. range:
100–224 µg/kg, Ø conc.: 136 µg/kg, country:
Norway[224], *ncac
see also feed components, feed (poultry),
maize, maize ears, maize germ/bran, maize
gluten, maize kernels, maize meal, maize
screenings, oats, wheat

HT-2 Toxin
incidence: 5/24, conc. range: 10–370 µg/kg,
country: Poland[2]
incidence: 1/8, conc.: 10,000 µg/kg, country:
Germany[68]
incidence: 36/86, conc. range: 100–500 µg/kg,
country: Germany[68]
incidence: 1/7, conc. range: 10–20 µg/kg,
country: Finland[112]
incidence: 12/189, conc. range: 8–288 µg/kg,
Ø conc.: 47.7 µg/kg, country: Germany[123]
incidence: 2/44, conc. range: 10–32 µg/kg,
Ø conc.: 21 µg/kg, country: Germany[130]
incidence: 3/94, conc. range: 3–53 µg/kg,
Ø conc.: 25.7 µg/kg, country: Canada[240]
incidence: 10/29*, conc. range: ≤88 µg/kg,
Ø conc.: 41 µg/kg, country: Lithuania[399],
*ncac
see also feed components, feed (dog), feed
(fish), feed (pig), feed (poultry), feed
(reindeer), grains, mixed, grains, mixed feed,
grains (no specification), maize, maize
germ/bran, maize gluten, maize meal, maize
screenings, oats, rye, silage, wheat

Moniliformin
incidence: 5/19*, conc. range: tr–43 µg/kg,
Ø conc.: <40 µg/kg, country:
Norway[204], *ncac
incidence: 15/23*, conc. range: tr–380 µg/kg,
Ø conc.: 46 µg/kg, country: Norway[204], *ncac
incidence: 33/33*, conc. range: tr–230 µg/kg,
Ø conc.: 48 µg/kg, country: Norway[204], *ncac
incidence: 11/14*, conc. range: tr–290 µg/kg,
country: Finland[205], *ncac
incidence: 6/8*, conc. range: 35–750 µg/kg,
Ø conc.: 390.8 µg/kg, country: Finland[205],
*ncac
incidence: 11/14*, conc.
range: <20–290 µg/kg, country: Finland[312],
*ncac

see also feed, mixed, feed (poultry), maize,
maize flakes, maize germ, maize germ/bran,
maize gluten, maize meal, maize screenings,
maize, "Baby", oats, rice bran, triticale,
wheat, wheat, summer, wheat, winter

Nivalenol
incidence: 14/222, conc. range: 60–340 µg/kg,
Ø conc.: 136.4 µg/kg, country: Hungary[6]
incidence: 1/15*, conc.: 51 µg/kg, country:
Finland[9], *imported?
incidence: 5/45, conc. range: ≥50–80 µg/kg,
Ø conc.: 61 µg/kg, country: Sweden[65]
incidence: 1/32, conc.: 700 µg/kg, country:
Sweden[65]
incidence: 15/67, conc. range: ≥50–130 µg/kg,
Ø conc.: 78 µg/kg, country: Sweden[65]
incidence: 6/39, conc. range: ≥50–215 µg/kg,
Ø conc.: 169 µg/kg, country: Sweden[65]
incidence: 52/240, conc. range: 2–333 µg/kg,
Ø conc.: 21.1 µg/kg, country: Germany[123]
incidence: 5/44, conc. range: 3–10 µg/kg,
Ø conc.: 5 µg/kg, country: Germany[130]
incidence: 25/25*, conc. range: 13–156 µg/kg,
Ø conc.: 44.4 µg/kg, country: Norway[132],
*probably feed
incidence: 25/25*, conc. range: 22–258 µg/kg,
Ø conc.: 52.1 µg/kg, country: Norway[132],
*probably feed
incidence: 3/6*, conc. range: 56–91 µg/kg,
Ø conc.: 78.3 µg/kg, country: Poland[214], *ncac
incidence: 91/1095*, conc. range:
50–770 µg/kg, Ø conc.: 121 µg/kg, country:
Norway[224], *ncac
incidence: 2/2*, conc. range: 600–1240 µg/kg,
Ø conc.: 920 µg/kg, country: Japan[230], *ncac
incidence: 19/25*, conc. range:
tr–36,900 µg/kg, Ø conc.: 3900 µg/kg,
country: Japan[231], *ncac
incidence: 1/94, conc.: 65 µg/kg, country:
Canada[240]
incidence: 78/94*, conc. range: 5–3900 µg/kg,
Ø conc.: 218 µg/kg, country: Japan[265], *ncac
incidence: 18/29*, conc. range: ≤571 µg/kg,
Ø conc.: 101 µg/kg, country: Lithuania[399],
*ncac
incidence: 18/53, conc. range: ≤360 µg/kg,
Ø conc.: 170 µg/kg, country: Canada[410]

see also barley, husked, barley, unhusked (naked), barley (pressed), bran, feed, feed components, feed (cattle), feed (poultry), feed (reindeer), feeds, industrial, maize, maize ears, maize germ, maize germ/bran, maize gluten, maize kernels, maize meal, maize powder, maize screenings, maize, "Baby", oats, rye, silage, triticale, wheat

OCHRATOXIN A

incidence: 6/222, conc. range: 50–250 μg/kg, Ø conc.: 76.7 μg/kg, country: Hungary[6]

incidence: 10/68 *, conc. range: 0.1–206 μg/kg, Ø conc.: 58.8 μg/kg, country: Germany[13], *ncac

incidence: 23/85, max. conc.: 17.8 μg/kg, Ø conc.: 2.6 μg/kg, country: UK[15]

incidence: 17/17, conc. range: 9–27,520 μg/kg, Ø conc.: 2907.9 μg/kg, country: Denmark[24]

incidence: 23/79, conc. range: ≤17.8 μg/kg, Ø conc.: 3 μg/kg, country: UK[43]

incidence: 1/1, conc.: 1900 μg/kg, country: Denmark[82]

incidence: 19/137, conc. range: 2–200 μg/kg, country: Poland[84]

incidence: 12/45*, conc. range: 1–102 μg/kg, country: UK[94], *imported?

incidence: 2/8, conc. range: 6–120 μg/kg, Ø conc.: 63 μg/kg, country: The Netherlands[103]

incidence: 22/159, conc. range: 10–29 μg/kg, Ø conc.: 18 μg/kg, country: Denmark[193]

incidence: 14/68, conc. range: 0.3–9.9 μg/kg (11 sa), 10–99.9 μg/kg (2 sa), 117 μg/kg (1 sa), country: UK[203]

incidence: 1/7345, conc.: nc, country: Hungary[209]

incidence: 18/127, conc. range: tr–38 μg/kg, country: USA[219]

incidence: 1/94, conc.: 60 μg/kg, country: Canada[240]

incidence: 17/269*, conc. range: 2–20 μg/kg, country: Sweden[261], *ncac

incidence: 1/1, conc.: 68 μg/kg, country: UK?[288]

incidence: 3/3, conc. range: 15.8–24.3 μg/kg, Ø conc.: 20.2 μg/kg, country: Denmark[297]

incidence: 2/36* **, conc. range: 1.2–9.7 μg/kg, Ø conc.: 5.5 μg/kg, country: Poland[339], *ncac, **cg

incidence: 2/17* **, conc. range: 1.4–35.3 μg/kg, Ø conc.: 18.4 μg/kg, country: Poland[339], *ncac, **eg

incidence: 1/26* **, conc.: 0.3 μg/kg, country: Poland[341], *ncac, **cg

incidence: 3/40* **, conc. range: 6.7–57 μg/kg, Ø conc.: 25.7 μg/kg, country: Poland[341], *ncac, **eg

incidence: 6/116, conc. range: 12–26 μg/kg, country: Canada[410]

incidence: 21/40*, conc. range: 3–934 μg/kg, Ø conc.: 119.3 μg/kg, country: Sweden[418], *ncac

see also alfalfa, barley, oats, barley (high moisture), barley-soybean diet, bird food, domestic, bird food, wild, broilers feed, cereal grains, citrus pulp, coconut, expeller, corn cob mix silage, diet (dairy cow), diet (poultry), diet (starter), dog food, eat, egg production mixed feed, feed, feed wheat, oat and barley, feed, commercial mix, feed, mixed, feed, mixed (pelleted), feed (broilers), feed (cat), feed (cattle), feed (cereals), feed (pig), feed ec (pig), feed (poultry), feed ec (poultry), feed (poultry, pig), feed ec (rabbit), feed (trout), grain, mixed feed, grains, mixed, grains (heated), hay, horse bean, maize, maize feed, milo, maize gluten, maize meal, maize, white, *Makhana* (*Euryale ferox* Salisb) puffs, milk production mixed feed, millet, oat and barley (hammer-milled), oats, palm products, peanut cake, peas, peas and beans, pet food, pig feedstuffs, pig grower diet, pig meal, piglet diet, poultry feedstuffs, rice bran, rice germ, rice germ cake, rye, sorghum, soybean groats, sunflower, sunflower seeds, extracted, tapioca, triticale, *Vicia faba*, wheat, wheat and barley, wheat bran, wheat hay, wheat, oats

T-2 TETRAOL

incidence: 2/24, conc. range: 10–210 μg/kg, country: Poland[2]

see also feed components, maize, silage, wheat

T-2 TOXIN

incidence: 12/24, conc. range: 20–2400 μg/kg, country: Poland[2]

incidence: 16/222, conc. range: 50–310 μg/kg, Ø conc.: 220 μg/kg, country: Hungary[6]

incidence: 2/43, conc. range:
300–14,000 µg/kg, country: Germany[68]
incidence: 4/89, conc. range: 200–500 µg/kg,
country: Germany[68]
incidence: 34/240, conc. range: 2–305 µg/kg,
Ø conc.: 25.5 µg/kg, country: Germany[123]
incidence: 4/44, conc. range: 3–9 µg/kg,
Ø conc.: 5 µg/kg, country: Germany[130]
incidence: 2/25*, conc. range: 22–46 µg/kg,
Ø conc.: 34 µg/kg, country: Norway[132],
*probably feed
incidence: 1/1, conc.: 25,000 µg/kg, country:
Canada[180]
incidence: 1/94, conc.: 14 µg/kg, country:
Canada[240]
incidence: 4/29*, conc. range: ≤76 µg/kg,
Ø conc.: 40 µg/kg, country: Lithuania[399],
*ncac
incidence: 1/116, conc.: 1000 µg/kg, country:
Canada[410]
see also alfalfa, bran, diet (grower), diet
(poultry), feed, feed components, feed, layer,
feed, mixed, feed (dog), feed (fish), feed
(mink), feed (pig), feed (poultry), feedstuff,
forage grass, grain, mixed feed, grains (no
specification), hay, maize, maize germ/bran,
maize gluten, maize meal, maize screenings,
maize stalks (pith), oat and barley (hammer-
milled), oats, peanuts, pig grower diet, piglet
diet, rye, sorghum, triticale, wheat

T-2 TRIOL
incidence: 5/86, conc. range: 100–300 µg/kg,
country: Germany[68]
see also feed components, grain, mixed feed,
grains (no specification), maize, oats, wheat

VIOMELLEIN
incidence: 1/1, conc.: 1000 µg/kg, country:
Denmark[82]

ZEARALENONE
incidence: 107/222, conc. range:
50–840 µg/kg, Ø conc.: 139.8 µg/kg, country:
Hungary[6]
incidence: 5/40, conc. range: 10–20 µg/kg,
country: Germany[68]
incidence: 2/45*, conc. range: 5–16 µg/kg,
country: UK[94], *imported?

incidence: 1/8, conc.: 43 µg/kg, country: The
Netherlands[103]
incidence: 53/240, conc. range: 1–311 µg/kg,
Ø conc.: 13.5 µg/kg, country: Germany[123]
incidence: 1/1, conc.: ~100 µg/kg, country:
Germany[124]
incidence: 30/44, conc. range: 1–310 µg/kg,
Ø conc.: 35 µg/kg, country: Germany[130]
incidence: 3/25*, conc. range: 1–5 µg/kg,
Ø conc.: 2.7 µg/kg, country: Norway[132],
*probably feed
incidence: 14/25*, conc. range: 1–4 µg/kg,
Ø conc.: 2.6 µg/kg, country: Norway[132],
*probably feed
incidence: 5/7*, conc. range: 24–450 µg/kg,
Ø conc.: 166.2 µg/kg, country: Canada[202],
*ncac
incidence: 10/30*, conc. range: 14–171 µg/kg,
Ø conc.: 36 µg/kg, country: Korea[364], *ncac
incidence: 17/116, conc. range: 30–840 µg/kg,
Ø conc.: 250 µg/kg, country: Canada[410]
see also alfalfa, barley, husked, barley,
unhusked (naked), barley and feed, bone
meal, bran, broilers feed, *Carthamus* cake,
chick pea, concentrate, mixed, corn cob mix
silage, cotton cake, cottonseed, cottonseed
cake, diet (dairy cow), diet (poultry), diets
(mixed), feed, feed components, feed, mixed,
feed, mixed (primarily maize also maize, oats,
wheat), feed (bran), feed (broiler chicken),
feed (cattle), feed (chicken), feed (dairy), feed
(developing pig), feed (mill run, from wheat),
feed (miscellaneous), feed (pig), feed
(poultry), feed (poultry, pig), feed (starter
chicken), feedstuff, fish meal, forage grass,
grain, bruised, grain, mixed feed, grains (no
specification), hay, maize, maize ears, maize
flakes, maize germ, maize germ/bran, maize
gluten, maize kernels, maize meal, maize oil
cake, maize screenings, maize stalks (pith),
maize, "Baby", maize, hybrid, maize, shelled,
maize, unshelled, maize, white, maize grain,
artificially dried, maize grain, crib dried,
maize grain, ensiled, milk production mixed
feed, oats, *Paspalum palidosum*, straw, peanut
hulls/skins, rice bran, rice germ, rice germ
cake, rye, silage, sorghum, soybeans, soybeans,
extracted, sunflower cake, tapioca, triticale,

wheat, wheat bran, wheat bran and chana testa, wheat soya meal

Barley, husked   for feed may contain the following mycotoxins:

DEOXYNIVALENOL
incidence: 5/6, conc. range: 3–65 µg/kg,
Ø conc.: 19 µg/kg, country: Korea[108]
incidence: 5/5*, conc. range: 26–223 µg/kg,
Ø conc.: 108.8 µg/kg, country: Korea[227], *ncac
incidence: 4/14*, conc. range: 25–85 µg/kg,
Ø conc.: 59 µg/kg, country: Korea[227], *ncac
see also barley, barley, unhusked (naked),
barley (pressed), bone meal, bran, broilers
feed, calf fattening mixed feed, coconut,
expeller, corn cob mix silage, cottonseed,
cottonseed cake, dairy cattle feed, egg
production mixed feed, feed, feed components,
feed, commercial mix, feed, mixed, feed,
mixed (primarily maize), feed (barley), feed
(cattle), feed (chicken), feed (dog), feed
(fish), feed (mill run, from wheat), feed
(mink), feed (pig), feed (poultry), feed
(reindeer), feeds, grain, feeds, industrial,
feedstuff, feedstuffs (rapeseed, turnip, fish
meal, concentrates), fish meal, grain, mixed
feed, grains, mixed, grains (no specification),
maize, maize ears, maize fibre, maize germ,
maize germ/bran, maize germ meal, maize
gluten, maize kernels, maize meal, maize
powder, maize screenings, maize stalks (pith),
maize, "Baby", maize, hybrid, maize, white,
oats, rice bran, rice germ cake, rye, silage,
sorghum, soybeans, triticale, wheat, wheat
and barley, wheat, red hard winter, wheat,
soft white winter, wheat, spring, wheat,
winter

NIVALENOL
incidence: 6/6, conc. range: 39–228 µg/kg,
Ø conc.: 112 µg/kg, country: Korea[108]
incidence: 5/5*, conc. range: 12–321 µg/kg,
Ø conc.: 222.4 µg/kg, country: Korea[227], *ncac
incidence: 10/14*, conc. range: 4–120 µg/kg,
Ø conc.: 48.2 µg/kg, country: Korea[227], *ncac
see also barley, barley, unhusked (naked),
barley (pressed), bran, feed, feed components,
feed (cattle), feed (poultry), feed (reindeer),

feeds, industrial, maize, maize ears, maize
germ/bran, maize gluten, maize kernels,
maize meal, maize powder, maize screenings,
maize, "Baby", oats, rye, silage, triticale,
wheat

ZEARALENONE
incidence: 3/6, conc. range: 1–2 µg/kg,
country: Korea[108]
incidence: 1/5*, conc.: 28 µg/kg, country:
Korea[227], *ncac
incidence: 3/14*, conc. range: 27–49 µg/kg,
Ø conc.: 40.7 µg/kg, country: Korea[227], *ncac
see also alfalfa, barley, barley, unhusked
(naked), barley and feed, bone meal, bran,
broilers feed, *Carthamus* cake, chick pea,
concentrate, mixed, corn cob mix silage,
cotton cake, cottonseed, cottonseed cake, diet
(dairy cow), diet (poultry), diets (mixed),
feed, feed components, feed, mixed, feed,
mixed (primarily maize also maize, oats,
wheat), feed (bran), feed (broiler chicken),
feed (cattle), feed (chicken), feed (dairy), feed
(developing pig), feed (mill run, from wheat),
feed (miscellaneous), feed (pig), feed
(poultry), feed (poultry, pig), feed (starter
chicken), feedstuff, fish meal, forage grass,
grain, bruised, grain, mixed feed, grains (no
specification), hay, maize, maize ears, maize
flakes, maize germ, maize germ/bran, maize
gluten, maize kernels, maize meal, maize oil
cake, maize screenings, maize stalks (pith),
maize, "Baby", maize, hybrid, maize, shelled,
maize, unshelled, maize, white, maize grain,
artificially dried, maize grain, crib dried,
maize grain, ensiled, milk production mixed
feed, oats, *Paspalum palidosum*, straw, peanut
hulls/skins, rice bran, rice germ, rice germ
cake, rye, silage, sorghum, soybeans, soybeans,
extracted, sunflower cake, tapioca, triticale,
wheat, wheat bran, wheat bran and chana
testa, wheat soya meal

Barley, oats   Barley and oats for feed may
contain the following mycotoxins:

ALTERNARIOL METHYL ETHER
incidence: 2/7, conc. range: 80–160 µg/kg,
Ø conc.: 120 µg/kg, country: Germany[294]

see also barley, cereals, mixture, oats, sunflower, wheat, wheat, oats

CITRININ
incidence: 1/1, conc.: 2000 μg/kg, country: Denmark[24]
see also barley, barley-soybean diet, cotton-seed cake, feed, feed, mixed, feed (cattle), feed (pig), fish meal, hay, maize, maize, white, *Makhana* (*Euryale ferox* Salisb) puffs, oats, palm products, peas and beans, rice bran, rice germ, wheat, wheat and other grains (moldy), wheat bran

OCHRATOXIN A
incidence: 7/84, conc. range: 16–409 μg/kg, country: Sweden[23]
incidence: 19/33, conc. range: 28–27,500 μg/kg, country: Denmark[24]
incidence: 4/4, conc.range: 69–6190 μg/kg, Ø conc.: 1805.5 μg/kg, country: Denmark[24]
incidence: 3/3, conc. range: 4–141 μg/kg, Ø conc.: 72.3 μg/kg, country: Finland[281]
see also alfalfa, barley, barley (high moisture), barley-soybean diet, bird food, domestic, bird food, wild, broilers feed, cereal grains, citrus pulp, coconut, expeller, corn cob mix silage, diet (dairy cow), diet (poultry), diet (starter), dog food, eat, egg production mixed feed, feed, feed wheat, oat and barley, feed, commercial mix, feed, mixed, feed, mixed (pelleted), feed (broilers), feed (cat), feed (cattle), feed (cereals), feed (pig), feed ec (pig), feed (poultry), feed ec (poultry), feed (poultry, pig), feed ec (rabbit), feed (trout), grain, mixed feed, grains, mixed, grains (heated), hay, horse bean, maize, maize feed, milo, maize gluten, maize meal, maize, white, *Makhana* (*Euryale ferox* Salisb) puffs, milk production mixed feed, millet, oat and barley (hammer-milled), oats, palm products, peanut cake, peas, peas and beans, pet food, pig feedstuffs, pig grower diet, pig meal, piglet diet, poultry feedstuffs, rice bran, rice germ, rice germ cake, rye, sorghum, soybean groats, sunflower, sunflower seeds, extracted, tapioca, triticale, *Vicia faba*, wheat, wheat and barley, wheat bran, wheat hay, wheat, oats

Barley, unhusked (naked) for feed may contain the following mycotoxins:

DEOXYNIVALENOL
incidence: 31/31, conc. range: 12–901 μg/kg, Ø conc.: 124.7 μg/kg, country: Korea[108]
incidence: 17/18*, conc. range: 8–495 μg/kg, Ø conc.: 115.9 μg/kg, country: Korea[227], *ncac
incidence: 9/20*, conc. range: 10–161 μg/kg, Ø conc.: 52.9 μg/kg, country: Korea[227], *ncac
incidence: 2/2*, conc. range: 180–330 μg/kg, Ø conc.: 255 μg/kg, country: Japan[230], *ncac
see also barley, barley, husked, barley (pressed), bone meal, bran, broilers feed, calf fattening mixed feed, coconut, expeller, corn cob mix silage, cottonseed, cottonseed cake, dairy cattle feed, egg production mixed feed, feed, feed components, feed, commercial mix, feed, mixed, feed, mixed (primarily maize), feed (barley), feed (cattle), feed (chicken), feed (dog), feed (fish), feed (mill run, from wheat), feed (mink), feed (pig), feed (poultry), feed (reindeer), feeds, grain, feeds, industrial, feedstuff, feedstuffs (rapeseed, turnip, fish meal, concentrates), fish meal, grain, mixed feed, grains, mixed, grains (no specification), maize, maize ears, maize fibre, maize germ, maize germ/bran, maize germ meal, maize gluten, maize kernels, maize meal, maize powder, maize screenings, maize stalks (pith), maize, "Baby", maize, hybrid, maize, white, oats, rice bran, rice germ cake, rye, silage, sorghum, soybeans, triticale, wheat, wheat and barley, wheat, red hard winter, wheat, soft white winter, wheat, spring, wheat, winter

NIVALENOL
incidence: 31/31, conc. range: 180–1145 μg/kg, Ø conc.: 489.5 μg/kg, country: Korea[108]
incidence: 17/18*, conc. range: 28–1109 μg/kg, Ø conc.: 330.8 μg/kg, country: Korea[227], *ncac
incidence: 14/20*, conc. range: 3–904 μg/kg, Ø conc.: 242.5 μg/kg, country: Korea[227], *ncac
incidence: 2/2*, conc. range: 920–1670 μg/kg, Ø conc.: 1295 μg/kg, country: Japan[230], *ncac
see also barley, barley, husked, barley (pressed), bran, feed, feed components, feed (cattle),

feed (poultry), feed (reindeer), feeds, industrial, maize, maize ears, maize germ/ bran, maize gluten, maize kernels, maize meal, maize powder, maize screenings, maize, "Baby", oats, rye, silage, triticale, wheat

Zearalenone
incidence: 29/31, conc. range: 1–388 μg/kg, Ø conc.: 26.5 μg/kg, country: Korea[108]
incidence: 8/18*, conc. range: 43–1132 μg/kg, Ø conc.: 280.6 μg/kg, country: Korea[227], *ncac
incidence: 6/20*, conc. range: 81–580 μg/kg, Ø conc.: 216 μg/kg, country: Korea[227], *ncac
see also alfalfa, barley, barley and feed, barley, husked, bone meal, bran, broilers feed, *Carthamus* cake, chick pea, concentrate, mixed, corn cob mix silage, cotton cake, cottonseed, cottonseed cake, diet (dairy cow), diet (poultry), diets (mixed), feed, feed components, feed, mixed, feed, mixed (primarily maize also maize, oats, wheat), feed (bran), feed (broiler chicken), feed (cattle), feed (chicken), feed (dairy), feed (developing pig), feed (mill run, from wheat), feed (miscellaneous), feed (pig), feed (poultry), feed (poultry, pig), feed (starter chicken), feedstuff, fish meal, forage grass, grain, bruised, grain, mixed feed, grains (no specification), hay, maize, maize ears, maize flakes, maize germ, maize germ/bran, maize gluten, maize kernels, maize meal, maize oil cake, maize screenings, maize stalks (pith), maize, "Baby", maize, hybrid, maize, shelled, maize, unshelled, maize, white, maize grain, artificially dried, maize grain, crib dried, maize grain, ensiled, milk production mixed feed, oats, *Paspalum palidosum*, straw, peanut hulls/skins, rice bran, rice germ, rice germ cake, rye, silage, sorghum, soybeans, soybeans, extracted, sunflower cake, tapioca, triticale, wheat, wheat bran, wheat bran and chana testa, wheat soya meal

Barley (high moisture)   Barley for feed may contain the following mycotoxins:

Ochratoxin A
incidence: 1/51, conc.: 48 μg/kg, country: Canada[133]

see also alfalfa, barley, barley, oats, barley-soybean diet, bird food, domestic, bird food, wild, broilers feed, cereal grains, citrus pulp, coconut, expeller, corn cob mix silage, diet (dairy cow), diet (poultry), diet (starter), dog food, eat, egg production mixed feed, feed, feed wheat, oat and barley, feed, commercial mix, feed, mixed, feed, mixed (pelleted), feed (broilers), feed (cat), feed (cattle), feed (cereals), feed (pig), feed ec (pig), feed (poultry), feed ec (poultry), feed (poultry, pig), feed ec (rabbit), feed (trout), grain, mixed feed, grains, mixed, grains (heated), hay, horse bean, maize, maize feed, milo, maize gluten, maize meal, maize, white, *Makhana* (*Euryale ferox* Salisb) puffs, milk production mixed feed, millet, oat and barley (hammer-milled), oats, palm products, peanut cake, peas, peas and beans, pet food, pig feedstuffs, pig grower diet, pig meal, piglet diet, poultry feedstuffs, rice bran, rice germ, rice germ cake, rye, sorghum, soybean groats, sunflower, sunflower seeds, extracted, tapioca, triticale, *Vicia faba*, wheat, wheat and barley, wheat bran, wheat hay, wheat, oats

Barley (pressed)   may contain the following mycotoxins:

Deoxynivalenol
incidence: 9/10*, conc. range: 3–23 μg/kg, Ø conc.: 11 μg/kg, country: Japan[230], *ncac
see also barley, barley, husked, barley, unhusked (naked), bone meal, bran, broilers feed, calf fattening mixed feed, coconut, expeller, corn cob mix silage, cottonseed, cottonseed cake, dairy cattle feed, egg production mixed feed, feed, feed components, feed, commercial mix, feed, mixed, feed, mixed (primarily maize), feed (barley), feed (cattle), feed (chicken), feed (dog), feed (fish), feed (mill run, from wheat), feed (mink), feed (pig), feed (poultry), feed (reindeer), feeds, grain, feeds, industrial, feedstuff, feedstuffs (rapeseed, turnip, fish meal, concentrates), fish meal, grain, mixed feed, grains, mixed, grains (no specification), maize, maize ears, maize fibre, maize germ, maize germ/bran, maize germ meal, maize gluten, maize kernels, maize

meal, maize powder, maize screenings, maize stalks (pith), maize, "Baby", maize, hybrid, maize, white, oats, rice bran, rice germ cake, rye, silage, sorghum, soybeans, triticale, wheat, wheat and barley, wheat, red hard winter, wheat, soft white winter, wheat, spring, wheat, winter

NIVALENOL
incidence: 2/2*, conc. range: 16–56 μg/kg, Ø conc.: 33 μg/kg, country: Japan[230], *ncac
see also barley, barley, husked, barley, unhusked (naked), bran, feed, feed components, feed (cattle), feed (poultry), feed (reindeer), feeds, industrial, maize, maize ears, maize germ/bran, maize gluten, maize kernels, maize meal, maize powder, maize screenings, maize, "Baby", oats, rye, silage, triticale, wheat

Barley and feed   may contain the following mycotoxins:

ZEARALENONE
incidence: 2/12, conc. range: 500–750 μg/kg, Ø conc.: 625 μg/kg, country: Scotland[178]
see also alfalfa, barley, barley, husked, barley, unhusked (naked), bone meal, bran, broilers feed, *Carthamus* cake, chick pea, concentrate, mixed, corn cob mix silage, cotton cake, cottonseed, cottonseed cake, diet (dairy cow), diet (poultry), diets (mixed), feed, feed components, feed, mixed, feed, mixed (primarily maize also maize, oats, wheat), feed (bran), feed (broiler chicken), feed (cattle), feed (chicken), feed (dairy), feed (developing pig), feed (mill run, from wheat), feed (miscellaneous), feed (pig), feed (poultry), feed (poultry, pig), feed (starter chicken), feedstuff, fish meal, forage grass, grain, bruised, grain, mixed feed, grains (no specification), hay, maize, maize ears, maize flakes, maize germ, maize germ/bran, maize gluten, maize kernels, maize meal, maize oil cake, maize screenings, maize stalks (pith), maize, "Baby", maize, hybrid, maize, shelled, maize, unshelled, maize, white, maize grain, artificially dried, maize grain, crib dried, maize grain, ensiled, milk production mixed

feed, rice bran, rice germ, peanut hulls/skins, rice germ cake, rye, sorghum, soybeans, soybeans, extracted, sunflower cake, tapioca, triticale, wheat, wheat bran, wheat bran and chana testa, wheat soya meal

Barley-soybean diet   may contain the following mycotoxins:

CITRININ
incidence: 1/1, conc.: 650 μg/kg, country: Denmark[295]
see also barley, barley, oats, cottonseed cake, feed, feed, mixed, feed (cattle), feed (pig), fish meal, hay, maize, maize, white, *Makhana* (*Euryale ferox* Salisb) puffs, oats, palm products, peas and beans, rice bran, rice germ, wheat, wheat and other grains (moldy), wheat bran

OCHRATOXIN A
incidence: 1/1, conc.: 1400 μg/kg, country: Denmark[295]
see also alfalfa, barley, barley, oats, barley (high moisture), bird food, domestic, bird food, wild, broilers feed, cereal grains, citrus pulp, coconut, expeller, corn cob mix silage, diet (dairy cow), diet (poultry), diet (starter), dog food, eat, egg production mixed feed, feed, feed wheat, oat and barley, feed, commercial mix, feed, mixed, feed, mixed (pelleted), feed (broilers), feed (cat), feed (cattle), feed (cereals), feed (pig), feed ec (pig), feed (poultry), feed ec (poultry), feed (poultry, pig), feed ec (rabbit), feed (trout), grain, mixed feed, grains, mixed, grains (heated), hay, horse bean, maize, maize feed, milo, maize gluten, maize meal, maize, white, *Makhana* (*Euryale ferox* Salisb) puffs, milk production mixed feed, millet, oat and barley (hammer-milled), oats, palm products, peanut cake, peas, peas and beans, pet food, pig feedstuffs, pig grower diet, pig meal, piglet diet, poultry feedstuffs, rice bran, rice germ, rice germ cake, rye, sorghum, soybean groats, sunflower, sunflower seeds, extracted, tapioca, triticale, *Vicia faba*, wheat, wheat and barley, wheat bran, wheat hay, wheat, oats

Bengalgram husk　may contain the following mycotoxins:

Aflatoxin B$_1$
incidence: 2/8, conc. range: nd–17 µg/kg, country: India[247]
see also alfalfa, *Ambadi* cake, animal feed-stuffs (dairy cake), bagasse, barley, bird food, bird food, wild, biri testa, blackgram, blackgram husk, bran, broiler mixed feed, calf fattening mixed feed, calf fattening mixed feed (containing 4–20 % peanut products), *Carthamus* cake, castor cake, cereals, cereal products, chick pea, coconut cake, cocos, concentrate, mixed, concentrates, cotton cake, cottonseed, cottonseed (dehulled), cottonseed cake, cottonseed extract, cotton-seed meal, cottonseed meal (ammoniated), cottonseed meal (decorticated), cottonseed meats, cottonseed products, crumbles, crumbles, grower, cycad meal, dairy cattle feed, dairy cattle feed (containing 2–5 % peanut products), dairy cattle feed (containing 6–10 % peanut products), dairy cattle feed (containing 6–12 % peanut products), dairy cattle feed (containing more than 20 % peanut products), diets, mixed, dog food, egg production mixed feed, feed, feed and ingredients, feed, compound, feed, layer, feed, mixed, feed (beef), feed (broilers), feed (calf), feed (cat), feed (cattle), feed (chicken), feed (dairy), feed (dog), feed (dug), feed (fish), feed (gluten), feed (horse), feed (miscellaneous), feed (pig), feed (poultry), feed (poultry, pig), feed (rabbit), feed (sheep), feed (50–60 % maize), fish meal, grain by-products, grains (no specification), greengram, hay/silage, horsegram, husk, *Jagni* cake, legume mixture, linseed, linseed cake, livol, *Mahua* cake, maize, maize germ, maize gluten, maize grits, maize husk, maize meal, maize oil cake, maize powder, maize screenings, maize, ground, maize, hybrid, maize, preharvest, maize, yellow, maize (dark grains), *Makhana* (*Euryale ferox* Salisb) puffs, manioc, milk production mixed feed, mung testa, murkool, mustard cake, neem cake, niger cake, oats, palm kernel expeller cake, palm kernels, palm products, peanut cake, peanut cake (deoiled), peanut expeller, peanut hay, peanut meal, peanut, kernels, peanut, shells, peanuts, pellets, finisher, pig meal and pellets, pigeon pea, poultry feeds (peanut containing), rapeseed cake, redgram husk, rice, rice bran, rice bran (deoiled), rice chaff, rice crack, rice germ, rice germ cake, rice meal, rice straw, rice (damaged), rice (polish), safflower cake, sal seed cake, sesame, sesame cake, sorghum, soybean meal, soybeans, sunflower, sunflower cake, sunflower flour, tapioca, wheat, wheat bran, wheat bran and chana testa

Bird food　may contain the following mycotoxins:

Aflatoxin B$_1$
incidence: 8/30, conc. range: 23–256 µg/kg, Ø conc.: 109.75 µg/kg, country: Brazil[4]
see also alfalfa, *Ambadi* cake, animal feedstuffs (dairy cake), bagasse, barley, bengalgram husk, bird food, wild, biri testa, blackgram, blackgram husk, bran, broiler mixed feed, calf fattening mixed feed, calf fattening mixed feed (containing 4–20 % peanut products), *Carthamus* cake, castor cake, cereals, cereal products, chick pea, coconut cake, cocos, concentrate, mixed, concentrates, cotton cake, cottonseed, cottonseed (dehulled), cottonseed cake, cottonseed extract, cotton-seed meal, cottonseed meal (ammoniated), cottonseed meal (decorticated), cottonseed meats, cottonseed products, crumbles, crumbles, grower, cycad meal, dairy cattle feed, dairy cattle feed (containing 2–5 % peanut products), dairy cattle feed (containing 6–10 % peanut products), dairy cattle feed (containing 6–12 % peanut products), dairy cattle feed (containing more than 20 % peanut prod ucts), diets, mixed, dog food, egg production mixed feed, feed, feed and ingredients, feed, compound, feed, layer, feed, mixed, feed (beef), feed (broilers), feed (calf), feed (cat), feed (cattle), feed (chicken), feed (dairy), feed (dog), feed (dug), feed (fish), feed (gluten), feed (horse), feed (miscellaneous), feed (pig), feed (poultry), feed (poultry, pig), feed (rabbit),

feed (sheep), feed (50–60 % maize), fish meal, grain by-products, grains (no specification), greengram, hay/silage, horsegram, husk, *Jagni* cake, legume mixture, linseed, linseed cake, livol, *Mahua* cake, maize, maize germ, maize gluten, maize grits, maize husk, maize meal, maize oil cake, maize powder, maize screenings, maize, ground, maize, hybrid, maize, preharvest, maize, yellow, maize (dark grains), *Makhana* (*Euryale ferox* Salisb) puffs, manioc, milk production mixed feed, mung testa, murkool, mustard cake, neem cake, niger cake, oats, palm kernel expeller cake, palm kernels, palm products, peanut cake, peanut cake (deoiled), peanut expeller, peanut hay, peanut meal, peanut, kernels, peanut, shells, peanuts, pellets, finisher, pig meal and pellets, pigeon pea, poultry feeds (peanut containing), rapeseed cake, redgram husk, rice, rice bran, rice bran (deoiled), rice chaff, rice crack, rice germ, rice germ cake, rice meal, rice straw, rice (damaged), rice (polish), safflower cake, sal seed cake, sesame, sesame cake, sorghum, soybean meal, soybeans, sunflower, sunflower cake, sunflower flour, tapioca, wheat, wheat bran, wheat bran and chana testa

Aflatoxin B_2
incidence: 6/30, conc. range: 12–114 µg/kg, Ø conc.: 70.5 µg/kg, country: Brazil[4]
see also animal feedstuff (dairy cake), bird food, wild, biri testa, blackgram, blackgram husk, cottonseed, cottonseed cake, cottonseed extract, cottonseed meal, cottonseed meal (ammoniated), cottonseed meats, dog food, egg production mixed feed, feed, compound, feed (cat), feed (cattle), feed (dog), feed (pig), feed (poultry), feed (rabbit), feed (sheep), fish meal, horsegram, maize, maize gluten, maize husk, maize, ground, maize, preharvest, mung testa, mustard cake, niger cake, peanut cake, peanut cake (deoiled), peanut expeller, peanut hay, peanut meal, peanuts, peanut, kernels, redgram husk, rice bran, rice bran (deoiled), rice chaff, rice meal, rice (polish), sal seed cake, sesame cake, sorghum, soybean meal, soybeans, wheat, wheat bran

Aflatoxin G_1
incidence: 2/30, conc. range: 43–87 µg/kg, Ø conc.: 65 µg/kg, country: Brazil[4]
see also animal feedstuffs (dairy cake), bird food, wild, concentrate, mixed, cottonseed, cottonseed cake, feed (cat), feed (chicken), feed (dog), maize, maize, ground, maize, preharvest, meat meal, milk production mixed feed, murkool, peanut cake, peanut expeller, peanut hay, peanut meal, peanuts, rice bran, rice germ, sorghum, soybeans, wheat, wheat bran, wheat bran and chana testa

Aflatoxin G_2
incidence: 2/30, conc. range: 18–36 µg/kg, Ø conc.: 27 µg/kg, country: Brazil[4]
see also animal feedstuffs (dairy cake), feed (cat), feed (dog), maize, maize, preharvest, peanut expeller, peanut hay, peanut meal, peanuts, soybeans

Bird food, domestic may contain the following mycotoxins:

Ochratoxin A
incidence: 4/15, conc. range: 1.5–7 µg/kg, Ø conc.: 4 µg/kg, country: UK[42]
see also alfalfa, barley, barley, oats, barley (high moisture), barley-soybean diet, bird food, wild, broilers feed, cereal grains, citrus pulp, coconut, expeller, corn cob mix silage, diet (dairy cow), diet (poultry), diet (starter), dog food, eat, egg production mixed feed, feed, feed wheat, oat and barley, feed, commercial mix, feed, mixed, feed, mixed (pelleted), feed (broilers), feed (cat), feed (cattle), feed (cereals), feed (pig), feed ec (pig), feed (poultry), feed ec (poultry), feed (poultry, pig), feed ec (rabbit), feed (trout), grain, mixed feed, grains, mixed, grains (heated), hay, horse bean, maize, maize feed, milo, maize gluten, maize meal, maize, white, *Makhana* (*Euryale ferox* Salisb) puffs, milk production mixed feed, millet, oat and barley (hammer-milled), oats, palm products, peanut cake, peas, peas and beans, pet food, pig feedstuffs, pig grower diet, pig meal, piglet diet, poultry feedstuffs, rice bran, rice

germ, rice germ cake, rye, sorghum, soybean groats, sunflower, sunflower seeds, extracted, tapioca, triticale, *Vicia faba*, wheat, wheat and barley, wheat bran, wheat hay, wheat, oats

Bird food, wild   may contain the following mycotoxins:

Aflatoxin B_1
incidence: 1/15, conc.: 370 μg/kg, country: UK[42]
see also alfalfa, *Ambadi* cake, animal feed-stuffs (dairy cake), bagasse, barley, bengalgram husk, bird food, biri testa, blackgram, blackgram husk, bran, broiler mixed feed, calf fattening mixed feed, calf fattening mixed feed (containing 4–20 % peanut products), *Carthamus* cake, castor cake, cereals, cereal products, chick pea, coconut cake, cocos, concentrate, mixed, concentrates, cotton cake, cottonseed, cottonseed (dehulled), cottonseed cake, cottonseed extract, cotton-seed meal, cottonseed meal (ammoniated), cottonseed meal (decorticated), cottonseed meats, cottonseed products, crumbles, crumbles, grower, cycad meal, dairy cattle feed, dairy cattle feed (containing 2–5 % peanut products), dairy cattle feed (containing 6–10 % peanut products), dairy cattle feed (containing 6–12 % peanut products), dairy cattle feed (containing more than 20 % peanut products), diets, mixed, dog food, egg production mixed feed, feed, feed and ingredients, feed, compound, feed, layer, feed, mixed, feed (beef), feed (broilers), feed (calf), feed (cat), feed (cattle), feed (chicken), feed (dairy), feed (dog), feed (dug), feed (fish), feed (gluten), feed (horse), feed (miscellaneous), feed (pig), feed (poultry), feed (poultry, pig), feed (rabbit), feed (sheep), feed (50–60 % maize), fish meal, grain by-products, grains (no specification), greengram, hay/silage, horsegram, husk, *Jagni* cake, legume mixture, linseed, linseed cake, livol, *Mahua* cake, maize, maize germ, maize gluten, maize grits, maize husk, maize meal, maize oil cake, maize powder, maize screenings, maize, ground, maize, hybrid, maize, preharvest,

maize, yellow, maize (dark grains), *Makhana* (*Euryale ferox* Salisb) puffs, manioc, milk production mixed feed, mung testa, murkool, mustard cake, neem cake, niger cake, oats, palm kernel expeller cake, palm kernels, palm products, peanut cake, peanut cake (deoiled), peanut expeller, peanut hay, peanut meal, peanut, kernels, peanut, shells, peanuts, pellets, finisher, pig meal and pellets, pigeon pea, poultry feeds (peanut containing), rapeseed cake, redgram husk, rice, rice bran, rice bran (deoiled), rice chaff, rice crack, rice germ, rice germ cake, rice meal, rice straw, rice (damaged), rice (polish), safflower cake, sal seed cake, sesame, sesame cake, sorghum, soybean meal, soybeans, sunflower, sunflower cake, sunflower flour, tapioca, wheat, wheat bran, wheat bran and chana testa

Aflatoxin B_2
incidence: 1/15, conc.: 29 μg/kg, country: UK[42]
see also animal feedstuff (dairy cake), bird food, biri testa, blackgram, blackgram husk, cottonseed, cottonseed cake, cottonseed extract, cottonseed meal, cottonseed meal (ammoniated), cottonseed meats, dog food, egg production mixed feed, feed, compound, feed (cat), feed (cattle), feed (dog), feed (pig), feed (poultry), feed (rabbit), feed (sheep), fish meal, horsegram, maize, maize gluten, maize husk, maize, ground, maize, preharvest, mung testa, mustard cake, niger cake, peanut cake, peanut cake (deoiled), peanut expeller, peanut hay, peanut meal, peanuts, peanut, kernels, redgram husk, rice bran, rice bran (deoiled), rice chaff, rice meal, rice (polish), sal seed cake, sesame cake, sorghum, soybean meal, soybeans, wheat, wheat bran

Aflatoxin G_1
incidence: 1/15, conc.: 3.3 μg/kg, country: UK[42]
see also animal feedstuffs (dairy cake), bird food, concentrate, mixed, cottonseed, cottonseed cake, feed (cat), feed (chicken), feed (dog), maize, maize, ground, maize, preharvest, meat meal, milk production mixed feed, murkool, peanut cake, peanut

expeller, peanut hay, peanut meal, peanuts, rice bran, rice germ, sorghum, soybeans, wheat, wheat bran, wheat bran and chana testa

Fumonisin B$_1$
incidence: 2/15, conc. range: 240–410 µg/kg, Ø conc.: 325 µg/kg, country: UK[42]
see also barley, dog food, feed, feed, complete ration, feed, general, feed, layer, feed, maize-based, feed, mixed, feed, pelleted ration, feed, screenings, feed, sweet, feed (broilers), feed (cat), feed (chicken), feed (dog), feed (gluten), feed (horse), feed (maize), feed (pig), feed (poultry), feed (rat), feed (rodent), forage grass, maize, maize and maize screenings, maize bran, maize ears, maize fine fractions, maize flakes, maize germ, maize germ/bran, maize germ meal, maize gluten, maize grits, maize kernels, maize meal, maize powder, maize screenings, maize, "Baby", maize, ground, maize, preharvest, maize, sweet feed, maize/oats mix, rat chow, silage, sorghum, soybeans, wheat

Fumonisin B$_2$
incidence: 2/15, conc.: 50 µg/kg, Ø conc.: 50 µg/kg, country: UK[42]
see also barley, dog food, feed, feed, maize-based, feed, mixed, feed (cat), feed (dog), feed (gluten), feed (horse), feed (maize), feed (poultry), feed (rodent), maize, maize bran, maize fine fractions, maize flakes, maize germ, maize germ/bran, maize germ meal, maize gluten, maize grits, maize kernels, maize meal, maize powder, maize screenings, maize, "Baby", maize, ground, maize, preharvest, rat chow, wheat

Ochratoxin A
incidence: 1/15, conc.: 6 µg/kg, country: UK[42]
see also alfalfa, barley, barley, oats, barley (high moisture), barley-soybean diet, bird food, domestic, broilers feed, cereal grains, citrus pulp, coconut, expeller, corn cob mix silage, diet (dairy cow), diet (poultry), diet (starter), dog food, eat, egg production mixed feed, feed, feed wheat, oat and barley, feed, commercial mix, feed, mixed, feed, mixed (pelleted), feed (broilers), feed (cat),

feed (cattle), feed (cereals), feed (pig), feed ec (pig), feed (poultry), feed ec (poultry), feed (poultry, pig), feed ec (rabbit), feed (trout), grain, mixed feed, grains, mixed, grains (heated), hay, horse bean, maize, maize feed, milo, maize gluten, maize meal, maize, white, *Makhana* (*Euryale ferox* Salisb) puffs, milk production mixed feed, millet, oat and barley (hammer-milled), oats, palm products, peanut cake, peas, peas and beans, pet food, pig feedstuffs, pig grower diet, pig meal, piglet diet, poultry feedstuffs, rice bran, rice germ, rice germ cake, rye, sorghum, soybean groats, sunflower, sunflower seeds, extracted, tapioca, triticale, *Vicia faba*, wheat, wheat and barley, wheat bran, wheat hay, wheat, oats

Biri testa   may contain the following mycotoxins:

Aflatoxin B$_1$
incidence: 1/1, conc.: 22 µg/kg, country: India[129]
see also alfalfa, *Ambadi* cake, animal feedstuffs (dairy cake), bagasse, barley, bengalgram husk, bird food, bird food, wild, blackgram, blackgram husk, bran, broiler mixed feed, calf fattening mixed feed, calf fattening mixed feed (containing 4–20 % peanut products), *Carthamus* cake, castor cake, cereals, cereal products, chick pea, coconut cake, cocos, concentrate, mixed, concentrates, cotton cake, cottonseed, cottonseed (dehulled), cottonseed cake, cottonseed extract, cottonseed meal, cottonseed meal (ammoniated), cottonseed meal (decorticated), cottonseed meats, cottonseed products, crumbles, crumbles, grower, cycad meal, dairy cattle feed, dairy cattle feed (containing 2–5 % peanut products), dairy cattle feed (containing 6–10 % peanut products), dairy cattle feed (containing 6–12 % peanut products), dairy cattle feed (containing more than 20 % peanut products), diets, mixed, dog food, egg production mixed feed, feed, feed and ingredients, feed, compound, feed, layer, feed, mixed, feed (beef), feed (broilers), feed (calf), feed (cat), feed (cattle), feed (chicken), feed (dairy), feed (dog), feed

(dug), feed (fish), feed (gluten), feed (horse), feed (miscellaneous), feed (pig), feed (poultry), feed (poultry, pig), feed (rabbit), feed (sheep), feed (50–60 % maize), fish meal, grain by-products, grains (no specification), greengram, hay/silage, horsegram, husk, *Jagni* cake, legume mixture, linseed, linseed cake, livol, *Mahua* cake, maize, maize germ, maize gluten, maize grits, maize husk, maize meal, maize oil cake, maize powder, maize screenings, maize, ground, maize, hybrid, maize, preharvest, maize, yellow, maize (dark grains), *Makhana* (*Euryale ferox* Salisb) puffs, manioc, milk production mixed feed, mung testa, murkool, mustard cake, neem cake, niger cake, oats, palm kernel expeller cake, palm kernels, palm products, peanut cake, peanut cake (deoiled), peanut expeller, peanut hay, peanut meal, peanut, kernels, peanut, shells, peanuts, pellets, finisher, pig meal and pellets, pigeon pea, poultry feeds (peanut containing), rapeseed cake, redgram husk, rice, rice bran, rice bran (deoiled), rice chaff, rice crack, rice germ, rice germ cake, rice meal, rice straw, rice (damaged), rice (polish), safflower cake, sal seed cake, sesame, sesame cake, sorghum, soybean meal, soybeans, sunflower, sunflower cake, sunflower flour, tapioca, wheat, wheat bran, wheat bran and chana testa

AFLATOXIN B$_2$
incidence: 1/1, conc.: 4 μg/kg, country: India[129]
see also animal feedstuff (dairy cake), bird food, bird food, wild, blackgram, blackgram husk, cottonseed, cottonseed cake, cottonseed extract, cottonseed meal, cottonseed meal (ammoniated), cottonseed meats, dog food, egg production mixed feed, feed, compound, feed (cat), feed (cattle), feed (dog), feed (pig), feed (poultry), feed (rabbit), feed (sheep), fish meal, horsegram, maize, maize gluten, maize husk, maize, ground, maize, preharvest, mung testa, mustard cake, niger cake, peanut cake, peanut cake (deoiled), peanut expeller, peanut hay, peanut meal, peanuts, peanut, kernels, redgram husk, rice bran, rice bran (deoiled), rice chaff, rice meal, rice (polish),

sal seed cake, sesame cake, sorghum, soybean meal, soybeans, wheat, wheat bran

Blackgram   Blackgram for feed may contain the following mycotoxins:

AFLATOXIN B$_1$
incidence: ?/10, conc. range: 10–15 μg/kg, country: India[183]
see also alfalfa, *Ambadi* cake, animal feedstuffs (dairy cake), bagasse, barley, bengalgram husk, bird food, bird food, wild, biri testa, blackgram husk, bran, broiler mixed feed, calf fattening mixed feed, calf fattening mixed feed (containing 4–20 % peanut products), *Carthamus* cake, castor cake, cereals, cereal products, chick pea, coconut cake, cocos, concentrate, mixed, concentrates, cotton cake, cottonseed, cottonseed (dehulled), cottonseed cake, cottonseed extract, cottonseed meal, cottonseed meal (ammoniated), cottonseed meal (decorticated), cottonseed meats, cottonseed products, crumbles, crumbles, grower, cycad meal, dairy cattle feed, dairy cattle feed (containing 2–5 % peanut products), dairy cattle feed (containing 6–10 % peanut products), dairy cattle feed (containing 6–12 % peanut products), dairy cattle feed (containing more than 20 % peanut products), diets, mixed, dog food, egg production mixed feed, feed, feed and ingredients, feed, compound, feed, layer, feed, mixed, feed (beef), feed (broilers), feed (calf), feed (cat), feed (cattle), feed (chicken), feed (dairy), feed (dog), feed (dug), feed (fish), feed (gluten), feed (horse), feed (miscellaneous), feed (pig), feed (poultry), feed (poultry, pig), feed (rabbit), feed (sheep), feed (50–60 % maize), fish meal, grain by-products, grains (no specification), greengram, hay/silage, horsegram, husk, *Jagni* cake, legume mixture, linseed, linseed cake, livol, *Mahua* cake, maize, maize germ, maize gluten, maize grits, maize husk, maize meal, maize oil cake, maize powder, maize screenings, maize, ground, maize, hybrid, maize, preharvest, maize, yellow, maize (dark grains), *Makhana* (*Euryale ferox* Salisb) puffs, manioc, milk production mixed feed, mung

testa, murkool, mustard cake, neem cake, niger cake, oats, palm kernel expeller cake, palm kernels, palm products, peanut cake, peanut cake (deoiled), peanut expeller, peanut hay, peanut meal, peanut, kernels, peanut, shells, peanuts, pellets, finisher, pig meal and pellets, pigeon pea, poultry feeds (peanut containing), rapeseed cake, redgram husk, rice, rice bran, rice bran (deoiled), rice chaff, rice crack, rice germ, rice germ cake, rice meal, rice straw, rice (damaged), rice (polish), safflower cake, sal seed cake, sesame, sesame cake, sorghum, soybean meal, soybeans, sunflower, sunflower cake, sunflower flour, tapioca, wheat, wheat bran, wheat bran and chana testa

AFLATOXIN B$_2$
incidence: ?/10, conc. range: 10 μg/kg, country: India[183]
see also animal feedstuff (dairy cake), bird food, bird food, wild, biri testa, blackgram husk, cottonseed, cottonseed cake, cottonseed extract, cottonseed meal, cottonseed meal (ammoniated), cottonseed meats, dog food, egg production mixed feed, feed, compound, feed (cat), feed (cattle), feed (dog), feed (pig), feed (poultry), feed (rabbit), feed (sheep), fish meal, horsegram, maize, maize gluten, maize husk, maize, ground, maize, preharvest, mung testa, mustard cake, niger cake, peanut cake, peanut cake (deoiled), peanut expeller, peanut hay, peanut meal, peanuts, peanut, kernels, redgram husk, rice bran, rice bran (deoiled), rice chaff, rice meal, rice (polish), sal seed cake, sesame cake, sorghum, soybean meal, soybeans, wheat, wheat bran

Blackgram husk   may contain the following mycotoxins:

AFLATOXIN B$_1$
incidence: 2/17, conc. range: $\leq$25 μg/kg (1 sa), 26–50 μg/kg (1 sa), $\varnothing$ conc.: 22.5 μg/kg, country: India[247]
see also alfalfa, *Ambadi* cake, animal feedstuffs (dairy cake), bagasse, barley, bengalgram husk, bird food, bird food, wild, biri testa, blackgram, bran, broiler mixed feed, calf

fattening mixed feed, calf fattening mixed feed (containing 4–20 % peanut products), *Carthamus* cake, castor cake, cereals, cereal products, chick pea, coconut cake, cocos, concentrate, mixed, concentrates, cotton cake, cottonseed, cottonseed (dehulled), cottonseed cake, cottonseed extract, cotton-seed meal, cottonseed meal (ammoniated), cottonseed meal (decorticated), cottonseed meats, cottonseed products, crumbles, crumbles, grower, cycad meal, dairy cattle feed, dairy cattle feed (containing 2–5 % peanut products), dairy cattle feed (containing 6–10 % peanut products), dairy cattle feed (containing 6–12 % peanut products), dairy cattle feed (containing more than 20 % peanut products), diets, mixed, dog food, egg production mixed feed, feed, feed and ingredients, feed, compound, feed, layer, feed, mixed, feed (beef), feed (broilers), feed (calf), feed (cat), feed (cattle), feed (chicken), feed (dairy), feed (dog), feed (dug), feed (fish), feed (gluten), feed (horse), feed (miscellaneous), feed (pig), feed (poultry), feed (poultry, pig), feed (rabbit), feed (sheep), feed (50–60 % maize), fish meal, grain by-products, grains (no specification), greengram, hay/silage, horsegram, husk, *Jagni* cake, legume mixture, linseed, linseed cake, livol, *Mahua* cake, maize, maize germ, maize gluten, maize grits, maize husk, maize meal, maize oil cake, maize powder, maize screenings, maize, ground, maize, hybrid, maize, preharvest, maize, yellow, maize (dark grains), *Makhana* (*Euryale ferox* Salisb) puffs, manioc, milk production mixed feed, mung testa, murkool, mustard cake, neem cake, niger cake, oats, palm kernel expeller cake, palm kernels, palm products, peanut cake, peanut cake (deoiled), peanut expeller, peanut hay, peanut meal, peanut, kernels, peanut, shells, peanuts, pellets, finisher, pig meal and pellets, pigeon pea, poultry feeds (peanut containing), rapeseed cake, redgram husk, rice, rice bran, rice bran (deoiled), rice chaff, rice crack, rice germ, rice germ cake, rice meal, rice straw, rice (damaged), rice (polish), safflower cake, sal seed cake, sesame, sesame cake, sorghum, soybean meal, soybeans, sunflower, sunflower cake, sunflower flour, tapioca, wheat, wheat bran, wheat bran and chana testa

AFLATOXIN B$_2$
incidence: 1/17, conc.: 3 µg/kg, country: India[247]
see also animal feedstuff (dairy cake), bird food, bird food, wild, biri testa, blackgram, cottonseed, cottonseed cake, cottonseed extract, cottonseed meal, cottonseed meal (ammoniated), cottonseed meats, dog food, egg production mixed feed, feed (cat), feed (cattle), feed (dog), feed (pig), feed (poultry), feed (rabbit), feed (sheep), fish meal, horsegram, maize, maize gluten, maize husk, maize, ground, maize, preharvest, mung testa, mustard cake, niger cake, peanut cake, peanut cake (deoiled), peanut expeller, peanut hay, peanut meal, peanuts, peanut, kernels, redgram husk, rice bran, rice bran (deoiled), rice chaff, rice meal, rice (polish), sal seed cake, sesame cake, sorghum, soybean meal, soybeans, wheat, wheat bran

AFLATOXIN
incidence: 2/17, conc. range: 9–39 µg/kg, country: India[247]
see also bread crumbs, broiler finisher, broiler starter, cotton cake, cottonseed, cottonseed cake, cottonseed extract, cottonseed meal, feed, feed (cattle), feed (cow), feed (dog), feed (horse), feed (maize, gluten), feed (pig), feed (poultry), feed (rabbit), feed (rat/mice), feed (sheep), feeds, grain, fish meal, flour (wheat), groundnut cake, grower's mash, horsegram, layer's mash, maize, maize gluten, maize, white, milo, peanut cake, peanut cake (deoiled), peanut, kernels, peanut (oil cake), pearlmillet, pig breeder's mash, pig finisher, pig starter, pod with haulms, poultry breeder's mash, rabbit pellets, redgram husk, rice, rice bran (deoiled), rice, broken, rice (polish), sesame cake, silk worm pupae, sorghum, soybean cake, soybean meal, wheat, wheat bran

Bone meal may contain the following mycotoxins:

DEOXYNIVALENOL
incidence: 2/4, conc. range: 100–150 µg/kg, Ø conc.: 125 µg/kg, country: Egypt[16]
see also barley, barley, husked, barley, unhusked (naked), barley (pressed), bran, broilers feed, calf fattening mixed feed, coconut, expeller, corn cob mix silage, cottonseed, cottonseed cake, dairy cattle feed, egg production mixed feed, feed, feed components, feed, commercial mix, feed, mixed, feed, mixed (primarily maize), feed (barley), feed (cattle), feed (chicken), feed (dog), feed (fish), feed (mill run, from wheat), feed (mink), feed (pig), feed (poultry), feed (reindeer), feeds, grain, feeds, industrial, feedstuff, feedstuffs (rapeseed, turnip, fish meal, concentrates), fish meal, grain, mixed feed, grains, mixed, grains (no specification), maize, maize ears, maize fibre, maize germ, maize germ/bran, maize germ meal, maize gluten, maize kernels, maize meal, maize powder, maize screenings, maize stalks (pith), maize, "Baby", maize, hybrid, maize, white, oats, rice bran, rice germ cake, rye, silage, sorghum, soybeans, triticale, wheat, wheat and barley, wheat, red hard winter, wheat, spring, wheat, soft white winter, wheat, winter

ZEARALENONE
incidence: 4/4, conc. range: 12–55 µg/kg, Ø conc.: 30 µg/kg, country: Egypt[16]
see also alfalfa, barley, barley, husked, barley, unhusked (naked), barley and feed, bran, broilers feed, *Carthamus* cake, chick pea, concentrate, mixed, corn cob mix silage, cotton cake, cottonseed, cottonseed cake, diet (dairy cow), diet (poultry), diets (mixed), feed, feed components, feed, mixed, feed, mixed (primarily maize also maize, oats, wheat), feed (bran), feed (broiler chicken), feed (cattle), feed (chicken), feed (dairy), feed (developing pig), feed (mill run, from wheat), feed (miscellaneous), feed (pig), feed (poultry), feed (poultry, pig), feed (starter chicken), feedstuff, fish meal, forage grass, grain, bruised, grain, mixed feed, grains (no specification), hay, maize, maize ears, maize flakes, maize germ, maize germ/bran, maize

gluten, maize kernels, maize meal, maize oil cake, maize screenings, maize stalks (pith), maize, "Baby", maize, hybrid, maize, shelled, maize, unshelled, maize, white, maize grain, artificially dried, maize grain, crib dried, maize grain, ensiled, milk production mixed feed, oats, *Paspalum palidosum*, straw, peanut hulls/skins, rice bran, rice germ, rice germ cake, rye, silage, sorghum, soybeans, soybeans, extracted, sunflower cake, tapioca, triticale, wheat, wheat bran, wheat bran and chana testa, wheat soya meal

Bran   Bran for feed may contain the following mycotoxins:

Aflatoxin B$_1$
incidence: 1/1, conc.: 70 µg/kg, country: Australia[121]
see also alfalfa, *Ambadi* cake, animal feedstuffs (dairy cake), bagasse, barley, bengalgram husk, bird food, bird food, wild, biri testa, blackgram, blackgram husk, broiler mixed feed, calf fattening mixed feed, calf fattening mixed feed (containing 4–20 % peanut products), *Carthamus* cake, castor cake, cereals, cereal products, chick pea, coconut cake, cocos, concentrate, mixed, concentrates, cotton cake, cottonseed, cottonseed (dehulled), cottonseed cake, cottonseed extract, cottonseed meal, cottonseed meal (ammoniated), cottonseed meal (decorticated), cottonseed meats, cottonseed products, crumbles, crumbles, grower, cycad meal, dairy cattle feed, dairy cattle feed (containing 2–5 % peanut products), dairy cattle feed (containing 6–10 % peanut products), dairy cattle feed (containing 6–12 % peanut products), dairy cattle feed (containing more than 20 % peanut products), diets, mixed, dog food, egg production mixed feed, feed, feed and ingredients, feed, compound, feed, layer, feed, mixed, feed (beef), feed (broilers), feed (calf), feed (cat), feed (cattle), feed (chicken), feed (dairy), feed (dog), feed (dug), feed (fish), feed (gluten), feed (horse), feed (miscellaneous), feed (pig), feed (poultry), feed (poultry, pig), feed (rabbit), feed

(sheep), feed (50–60 % maize), fish meal, grain by-products, grains (no specification), greengram, hay/silage, horsegram, husk, *Jagni* cake, legume mixture, linseed, linseed cake, livol, *Mahua* cake, maize, maize germ, maize gluten, maize grits, maize husk, maize meal, maize oil cake, maize powder, maize screenings, maize, ground, maize, hybrid, maize, preharvest, maize, yellow, maize (dark grains), *Makhana* (*Euryale ferox* Salisb) puffs, manioc, milk production mixed feed, mung testa, murkool, mustard cake, neem cake, niger cake, oats, palm kernel expeller cake, palm kernels, palm products, peanut cake, peanut cake (deoiled), peanut expeller, peanut hay, peanut meal, peanut, kernels, peanut, shells, peanuts, pellets, finisher, pig meal and pellets, pigeon pea, poultry feeds (peanut containing), rapeseed cake, redgram husk, rice, rice bran, rice bran (deoiled), rice chaff, rice crack, rice germ, rice germ cake, rice meal, rice straw, rice (damaged), rice (polish), safflower cake, sal seed cake, sesame, sesame cake, sorghum, soybean meal, soybeans, sunflower, sunflower cake, sunflower flour, tapioca, wheat, wheat bran, wheat bran and chana testa

Deoxynivalenol
incidence: 78/85, conc. range: 50–960 µg/kg, Ø conc.: 292.9 µg/kg, country: Hungary[6]
see also barley, barley, husked, barley, unhusked (naked), barley (pressed), bone meal, broilers feed, calf fattening mixed feed, coconut, expeller, corn cob mix silage, cottonseed, cottonseed cake, dairy cattle feed, egg production mixed feed, feed, feed components, feed, commercial mix, feed, mixed, feed, mixed (primarily maize), feed (barley), feed (cattle), feed (chicken), feed (dog), feed (fish), feed (mill run, from wheat), feed (mink), feed (pig), feed (poultry), feed (reindeer), feeds, grain, feeds, industrial, feedstuff, feedstuffs (rapeseed, turnip, fish meal, concentrates), fish meal, grain, mixed feed, grains, mixed, grains (no specification), maize, maize ears, maize fibre, maize germ, maize germ/bran, maize germ meal, maize gluten, maize kernels, maize

meal, maize powder, maize screenings, maize stalks (pith), maize, "Baby", maize, hybrid, maize, white, oats, rice bran, rice germ cake, rye, silage, sorghum, soybeans, triticale, wheat, wheat and barley, wheat, red hard winter, wheat, soft white winter, wheat, spring, wheat, winter

NIVALENOL
incidence: 11/85, conc. range: 60–180 μg/kg, Ø conc.: 137.3 μg/kg, country: Hungary[6]
see also barley, barley, husked, barley, unhusked (naked), barley (pressed), feed, feed components, feed (cattle), feed (poultry), feed (reindeer), feeds, industrial, maize, maize ears, maize germ, maize germ/bran, maize gluten, maize kernels, maize meal, maize powder, maize screenings, maize, "Baby", oats, rye, silage, triticale, wheat

T-2 TOXIN
incidence: 15/85, conc. range: 50–950 μg/kg, Ø conc.: 234.7 μg/kg, country: Hungary[6]
see also alfalfa, barley, diet (grower), diet (poultry), feed, feed components, feed, layer, feed, mixed, feed (dog), feed (fish), feed (mink), feed (pig), feed (poultry), feedstuff, grain, mixed feed, grains (no specification), hay, maize, maize germ/bran, maize gluten, maize meal, maize screenings, maize stalks (pith), oat and barley (hammer-milled), oats, peanuts, pig grower diet, piglet diet, rye, sorghum, triticale, wheat

ZEARALENONE
incidence: 53/85, conc. range: 50–1560 μg/kg, Ø conc.: 308.5 μg/kg, country: Hungary[6]
see also alfalfa, barley, barley, husked, barley, unhusked (naked), barley and feed, bone meal, broilers feed, *Carthamus* cake, chick pea, concentrate, mixed, corn cob mix silage, cotton cake, cottonseed, cottonseed cake, diet (dairy cow), diet (poultry), diets (mixed), feed, feed components, feed, mixed, feed, mixed (primarily maize also maize, oats, wheat), feed (bran), feed (broiler chicken), feed (cattle), feed (chicken), feed (dairy), feed (developing pig), feed (mill run, from wheat), feed (miscellaneous), feed (pig), feed (poultry), feed (poultry, pig), feed (starter chicken), feedstuff, fish meal, forage grass,

grain, bruised, grain, mixed feed, grains (no specification), hay, maize, maize ears, maize flakes, maize germ, maize germ/bran, maize gluten, maize kernels, maize meal, maize oil cake, maize screenings, maize stalks (pith), maize, "Baby", maize, hybrid, maize, shelled, maize, unshelled, maize, white, maize grain, artificially dried, maize grain, crib dried, maize grain, ensiled, milk production mixed feed, oats, *Paspalum palidosum*, straw, peanut hulls/skins, rice bran, rice germ, rice germ cake, rye, silage, sorghum, soybeans, soybeans, extracted, sunflower cake, tapioca, triticale, wheat, wheat bran, wheat bran and chana testa, wheat soya meal

Bread crumbs   may contain the following mycotoxins:

AFLATOXIN
incidence: 1/1, conc.: <30 μg/kg, country: India[380]
see also blackgram husk, broiler finisher, broiler starter, cotton cake, cottonseed, cottonseed cake, cottonseed extract, cottonseed meal, feed, feed (cattle), feed (cow), feed (dog), feed (horse), feed (maize, gluten), feed (pig), feed (poultry), feed (rabbit), feed (rat/mice), feed (sheep), feeds, grain, fish meal, flour (wheat), groundnut cake, grower's mash, horsegram, layer's mash, maize, maize gluten, maize, white, milo, peanut cake, peanut cake (deoiled), peanut, kernels, peanut (oil cake), pearlmillet, pig breeder's mash, pig finisher, pig starter, pod with haulms, poultry breeder's mash, rabbit pellets, redgram husk, rice, rice bran (deoiled), rice, broken, rice (polish), sesame cake, silk worm pupae, sorghum, soybean cake, soybean meal, wheat, wheat bran

Broiler finisher   may contain the following mycotoxins:

AFLATOXIN
incidence: 14/25, conc. range: 0–180 μg/kg, country: Nigeria[109]
see also blackgram husk, bread crumbs, broiler starter, cotton cake, cottonseed,

cottonseed cake, cottonseed extract, cottonseed meal, feed, feed (cattle), feed (cow), feed (dog), feed (horse), feed (maize, gluten), feed (pig), feed (poultry), feed (rabbit), feed (rat/mice), feed (sheep), feeds, grain, fish meal, flour (wheat), groundnut cake, grower's mash, horsegram, layer's mash, maize, maize gluten, maize, white, milo, peanut cake, peanut cake (deoiled), peanut, kernels, peanut (oil cake), pearlmillet, pig breeder's mash, pig finisher, pig starter, pod with haulms, poultry breeder's mash, rabbit pellets, redgram husk, rice, rice bran (deoiled), rice, broken, rice (polish), sesame cake, silk worm pupae, sorghum, soybean cake, soybean meal, wheat, wheat bran

Broiler mixed feed   may contain the following mycotoxins:

Aflatoxin B$_1$
incidence: 2/5, Ø conc.: 10 µg/kg, country: Egypt[16]
see also alfalfa, *Ambadi* cake, animal feedstuffs (dairy cake), bagasse, barley, bengalgram husk, bird food, bird food, wild, biri testa, blackgram, blackgram husk, bran, calf fattening mixed feed, calf fattening mixed feed (containing 4–20% peanut products), *Carthamus* cake, castor cake, cereals, cereal products, chick pea, coconut cake, cocos, concentrate, mixed, concentrates, cotton cake, cottonseed, cottonseed (dehulled), cottonseed cake, cottonseed extract, cottonseed meal, cottonseed meal (ammoniated), cottonseed meal (decorticated), cottonseed meats, cottonseed products, crumbles, crumbles, grower, cycad meal, dairy cattle feed, dairy cattle feed (containing 2–5% peanut products), dairy cattle feed (containing 6–10% peanut products), dairy cattle feed (containing 6–12% peanut products), dairy cattle feed (containing more than 20% peanut products), diets, mixed, dog food, egg production mixed feed, feed and ingredients, feed, compound, feed, layer, feed, mixed, feed (beef), feed (broilers), feed (calf), feed (cat), feed (cattle), feed (chicken), feed (dairy), feed

(dog), feed (dug), feed (fish), feed (gluten), feed (horse), feed (miscellaneous), feed (pig), feed (poultry), feed (poultry, pig), feed (rabbit), feed (sheep), feed (50–60% maize), fish meal, grain by-products, grains (no specification), greengram, hay/silage, horsegram, husk, *Jagni* cake, legume mixture, linseed, linseed cake, livol, *Mahua* cake, maize, maize germ, maize gluten, maize grits, maize husk, maize meal, maize oil cake, maize powder, maize screenings, maize, ground, maize, hybrid, maize, preharvest, maize, yellow, maize (dark grains), *Makhana* (*Euryale ferox* Salisb) puffs, manioc, milk production mixed feed, mung testa, murkool, mustard cake, neem cake, niger cake, oats, palm kernel expeller cake, palm kernels, palm products, peanut cake, peanut cake (deoiled), peanut expeller, peanut hay, peanut meal, peanut, kernels, peanut, shells, peanuts, pellets, finisher, pig meal and pellets, pigeon pea, poultry feeds (peanut containing), rapeseed cake, redgram husk, rice, rice bran, rice bran (deoiled), rice chaff, rice crack, rice germ, rice germ cake, rice meal, rice straw, rice (damaged), rice (polish), safflower cake, sal seed cake, sesame, sesame cake, sorghum, soybean meal, soybeans, sunflower, sunflower cake, sunflower flour, tapioca, wheat, wheat bran, wheat bran and chana testa

Broiler starter   may contain the following mycotoxins:

Aflatoxin
incidence: 13/24, conc. range: 0–322 µg/kg, country: Nigeria[109]
see also blackgram husk, bread crumbs, broiler finisher, cotton cake, cottonseed, cottonseed cake, cottonseed extract, cottonseed meal, feed, feed (cattle), feed (cow), feed (dog), feed (horse), feed (maize, gluten), feed (pig), feed (poultry), feed (rabbit), feed (rat/mice), feed (sheep), feeds, grain, fish meal, flour (wheat), groundnut cake, grower's mash, horsegram, layer's mash, maize, maize gluten, maize, white, milo, peanut cake, peanut cake (deoiled), peanut, kernels, peanut (oil cake), pearlmillet, pig

breeder's mash, pig finisher, pig starter, pod with haulms, poultry breeder's mash, rabbit pellets, redgram husk, rice, rice bran (deoiled), rice, broken, rice (polish), sesame cake, silk worm pupae, sorghum, soybean cake, soybean meal, wheat, wheat bran

Broilers feed   may contain the following mycotoxins:

DEOXYNIVALENOL
incidence: 2/4, conc. range: 680–960 µg/kg, Ø conc.: 820 µg/kg, country: Egypt[16]
see also barley, barley, husked, barley, unhusked (naked), barley (pressed), bone meal, bran, calf fattening mixed feed, coconut, expeller, corn cob mix silage, cottonseed, cottonseed cake, dairy cattle feed, egg production mixed feed, feed, feed components, feed, commercial mix, feed, mixed, feed, mixed (primarily maize), feed (barley), feed (cattle), feed (chicken), feed (dog), feed (fish), feed (mill run, from wheat), feed (mink), feed (pig), feed (poultry), feed (reindeer), feeds, grain, feeds, industrial, feedstuff, feedstuffs (rapeseed, turnip, fish meal, concentrates), fish meal, grain, mixed feed, grains, mixed, grains (no specification), maize, maize ears, maize fibre, maize germ, maize germ/bran, maize germ meal, maize gluten, maize kernels, maize meal, maize powder, maize screenings, maize stalks (pith), maize, „Baby", maize, hybrid, maize, white, oats, rice bran, rice germ cake, rye, silage, sorghum, soybeans, triticale, wheat, wheat and barley, wheat, red hard winter, wheat, soft white winter, wheat, spring, wheat, winter

OCHRATOXIN A
incidence: 1/3, conc.: 13 µg/kg, country: Egypt[16]
see also alfalfa, barley, barley, oats, barley (high moisture), barley-soybean diet, bird food, domestic, bird food, wild, cereal grains, citrus pulp, coconut, expeller, corn cob mix silage, diet (dairy cow), diet (poultry), diet (starter), dog food, eat, egg production mixed feed, feed, feed wheat, oat and barley,

feed, commercial mix, feed, mixed, feed, mixed (pelleted), feed (broilers), feed (cat), feed (cattle), feed (cereals), feed (pig), feed ec (pig), feed (poultry), feed ec (poultry), feed (poultry, pig), feed ec (rabbit), feed (trout), grain, mixed feed, grains, mixed, grains (heated), hay, horse bean, maize, maize feed, milo, maize gluten, maize meal, maize, white, *Makhana* (*Euryale ferox* Salisb) puffs, milk production mixed feed, millet, oat and barley (hammer-milled), oats, palm products, peanut cake, peas, peas and beans, pet food, pig feedstuffs, pig grower diet, pig meal, piglet diet, poultry feedstuffs, rice bran, rice germ, rice germ cake, rye, sorghum, soybean groats, sunflower, sunflower seeds, extracted, tapioca, triticale, *Vicia faba*, wheat, wheat and barley, wheat bran, wheat hay, wheat, oats

ZEARALENONE
incidence: 4/4, conc. range: 3–426 µg/kg, Ø conc.: 143 µg/kg, country: Egypt[16]
see also alfalfa, barley, barley, husked, barley, unhusked (naked), barley and feed, bone meal, bran, *Carthamus* cake, chick pea, concentrate, mixed, corn cob mix silage, cotton cake, cottonseed, cottonseed cake, diet (dairy cow), diet (poultry), diets (mixed), feed, feed components, feed, mixed, feed, mixed (primarily maize also maize, oats, wheat), feed (bran), feed (broiler chicken), feed (cattle), feed (chicken), feed (dairy), feed (developing pig), feed (mill run, from wheat), feed (miscellaneous), feed (pig), feed (poultry), feed (poultry, pig), feed (starter chicken), feedstuff, fish meal, forage grass, grain, bruised, grain, mixed feed, grains (no specification), hay, maize, maize ears, maize flakes, maize germ, maize germ/bran, maize gluten, maize kernels, maize meal, maize oil cake, maize screenings, maize stalks (pith), maize, „Baby", maize, hybrid, maize, shelled, maize, unshelled, maize, white, maize grain, artificially dried, maize grain, crib dried, maize grain, ensiled, milk production mixed feed, oats, *Paspalum palidosum*, straw, peanut hulls/skins, rice bran, rice germ, rice germ cake, rye, silage, sorghum, soybeans, soybeans, extracted, sunflower cake, tapioca,

triticale, wheat, wheat bran, wheat bran and chana testa, wheat soya meal

CALF FEED
see Feed (calf)

Calf fattening mixed feed   may contain the following mycotoxins:

AFLATOXIN B$_1$
incidence: 3/5, Ø conc.: 50 µg/kg, country: Egypt[16]
incidence: 37/37, conc. range: <5 µg/kg (30 sa), 6–10 µg/kg (5 sa), 11–20 µg/kg (2 sa), country: Germany[298]
see also alfalfa, *Ambadi* cake, animal feedstuffs (dairy cake), bagasse, barley, bengalgram husk, bird food, bird food, wild, biri testa, blackgram, blackgram husk, bran, broiler mixed feed, calf fattening mixed feed (containing 4–20% peanut products), *Carthamus* cake, castor cake, cereals, cereal products, chick pea, coconut cake, cocos, concentrate, mixed, concentrates, cotton cake, cottonseed, cottonseed (dehulled), cottonseed cake, cottonseed extract, cottonseed meal, cottonseed meal (ammoniated), cottonseed meal (decorticated), cottonseed meats, cottonseed products, crumbles, crumbles, grower, cycad meal, dairy cattle feed, dairy cattle feed (containing 2–5% peanut products), dairy cattle feed (containing 6–10% peanut products), dairy cattle feed (containing 6–12% peanut products), dairy cattle feed (containing more than 20% peanut products), diets, mixed, dog food, egg production mixed feed, feed and ingredients, feed, compound, feed, layer, feed, mixed, feed (beef), feed (broilers), feed (calf), feed (cat), feed (cattle), feed (chicken), feed (dairy), feed (dog), feed (dug), feed (fish), feed (gluten), feed (horse), feed (miscellaneous), feed (pig), feed (poultry), feed (poultry, pig), feed (rabbit), feed (sheep), feed (50–60% maize), fish meal, grain by-products, grains (no specification), greengram, hay/silage, horsegram, husk, *Jagni* cake, legume mixture, linseed, linseed cake, livol, *Mahua* cake, maize, maize germ, maize gluten, maize grits, maize husk, maize meal, maize oil cake, maize powder, maize screenings, maize, ground, maize, hybrid, maize, preharvest, maize, yellow, maize (dark grains), *Makhana* (*Euryale ferox* Salisb) puffs, manioc, milk production mixed feed, mung testa, murkool, mustard cake, neem cake, niger cake, oats, palm kernel expeller cake, palm kernels, palm products, peanut cake, peanut cake (deoiled), peanut expeller, peanut hay, peanut meal, peanut, kernels, peanut, shells, peanuts, pellets, finisher, pig meal and pellets, pigeon pea, poultry feeds (peanut containing), rapeseed cake, redgram husk, rice, rice bran, rice bran (deoiled), rice chaff, rice crack, rice germ, rice germ cake, rice meal, rice straw, rice (damaged), rice (polish), safflower cake, sal seed cake, sesame, sesame cake, sorghum, soybean meal, soybeans, sunflower, sunflower cake, sunflower flour, tapioca, wheat, wheat bran, wheat bran and chana testa

DEOXYNIVALENOL
incidence: 3/4, conc. range: 80–400 µg/kg, Ø conc.: 230 µg/kg, country: Egypt[16]
see also barley, barley, husked, barley, unhusked (naked), barley (pressed), bone meal, bran, broilers feed, coconut, expeller, corn cob mix silage, cottonseed, cottonseed cake, dairy cattle feed, egg production mixed feed, feed, feed components, feed, commercial mix, feed, mixed, feed, mixed (primarily maize), feed (barley), feed (cattle), feed (chicken), feed (dog), feed (fish), feed (mill run, from wheat), feed (mink), feed (pig), feed (poultry), feed (reindeer), feeds, grain, feeds, industrial, feedstuff, feedstuffs (rapeseed, turnip, fish meal, concentrates), fish meal, grain, mixed feed, grains, mixed, grains (no specification), maize, maize ears, maize fibre, maize germ, maize germ/bran, maize germ meal, maize gluten, maize kernels, maize meal, maize powder, maize screenings, maize stalks (pith), maize, „Baby", maize, hybrid, maize, white, oats, rice bran, rice germ cake, rye, silage, sorghum, soybeans, triticale, wheat, wheat and barley, wheat, red hard winter, wheat, soft white winter, wheat, spring, wheat, winter

Calf fattening mixed feed (containing 4–20% peanut products) may contain the following mycotoxins:

AFLATOXIN B_1
incidence: ?/8, conc. range: 10–300 μg/kg, country: Germany[298]
see also alfalfa, *Ambadi* cake, animal feedstuffs (dairy cake), bagasse, barley, bengalgram husk, bird food, bird food, wild, biri testa, blackgram, blackgram husk, bran, broiler mixed feed, calf fattening mixed feed, *Carthamus* cake, castor cake, cereals, cereal products, chick pea, coconut cake, cocos, concentrate, mixed, concentrates, cotton cake, cottonseed, cottonseed (dehulled), cottonseed cake, cottonseed extract, cottonseed meal, cottonseed meal (ammoniated), cottonseed meal (decorticated), cottonseed meats, cottonseed products, crumbles, crumbles, grower, cycad meal, dairy cattle feed, dairy cattle feed (containing 2–5% peanut products), dairy cattle feed (containing 6–10% peanut products), dairy cattle feed (containing 6–12% peanut products), dairy cattle feed (containing more than 20% peanut products), diets, mixed, dog food, egg production mixed feed, feed and ingredients, feed, compound, feed, layer, feed, mixed, feed (beef), feed (broilers), feed (calf), feed (cat), feed (cattle), feed (chicken), feed (dairy), feed (dog), feed (dug), feed (fish), feed (gluten), feed (horse), feed (miscellaneous), feed (pig), feed (poultry), feed (poultry, pig), feed (rabbit), feed (sheep), feed (50–60% maize), fish meal, grain by-products, grains (no specification), greengram, hay/silage, horsegram, husk, *Jagni* cake, legume mixture, linseed, linseed cake, livol, *Mahua* cake, maize, maize germ, maize gluten, maize grits, maize husk, maize meal, maize oil cake, maize powder, maize screenings, maize, ground, maize, hybrid, maize, preharvest, maize, yellow, maize (dark grains), *Makhana* (*Euryale ferox* Salisb) puffs, manioc, milk production mixed feed, mung testa, murkool, mustard cake, neem cake, niger cake, oats, palm kernel expeller cake, palm kernels, palm products, peanut cake, peanut cake (deoiled), peanut expeller, peanut hay, peanut meal, peanut, kernels, peanut, shells, peanuts, pellets, finisher, pig meal and pellets, pigeon pea, poultry feeds (peanut containing), rapeseed cake, redgram husk, rice, rice bran, rice bran (deoiled), rice chaff, rice crack, rice germ, rice germ cake, rice meal, rice straw, rice (damaged), rice (polish), safflower cake, sal seed cake, sesame, sesame cake, sorghum, soybean meal, soybeans, sunflower, sunflower cake, sunflower flour, tapioca, wheat, wheat bran, wheat bran and chana testa

Carthamus cake may contain the following mycotoxins:

AFLATOXIN B_1
incidence: ?/10, conc.: tr μg/kg, country: India[183]
see also alfalfa, *Ambadi* cake, animal feedstuffs (dairy cake), bagasse, barley, bengalgram husk, bird food, bird food, wild, biri testa, blackgram, blackgram husk, bran, broiler mixed feed, calf fattening mixed feed, calf fattening mixed feed (containing 4–20% peanut products), castor cake, cereals, cereal products, chick pea, coconut cake, cocos, concentrate, mixed, concentrates, cotton cake, cottonseed, cottonseed (dehulled), cottonseed cake, cottonseed extract, cottonseed meal, cottonseed meal (ammoniated), cottonseed meal (decorticated), cottonseed meats, cottonseed products, crumbles, crumbles, grower, cycad meal, dairy cattle feed, dairy cattle feed (containing 2–5% peanut products), dairy cattle feed (containing 6–10% peanut products), dairy cattle feed (containing 6–12% peanut products), dairy cattle feed (containing more than 20% peanut products), diets, mixed, dog food, egg production mixed feed, feed and ingredients, feed, compound, feed, layer, feed, mixed, feed (beef), feed (calf), feed (broilers), feed (cat), feed (cattle), feed (chicken), feed (dairy), feed (dog), feed (dug), feed (fish), feed (gluten), feed (horse), feed (miscellaneous), feed (pig), feed (poultry), feed (poultry, pig), feed (rabbit), feed (sheep), feed (50–60% maize), fish meal,

grain by-products, grains (no specification), greengram, hay/silage, horsegram, husk, *Jagni* cake, legume mixture, linseed, linseed cake, livol, *Mahua* cake, maize, maize germ, maize gluten, maize grits, maize husk, maize meal, maize oil cake, maize powder, maize screenings, maize, ground, maize, hybrid, maize, preharvest, maize, yellow, maize (dark grains), *Makhana* (*Euryale ferox* Salisb) puffs, manioc, milk production mixed feed, mung testa, murkool, mustard cake, neem cake, niger cake, oats, palm kernel expeller cake, palm kernels, palm products, peanut cake, peanut cake (deoiled), peanut expeller, peanut hay, peanut meal, peanut, kernels, peanut, shells, peanuts, pellets, finisher, pig meal and pellets, pigeon pea, poultry feeds (peanut containing), rapeseed cake, redgram husk, rice, rice bran, rice bran (deoiled), rice chaff, rice crack, rice germ, rice germ cake, rice meal, rice straw, rice (damaged), rice (polish), safflower cake, sal seed cake, sesame, sesame cake, sorghum, soybean meal, soybeans, sunflower, sunflower cake, sunflower flour, tapioca, wheat, wheat bran, wheat bran and chana testa

Zearalenone
incidence: ?/10, conc.: 20 μg/kg, country: India[183]
see also alfalfa, barley, barley, husked, barley, unhusked (naked), barley and feed, bone meal, bran, broilers feed, chick pea, concentrate, mixed, corn cob mix silage, cotton cake, cottonseed, cottonseed cake, diet (dairy cow), diet (poultry), diets (mixed), feed, feed components, feed, mixed, feed, mixed (primarily maize also maize, oats, wheat), feed (bran), feed (broiler chicken), feed (cattle), feed (chicken), feed (dairy), feed (developing pig), feed (mill run, from wheat), feed (miscellaneous), feed (pig), feed (poultry), feed (poultry, pig), feed (starter chicken), feedstuff, fish meal, forage grass, grain, bruised, grain, mixed feed, grains (no specification), hay, maize, maize ears, maize flakes, maize germ, maize germ/bran, maize gluten, maize kernels, maize meal, maize oil cake, maize screenings, maize, „Baby", maize,

hybrid, maize, shelled, maize, unshelled, maize, white, maize grain, artificially dried, maize grain, crib dried, maize grain, ensiled, maize stalks (pith), milk production mixed feed, oats, *Paspalum palidosum*, straw, peanut hulls/skins, rice bran, rice germ, rice germ cake, rye, silage, sorghum, soybeans, soybeans, extracted, sunflower cake, tapioca, triticale, wheat, wheat bran, wheat bran and chana testa, wheat soya meal

Castor cake   may contain the following mycotoxins:

Aflatoxin B$_1$
incidence: 4/6, conc. range: tr–259.1 μg/kg, Ø conc.: 76.3 μg/kg, country: India[253]
see also alfalfa, *Ambadi* cake, animal feedstuffs (dairy cake), bagasse, barley, bengalgram husk, bird food, bird food, wild, biri testa, blackgram, blackgram husk, bran, broiler mixed feed, calf fattening mixed feed, calf fattening mixed feed (containing 4–20% peanut products), *Carthamus* cake, cereals, cereal products, chick pea, coconut cake, cocos, concentrate, mixed, concentrates, cotton cake, cottonseed, cottonseed (dehulled), cottonseed cake, cottonseed extract, cottonseed meal, cottonseed meal (ammoniated), cottonseed meal (decorticated), cottonseed meats, cottonseed products, crumbles, crumbles, grower, cycad meal, dairy cattle feed, dairy cattle feed (containing 2–5% peanut products), dairy cattle feed (containing 6–10% peanut products), dairy cattle feed (containing 6–12% peanut products), dairy cattle feed (containing more than 20% peanut products), diets, mixed, dog food, egg production mixed feed, feed, feed and ingredients, feed, compound, feed, layer, feed, mixed, feed (beef), feed (broilers), feed (calf), feed (cat), feed (cattle), feed (chicken), feed (dairy), feed (dog), feed (dug), feed (fish), feed (gluten), feed (horse), feed (miscellaneous), feed (pig), feed (poultry), feed (poultry, pig), feed (rabbit), feed (sheep), feed (50–60% maize), fish meal, grain by-products, grains (no specification), greengram, hay/silage, horsegram, husk, *Jagni*

cake, legume mixture, linseed, linseed cake, livol, *Mahua* cake, maize, maize germ, maize gluten, maize grits, maize husk, maize meal, maize oil cake, maize powder, maize screenings, maize, ground, maize, hybrid, maize, preharvest, maize, yellow, maize (dark grains), *Makhana* (*Euryale ferox* Salisb) puffs, manioc, milk production mixed feed, mung testa, murkool, mustard cake, neem cake, niger cake, oats, palm kernel expeller cake, palm kernels, palm products, peanut cake, peanut cake (deoiled), peanut expeller, peanut hay, peanut meal, peanut, kernels, peanut, shells, peanuts, pellets, finisher, pig meal and pellets, pigeon pea, poultry feeds (peanut containing), rapeseed cake, redgram husk, rice, rice bran, rice bran (deoiled), rice chaff, rice crack, rice germ, rice germ cake, rice meal, rice straw, rice (damaged), rice (polish), safflower cake, sal seed cake, sesame, sesame cake, sorghum, soybean meal, soybeans, sunflower, sunflower cake, sunflower flour, tapioca, wheat, wheat bran, wheat bran and chana testa

CATTLE FEED
see Feed (cattle)

Cereal grains   Cereal grains for feed may contain the following mycotoxins:

OCHRATOXIN A
incidence: 20/296*, conc. range:
20–470 µg/kg, country: Poland[87], *ncac
see also alfalfa, barley, barley, oats, barley (high moisture), barley-soybean diet, bird food, domestic, bird food, wild, broilers feed, citrus pulp, coconut, expeller, corn cob mix silage, diet (dairy cow), diet (poultry), diet (starter), dog food, eat, egg production mixed feed, feed, feed wheat, oat and barley, feed, commercial mix, feed, mixed, feed, mixed (pelleted), feed (broilers), feed (cat), feed (cattle), feed (cereals), feed (pig), feed ec (pig), feed (poultry), feed ec (poultry), feed (poultry, pig), feed ec (rabbit), feed (trout), grain, mixed feed, grains, mixed, grains (heated), hay, horse bean, maize, maize feed, milo, maize gluten, maize meal, maize, white,

Makhana (*Euryale ferox* Salisb) puffs, milk production mixed feed, millet, oat and barley (hammer-milled), oats, palm products, peanut cake, peas, peas and beans, pet food, pig feedstuffs, pig grower diet, pig meal, piglet diet, poultry feedstuffs, rice bran, rice germ, rice germ cake, rye, sorghum, soybean groats, sunflower, sunflower seeds, extracted, tapioca, triticale, *Vicia faba*, wheat, wheat and barley, wheat bran, wheat hay, wheat, oats

Cereals, cereal products   Cereals and cereal products for feed may contain the following mycotoxins:

AFLATOXIN B$_1$
incidence: 1/29* (maize gluten), conc.:
27 µg/kg, country: UK[36], *imported?
see also alfalfa, *Ambadi* cake, animal feedstuffs (dairy cake), bagasse, barley, bengalgram husk, bird food, bird food, wild, biri testa, blackgram, blackgram husk, bran, broiler mixed feed, calf fattening mixed feed, calf fattening mixed feed (containing 4–20% peanut products), *Carthamus* cake, castor cake, chick pea, coconut cake, cocos, concentrate, mixed, concentrates, cotton cake, cottonseed, cottonseed (dehulled), cottonseed cake, cottonseed extract, cottonseed meal, cottonseed meal (ammoniated), cottonseed meal (decorticated), cottonseed meats, cottonseed products, crumbles, crumbles, grower, cycad meal, dairy cattle feed, dairy cattle feed (containing 2–5% peanut products), dairy cattle feed (containing 6–10% peanut products), dairy cattle feed (containing 6–12% peanut products), dairy cattle feed (containing more than 20% peanut products), diets, mixed, dog food, egg production mixed feed, feed and ingredients, feed, compound, feed, layer, feed, mixed, feed (beef), feed (broilers), feed (calf), feed (cat), feed (cattle), feed (chicken), feed (dairy), feed (dog), feed (dug), feed (fish), feed (gluten), feed (horse), feed (miscellaneous), feed (pig), feed (poultry), feed (poultry, pig), feed (rabbit), feed (sheep), feed (50–60% maize), fish meal, grain by-products, grains (no specification),

greengram, hay/silage, horsegram, husk, *Jagni* cake, legume mixture, linseed, linseed cake, livol, *Mahua* cake, maize, maize germ, maize gluten, maize grits, maize husk, maize meal, maize oil cake, maize powder, maize screenings, maize, ground, maize, hybrid, maize, preharvest, maize, yellow, maize (dark grains), *Makhana* (*Euryale ferox* Salisb) puffs, manioc, milk production mixed feed, mung testa, murkool, mustard cake, neem cake, niger cake, oats, palm kernel expeller cake, palm kernels, palm products, peanut cake, peanut cake (deoiled), peanut expeller, peanut hay, peanut meal, peanut, kernels, peanut, shells, peanuts, pellets, finisher, pig meal and pellets, pigeon pea, poultry feeds (peanut containing), rapeseed cake, redgram husk, rice, rice bran, rice bran (deoiled), rice chaff, rice crack, rice germ, rice germ cake, rice meal, rice straw, rice (damaged), rice (polish), safflower cake, sal seed cake, sesame, sesame cake, sorghum, soybean meal, soybeans, sunflower, sunflower cake, sunflower flour, tapioca, wheat, wheat bran, wheat bran and chana testa

Cereals, mixture may contain the following mycotoxins:

ALTERNARIOL
incidence: 1/5, conc.: 8 μg/kg, country: Germany[294]
see also oilseed rape meal, sunflower, sunflower seed meal, sunflower seeds, ensiling

ALTERNARIOL METHYL ETHER
incidence: 2/7, conc. range: 4–8 μg/kg, Ø conc.: 6 μg/kg, country: Germany[294]
see also barley, barley, oats, oats, sunflower, wheat, wheat, oats

Chick mash may contain the following mycotoxins:

CYCLOPIAZONIC ACID
incidence: 1/1, conc.: 1500 μg/kg (estimated), country: India[33]
see also feed, groundnut cake, maize, millet, little, peanuts, rice bran, wheat

Chick pea Chick pea for feed may contain the following mycotoxins:

AFLATOXIN B₁
incidence: ?/10, conc. range: tr–20 μg/kg, country: India[183]
see also alfalfa, *Ambadi* cake, animal feedstuffs (dairy cake), bagasse, barley, bengalgram husk, bird food, bird food, wild, biri testa, blackgram, blackgram husk, bran, broiler mixed feed, calf fattening mixed feed, calf fattening mixed feed (containing 4–20% peanut products), *Carthamus* cake, castor cake, cereals, cereal products, coconut cake, cocos, concentrate, mixed, concentrates, cotton cake, cottonseed, cottonseed (dehulled), cottonseed cake, cottonseed extract, cottonseed meal, cottonseed meal (ammoniated), cottonseed meal (decorticated), cottonseed meats, cottonseed products, crumbles, crumbles, grower, cycad meal, dairy cattle feed, dairy cattle feed (containing 2–5% peanut products), dairy cattle feed (containing 6–10% peanut products), dairy cattle feed (containing 6–12% peanut products), dairy cattle feed (containing more than 20 % peanut products), diets, mixed, dog food, egg production mixed feed, feed and ingredients, feed, compound, feed, layer, feed, mixed, feed (beef), feed (broilers), feed (calf), feed (cat), feed (cattle), feed (chicken), feed (dairy), feed (dog), feed (dug), feed (fish), feed (gluten), feed (horse), feed (miscellaneous), feed (pig), feed (poultry), feed (poultry, pig), feed (rabbit), feed (sheep), feed (50–60% maize), fish meal, grain by-products, grains (no specification), greengram, hay/silage, horsegram, husk, *Jagni* cake, legume mixture, linseed, linseed cake, livol, *Mahua* cake, maize, maize germ, maize gluten, maize grits, maize husk, maize meal, maize oil cake, maize powder, maize screenings, maize, ground, maize, hybrid, maize, preharvest, maize, yellow, maize (dark grains), *Makhana* (*Euryale ferox* Salisb) puffs, manioc, milk production mixed feed, mung testa, murkool, mustard cake, neem cake, niger cake, oats, palm kernel expeller cake, palm kernels, palm products, peanut cake, peanut cake (deoiled),

peanut expeller, peanut hay, peanut meal, peanut, kernels, peanut, shells, peanuts, pellets, finisher, pig meal and pellets, pigeon pea, poultry feeds (peanut containing), rapeseed cake, redgram husk, rice, rice bran, rice bran (deoiled), rice chaff, rice crack, rice germ, rice germ cake, rice meal, rice straw, rice (damaged), rice (polish), safflower cake, sal seed cake, sesame, sesame cake, sorghum, soybean meal, soybeans, sunflower, sunflower cake, sunflower flour, tapioca, wheat, wheat bran, wheat bran and chana testa

ZEARALENONE
incidence: ?/10, conc. range: 20–40 µg/kg, country: India[183]
see also alfalfa, barley, barley, husked, barley, unhusked (naked), barley and feed, bone meal, bran, broilers feed, *Carthamus* cake, concentrate, mixed, corn cob mix silage, cotton cake, cottonseed, cottonseed cake, diet (dairy cow), diet (poultry), diets (mixed), feed, feed components, feed, mixed, feed, mixed (primarily maize also maize, oats, wheat), feed (bran), feed (broiler chicken), feed (cattle), feed (chicken), feed (dairy), feed (developing pig), feed (mill run, from wheat), feed (miscellaneous), feed (pig), feed (poultry), feed (poultry, pig), feed (starter chicken), feedstuff, fish meal, forage grass, grain, bruised, grain, mixed feed, grains (no specification), hay, maize, maize ears, maize flakes, maize germ, maize germ/bran, maize gluten, maize kernels, maize meal, maize oil cake, maize screenings, maize stalks (pith), maize, „Baby", maize, hybrid, maize, shelled, maize, unshelled, maize, white, maize grain, artificially dried, maize grain, crib dried, maize grain, ensiled, milk production mixed feed, oats, *Paspalum palidosum*, straw, peanut hulls/skins, rice bran, rice germ, rice germ cake, rye, silage, sorghum, soybeans, soybeans, extracted, sunflower cake, tapioca, triticale, wheat, wheat bran, wheat bran and chana testa, wheat soya meal

Citrus pulp   may contain the following mycotoxins:

OCHRATOXIN A
incidence: 1/4*, conc.: 29 µg/kg, country: The Netherlands[103], *imported?
see also alfalfa, barley, barley, oats, barley (high moisture), barley-soybean diet, bird food, domestic, bird food, wild, broilers feed, cereal grains, coconut, expeller, corn cob mix silage, diet (dairy cow), diet (poultry), diet (starter), dog food, eat, egg production mixed feed, feed, feed wheat, oat and barley, feed, commercial mix, feed, mixed, feed, mixed (pelleted), feed (broilers), feed (cat), feed (cattle), feed (cereals), feed (pig), feed ec (pig), feed (poultry), feed ec (poultry), feed (poultry, pig), feed ec (rabbit), feed (trout), grain, mixed feed, grains, mixed, grains (heated), hay, horse bean, maize, maize feed, milo, maize gluten, maize meal, maize, white, *Makhana* (*Euryale ferox* Salisb) puffs, milk production mixed feed, millet, oat and barley (hammer-milled), oats, palm products, peanut cake, peas, peas and beans, pet food, pig feedstuffs, pig grower diet, pig meal, piglet diet, poultry feedstuffs, rice bran, rice germ, rice germ cake, rye, sorghum, soybean groats, sunflower, sunflower seeds, extracted, tapioca, triticale, *Vicia faba*, wheat, wheat and barley, wheat bran, wheat hay, wheat, oats

Cocoa (oilcake)   may contain the following mycotoxins:

AFLATOXIN B
incidence: 5/7, conc. range: ≥25–>100 µg/kg, country: France[46]
see also barley, cottonseed cake, feed, feed (cattle), feed (maize, gluten), feed (pig), feed (poultry), flour (wheat), lucern (dried), maize, maize gluten, maize grains, oats, peanut (oil cake), rice, broken, sorghum, soybean (oil cake), sunflower (oil cake), wheat

Coconut cake   may contain the following mycotoxins:

AFLATOXIN B$_1$
incidence: 7/10, conc. range: 10–60 µg/kg, country: India[183]

see also alfalfa, *Ambadi* cake, animal feedstuffs (dairy cake), bagasse, barley, bengalgram husk, bird food, bird food, wild, biri testa, blackgram, blackgram husk, bran, broiler mixed feed, calf fattening mixed feed, calf fattening mixed feed (containing 4–20 % peanut products), *Carthamus* cake, castor cake, cereals, cereal products, chick pea, cocos, concentrate, mixed, concentrates, cotton cake, cottonseed, cottonseed (dehulled), cottonseed cake, cottonseed extract, cottonseed meal, cottonseed meal (ammoniated), cottonseed meal (decorticated), cottonseed meats, cottonseed products, crumbles, crumbles, grower, cycad meal, dairy cattle feed, dairy cattle feed (containing 2–5 % peanut products), dairy cattle feed (containing 6–10 % peanut products), dairy cattle feed (containing 6–12 % peanut products), dairy cattle feed (containing more than 20 % peanut products), diets, mixed, dog food, egg production mixed feed, feed, feed and ingredients, feed, compound, feed, layer, feed, mixed, feed (beef), feed (broilers), feed (calf), feed (cat), feed (cattle), feed (chicken), feed (dairy), feed (dog), feed (dug), feed (fish), feed (gluten), feed (horse), feed (miscellaneous), feed (pig), feed (poultry), feed (poultry, pig), feed (rabbit), feed (sheep), feed (50–60 % maize), fish meal, grain by-products, grains (no specification), greengram, hay/silage, horsegram, husk, *Jagni* cake, legume mixture, linseed, linseed cake, livol, *Mahua* cake, maize, maize germ, maize gluten, maize grits, maize husk, maize meal, maize oil cake, maize powder, maize screenings, maize, ground, maize, hybrid, maize, preharvest, maize, yellow, maize (dark grains), *Makhana* (*Euryale ferox* Salisb) puffs, manioc, milk production mixed feed, mung testa, murkool, mustard cake, neem cake, niger cake, oats, palm kernel expeller cake, palm kernels, palm products, peanut cake, peanut cake (deoiled), peanut expeller, peanut hay, peanut meal, peanut, kernels, peanut, shells, peanuts, pellets, finisher, pig meal and pellets, pigeon pea, poultry feeds (peanut containing), rapeseed cake, redgram husk, rice, rice bran, rice bran (deoiled), rice chaff, rice crack, rice germ, rice germ cake, rice meal, rice straw, rice (damaged), rice (polish), safflower cake, sal seed cake, sesame, sesame cake, sorghum, soybean meal, soybeans, sunflower, sunflower cake, sunflower flour, tapioca, wheat, wheat bran, wheat bran and chana testa

Coconut, expeller   may contain the following mycotoxins:

DEOXYNIVALENOL
incidence: 1/4*, conc.: 620 µg/kg, country: The Netherlands[103], *imported?
see also barley, barley, husked, barley, unhusked (naked), barley (pressed), bone meal, bran, broilers feed, calf fattening mixed feed, corn cob mix silage, cottonseed, cottonseed cake, dairy cattle feed, egg production mixed feed, feed, feed components, feed, commercial mix, feed, mixed, feed, mixed (primarily maize), feed (barley), feed (cattle), feed (chicken), feed (dog), feed (fish), feed (mill run, from wheat), feed (mink), feed (pig), feed (poultry), feed (reindeer), feeds, grain, feeds, industrial, feedstuff, feedstuffs (rapeseed, turnip, fish meal, concentrates), fish meal, grain, mixed feed, grains, mixed, grains (no specification), maize, maize ears, maize fibre, maize germ, maize germ/bran, maize germ meal, maize gluten, maize kernels, maize meal, maize powder, maize screenings, maize stalks (pith), maize, "Baby", maize, hybrid, maize, white, oats, rice bran, rice germ cake, rye, silage, sorghum, soybeans, triticale, wheat, wheat and barley, wheat, red hard winter, wheat, soft white winter, wheat, spring, wheat, winter

OCHRATOXIN A
incidence: 3/4*, conc. range: 4–6 µg/kg, Ø conc.: 5 µg/kg, country: The Netherlands[103], *imported?
see also alfalfa, barley, barley, oats, barley (high moisture), barley-soybean diet, bird food, domestic, bird food, wild, broilers feed,

cereal grains, citrus pulp, corn cob mix silage, diet (dairy cow), diet (poultry), diet (starter), dog food, eat, egg production mixed feed, feed, feed wheat, oat and barley, feed, commercial mix, feed, mixed, feed, mixed (pelleted), feed (broilers), feed (cat), feed (cattle), feed (cereals), feed (pig), feed ec (pig), feed (poultry), feed ec (poultry), feed (poultry, pig), feed ec (rabbit), feed (trout), grain, mixed feed, grains, mixed, grains (heated), hay, horse bean, maize, maize feed, milo, maize gluten, maize meal, maize, white, *Makhana* (*Euryale ferox* Salisb) puffs, milk production mixed feed, millet, oat and barley (hammer-milled), oats, palm products, peanut cake, peas, peas and beans, pet food, pig feedstuffs, pig grower diet, pig meal, piglet diet, poultry feedstuffs, rice bran, rice germ, rice germ cake, rye, sorghum, soybean groats, sunflower, sunflower seeds, extracted, tapioca, triticale, *Vicia faba*, wheat, wheat and barley, wheat bran, wheat hay, wheat, oats

Cocos Cocos for feed may contain the following mycotoxins:

Aflatoxin B$_1$
incidence: 7/8, conc. range: 15–59 µg/kg, Ø conc.: 30 µg/kg, country: investigated in Germany[298]
see also alfalfa, *Ambadi* cake, animal feedstuffs (dairy cake), bagasse, barley, bengalgram husk, bird food, bird food, wild, biri testa, blackgram, blackgram husk, bran, broiler mixed feed, calf fattening mixed feed, calf fattening mixed feed (containing 4–20 % peanut products), *Carthamus* cake, castor cake, cereals, cereal products, chick pea, coconut cake, concentrate, mixed, concentrates, cotton cake, cottonseed, cottonseed (dehulled), cottonseed cake, cottonseed extract, cottonseed meal, cottonseed meal (ammoniated), cottonseed meal (decorticated), cottonseed meats, cottonseed products, crumbles, crumbles, grower, cycad meal, dairy cattle feed, dairy cattle feed (containing 2–5 % peanut products), dairy cattle feed (containing 6–10 % peanut products), dairy cattle feed (containing 6–12 % peanut products), dairy cattle feed (containing more than 20 % peanut products), diets, mixed, dog food, egg production mixed feed, feed, feed and ingredients, feed, compound, feed, layer, feed, mixed, feed (beef), feed (broilers), feed (calf), feed (cat), feed (cattle), feed (chicken), feed (dairy), feed (dog), feed (dug), feed (fish), feed (gluten), feed (horse), feed (miscellaneous), feed (pig), feed (poultry), feed (poultry, pig), feed (rabbit), feed (sheep), feed (50–60 % maize), fish meal, grain by-products, grains (no specification), greengram, hay/silage, horsegram, husk, *Jagni* cake, legume mixture, linseed, linseed cake, livol, *Mahua* cake, maize, maize germ, maize gluten, maize grits, maize husk, maize meal, maize oil cake, maize powder, maize screenings, maize, ground, maize, hybrid, maize, preharvest, maize, yellow, maize (dark grains), *Makhana* (*Euryale ferox* Salisb) puffs, manioc, milk production mixed feed, mung testa, murkool, mustard cake, neem cake, niger cake, oats, palm kernel expeller cake, palm kernels, palm products, peanut cake, peanut cake (deoiled), peanut expeller, peanut hay, peanut meal, peanut, kernels, peanut, shells, peanuts, pellets, finisher, pig meal and pellets, pigeon pea, poultry feeds (peanut containing), rapeseed cake, redgram husk, rice, rice bran, rice bran (deoiled), rice chaff, rice crack, rice germ, rice germ cake, rice meal, rice straw, rice (damaged), rice (polish), safflower cake, sal seed cake, sesame, sesame cake, sorghum, soybean meal, soybeans, sunflower, sunflower cake, sunflower flour, tapioca, wheat, wheat bran, wheat bran and chana testa

Concentrate, mixed may contain the following mycotoxins:

Aflatoxin B$_1$
incidence: 3/5, conc. range: 7–13 µg/kg, Ø conc.: 5.4 µg/kg, country: India[129]
see also alfalfa, *Ambadi* cake, animal feedstuffs (dairy cake), bagasse, barley, bengalgram husk, bird food, bird food, wild, biri testa, blackgram, blackgram husk, bran, broiler

mixed feed, calf fattening mixed feed, calf fattening mixed feed (containing 4–20 % peanut products), *Carthamus* cake, castor cake, cereals, cereal products, chick pea, coconut cake, cocos, concentrates, cotton cake, cottonseed, cottonseed (dehulled), cottonseed cake, cottonseed extract, cottonseed meal, cottonseed meal (ammoniated), cottonseed meal (decorticated), cottonseed meats, cottonseed products, crumbles, crumbles, grower, cycad meal, dairy cattle feed, dairy cattle feed (containing 2–5 % peanut products), dairy cattle feed (containing 6–10 % peanut products), dairy cattle feed (containing 6–12 % peanut products), dairy cattle feed (containing more than 20 % peanut products), diets, mixed, dog food, egg production mixed feed, feed, feed and ingredients, feed, compound, feed, layer, feed, mixed, feed (beef), feed (broilers), feed (calf), feed (cat), feed (cattle), feed (chicken), feed (dairy), feed (dog), feed (dug), feed (fish), feed (gluten), feed (horse), feed (miscellaneous), feed (pig), feed (poultry), feed (poultry, pig), feed (rabbit), feed (sheep), feed (50–60 % maize), fish meal, grain by-products, grains (no specification), greengram, hay/silage, horsegram, husk, *Jagni* cake, legume mixture, linseed, linseed cake, livol, *Mahua* cake, maize, maize germ, maize gluten, maize grits, maize husk, maize meal, maize oil cake, maize powder, maize screenings, maize, ground, maize, hybrid, maize, preharvest, maize, yellow, maize (dark grains), *Makhana* (*Euryale ferox* Salisb) puffs, manioc, milk production mixed feed, mung testa, murkool, mustard cake, neem cake, niger cake, oats, palm kernel expeller cake, palm kernels, palm products, peanut cake, peanut cake (deoiled), peanut expeller, peanut hay, peanut meal, peanut, kernels, peanut, shells, peanuts, pellets, finisher, pig meal and pellets, pigeon pea, poultry feeds (peanut containing), rapeseed cake, redgram husk, rice, rice bran, rice bran (deoiled), rice chaff, rice crack, rice germ, rice germ cake, rice meal, rice straw, rice (damaged), rice (polish), safflower cake, sal seed cake, sesame, sesame cake, sorghum, soybean meal, soybeans, sunflower, sunflower cake, sunflower flour, tapioca, wheat, wheat bran, wheat bran and chana testa

Aflatoxin G_1

incidence: 1/1, conc.: 3 µg/kg, country: India[129]

see also animal feedstuffs (dairy cake), bird food, bird food, wild, cottonseed, cottonseed cake, feed (cat), feed (chicken), feed (dog), maize, maize, ground, maize, preharvest, meat meal, milk production mixed feed, murkool, peanut cake, peanut expeller, peanut hay, peanut meal, peanuts, rice bran, rice germ, sorghum, soybeans, wheat, wheat bran, wheat bran and chana testa

Zearalenone

incidence: 1/5, conc.: 843 µg/kg, country: India[129]

see also alfalfa, barley, barley, husked, barley, unhusked (naked), barley and feed, bone meal, bran, broilers feed, *Carthamus* cake, chick pea, corn cob mix silage, cotton cake, cottonseed, cottonseed cake, diet (dairy cow), diet (poultry), diets (mixed), feed, feed components, feed, mixed, feed, mixed (primarily maize also maize, oats, wheat), feed (bran), feed (broiler chicken), feed (cattle), feed (chicken), feed (dairy), feed (developing pig), feed (mill run, from wheat), feed (miscellaneous), feed (pig), feed (poultry), feed (poultry, pig), feed (starter chicken), feedstuff, fish meal, forage grass, grain, bruised, grain, mixed feed, grains (no specification), hay, maize, maize ears, maize flakes, maize germ, maize germ/bran, maize gluten, maize kernels, maize meal, maize oil cake, maize screenings, maize stalks (pith), maize, "Baby", maize, hybrid, maize, shelled, maize, unshelled, maize, white, maize grain, artificially dried, maize grain, crib dried, maize grain, ensiled, milk production mixed feed, oats, *Paspalum palidosum*, straw, peanut hulls/skins, rice bran, rice germ, rice germ cake, rye, silage, sorghum, soybeans, soybeans, extracted, sunflower cake, tapioca, triticale, wheat, wheat bran, wheat bran and chana testa, wheat soya meal

Concentrates   may contain the following mycotoxins:

A F L A T O X I N B$_1$
incidence: 312/1545, conc. range:
1–2000 µg/kg, country: Cuba[106]
see also alfalfa, *Ambadi* cake, animal feedstuffs (dairy cake), bagasse, barley, bengalgram husk, bird food, bird food, wild, biri testa, blackgram, blackgram husk, bran, broiler mixed feed, calf fattening mixed feed, calf fattening mixed feed (containing 4–20 % peanut products), *Carthamus* cake, castor cake, cereals, cereal products, chick pea, coconut cake, cocos, concentrate, mixed, cotton cake, cottonseed, cottonseed (dehulled), cottonseed cake, cottonseed extract, cottonseed meal, cottonseed meal (ammoniated), cottonseed meal (decorticated), cottonseed meats, cottonseed products, crumbles, crumbles, grower, cycad meal, dairy cattle feed, dairy cattle feed (containing 2–5 % peanut products), dairy cattle feed (containing 6–10 % peanut products), dairy cattle feed (containing 6–12 % peanut products), dairy cattle feed (containing more than 20 % peanut products), diets, mixed, dog food, egg production mixed feed, feed, feed and ingredients, feed, compound, feed, layer, feed, mixed, feed (beef), feed (broilers), feed (calf), feed (cat), feed (cattle), feed (chicken), feed (dairy), feed (dog), feed (dug), feed (fish), feed (gluten), feed (horse), feed (miscellaneous), feed (pig), feed (poultry), feed (poultry, pig), feed (rabbit), feed (sheep), feed (50–60 % maize), fish meal, grain by-products, grains (no specification), greengram, hay/silage, horsegram, husk, *Jagni* cake, legume mixture, linseed, linseed cake, livol, *Mahua* cake, maize, maize germ, maize gluten, maize grits, maize husk, maize meal, maize oil cake, maize powder, maize screenings, maize, ground, maize, hybrid, maize, preharvest, maize, yellow, maize (dark grains), *Makhana* (*Euryale ferox* Salisb) puffs, manioc, milk production mixed feed, mung testa, murkool, mustard cake, neem cake, niger cake, oats, palm kernel expeller cake, palm kernels, palm products, peanut cake, peanut cake (deoiled), peanut expeller, peanut hay, peanut meal, peanut, kernels, peanut, shells, peanuts, pellets, finisher, pig meal and pellets, pigeon pea, poultry feeds (peanut containing), rapeseed cake, redgram husk, rice, rice bran, rice bran (deoiled), rice chaff, rice crack, rice germ, rice germ cake, rice meal, rice straw, rice (damaged), rice (polish), safflower cake, sal seed cake, sesame, sesame cake, sorghum, soybean meal, soybeans, sunflower, sunflower cake, sunflower flour, tapioca, wheat, wheat bran, wheat bran and chana testa

Copra meal   may contain the following mycotoxins:

A F L A T O X I N S
incidence: 1/1*, conc.: 16 µg/kg, country: USA[280], *imported
see also animal feed (maize), animal feed (mixed), barley, cottonseed, cottonseed fines, cottonseed meal, cottonseed meats, feed, feed ingredients (miscellaneous), feed, mixed, feed (excluding peanuts, suspect), feed (goat), feed (pig), feed (poultry), feedstuff, grain, grain, mixed feed, maize, maize, shelled, millet, peanut cake, peanut meal, peanut meal and by-products, peanuts, protein concentrates, rice, rice bran, sorghum, soybean meal, sunflower

C O R N
see Maize

Corn cob mix silage   may contain the following mycotoxins:

D E O X Y N I V A L E N O L
incidence: 1/2*, conc.: 130 µg/kg, country: The Netherlands[103], *not molded
see also barley, barley, husked, barley, unhusked (naked), barley (pressed), bone meal, bran, broilers feed, calf fattening mixed feed, coconut, expeller, cottonseed, cottonseed cake, dairy cattle feed, egg production mixed feed, feed, feed components, feed, commercial mix, feed, mixed, feed, mixed (primarily maize), feed (barley), feed

(cattle), feed (chicken), feed (dog), feed (fish), feed (mill run, from wheat), feed (mink), feed (pig), feed (poultry), feed (reindeer), feeds, grain, feeds, industrial, feedstuff, feedstuffs (rapeseed, turnip, fish meal, concentrates), fish meal, grain, mixed feed, grains, mixed, grains (no specification), maize, maize ears, maize fibre, maize germ, maize germ/bran, maize germ meal, maize gluten, maize kernels, maize meal, maize powder, maize screenings, maize stalks (pith), maize, "Baby", maize, hybrid, maize, white, oats, rice bran, rice germ cake, rye, silage, sorghum, soybeans, triticale, wheat, wheat and barley, wheat, red hard winter, wheat, soft white winter, wheat, spring, wheat, winter

OCHRATOXIN A
incidence: 3/4*, conc. range: 4–5 µg/kg, Ø conc.: 5 µg/kg, country: The Netherlands[103], *molded
incidence: 2/2*, conc. range: 3–6 µg/kg, Ø conc.: 5 µg/kg, country: The Netherlands[103], *not molded
see also alfalfa, barley, barley, oats, barley (high moisture), barley-soybean diet, bird food, domestic, bird food, wild, broilers feed, cereal grains, citrus pulp, coconut, expeller, diet (dairy cow), diet (poultry), diet (starter), dog food, eat, egg production mixed feed, feed, feed wheat, oat and barley, feed, commercial mix, feed, mixed, feed, mixed (pelleted), feed (broilers), feed (cat), feed (cattle), feed (cereals), feed (pig), feed ec (pig), feed (poultry), feed ec (poultry), feed (poultry, pig), feed ec (rabbit), feed (trout), grain, mixed feed, grains, mixed, grains (heated), hay, horse bean, maize, maize feed, milo, maize gluten, maize meal, maize, white, *Makhana* (*Euryale ferox* Salisb) puffs, milk production mixed feed, millet, oat and barley (hammer-milled), oats, palm products, peanut cake, peas, peas and beans, pet food, pig feedstuffs, pig grower diet, pig meal, piglet diet, poultry feedstuffs, rice bran, rice germ, rice germ cake, rye, sorghum, soybean groats, sunflower, sunflower seeds, extracted, tapioca, triticale, *Vicia faba*, wheat, wheat and barley, wheat bran, wheat hay, wheat, oats

ZEARALENONE
incidence: 4/4*, conc. range: 47–3100 µg/kg, Ø conc.: 890 µg/kg, country: The Netherlands[103], *molded
incidence: 2/2*, conc. range: 33–140 µg/kg, Ø conc.: 87 µg/kg, country: The Netherlands[103], *not molded
see also alfalfa, barley, barley, husked, barley, unhusked (naked), barley and feed, bone meal, bran, broilers feed, *Carthamus* cake, chick pea, concentrate, mixed, cotton cake, cottonseed, cottonseed cake, diet (dairy cow), diet (poultry), diets (mixed), feed, feed components, feed, mixed, feed, mixed (primarily maize also maize, oats, wheat), feed (bran), feed (broiler chicken), feed (cattle), feed (chicken), feed (dairy), feed (developing pig), feed (mill run, from wheat), feed (miscellaneous), feed (pig), feed (poultry), feed (poultry, pig), feed (starter chicken), feedstuff, fish meal, forage grass, grain, bruised, grain, mixed feed, grains (no specification), hay, maize, maize ears, maize flakes, maize germ, maize germ/bran, maize gluten, maize kernels, maize meal, maize oil cake, maize screenings, maize stalks (pith), maize, "Baby", maize, hybrid, maize, shelled, maize, unshelled, maize, white, maize grain, artificially dried, maize grain, crib dried, maize grain, ensiled, milk production mixed feed, oats, *Paspalum palidosum*, straw, peanut hulls/skins, rice bran, rice germ, rice germ cake, rye, silage, sorghum, soybeans, soybeans, extracted, sunflower cake, tapioca, triticale, wheat, wheat bran, wheat bran and chana testa, wheat soya meal

CORN POWDER
see Maize powder

Cotton cake   may contain the following mycotoxins:

AFLATOXIN B$_1$
incidence: ?/10, conc.: 10–40 µg/kg, country: India[183]

incidence: 8/10, conc.: tr–20 µg/kg, country: India[183]

see also alfalfa, *Ambadi* cake, animal feedstuffs (dairy cake), bagasse, barley, bengalgram husk, bird food, bird food, wild, biri testa, blackgram, blackgram husk, bran, broiler mixed feed, calf fattening mixed feed, calf fattening mixed feed (containing 4–20 % peanut products), *Carthamus* cake, castor cake, cereals, cereal products, chick pea, coconut cake, cocos, concentrate, mixed, concentrates, alfalfa, cottonseed, cottonseed (dehulled), cottonseed cake, cottonseed extract, cottonseed meal, cottonseed meal (ammoniated), cottonseed meal (decorticated), cottonseed meats, cottonseed products, crumbles, crumbles, grower, cycad meal, dairy cattle feed, dairy cattle feed (containing 2–5 % peanut products), dairy cattle feed (containing 6–10 % peanut products), dairy cattle feed (containing 6–12 % peanut products), dairy cattle feed (containing more than 20 % peanut products), diets, mixed, dog food, egg production mixed feed, feed, feed and ingredients, feed, compound, feed, layer, feed, mixed, feed (beef), feed (broilers), feed (calf), feed (cat), feed (cattle), feed (chicken), feed (dairy), feed (dog), feed (dug), feed (fish), feed (gluten), feed (horse), feed (miscellaneous), feed (pig), feed (poultry), feed (poultry, pig), feed (rabbit), feed (sheep), feed (50–60 % maize), fish meal, grain by-products, grains (no specification), greengram, hay/silage, horsegram, husk, *Jagni* cake, legume mixture, linseed, linseed cake, livol, *Mahua* cake, maize, maize germ, maize gluten, maize grits, maize husk, maize meal, maize oil cake, maize powder, maize screenings, maize, ground, maize, hybrid, maize, preharvest, maize, yellow, maize (dark grains), *Makhana* (*Euryale ferox* Salisb) puffs, manioc, milk production mixed feed, mung testa, murkool, mustard cake, neem cake, niger cake, oats, palm kernel expeller cake, palm kernels, palm products, peanut cake, peanut cake (deoiled), peanut expeller, peanut hay, peanut meal, peanut, kernels, peanut, shells, peanuts, pellets, finisher, pig meal and pellets, pigeon pea, poultry feeds (peanut containing), rapeseed cake, redgram husk, rice, rice bran, rice bran (deoiled), rice chaff, rice crack, rice germ, rice germ cake, rice meal, rice straw, rice (damaged), rice (polish), safflower cake, sal seed cake, sesame, sesame cake, sorghum, soybean meal, soybeans, sunflower, sunflower cake, sunflower flour, tapioca, wheat, wheat bran, wheat bran and chana testa

AFLATOXIN

incidence: 9/9, conc. range: 11–30 µg/kg (8 sa), 31–100 µg/kg (1 sa), country: India[311]

see also blackgram husk, bread crumbs, broiler finisher, broiler starter, cottonseed, cottonseed cake, cottonseed extract, cottonseed meal, feed, feed (cattle), feed (cow), feed (dog), feed (horse), feed (maize, gluten), feed (pig), feed (poultry), feed (rabbit), feed (rat/mice), feed (sheep), feeds, grain, fish meal, flour (wheat), groundnut cake, grower's mash, horsegram, layer's mash, maize, maize gluten, maize, white, milo, peanut cake, peanut cake (deoiled), peanut, kernels, peanut (oil cake), pearlmillet, pig breeder's mash, pig finisher, pig starter, pod with haulms, poultry breeder's mash, rabbit pellets, redgram husk, rice, rice bran (deoiled), rice, broken, rice (polish), sesame cake, silk worm pupae, sorghum, soybean cake, soybean meal, wheat, wheat bran

ZEARALENONE

incidence: ?/10, conc.: 20 µg/kg, country: India[183]

see also alfalfa, barley, barley, husked, barley, unhusked (naked), barley and feed, bone meal, bran, broilers feed, *Carthamus* cake, chick pea, concentrate, mixed, corn cob mix silage, cottonseed, cottonseed cake, diet (dairy cow), diet (poultry), diets (mixed), feed, feed components, feed, mixed, feed, mixed (primarily maize also maize, oats, wheat), feed (bran), feed (broiler chicken), feed (cattle), feed (chicken), feed (dairy), feed (developing pig), feed (mill run, from wheat), feed (miscellaneous), feed (pig), feed (poultry), feed (poultry, pig), feed (starter chicken), feedstuff, fish meal, forage grass, grain, bruised, grain, mixed feed, grains (no specification), hay, maize, maize ears, maize flakes, maize germ, maize germ/bran, maize gluten, maize kernels, maize meal, maize oil cake, maize screenings, maize stalks (pith), maize, "Baby", maize, hybrid, maize, shelled, maize, unshelled, maize, white, maize grain, artificially dried, maize grain, crib dried, maize grain, ensiled, milk production mixed feed, oats, *Paspalum palidosum*, straw, peanut hulls/skins, rice bran, rice germ, rice germ cake, rye, silage, sorghum, soybeans, soybeans, extracted, sunflower cake, tapioca, triticale, wheat, wheat bran, wheat bran and chana testa, wheat soya meal

Cottonseed   may contain the following mycotoxins:

AFLATOXIN B$_1$
incidence: 4/5, Ø conc. range: 15 μg/kg, country: Egypt[16]
incidence: 15/21*, conc. range: 5–20 μg/kg, country: UK[94], *imported?
incidence: 179?/388, conc. range: 1–50 μg/kg (76 sa), 51–500 μg/kg (58 sa), 501–1000 μg/kg (18 sa), >1000 μg/kg (29 sa), country: India[210]
incidence: 1/2, conc.: 210 μg/kg, country: USA[246]
incidence: 1/3, conc.: 404 μg/kg, country: USA[246]
incidence: 1/9, conc.: 82 μg/kg, country: USA[246]
incidence: 1/5, conc.: 910 μg/kg, country: USA[246]
incidence: 5/5, conc. range: <20 μg/kg (3 sa), 51–100 μg/kg (1 sa), 101–1000 μg/kg (1 sa), country: UK[267]
incidence: 7?/7?, conc. range: 5–560 μg/kg, Ø conc.: 163.1 μg/kg, country: USA[287]
incidence: 12/12, conc. range: 5–1500 μg/kg, Ø conc.: 197.9 μg/kg, country: USA[304]
incidence: 6/7, conc. range: 10–380 μg/kg, Ø conc.: 110 μg/kg, country: USA[304]
incidence: 62/106*, conc. range: 1–20 μg/kg (36 sa), 21–100 μg/kg (14 sa), 101–300 μg/kg (5 sa), >300 μg/kg (7 sa), country: USA[370],

*includes 73 cottonseed and 33 cottonseed meal sa
see also alfalfa, *Ambadi* cake, animal feedstuffs (dairy cake), bagasse, barley, bengalgram husk, bird food, bird food, wild, biri testa, blackgram, blackgram husk, bran, broiler mixed feed, calf fattening mixed feed, calf fattening mixed feed (containing 4–20 % peanut products), *Carthamus* cake, castor cake, cereals, cereal products, chick pea, coconut cake, cocos, concentrate, mixed, concentrates, cotton cake, cottonseed (dehulled), cottonseed cake, cottonseed extract, cottonseed meal, cottonseed meal (ammoniated), cottonseed meal (decorticated), cottonseed meats, cottonseed products, crumbles, crumbles, grower, cycad meal, dairy cattle feed, dairy cattle feed (containing 2–5 % peanut products), dairy cattle feed (containing 6–10 % peanut products), dairy cattle feed (containing 6–12 % peanut products), dairy cattle feed (containing more than 20 % peanut products), diets, mixed, dog food, egg production mixed feed, feed, feed and ingredients, feed, compound, feed, layer, feed, mixed, feed (beef), feed (broilers), feed (calf), feed (cat), feed (cattle), feed (chicken), feed (dairy), feed (dog), feed (dug), feed (fish), feed (gluten), feed (horse), feed (miscellaneous), feed (pig), feed (poultry), feed (poultry, pig), feed (rabbit), feed (sheep), feed (50–60 % maize), fish meal, grain by-products, grains (no specification), greengram, hay/silage, horsegram, husk, *Jagni* cake, legume mixture, linseed, linseed cake, livol, *Mahua* cake, maize, maize germ, maize gluten, maize grits, maize husk, maize meal, maize oil cake, maize powder, maize screenings, maize, ground, maize, hybrid, maize, preharvest, maize, yellow, maize (dark grains), *Makhana* (*Euryale ferox* Salisb) puffs, manioc, milk production mixed feed, mung testa, murkool, mustard cake, neem cake, niger cake, oats, palm kernel expeller cake, palm kernels, palm products, peanut cake, peanut cake (deoiled), peanut expeller, peanut hay, peanut meal, peanut, kernels, peanut, shells, peanuts,

pellets, finisher, pig meal and pellets, pigeon pea, poultry feeds (peanut containing), rapeseed cake, redgram husk, rice, rice bran, rice bran (deoiled), rice chaff, rice crack, rice germ, rice germ cake, rice meal, rice straw, rice (damaged), rice (polish), safflower cake, sal seed cake, sesame, sesame cake, sorghum, soybean meal, soybeans, sunflower, sunflower cake, sunflower flour, tapioca, wheat, wheat bran, wheat bran and chana testa

Aflatoxin B$_2$
incidence: 1/5, conc.: 810 µg/kg, country: USA[246]
incidence: 3/7, conc. range: 5–90 µg/kg, Ø conc.: 53.3 µg/kg, country: USA[304]
see also animal feedstuff (dairy cake), bird food, bird food, wild, biri testa, blackgram, blackgram husk, cottonseed cake, cottonseed extract, cottonseed meal, cottonseed meal (ammoniated), cottonseed meats, dog food, egg production mixed feed, feed, compound, feed (cat), feed (cattle), feed (dog), feed (pig), feed (poultry), feed (rabbit), feed (sheep), fish meal, horsegram, maize, maize gluten, maize husk, maize, ground, maize, preharvest, mung testa, mustard cake, niger cake, peanut cake, peanut cake (deoiled), peanut expeller, peanut hay, peanut meal, peanuts, peanut, kernels, redgram husk, rice bran, rice bran (deoiled), rice chaff, rice meal, rice (polish), sal seed cake, sesame cake, sorghum, soybean meal, soybeans, wheat, wheat bran

Aflatoxin B$_1$ + B$_2$
incidence: 4/24*, conc.: 100–200 µg/kg, country: Egypt[361], *and cottonseed products
see also feed (poultry), maize, sorghum

Aflatoxin G$_1$
incidence: 1/2, conc.: 130 µg/kg, country: USA[246]
incidence: 1/3, conc.: 207 µg/kg, country: USA[246]
see also animal feedstuffs (dairy cake), bird food, bird food, wild, concentrate, mixed, cottonseed cake, feed (cat), feed (chicken), feed (dog), maize, maize, ground, maize, preharvest, meat meal, milk production mixed feed, murkool, peanut cake, peanut

expeller, peanut hay, peanut meal, peanuts, rice bran, rice germ, sorghum, soybeans, wheat, wheat bran, wheat bran and chana testa

Aflatoxin
incidence: 60/928, highest conc.: >1500 µg/kg, Ø conc.: 220 µg/kg, country: unknown[193]
incidence: 108/1319, conc. range: ≤500 µg/kg, Ø conc.: 44 µg/kg, country: unknown[193]
incidence: 83/943, conc. range: ≤1500 µg/kg, Ø conc.: 228 µg/kg, country: unknown[193]
incidence: 11/18, conc. range: ≤111 µg/kg, Ø conc.: 37 µg/kg, country: USA[280]
see also blackgram husk, bread crumbs, broiler finisher, broiler starter, cotton cake, cottonseed cake, cottonseed extract, cottonseed meal, feed, feed (cattle), feed (cow), feed (dog), feed (horse), feed (maize, gluten), feed (pig), feed (poultry), feed (rabbit), feed (rat/mice), feed (sheep), feeds, grain, fish meal, flour (wheat), groundnut cake, grower's mash, horsegram, layer's mash, maize, maize gluten, maize, white, milo, peanut cake, peanut cake (deoiled), peanut, kernels, peanut (oil cake), pearlmillet, pig breeder's mash, pig finisher, pig starter, pod with haulms, poultry breeder's mash, rabbit pellets, redgram husk, rice, rice bran (deoiled), rice, broken, rice (polish), sesame cake, silk worm pupae, sorghum, soybean cake, soybean meal, wheat, wheat bran

Aflatoxins
incidence: 1/1*, conc.: 11 µg/kg, country: USA[280], *imported
incidence: 9/73, conc.: 20 µg/kg (3 sa), >300 µg/kg (1 sa), country: USA[195]
incidence: 17/45, conc. range: >20 µg/kg (12 sa), >300 µg/kg (5 sa), country: USA[195]
see also maize germ, maize gluten, maize (dark grains), palm products, rice bran, soybeans, sunflower

Aflatoxins (total)
incidence: 15/21*, conc. range: 5–25 µg/kg, country: UK[94], *imported?
incidence: 4/37, conc. range: >20 µg/kg (3 sa), >300 µg/kg (1 sa), country: USA[195]

incidence: 17/45, conc. range: >20 µg/kg
(12 sa), >300 µg/kg (5 sa), country: USA[195]
see also maize germ, maize gluten, maize
(dark grains), palm products, rice bran,
soybeans, sunflower

DEOXYNIVALENOL
incidence: 2/4, conc. range: 830–1100 µg/kg,
Ø conc.: 965 µg/kg, country: Egypt[16]
see also barley, barley, husked, barley,
unhusked (naked), barley (pressed), bone
meal, bran, broilers feed, calf fattening mixed
feed, coconut, expeller, corn cob mix silage,
cottonseed cake, dairy cattle feed, egg
production mixed feed, feed, feed
components, feed, commercial mix, feed,
mixed, feed, mixed (primarily maize), feed
(barley), feed (cattle), feed (chicken), feed
(dog), feed (fish), feed (mill run, from
wheat), feed (mink), feed (pig), feed
(poultry), feed (reindeer), feeds, grain, feeds,
industrial, feedstuff, feedstuffs (rapeseed,
turnip, fish meal, concentrates), fish meal,
grain, mixed feed, grains, mixed, grains (no
specification), maize, maize ears, maize fibre,
maize germ, maize germ/bran, maize germ
meal, maize gluten, maize kernels, maize
meal, maize powder, maize screenings, maize
stalks (pith), maize, "Baby", maize, hybrid,
maize, white, oats, rice bran, rice germ cake,
rye, silage, sorghum, soybeans, triticale,
wheat, wheat and barley, wheat, red hard
winter, wheat, soft white winter, wheat,
spring, wheat, winter

ZEARALENONE
incidence: 2/4, conc. range: 20–46 µg/kg,
Ø conc.: 33 µg/kg, country: Egypt[16]
incidence: 3/15*, conc. range: 70–150 µg/kg
(3 sa), country: USA[370], *includes 10
cottonseed and 5 cottonseed meal sa
see also alfalfa, barley, barley, husked, barley,
unhusked (naked), barley and feed, bone
meal, bran, broilers feed, Carthamus cake,
chick pea, concentrate, mixed, corn cob mix
silage, cotton cake, cottonseed cake, diet
(dairy cow), diet (poultry), diets (mixed),
feed, feed components, feed, mixed, feed,
mixed (primarily maize also maize, oats,

wheat), feed (bran), feed (broiler chicken),
feed (cattle), feed (chicken), feed (dairy), feed
(developing pig), feed (mill run, from
wheat), feed (miscellaneous), feed (pig), feed
(poultry), feed (poultry, pig), feed (starter
chicken), feedstuff, fish meal, forage grass,
grain, bruised, grain, mixed feed, grains (no
specification), hay, maize, maize ears, maize
flakes, maize germ, maize germ/bran, maize
gluten, maize kernels, maize meal, maize oil
cake, maize screenings, maize stalks (pith),
maize, "Baby", maize, hybrid, maize, shelled,
maize, unshelled, maize, white, maize grain,
artificially dried, maize grain, crib dried,
maize grain, ensiled, milk production mixed
feed, oats, *Paspalum palidosum*, straw, peanut
hulls/skins, rice bran, rice germ, rice germ
cake, rye, silage, sorghum, soybeans,
soybeans, extracted, sunflower cake, tapioca,
triticale, wheat, wheat bran, wheat bran and
chana testa, wheat soya meal

Cottonseed cake   may contain the following
mycotoxins:

AFLATOXIN B₁
incidence: 44/55, conc. range:
44.5–397.4 µg/kg, Ø conc.: 138.6 µg/kg,
country: India[253]
incidence: 1/1, conc.: ≈500 µg/kg, country:
UK[284]
incidence: 3/3, conc. range: 3.2–4.4 µg/kg,
Ø conc.: 3.6 µg/kg, country: Egypt[285]
incidence: 3/3, conc. range: 4.2–6.1 µg/kg,
Ø conc.: 5.2 µg/kg, country: Egypt[285]
incidence: 4/13, conc. range: 40–500 µg/kg,
country: India[358]
see also alfalfa, *Ambadi* cake, animal feedstuffs
(dairy cake), bagasse, barley, bengalgram
husk, bird food, bird food, wild, biri testa,
blackgram, blackgram husk, bran, broiler
mixed feed, calf fattening mixed feed, calf
fattening mixed feed (containing 4–20 %
peanut products), *Carthamus* cake, castor
cake, cereals, cereal products, chick pea,
coconut cake, cocos, concentrate, mixed,
concentrates, cotton cake, cottonseed,
cottonseed (dehulled), cottonseed extract,
cottonseed meal, cottonseed meal

(ammoniated), cottonseed meal (decorticated), cottonseed meats, cottonseed products, crumbles, crumbles, grower, cycad meal, dairy cattle feed, dairy cattle feed (containing 2–5 % peanut products), dairy cattle feed (containing 6–10 % peanut products), dairy cattle feed (containing 6–12 % peanut products), dairy cattle feed (containing more than 20 % peanut products), diets, mixed, dog food, egg production mixed feed, feed, feed and ingredients, feed, compound, feed, layer, feed, mixed, feed (beef), feed (broilers), feed (calf), feed (cat), feed (cattle), feed (chicken), feed (dairy), feed (dog), feed (dug), feed (fish), feed (gluten), feed (horse), feed (miscellaneous), feed (pig), feed (poultry), feed (poultry, pig), feed (rabbit), feed (sheep), feed (50–60 % maize), fish meal, grain by-products, grains (no specification), greengram, hay/silage, horsegram, husk, *Jagni* cake, legume mixture, linseed, linseed cake, livol, *Mahua* cake, maize, maize germ, maize gluten, maize grits, maize husk, maize meal, maize oil cake, maize powder, maize screenings, maize, ground, maize, hybrid, maize, preharvest, maize, yellow, maize (dark grains), *Makhana* (*Euryale ferox* Salisb) puffs, manioc, milk production mixed feed, mung testa, murkool, mustard cake, neem cake, niger cake, oats, palm kernel expeller cake, palm kernels, palm products, peanut cake, peanut cake (deoiled), peanut expeller, peanut hay, peanut meal, peanut, kernels, peanut, shells, peanuts, pellets, finisher, pig meal and pellets, pigeon pea, poultry feeds (peanut containing), rapeseed cake, redgram husk, rice, rice bran, rice bran (deoiled), rice chaff, rice crack, rice germ, rice germ cake, rice meal, rice straw, rice (damaged), rice (polish), safflower cake, sal seed cake, sesame, sesame cake, sorghum, soybean meal, soybeans, sunflower, sunflower cake, sunflower flour, tapioca, wheat, wheat bran, wheat bran and chana testa

Aflatoxin B_2
incidence: 3/3, conc. range: 1.6–3.1 µg/kg, Ø conc.: 2.2 µg/kg, country: Egypt[285]

incidence: 3/3, conc. range: 2.7–3 µg/kg, Ø conc.: 2.8 µg/kg, country: Egypt[285]
see also animal feedstuff (dairy cake), bird food, bird food, wild, biri testa, blackgram, blackgram husk, cottonseed, cottonseed extract, cottonseed meal, cottonseed meal (ammoniated), cottonseed meats, dog food, egg production mixed feed, feed, compound, feed (cat), feed (cattle), feed (dog), feed (pig), feed (poultry), feed (rabbit), feed (sheep), fish meal, horsegram, maize, maize gluten, maize husk, maize, ground, maize, preharvest, mung testa, mustard cake, niger cake, peanut cake, peanut cake (deoiled), peanut expeller, peanut hay, peanut meal, peanuts, peanut, kernels, redgram husk, rice bran, rice bran (deoiled), rice chaff, rice meal, rice (polish), sal seed cake, sesame cake, sorghum, soybean meal, soybeans, wheat, wheat bran

Aflatoxin G_1
incidence: 2/3, conc. range: 1.8–2.6 µg/kg, Ø conc.: 2.2 µg/kg, country: Egypt[285]
incidence: 3/3, conc. range: 1.9–2.9 µg/kg, Ø conc.: 2.4 µg/kg, country: Egypt[285]
incidence: 1/13, conc. range: <1 µg/kg, country: India[358]
see also animal feedstuffs (dairy cake), bird food, bird food, wild, concentrate, mixed, cottonseed, feed (cat), feed (chicken), feed (dog), maize, maize, ground, maize, preharvest, meat meal, milk production mixed feed, murkool, peanut cake, peanut expeller, peanut hay, peanut meal, peanuts, rice bran, rice germ, sorghum, soybeans, wheat, wheat bran, wheat bran and chana testa

Aflatoxin B
incidence: ?/13, conc. range: 6–10 µg/kg, Ø conc.: 7.3 µg/kg, country: Pakistan[162]
incidence: 5/19, conc. range: 4–19 µg/kg, Ø conc.: 9 µg/kg, country: Pakistan[412]
see also barley, cocoa (oil cake), feed, feed (cattle), feed (maize, gluten), feed (pig), feed (poultry), flour (wheat), lucern (dried), maize, maize gluten, maize grains, oats, peanut (oil cake), rice, broken, sorghum, soybean (oil cake), sunflower (oil cake), wheat

Aflatoxin G
incidence: ?/13, conc. range: 4–8 µg/kg,
Ø conc.: 5.7 µg/kg, country: Pakistan[162]
incidence: 5/19, conc. range: 0–5 µg/kg,
Ø conc.: 3.6 µg/kg, country: Pakistan[412]
see also feed (maize, gluten), maize gluten,
maize grains, rice, broken

Aflatoxin
incidence: 3/6, conc. range: 0–175 µg/kg,
country: Nigeria[109]
see also blackgram husk, bread crumbs,
broiler finisher, broiler starter, cotton cake,
cottonseed, cottonseed extract, cottonseed
meal, feed, feed (cattle), feed (cow), feed
(dog), feed (horse), feed (maize, gluten), feed
(pig), feed (poultry), feed (rabbit), feed
(rat/mice), feed (sheep), feeds, grain, fish
meal, flour (wheat), groundnut cake,
grower's mash, horsegram, layer's mash,
maize, maize gluten, maize, white, milo,
peanut cake, peanut cake (deoiled), peanut,
kernels, peanut (oil cake), pearlmillet, pig
breeder's mash, pig finisher, pig starter, pod
with haulms, poultry breeder's mash, rabbit
pellets, redgram husk, rice, rice bran
(deoiled), rice, broken, rice (polish), sesame
cake, silk worm pupae, sorghum, soybean
cake, soybean meal, wheat, wheat bran

Citrinin
incidence: 1/4, conc.: 50 µg/kg, country:
Egypt[16]
see also barley, barley, oats, barley-soybean
diet, feed, feed, mixed, feed (cattle), feed
(pig), fish meal, hay, maize, maize, white,
Makhana (Euryale ferox Salisb) puffs, oats,
palm products, peas and beans, rice bran,
rice germ, wheat, wheat and other grains
(moldy), wheat bran

Deoxynivalenol
incidence: 2/4, conc. range: 310–528 µg/kg,
Ø conc.: 419 µg/kg, country: Egypt[16]
see also barley, barley, husked, barley,
unhusked (naked), barley (pressed), bone
meal, bran, broilers feed, calf fattening mixed
feed, coconut, expeller, corn cob mix silage,
cottonseed, dairy cattle feed, egg production
mixed feed, feed, feed components, feed,
commercial mix, feed, mixed, feed, mixed
(primarily maize), feed (barley), feed (cattle),
feed (chicken), feed (dog), feed (fish), feed
(mill run, from wheat), feed (mink), feed
(pig), feed (poultry), feed (reindeer), feeds,
grain, feeds, industrial, feedstuff, feedstuffs
(rapeseed, turnip, fish meal, concentrates),
fish meal, grain, mixed feed, grains, mixed,
grains (no specification), maize, maize ears,
maize fibre, maize germ, maize germ/bran,
maize germ meal, maize gluten, maize
kernels, maize meal, maize powder, maize
screenings, maize stalks (pith), maize,
„Baby“, maize, hybrid, maize, white, oats,
rice bran, rice germ cake, rye, silage,
sorghum, soybeans, triticale, wheat, wheat
and barley, wheat, red hard winter, wheat,
soft white winter, wheat, spring, wheat,
winter

Zearalenone
incidence: 4/4, conc. range: 8–22 µg/kg,
Ø conc.: 13 µg/kg, country: Egypt[16]
see also alfalfa, barley, barley, husked, barley,
unhusked (naked), barley and feed, bone
meal, bran, broilers feed, Carthamus cake,
chick pea, concentrate, mixed, corn cob mix
silage, cotton cake, cottonseed, diet (dairy
cow), diet (poultry), diets (mixed), feed, feed
components, feed, mixed, feed, mixed
(primarily maize also maize, oats, wheat),
feed (bran), feed (broiler chicken), feed
(cattle), feed (chicken), feed (dairy), feed
(developing pig), feed (mill run, from
wheat), feed (miscellaneous), feed (pig), feed
(poultry), feed (poultry, pig), feed (starter
chicken), feedstuff, fish meal, forage grass,
grain, bruised, grain, mixed feed, grains (no
specification), hay, maize, maize ears, maize
flakes, maize germ, maize germ/bran, maize
gluten, maize kernels, maize meal, maize oil
cake, maize screenings, maize stalks (pith),
maize, „Baby“, maize, hybrid, maize, shelled,
maize, unshelled, maize, white, maize grain,
artificially dried, maize grain, crib dried,
maize grain, ensiled, milk production mixed
feed, oats, Paspalum palidosum, straw, peanut
hulls/skins, rice bran, rice germ, rice germ
cake, rye, silage, sorghum, soybeans,

soybeans, extracted, sunflower cake, tapioca, triticale, wheat, wheat bran, wheat bran and chana testa, wheat soya meal

Cottonseed extract may contain the following mycotoxins:

AFLATOXIN B$_1$
incidence: 3/7, conc. range: $\leq$25 μg/kg (3 sa), Ø conc.: 8.7 μg/kg, country: India[247]
see also alfalfa, *Ambadi* cake, animal feedstuffs (dairy cake), bagasse, barley, bengalgram husk, bird food, bird food, wild, biri testa, blackgram, blackgram husk, bran, broiler mixed feed, calf fattening mixed feed, calf fattening mixed feed (containing 4–20% peanut products), *Carthamus* cake, castor cake, cereals, cereal products, chick pea, coconut cake, cocos, concentrate, mixed, concentrates, cotton cake, cottonseed, cottonseed (dehulled), cottonseed cake, cottonseed meal, cottonseed meal (ammoniated), cottonseed meal (decorticated), cottonseed meats, cottonseed products, crumbles, crumbles, grower, cycad meal, dairy cattle feed, dairy cattle feed (containing 2–5% peanut products), dairy cattle feed (containing 6–10% peanut products), dairy cattle feed (containing 6–12% peanut products), dairy cattle feed (containing more than 20% peanut products), diets, mixed, dog food, egg production mixed feed, feed and ingredients, feed, compound, feed, layer, feed, mixed, feed (beef), feed (broilers), feed (calf), feed (cat), feed (cattle), feed (chicken), feed (dairy), feed (dog), feed (dug), feed (fish), feed (gluten), feed (horse), feed (miscellaneous), feed (pig), feed (poultry), feed (poultry, pig), feed (rabbit), feed (sheep), feed (50–60% maize), fish meal, grain by-products, grains (no specification), greengram, hay/silage, horsegram, husk, *Jagni* cake, legume mixture, linseed, linseed cake, livol, *Mahua* cake, maize, maize germ, maize gluten, maize grits, maize husk, maize meal, maize oil cake, maize powder, maize screenings, maize, ground, maize, hybrid, maize, preharvest, maize, yellow, maize (dark grains), *Makhana* (*Euryale ferox* Salisb) puffs,

manioc, milk production mixed feed, mung testa, murkool, mustard cake, neem cake, niger cake, oats, palm kernel expeller cake, palm kernels, palm products, peanut cake, peanut cake (deoiled), peanut expeller, peanut hay, peanut meal, peanut, kernels, peanut, shells, peanuts, pellets, finisher, pig meal and pellets, pigeon pea, poultry feeds (peanut containing), rapeseed cake, redgram husk, rice, rice bran, rice bran (deoiled), rice chaff, rice crack, rice germ, rice germ cake, rice meal, rice straw, rice (damaged), rice (polish), safflower cake, sal seed cake, sesame, sesame cake, sorghum, soybean meal, soybeans, sunflower, sunflower cake, sunflower flour, tapioca, wheat, wheat bran, wheat bran and chana testa

AFLATOXIN B$_2$
incidence: 1/7, conc.: 5 μg/kg, country: India[247]
see also animal feedstuff (dairy cake), bird food, bird food, wild, biri testa, blackgram, blackgram husk, cottonseed, cottonseed cake, cottonseed meal, cottonseed meal (ammoniated), cottonseed meats, dog food, egg production mixed feed, feed, compound, feed (cat), feed (cattle), feed (dog), feed (pig), feed (poultry), feed (rabbit), feed (sheep), fish meal, horsegram, maize, maize gluten, maize husk, maize, ground, maize, preharvest, mung testa, mustard cake, niger cake, peanut cake, peanut cake (deoiled), peanut expeller, peanut hay, peanut meal, peanuts, peanut, kernels, redgram husk, rice bran, rice bran (deoiled), rice chaff, rice meal, rice (polish), sal seed cake, sesame cake, sorghum, soybean meal, soybeans, wheat, wheat bran

AFLATOXIN
incidence: 3/7, conc. range: 6–19 μg/kg, country: India[247]
see also blackgram husk, bread crumbs, broiler finisher, broiler starter, cotton cake, cottonseed, cottonseed cake, cottonseed meal, feed, feed (cattle), feed (cow), feed (dog), feed (horse), feed (maize, gluten), feed (pig), feed (poultry), feed (rabbit), feed (rat/mice), feed (sheep), feeds, grain, fish meal, flour

(wheat), groundnut cake, grower's mash, horsegram, layer's mash, maize, maize gluten, maize, white, milo, peanut cake, peanut cake (deoiled), peanut, kernels, peanut (oil cake), pearlmillet, pig breeder's mash, pig finisher, pig starter, pod with haulms, poultry breeder's mash, rabbit pellets, redgram husk, rice, rice bran (deoiled), rice, broken, rice (polish), sesame cake, silk worm pupae, sorghum, soybean cake, soybean meal, wheat, wheat bran

Cottonseed fines   may contain the following mycotoxins:

AFLATOXINS
incidence: 9/9, conc. range: 300–9620 µg/kg, Ø conc.: 4023.7 µg/kg, country: USA[324]
see also animal feed (maize), animal feed (mixed), barley, copra meal, cottonseed meal, cottonseed meats, feed, feed ingredients (miscellaneous), feed, mixed, feed (excluding peanuts, suspect), feed (goat), feed (pig), feed (poultry), feedstuff, grain, grain, mixed feed, maize, maize, shelled, millet, peanut cake, peanut meal, peanut meal and by-products, peanuts, protein concentrates, rice, rice bran, sorghum, soybean meal, sunflower

Cottonseed meal   may contain the following mycotoxins:

AFLATOXIN B$_1$
incidence: 15/17, conc. range: 1.5–11.4 µg/kg, Ø conc.: 5.3 µg/kg, country: Colombia[125]
incidence: 604/3218, conc. range: 151–1500 µg/kg, country: USA[163,164]
incidence: 10/19, conc. range: 1.1–210 µg/kg, Ø conc.: 71.3 µg/kg, country: USA[245]
incidence: 6/6, conc. range: 11.3–117.1 µg/kg, Ø conc.: 41.9 µg/kg, country: USA[274]
incidence: 12/12, conc. range: 90–650 µg/kg, Ø conc.: 231.7 µg/kg, country: USA[304]
incidence: 7/7, conc. range: 40–190 µg/kg, Ø conc.: 114.3 µg/kg, country: USA[304]
see also alfalfa, *Ambadi* cake, animal feedstuffs (dairy cake), bagasse, barley, bengalgram husk, bird food, bird food, wild, biri testa, blackgram, blackgram husk, bran, broiler mixed feed, calf fattening mixed feed, calf fattening mixed feed (containing 4–20% peanut products), *Carthamus* cake, castor cake, cereals, cereal products, chick pea, coconut cake, cocos, concentrate, mixed, concentrates, cotton cake, cottonseed, cottonseed (dehulled), cottonseed cake, cottonseed extract, cottonseed meal (ammoniated), cottonseed meal (decorticated), cottonseed meats, cottonseed products, crumbles, crumbles, grower, cycad meal, dairy cattle feed, dairy cattle feed (containing 2–5% peanut products), dairy cattle feed (containing 6–10% peanut products), dairy cattle feed (containing 6–12% peanut products), dairy cattle feed (containing more than 20% peanut products), diets, mixed, dog food, egg production mixed feed, feed and ingredients, feed, compound, feed, layer, feed, mixed, feed (beef), feed (broilers), feed (calf), feed (cat), feed (cattle), feed (chicken), feed (dairy), feed (dog), feed (dug), feed (fish), feed (gluten), feed (horse), feed (miscellaneous), feed (pig), feed (poultry), feed (poultry, pig), feed (rabbit), feed (sheep), feed (50–60% maize), fish meal, grain by-products, grains (no specification), greengram, hay/silage, horsegram, husk, *Jagni* cake, legume mixture, linseed, linseed cake, livol, *Mahua* cake, maize, maize germ, maize gluten, maize grits, maize husk, maize meal, maize oil cake, maize powder, maize screenings, maize, ground, maize, hybrid, maize, preharvest, maize, yellow, maize (dark grains), *Makhana* (*Euryale ferox* Salisb) puffs, manioc, milk production mixed feed, mung testa, murkool, mustard cake, neem cake, niger cake, oats, palm kernel expeller cake, palm kernels, palm products, peanut cake, peanut cake (deoiled), peanut expeller, peanut hay, peanut meal, peanut, kernels, peanut, shells, peanuts, pellets, finisher, pig meal and pellets, pigeon pea, poultry feeds (peanut containing), rapeseed cake, redgram husk, rice, rice bran, rice bran (deoiled), rice chaff, rice crack, rice germ, rice germ cake, rice meal, rice straw, rice (damaged), rice (polish), safflower cake, sal seed cake, sesame, sesame cake, sorghum,

soybean meal, soybeans, sunflower, sunflower cake, sunflower flour, tapioca, wheat, wheat bran, wheat bran and chana testa

AFLATOXIN B_2
incidence: 6/17, conc. range: 1.1–2 µg/kg, Ø conc.: 1.6 µg/kg, country: Colombia[125]
incidence: 6/6, conc. range: 1.7–18.9 µg/kg, Ø conc.: 6.8 µg/kg, country: USA[274]
incidence: 7/7, conc. range: 10–55 µg/kg, Ø conc.: 32.9 µg/kg, country: USA[304]
see also animal feedstuffs (dairy cake), bird food, bird food, wild, biri testa, blackgram, blackgram husk, cottonseed, cottonseed cake, cottonseed extract, cottonseed meal (ammoniated), cottonseed meats, dog food, egg production mixed feed, feed, compound, feed (cat), feed (cattle), feed (dog), feed (pig), feed (poultry), feed (rabbit), feed (sheep), fish meal, horsegram, maize, maize gluten, maize husk, maize, ground, maize, preharvest, mung testa, mustard cake, niger cake, peanut cake, peanut cake (deoiled), peanut expeller, peanut hay, peanut meal, peanuts, peanut, kernels, redgram husk, rice bran, rice bran (deoiled), rice chaff, rice meal, rice (polish), sal seed cake, sesame cake, sorghum, soybean meal, soybeans, wheat, wheat bran

AFLATOXIN
incidence: 203/964, conc. range: ≤1500 µg/kg, Ø conc.: 136 µg/kg, country: unknown[193]
incidence: 278/1293, conc. range: ≤500 µg/kg, Ø conc.: 33 µg/kg, country: unknown[193]
incidence: 123/961, conc. range: ≤1500 µg/kg, Ø conc.: 150 µg/kg, country: unknown[193]
incidence: 12/15, conc. range: ≤143 µg/kg, Ø conc.: 43 µg/kg, country: USA[280]
incidence: 3/12, conc. range: ≤55 µg/kg, Ø conc.: 28 µg/kg, country: USA[280]
see also blackgram husk, bread crumbs, broiler finisher, broiler starter, cotton cake, cottonseed, cottonseed cake, cottonseed extract, feed, feed (cattle), feed (cow), feed (dog), feed (horse), feed (maize, gluten), feed (pig), feed (poultry), feed (rabbit), feed

(rat/mice), feed (sheep), feeds, grain, fish meal, flour (wheat), groundnut cake, grower's mash, horsegram, layer's mash, maize, maize gluten, maize, white, milo, peanut cake, peanut cake (deoiled), peanut, kernels, peanut (oil cake), pearlmillet, pig breeder's mash, pig finisher, pig starter, pod with haulms, poultry breeder's mash, rabbit pellets, redgram husk, rice, rice bran (deoiled), rice, broken, rice (polish), sesame cake, silk worm pupae, sorghum, soybean cake, soybean meal, wheat, wheat bran

AFLATOXINS
incidence: 8/27, conc. range: >20 µg/kg (6 sa), >300 µg/kg (2 sa), country: USA[195]
incidence: 3/17, conc. range: >20 µg/kg (3 sa), country: USA[195]
see also animal feed (mixed), animal feed (maize), barley, copra meal, cottonseed, cottonseed fines, cottonseed meats, feed, feed ingredients (miscellaneous), feed, mixed, feed (excluding peanuts, suspect), feed (goat), feed (pig), feed (poultry), feedstuff, grain, grain, mixed feed, maize, maize, shelled, millet, peanut cake, peanut meal, peanut meal and by-products, peanuts, protein concentrates, rice, rice bran, sorghum, soybean meal, sunflower

Cottonseed meal (ammoniated)   may contain the following mycotoxins:

AFLATOXIN B_1
incidence: 6/6, conc. range: 8.7–19.5 µg/kg, Ø conc.: 14.1 µg/kg, country: USA[274]
see also alfalfa, *Ambadi* cake, animal feedstuffs (dairy cake), bagasse, barley, bengalgram husk, bird food, bird food, wild, biri testa, blackgram, blackgram husk, bran, broiler mixed feed, calf fattening mixed feed, calf fattening mixed feed (containing 4–20 % peanut products), *Carthamus* cake, castor cake, cereals, cereal products, chick pea, coconut cake, cocos, concentrate, mixed, concentrates, cotton cake, cottonseed, cottonseed (dehulled), cottonseed cake, cottonseed extract, cottonseed meal, cottonseed meal (decorticated), cottonseed

meats, cottonseed products, crumbles, crumbles, grower, cycad meal, dairy cattle feed, dairy cattle feed (containing 2–5 % peanut products), dairy cattle feed (containing 6–10 % peanut products), dairy cattle feed (containing 6–12 % peanut products), dairy cattle feed (containing more than 20 % peanut products), diets, mixed, dog food, egg production mixed feed, feed, feed and ingredients, feed, compound, feed, layer, feed, mixed, feed (beef), feed (broilers), feed (calf), feed (cat), feed (cattle), feed (chicken), feed (dairy), feed (dog), feed (dug), feed (fish), feed (gluten), feed (horse), feed (miscellaneous), feed (pig), feed (poultry), feed (poultry, pig), feed (rabbit), feed (sheep), feed (50–60 % maize), fish meal, grain by-products, grains (no specification), greengram, hay/silage, horsegram, husk, *Jagni* cake, legume mixture, linseed, linseed cake, livol, *Mahua* cake, maize, maize germ, maize gluten, maize grits, maize husk, maize meal, maize oil cake, maize powder, maize screenings, maize, ground, maize, hybrid, maize, preharvest, maize, yellow, maize (dark grains), *Makhana* (*Euryale ferox* Salisb) puffs, manioc, milk production mixed feed, mung testa, murkool, mustard cake, neem cake, niger cake, oats, palm kernel expeller cake, palm kernels, palm products, peanut cake, peanut cake (deoiled), peanut expeller, peanut hay, peanut meal, peanut, kernels, peanut, shells, peanuts, pellets, finisher, pig meal and pellets, pigeon pea, poultry feeds (peanut containing), rapeseed cake, redgram husk, rice, rice bran, rice bran (deoiled), rice chaff, rice crack, rice germ, rice germ cake, rice meal, rice straw, rice (damaged), rice (polish), safflower cake, sal seed cake, sesame, sesame cake, sorghum, soybean meal, soybeans, sunflower, sunflower cake, sunflower flour, tapioca, wheat, wheat bran, wheat bran and chana testa

Aflatoxin B$_2$
incidence: 6/6, conc. range: 2.5–6.3 μg/kg, Ø conc.: 4.4 μg/kg, country: USA[274]
see also animal feedstuffs (dairy cake), bird food, bird food, wild, biri testa, blackgram,

blackgram husk, cottonseed, cottonseed cake, cottonseed extract,cottonseed meal,cottonseed meats, dog food, egg production mixed feed, feed, compound, feed (cat), feed (cattle), feed (dog), feed (pig), feed (poultry), feed (rabbit), feed (sheep), fish meal, horsegram, maize, maize gluten, maize husk, maize, ground, maize, preharvest, mung testa, mustard cake, niger cake, peanut cake, peanut cake (deoiled), peanut expeller, peanut hay, peanut meal, peanuts, peanut, kernels, redgram husk, rice bran, rice bran (deoiled), rice chaff, rice meal, rice (polish), sal seed cake, sesame cake, sorghum, soybean meal, soybeans, wheat, wheat bran

Cottonseed meal (decorticated)   may contain the following mycotoxins:

Aflatoxin B$_1$
incidence: 3/5, Ø conc.: 60 μg/kg, country: Egypt[16]
see also alfalfa, *Ambadi* cake, animal feedstuffs (dairy cake), bagasse, barley, bengalgram husk, bird food, bird food, wild, biri testa, blackgram, blackgram husk, bran, broiler mixed feed, calf fattening mixed feed, calf fattening mixed feed (containing 4–20 % peanut products), *Carthamus* cake, castor cake, cereals, cereal products, chick pea, coconut cake, cocos, concentrate, mixed, concentrates, cotton cake, cottonseed, cottonseed (dehulled), cottonseed cake, cottonseed extract, cottonseed meal, cottonseed meal (ammoniated), cottonseed meats, cottonseed products, crumbles, crumbles, grower, cycad meal, dairy cattle feed, dairy cattle feed (containing 2–5 % peanut products), dairy cattle feed (containing 6–10 % peanut products), dairy cattle feed (containing 6–12 % peanut products), dairy cattle feed (containing more than 20 % peanut products), diets, mixed, dog food, egg production mixed feed, feed, feed and ingredients, feed, compound, feed, layer, feed, mixed, feed (beef), feed (broilers), feed (calf), feed (cat), feed (cattle), feed (chicken), feed (dairy), feed (dog), feed (dug), feed (fish), feed (gluten), feed (horse),

feed (miscellaneous), feed (pig), feed (poultry), feed (poultry, pig), feed (rabbit), feed (sheep), feed (50–60 % maize), fish meal, grain by-products, grains (no specification), greengram, hay/silage, horsegram, husk, *Jagni* cake, legume mixture, linseed, linseed cake, livol, *Mahua* cake, maize, maize germ, maize gluten, maize grits, maize husk, maize meal, maize oil cake, maize powder, maize screenings, maize, ground, maize, hybrid, maize, preharvest, maize, yellow, maize (dark grains), *Makhana* (*Euryale ferox* Salisb) puffs, manioc, milk production mixed feed, mung testa, murkool, mustard cake, neem cake, niger cake, oats, palm kernel expeller cake, palm kernels, palm products, peanut cake, peanut cake (deoiled), peanut expeller, peanut hay, peanut meal, peanut, kernels, peanut, shells, peanuts, pellets, finisher, pig meal and pellets, pigeon pea, poultry feeds (peanut containing), rapeseed cake, redgram husk, rice, rice bran, rice bran (deoiled), rice chaff, rice crack, rice germ, rice germ cake, rice meal, rice straw, rice (damaged), rice (polish), safflower cake, sal seed cake, sesame, sesame cake, sorghum, soybean meal, soybeans, sunflower, sunflower cake, sunflower flour, tapioca, wheat, wheat bran, wheat bran and chana testa

Cottonseed meats   may contain the following mycotoxins:

AFLATOXIN B$_1$
incidence: 4/4, conc. range: 26.8–190.5 µg/kg, Ø conc.: 69.7 µg/kg, country: USA[274]
see also alfalfa, *Ambadi* cake, animal feedstuffs (dairy cake), bagasse, barley, bengalgram husk, bird food, bird food, wild, biri testa, blackgram, blackgram husk, bran, broiler mixed feed, calf fattening mixed feed, calf fattening mixed feed (containing 4–20 % peanut products), *Carthamus* cake, castor cake, cereals, cereal products, chick pea, coconut cake, cocos, concentrate, mixed, concentrates, cotton cake, cottonseed, cottonseed (dehulled), cottonseed cake, cottonseed extract, cottonseed meal, cottonseed meal (ammoniated), cottonseed

meal (decorticated), cottonseed products, crumbles, crumbles, grower, cycad meal, dairy cattle feed, dairy cattle feed (containing 2–5 % peanut products), dairy cattle feed (containing 6–10 % peanut products), dairy cattle feed (containing 6–12 % peanut products), dairy cattle feed (containing more than 20 % peanut products), diets, mixed, dog food, egg production mixed feed, feed, feed and ingredients, feed, compound, feed, layer, feed, mixed, feed (beef), feed (broilers), feed (calf), feed (cat), feed (cattle), feed (chicken), feed (dairy), feed (dog), feed (dug), feed (fish), feed (gluten), feed (horse), feed (miscellaneous), feed (pig), feed (poultry), feed (poultry, pig), feed (rabbit), feed (sheep), feed (50–60 % maize), fish meal, grain by-products, grains (no specification), greengram, hay/silage, horsegram, husk, *Jagni* cake, legume mixture, linseed, linseed cake, livol, *Mahua* cake, maize, maize germ, maize gluten, maize grits, maize husk, maize meal, maize oil cake, maize powder, maize screenings, maize, ground, maize, hybrid, maize, preharvest, maize, yellow, maize (dark grains), *Makhana* (*Euryale ferox* Salisb) puffs, manioc, milk production mixed feed, mung testa, murkool, mustard cake, neem cake, niger cake, oats, palm kernel expeller cake, palm kernels, palm products, peanut cake, peanut cake (deoiled), peanut expeller, peanut hay, peanut meal, peanut, kernels, peanut, shells, peanuts, pellets, finisher, pig meal and pellets, pigeon pea, poultry feeds (peanut containing), rapeseed cake, redgram husk, rice, rice bran, rice bran (deoiled), rice chaff, rice crack, rice germ, rice germ cake, rice meal, rice straw, rice (damaged), rice (polish), safflower cake, sal seed cake, sesame, sesame cake,sorghum, soybean meal, soybeans, sunflower, sunflower cake, sunflower flour, tapioca, wheat, wheat bran, wheat bran and chana testa

AFLATOXIN B$_2$
incidence: 4/4, conc. range: 4.7–29.7 µg/kg, Ø conc.: 11.2 µg/kg, country: USA[274]
see also animal feedstuff (dairy cake), bird food, bird food, wild, biri testa, blackgram,

blackgram husk, cottonseed, cottonseed cake, cottonseed extract, cottonseed meal, cottonseed meal (ammoniated), dog food, egg production mixed feed, feed, compound, feed (cat), feed (cattle), feed (dog), feed (pig), feed (poultry), feed (rabbit), feed (sheep), fish meal, horsegram, maize, maize gluten, maize husk, maize, ground, maize, preharvest, mung testa, mustard cake, niger cake, peanut cake, peanut cake (deoiled), peanut expeller, peanut hay, peanut meal, peanuts, peanut, kernels, redgram husk, rice bran, rice bran (deoiled), rice chaff, rice meal, rice (polish), sal seed cake, sesame cake, sorghum, soybean meal, soybeans, wheat, wheat bran

AFLATOXINS
incidence: 8/9, conc. range: 3–2560 µg/kg, Ø conc.: 318.2 µg/kg, country: USA[324]
see also animal feed (mixed), animal feed (maize), barley, copra meal, cottonseed fines, cottonseed meal, feed, feed ingredients (miscellaneous), feed, mixed, feed (excluding peanuts, suspect), feed (goat), feed (pig), feed (poultry), feedstuff, grain, grain, mixed feed, maize, maize, shelled, millet, peanut cake, peanut meal, peanut meal and by-products, peanuts, protein concentrates, rice, rice bran, sorghum, soybean meal, sunflower

Cottonseed products   may contain the following mycotoxins:

AFLATOXIN B$_1$
incidence: 40/120* conc. range: 5–120 µg/kg, country: Denmark[167],* imported
see also alfalfa, *Ambadi* cake, animal feedstuffs (dairy cake), bagasse, barley, bengalgram husk, bird food, bird food, wild, biri testa, blackgram, blackgram husk, bran, broiler mixed feed, calf fattening mixed feed, calf fattening mixed feed (containing 4–20 % peanut products), *Carthamus* cake, castor cake, cereals, cereal products, chick pea, coconut cake, cocos, concentrate, mixed, concentrates, cotton cake, cottonseed, cottonseed (dehulled), cottonseed cake, cottonseed extract, cottonseed meal, cottonseed meal (ammoniated), cottonseed meal (decorticated), cottonseed meats,

crumbles, crumbles, grower, cycad meal, dairy cattle feed, dairy cattle feed (containing 2–5 % peanut products), dairy cattle feed (containing 6–10 % peanut products), dairy cattle feed (containing 6–12 % peanut products), dairy cattle feed (containing more than 20 % peanut products), diets, mixed, dog food, egg production mixed feed, feed, feed and ingredients, feed, compound, feed, layer, feed, mixed, feed (beef), feed (broilers), feed (calf), feed (cat), feed (cattle), feed (chicken), feed (dairy), feed (dog), feed (dug), feed (fish), feed (gluten), feed (horse), feed (miscellaneous), feed (pig), feed (poultry), feed (poultry, pig), feed (rabbit), feed (sheep), feed (50–60 % maize), fish meal, grain by-products, grains (no specification), greengram, hay/silage, horsegram, husk, *Jagni* cake, legume mixture, linseed, linseed cake, livol, *Mahua* cake, maize, maize germ, maize gluten, maize grits, maize husk, maize meal, maize oil cake, maize powder, maize screenings, maize, ground, maize, hybrid, maize, preharvest, maize, yellow, maize (dark grains), *Makhana* (*Euryale ferox* Salisb) puffs, manioc, milk production mixed feed, mung testa, murkool, mustard cake, neem cake, niger cake, oats, palm kernel expeller cake, palm kernels, palm products, peanut cake, peanut cake (deoiled), peanut expeller, peanut hay, peanut meal, peanut, kernels, peanut, shells, peanuts, pellets, finisher, pig meal and pellets, pigeon pea, poultry feeds (peanut containing), rapeseed cake, redgram husk, rice, rice bran, rice bran (deoiled), rice chaff, rice crack, rice germ, rice germ cake, rice meal, rice straw, rice (damaged), rice (polish), safflower cake, sal seed cake, sesame, sesame cake, sorghum, soybean meal, soybeans, sunflower, sunflower cake, sunflower flour, tapioca, wheat, wheat bran, wheat bran and chana testa

Cottonseed (dehulled)   may contain the following mycotoxins:

AFLATOXIN B$_1$
incidence: 15/15, conc. range: 7–3258 µg/kg, Ø conc.: 1117.9 µg/kg, country: USA[136]

see also alfalfa, *Ambadi* cake, animal feedstuffs (dairy cake), bagasse, barley, bengalgram husk, bird food, bird food, wild, biri testa, blackgram, blackgram husk, bran, broiler mixed feed, calf fattening mixed feed, calf fattening mixed feed (containing 4–20 % peanut products), *Carthamus* cake, castor cake, cereals, cereal products, chick pea, coconut cake, cocos, concentrate, mixed, concentrates, cotton cake, cottonseed, cottonseed cake, cottonseed extract, cottonseed meal, cottonseed meal (ammoniated), cottonseed meal (decorticated), cottonseed meats, cottonseed products, crumbles, crumbles, grower, cycad meal, dairy cattle feed, dairy cattle feed (containing 2–5 % peanut products), dairy cattle feed (containing 6–10 % peanut products), dairy cattle feed (containing 6–12 % peanut products), dairy cattle feed (containing more than 20 % peanut products), diets, mixed, dog food, egg production mixed feed, feed, feed and ingredients, feed, compound, feed, layer, feed, mixed, feed (beef), feed (broilers), feed (calf), feed (cat), feed (cattle), feed (chicken), feed (dairy), feed (dog), feed (dug), feed (fish), feed (gluten), feed (horse), feed (miscellaneous), feed (pig), feed (poultry), feed (poultry, pig), feed (rabbit), feed (sheep), feed (50–60 % maize), fish meal, grain by-products, grains (no specification), greengram, hay/silage, horsegram, husk, *Jagni* cake, legume mixture, linseed, linseed cake, livol, *Mahua* cake, maize, maize germ, maize gluten, maize grits, maize husk, maize meal, maize oil cake, maize powder, maize screenings, maize, ground, maize, hybrid, maize, preharvest, maize, yellow, maize (dark grains), *Makhana* (*Euryale ferox* Salisb) puffs, manioc, milk production mixed feed, mung testa, murkool, mustard cake, neem cake, niger cake, oats, palm kernel expeller cake, palm kernels, palm products, peanut cake, peanut cake (deoiled), peanut expeller, peanut hay, peanut meal, peanut, kernels, peanut, shells, peanuts, pellets, finisher, pig meal and pellets, pigeon pea, poultry feeds (peanut containing), rapeseed cake, redgram husk, rice, rice bran, rice bran (deoiled), rice chaff, rice crack, rice germ, rice germ cake, rice meal, rice straw, rice (damaged), rice (polish), safflower cake, sal seed cake, sesame, sesame cake, sorghum, soybean meal, soybeans, sunflower, sunflower cake, sunflower flour, tapioca, wheat, wheat bran, wheat bran and chana testa

Crumbles   may contain the following mycotoxins:

Aflatoxin B$_1$
incidence: 1/2, conc.: 100 μg/kg, country: Australia[181]
see also alfalfa, *Ambadi* cake, animal feedstuffs (dairy cake), bagasse, barley, bengalgram husk, bird food, bird food, wild, biri testa, blackgram, blackgram husk, bran, broiler mixed feed, calf fattening mixed feed, calf fattening mixed feed (containing 4–20 % peanut products), *Carthamus* cake, castor cake, cereals, cereal products, chick pea, coconut cake, cocos, concentrate, mixed, concentrates, cotton cake, cottonseed, cottonseed (dehulled), cottonseed cake, cottonseed extract, cottonseed meal, cottonseed meal (ammoniated), cottonseed meal (decorticated), cottonseed meats, cottonseed products, crumbles, grower, cycad meal, dairy cattle feed, dairy cattle feed (containing 2–5 % peanut products), dairy cattle feed (containing 6–10 % peanut products), dairy cattle feed (containing 6–12 % peanut products), dairy cattle feed (containing more than 20 % peanut products), diets, mixed, dog food, egg production mixed feed, feed and ingredients, feed, compound, feed, layer, feed, mixed, feed (beef), feed (broilers), feed (calf), feed (cat), feed (cattle), feed (chicken), feed (dairy), feed (dog), feed (dug), feed (fish), feed (gluten), feed (horse), feed (miscellaneous), feed (pig), feed (poultry), feed (poultry, pig), feed (rabbit), feed (sheep), feed (50–60 % maize), fish meal, grain by-products, grains (no specification), greengram, hay/silage, horsegram, husk, *Jagni* cake, legume mixture, linseed, linseed cake, livol, *Mahua* cake,

maize, maize germ, maize gluten, maize grits, maize husk, maize meal, maize oil cake, maize powder, maize screenings, maize, ground, maize, hybrid, maize, preharvest, maize, yellow, maize (dark grains), *Makhana* (*Euryale ferox* Salisb) puffs, manioc, milk production mixed feed, mung testa, murkool, mustard cake, neem cake, niger cake, oats, palm kernel expeller cake, palm kernels, palm products, peanut cake, peanut cake (deoiled), peanut expeller, peanut hay, peanut meal, peanut, kernels, peanut, shells, peanuts, pellets, finisher, pig meal and pellets, pigeon pea, poultry feeds (peanut containing), rapeseed cake, redgram husk, rice, rice bran, rice bran (deoiled), rice chaff, rice crack, rice germ, rice germ cake, rice meal, rice straw, rice (damaged), rice (polish), safflower cake, sal seed cake, sesame, sesame cake, sorghum, soybean meal, soybeans, sunflower, sunflower cake, sunflower flour, tapioca, wheat, wheat bran, wheat bran and chana testa

Crumbles, grower   may contain the following mycotoxins:

AFLATOXIN B$_1$
incidence: 1/3, conc.: 20 µg/kg, country: Australia[121]
see also alfalfa, *Ambadi* cake, animal feed-stuffs (dairy cake), bagasse, barley, bengalgram husk, bird food, bird food, wild, biri testa, blackgram, blackgram husk, bran, broiler mixed feed, calf fattening mixed feed, calf fattening mixed feed (containing 4–20 % peanut products), *Carthamus* cake, castor cake, cereals, cereal products, chick pea, coconut cake, cocos, concentrate, mixed, concentrates, cotton cake, cottonseed, cottonseed (dehulled), cottonseed cake, cottonseed extract, cottonseed meal, cottonseed meal (ammoniated), cottonseed meal (decorticated), cottonseed meats, cottonseed products, crumbles, cycad meal, dairy cattle feed, dairy cattle feed (containing 2–5 % peanut products), dairy cattle feed (containing 6–10 % peanut products), dairy cattle feed (containing 6–12 % peanut products), dairy cattle feed (containing more

than 20 % peanut products), diets, mixed, dog food, egg production mixed feed, feed, feed and ingredients, feed, compound, feed, layer, feed, mixed, feed (beef), feed (broilers), feed (calf), feed (cat), feed (cattle), feed (chicken), feed (dairy), feed (dog), feed (dug), feed (fish), feed (gluten), feed (horse), feed (miscellaneous), feed (pig), feed (poultry), feed (poultry, pig), feed (rabbit), feed (sheep), feed (50–60 % maize), fish meal, grain by-products, grains (no specification), greengram, hay/silage, horsegram, husk, *Jagni* cake, legume mixture, linseed, linseed cake, livol, *Mahua* cake, maize, maize germ, maize gluten, maize grits, maize husk, maize meal, maize oil cake, maize powder, maize screenings, maize, ground, maize, hybrid, maize, preharvest, maize, yellow, maize (dark grains), *Makhana* (*Euryale ferox* Salisb) puffs, manioc, milk production mixed feed, mung testa, murkool, mustard cake, neem cake, niger cake, oats, palm kernel expeller cake, palm kernels, palm products, peanut cake, peanut cake (deoiled), peanut expeller, peanut hay, peanut meal, peanut, kernels, peanut, shells, peanuts, pellets, finisher, pig meal and pellets, pigeon pea, poultry feeds (peanut containing), rapeseed cake, redgram husk, rice, rice bran, rice bran (deoiled), rice chaff, rice crack, rice germ, rice germ cake, rice meal, rice straw, rice (damaged), rice (polish), safflower cake, sal seed cake, sesame, sesame cake, sorghum, soybean meal, soybeans, sunflower, sunflower cake, sunflower flour, tapioca, wheat, wheat bran, wheat bran and chana testa

Cycad meal   may contain the following mycotoxins:

AFLATOXIN B$_1$
incidence: 2/4, conc. range: 1.1–2.2 µg/kg, Ø conc.: 1.7 µg/kg, country: USA[245]
see also alfalfa, *Ambadi* cake, animal feedstuffs (dairy cake), bagasse, barley, bengalgram husk, bird food, bird food, wild, biri testa, blackgram, blackgram husk, bran, broiler mixed feed, calf fattening mixed feed,

calf fattening mixed feed (containing 4–20 % peanut products), *Carthamus* cake, castor cake, cereals, cereal products, chick pea, cocos, concentrate, mixed, concentrates, cotton cake, cottonseed, cottonseed (dehulled), cottonseed cake, cottonseed extract, cottonseed meal, cottonseed meal (ammoniated), cottonseed meal (decorticated), cottonseed meats, cottonseed products, crumbles, crumbles, grower, dairy cattle feed, dairy cattle feed (containing 2–5 % peanut products), dairy cattle feed (containing 6–10 % peanut products), dairy cattle feed (containing 6–12 % peanut products), dairy cattle feed (containing more than 20 % peanut products), diets, mixed, dog food, egg production mixed feed, feed, feed and ingredients, feed, compound, feed, layer, feed, mixed, feed (beef), feed (broilers), feed (calf), feed (cat), feed (cattle), feed (chicken), feed (dairy), feed (dog), feed (dug), feed (fish), feed (gluten), feed (horse), feed (miscellaneous), feed (pig), feed (poultry), feed (poultry, pig), feed (rabbit), feed (sheep), feed (50–60 % maize), fish meal, grain by-products, grains (no specification), greengram, hay/silage, horsegram, husk, *Jagni* cake, legume mixture, linseed, linseed cake, livol, *Mahua* cake, maize, maize germ, maize gluten, maize grits, maize husk, maize meal, maize oil cake, maize powder, maize screenings, maize, ground, maize, hybrid, maize, preharvest, maize, yellow, maize (dark grains), *Makhana* (*Euryale ferox* Salisb) puffs, manioc, milk production mixed feed, mung testa, murkool, mustard cake, neem cake, niger cake, oats, palm kernel expeller cake, palm kernels, palm products, peanut cake, peanut cake (deoiled), peanut expeller, peanut hay, peanut meal, peanut, kernels, peanut, shells, peanuts, pellets, finisher, pig meal and pellets, pigeon pea, poultry feeds (peanut containing), rapeseed cake, redgram husk, rice, rice bran, rice bran (deoiled), rice chaff, rice crack, rice germ, rice germ cake, rice meal, rice straw, rice (damaged), rice (polish), safflower cake, sal seed cake, sesame, sesame cake, sorghum, soybean meal, soybeans, sunflower, sunflower cake, sunflower flour, tapioca, wheat, wheat bran, wheat bran and chana testa

Dairy cattle feed may contain the following mycotoxins:

Aflatoxin B$_1$
incidence: 14/14, conc. range: <5 µg/kg (4 sa), 6–10 (1 sa), 11–20 µg/kg (5 sa), 21–30 µg/kg (3 sa), 37 µg/kg (1 sa), country: Germany[298]
incidence: 30/30, conc. range: <5 µg/kg (10 sa), 6–10 µg/kg (6 sa), 11–20 µg/kg (9 sa), 21–30 µg/kg (5 sa), country: Germany[298]
incidence: 16/16, conc. range: <5 µg/kg (6 sa), 6–10 µg/kg (6 sa), 11–20 µg/kg (4 sa), country: Germany[298]
see also alfalfa, *Ambadi* cake, animal feedstuffs (dairy cake), bagasse, barley, bengalgram husk, bird food, bird food, wild, biri testa, blackgram, blackgram husk, bran, broiler mixed feed, calf fattening mixed feed, calf fattening mixed feed (containing 4–20 % peanut products), *Carthamus* cake, castor cake, cereals, cereal products, chick pea, coconut cake, cocos, concentrate, mixed, concentrates, cotton cake, cottonseed, cottonseed (dehulled), cottonseed cake, cottonseed extract, cottonseed meal, cottonseed meal (ammoniated), cottonseed meal (decorticated), cottonseed meats, cottonseed products, crumbles, crumbles, grower, cycad meal, dairy cattle feed (containing 2–5 % peanut products), dairy cattle feed (containing 6–10 % peanut products), dairy cattle feed (containing 6–12 % peanut products), dairy cattle feed (containing more than 20 % peanut products), diets, mixed, dog food, egg production mixed feed, feed, feed and ingredients, feed, compound, feed, layer, feed, mixed, feed (beef), feed (broilers), feed (calf), feed (cat), feed (cattle), feed (chicken), feed (dairy), feed (dog), feed (dug), feed (fish), feed (gluten), feed (horse), feed (miscellaneous), feed (pig), feed (poultry), feed (poultry, pig), feed (rabbit), feed

(sheep), feed (50–60 % maize), fish meal, grain by-products, grains (no specification), greengram, hay/silage, horsegram, husk, *Jagni* cake, legume mixture, linseed, linseed cake, livol, *Mahua* cake, maize, maize germ, maize gluten, maize grits, maize husk, maize meal, maize oil cake, maize powder, maize screenings, maize, ground, maize, hybrid, maize, preharvest, maize, yellow, maize (dark grains), *Makhana* (*Euryale ferox* Salisb) puffs, manioc, milk production mixed feed, mung testa, murkool, mustard cake, neem cake, niger cake, oats, palm kernel expeller cake, palm kernels, palm products, peanut cake, peanut cake (deoiled), peanut expeller, peanut hay, peanut meal, peanut, kernels, peanut, shells, peanuts, pellets, finisher, pig meal and pellets, pigeon pea, poultry feeds (peanut containing), rapeseed cake, redgram husk, rice, rice bran, rice bran (deoiled), rice chaff, rice crack, rice germ, rice germ cake, rice meal, rice straw, rice (damaged), rice (polish), safflower cake, sal seed cake, sesame, sesame cake, sorghum, soybean meal, soybeans, sunflower, sunflower cake, sunflower flour, tapioca, wheat, wheat bran, wheat bran and chana testa

Deoxynivalenol
incidence: 2/4, conc. range: 950–1056 µg/kg, Ø conc.: 1003 µg/kg, country: Egypt[16]
see also barley, barley, husked, barley, unhusked (naked), barley (pressed), bone meal, bran, broilers feed, calf fattening mixed feed, coconut, expeller, corn cob mix silage, cottonseed, cottonseed cake, egg production mixed feed, feed, feed components, feed, commercial mix, feed, mixed, feed, mixed (primarily maize), feed (barley), feed (cattle), feed (chicken), feed (dog), feed (fish), feed (mill run, from wheat), feed (mink), feed (pig), feed (poultry), feed (reindeer), feeds, grain, feeds, industrial, feedstuff, feedstuffs (rapeseed, turnip, fish meal, concentrates), fish meal, grain, mixed feed, grains, mixed, grains (no specification), maize, maize ears, maize fibre, maize germ, maize germ/bran, maize germ meal, maize gluten, maize kernels, maize meal, maize powder, maize

screenings, maize stalks (pith), maize, "Baby", maize, hybrid, maize, white, oats, rice bran, rice germ cake, rye, silage, sorghum, soybeans, triticale, wheat, wheat and barley, wheat, red hard winter, wheat, soft white winter, wheat, spring, wheat, winter

Dairy cattle feed (containing 2–5 % peanut products)   may contain the following mycotoxins:

Aflatoxin B_1
incidence: ?/13, conc. range: 15–100 µg/kg, country: Germany[298]
incidence: ?/20, conc. range: ≤240 µg/kg, country: Germany[298]
incidence: ?/16, conc. range: 16–260 µg/kg, country: Germany[298]
see also alfalfa, *Ambadi* cake, animal feedstuffs (dairy cake), bagasse, barley, bengalgram husk, bird food, bird food, wild, biri testa, blackgram, blackgram husk, bran, broiler mixed feed, calf fattening mixed feed, calf fattening mixed feed (containing 4–20 % peanut products), *Carthamus* cake, castor cake, cereals, cereal products, chick pea, coconut cake, cocos, concentrate, mixed, concentrates, cotton cake, cottonseed, cottonseed (dehulled), cottonseed cake, cottonseed extract, cottonseed meal, cottonseed meal (ammoniated), cottonseed meal (decorticated), cottonseed meats, cottonseed products, crumbles, crumbles, grower, cycad meal, dairy cattle feed, dairy cattle feed (containing 6–10 % peanut products), dairy cattle feed (containing 6–12 % peanut products), dairy cattle feed (containing more than 20 % peanut products), diets, mixed, dog food, egg production mixed feed, feed, feed and ingredients, feed, compound, feed, layer, feed, mixed, feed (beef), feed (broilers), feed (calf), feed (cat), feed (cattle), feed (chicken), feed (dairy), feed (dog), feed (dug), feed (fish), feed (gluten), feed (horse), feed (miscellaneous), feed (pig), feed (poultry), feed (poultry, pig), feed (rabbit), feed (sheep), feed (50–60 % maize), fish meal, grain by-products, grains (no specification),

greengram, hay/silage, horsegram, husk, *Jagni* cake, legume mixture, linseed, linseed cake, livol, *Mahua* cake, maize, maize germ, maize gluten, maize grits, maize husk, maize meal, maize oil cake, maize powder, maize screenings, maize, ground, maize, hybrid, maize, preharvest, maize, yellow, maize (dark grains), *Makhana* (*Euryale ferox* Salisb) puffs, manioc, milk production mixed feed, mung testa, murkool, mustard cake, neem cake, niger cake, oats, palm kernel expeller cake, palm kernels, palm products, peanut cake, peanut cake (deoiled), peanut expeller, peanut hay, peanut meal, peanut, kernels, peanut, shells, peanuts, pellets, finisher, pig meal and pellets, pigeon pea, poultry feeds (peanut containing), rapeseed cake, redgram husk, rice, rice bran, rice bran (deoiled), rice chaff, rice crack, rice germ, rice germ cake, rice meal, rice straw, rice (damaged), rice (polish), safflower cake, sal seed cake, sesame, sesame cake, sorghum, soybean meal, soybeans, sunflower, sunflower cake, sunflower flour, tapioca, wheat, wheat bran, wheat bran and chana testa

Dairy cattle feed (containing 6–10 % peanut products) may contain the following mycotoxins:

AFLATOXIN B$_1$
incidence: ?/11, conc. range: 13–280 μg/kg, country: Germany[298]
incidence: ?/5, conc. range: ≤550 μg/kg, country: Germany[298]
see also alfalfa, *Ambadi* cake, animal feedstuffs (dairy cake), bagasse, barley, bengalgram husk, bird food, bird food, wild, biri testa, blackgram, blackgram husk, bran, broiler mixed feed, calf fattening mixed feed, calf fattening mixed feed (containing 4–20 % peanut products), *Carthamus* cake, castor cake, cereals, cereal products, chick pea, coconut cake, cocos, concentrate, mixed, concentrates, cotton cake, cottonseed, cottonseed (dehulled), cottonseed cake, cottonseed extract, cottonseed meal, cottonseed meal (ammoniated), cottonseed meal (decorticated), cottonseed meats, cottonseed products, crumbles, crumbles, grower, cycad meal, dairy cattle feed, dairy cattle feed (containing 2–5 % peanut products), dairy cattle feed (containing 6–12 % peanut products), dairy cattle feed (containing more than 20 % peanut products), diets, mixed, dog food, egg production mixed feed, feed, feed and ingredients, feed, compound, feed, layer, feed, mixed, feed (beef), feed (broilers), feed (calf), feed (cat), feed (cattle), feed (chicken), feed (dairy), feed (dog), feed (dug), feed (fish), feed (gluten), feed (horse), feed (miscellaneous), feed (pig), feed (poultry), feed (poultry, pig), feed (rabbit), feed (sheep), feed (50–60 % maize), fish meal, grain by-products, grains (no specification), greengram, hay/silage, horsegram, husk, *Jagni* cake, legume mixture, linseed, linseed cake, livol, *Mahua* cake, maize, maize germ, maize gluten, maize grits, maize husk, maize meal, maize oil cake, maize powder, maize screenings, maize, ground, maize, hybrid, maize, preharvest, maize, yellow, maize (dark grains), *Makhana* (*Euryale ferox* Salisb) puffs, manioc, milk production mixed feed, mung testa, murkool, mustard cake, neem cake, niger cake, oats, palm kernel expeller cake, palm kernels, palm products, peanut cake, peanut cake (deoiled), peanut expeller, peanut hay, peanut meal, peanut, kernels, peanut, shells, peanuts, pellets, finisher, pig meal and pellets, pigeon pea, poultry feeds (peanut containing), rapeseed cake, redgram husk, rice, rice bran, rice bran (deoiled), rice chaff, rice crack, rice germ, rice germ cake, rice meal, rice straw, rice (damaged), rice (polish), safflower cake, sal seed cake, sesame, sesame cake, sorghum, soybean meal, soybeans, sunflower, sunflower cake, sunflower flour, tapioca, wheat, wheat bran, wheat bran and chana testa

Dairy cattle feed (containing 6–12 % peanut products) may contain the following mycotoxins:

AFLATOXIN B$_1$
incidence: ?/2, conc. range: 25 µg/kg, country: Germany[298]
see also alfalfa, *Ambadi* cake, animal feedstuffs (dairy cake), bagasse, barley, bengalgram husk, bird food, bird food, wild, biri testa, blackgram, blackgram husk, bran, broiler mixed feed, calf fattening mixed feed, calf fattening mixed feed (containing 4–20 % peanut products), *Carthamus* cake, castor cake, cereals, cereal products, chick pea, coconut cake, cocos, concentrate, mixed, concentrates, cotton cake, cottonseed, cottonseed (dehulled), cottonseed cake, cottonseed extract, cottonseed meal, cottonseed meal (ammoniated), cottonseed meal (decorticated), cottonseed meats, cottonseed products, crumbles, crumbles, grower, cycad meal, dairy cattle feed, dairy cattle feed (containing 2–5 % peanut products), dairy cattle feed (containing 6–10 % peanut products), dairy cattle feed (containing more than 20 % peanut products), diets, mixed, dog food, egg production mixed feed, feed, feed and ingredients, feed, compound, feed, layer, feed, mixed, feed (beef), feed (broilers), feed (calf), feed (cat), feed (cattle), feed (chicken), feed (dairy), feed (dog), feed (dug), feed (fish), feed (gluten), feed (horse), feed (miscellaneous), feed (pig), feed (poultry), feed (poultry, pig), feed (rabbit), feed (sheep), feed (50–60 % maize), fish meal, grain by-products, grains (no specification), greengram, hay/silage, horsegram, husk, *Jagni* cake, legume mixture, linseed, linseed cake, livol, *Mahua* cake, maize, maize germ, maize gluten, maize grits, maize husk, maize meal, maize oil cake, maize powder, maize screenings, maize, ground, maize, hybrid, maize, preharvest, maize, yellow, maize (dark grains), *Makhana* (*Euryale ferox* Salisb) puffs, manioc, milk production mixed feed, mung testa, murkool, mustard cake, neem cake, niger cake, oats, palm kernel expeller cake, palm kernels, palm products, peanut cake, peanut cake (deoiled), peanut expeller, peanut hay, peanut meal, peanut, kernels, peanut, shells, peanuts, pellets, finisher, pig meal and pellets, pigeon pea, poultry feeds (peanut containing), rapeseed cake, redgram husk, rice, rice bran, rice bran (deoiled), rice chaff, rice crack, rice germ, rice germ cake, rice meal, rice straw, rice (damaged), rice (polish), safflower cake, sal seed cake, sesame, sesame cake, sorghum, soybean meal, soybeans, sunflower, sunflower cake, sunflower flour, tapioca, wheat, wheat bran, wheat bran and chana testa

Dairy cattle feed (containing more than 20 % peanut products)

may contain the following mycotoxins:

AFLATOXIN B$_1$
incidence: ?/3, conc. range: 80–1100 µg/kg, country: Germany[298]
see also alfalfa, *Ambadi* cake, animal feedstuffs (dairy cake), bagasse, barley, bengalgram husk, bird food, bird food, wild, biri testa, blackgram, blackgram husk, bran, broiler mixed feed, calf fattening mixed feed, calf fattening mixed feed (containing 4–20 % peanut products), *Carthamus* cake, castor cake, cereals, cereal products, chick pea, coconut cake, cocos, concentrate, mixed, concentrates, cotton cake, cottonseed, cottonseed (dehulled), cottonseed cake, cottonseed extract, cottonseed meal, cottonseed meal (ammoniated), cottonseed meal (decorticated), cottonseed meats, cottonseed products, crumbles, crumbles, grower, cycad meal, dairy cattle feed, dairy cattle feed (containing 2–5 % peanut products), dairy cattle feed (containing 6–10 % peanut products), dairy cattle feed (containing 6–12 % peanut products), diets, mixed, dog food, egg production mixed feed, feed, feed and ingredients, feed, compound, feed, layer, feed, mixed, feed (beef), feed (broilers), feed (calf), feed (cat), feed (cattle), feed (chicken), feed (dairy), feed (dog), feed (dug), feed (fish), feed (gluten), feed (horse), feed (miscellaneous), feed (pig), feed (poultry), feed (poultry, pig), feed (rabbit), feed (sheep), feed (50–60 % maize), fish meal, grain by-products, grains (no

specification), greengram, hay/silage, horsegram, husk, *Jagni* cake, legume mixture, linseed, linseed cake, livol, *Mahua* cake, maize, maize germ, maize gluten, maize grits, maize husk, maize meal, maize oil cake, maize powder, maize screenings, maize, ground, maize, hybrid, maize, preharvest, maize, yellow, maize (dark grains), *Makhana* (*Euryale ferox* Salisb) puffs, manioc, milk production mixed feed, mung testa, murkool, mustard cake, neem cake, niger cake, oats, palm kernel expeller cake, palm kernels, palm products, peanut cake, peanut cake (deoiled), peanut expeller, peanut hay, peanut meal, peanut, kernels, peanut, shells, peanuts, pellets, finisher, pig meal and pellets, pigeon pea, poultry feeds (peanut containing), rapeseed cake, redgram husk, rice, rice bran, rice bran (deoiled), rice chaff, rice crack, rice germ, rice germ cake, rice meal, rice straw, rice (damaged), rice (polish), safflower cake, sal seed cake, sesame, sesame cake, sorghum, soybean meal, soybeans, sunflower, sunflower cake, sunflower flour, tapioca, wheat, wheat bran, wheat bran and chana testa

Diet (dairy cow)   may contain the following mycotoxins:

Ochratoxin A
incidence: 1/7345, conc.: nc, country: Hungary[209]
see also alfalfa, barley, barley, oats, barley (high moisture), barley-soybean diet, bird food, domestic, bird food, wild, broilers feed, cereal grains, citrus pulp, coconut, expeller, corn cob mix silage, diet (poultry), diet (starter), dog food, eat, egg production mixed feed, feed, feed wheat, oat and barley, feed, commercial mix, feed, mixed, feed, mixed (pelleted), feed (broilers), feed (cat), feed (cattle), feed (cereals), feed (pig), feed ec (pig), feed (poultry), feed ec (poultry), feed (poultry, pig), feed ec(rabbit), feed (trout), grain, mixed feed, grains, mixed, grains (heated), hay, horse bean, maize, maize feed, milo, maize gluten, maize meal, maize, white, *Makhana* (*Euryale ferox* Salisb) puffs, milk production mixed feed, millet, oat and barley

(hammer-milled), oats, palm products, peanut cake, peas, peas and beans, pet food, pig feedstuffs, pig grower diet, pig meal, piglet diet, poultry feedstuffs, rice bran, rice germ, rice germ cake, rye, sorghum, soybean groats, sunflower, sunflower seeds, extracted, tapioca, triticale, *Vicia faba*, wheat, wheat and barley, wheat bran, wheat hay, wheat, oats

Zearalenone
incidence: 1/7345, conc.: nc, country: Hungary[209]
see also alfalfa, barley, barley, husked, barley, unhusked (naked), barley and feed, bone meal, bran, broilers feed, *Carthamus* cake, chick pea, concentrate, mixed, corn cob mix silage, cotton cake, cottonseed, cottonseed cake, diet (poultry), diets (mixed), feed, feed components, feed, mixed, feed, mixed (primarily maize also maize, oats, wheat), feed (bran), feed (broiler chicken), feed (cattle), feed (chicken), feed (dairy), feed (developing pig), feed (mill run, from wheat), feed (miscellaneous), feed (pig), feed (poultry), feed (poultry, pig), feed (starter chicken), feedstuff, fish meal, forage grass, grain, bruised, grain, mixed feed, grains (no specification), hay, maize, maize ears, maize flakes, maize germ, maize germ/bran, maize gluten, maize kernels, maize meal, maize oil cake, maize screenings, maize stalks (pith), maize, "Baby", maize, hybrid, maize, shelled, maize, unshelled, maize, white, maize grain, artificially dried, maize grain, crib dried, maize grain, ensiled, milk production mixed feed, oats, *Paspalum palidosum*, straw, peanut hulls/skins, rice bran, rice germ, rice germ cake, rye, silage, sorghum, soybeans, soybeans, extracted, sunflower cake, tapioca, triticale, wheat, wheat bran, wheat bran and chana testa, wheat soya meal

Diet (grower)   may contain the following mycotoxins:

Patulin
incidence: 1/7345, conc.: nc, country: Hungary[209]
see also hay, maize meal, piglet diet, wheat

RUBRATOXIN B
incidence: 1/7345, conc.: nc, country:
Hungary[209]
see also diet (poultry), diet (starter), pig diet
(fattening), pig grower diet, soybean groats,
wheat

T-2 TOXIN
incidence: 1/7345, conc.: nc, country:
Hungary[209]
see also alfalfa, barley, bran, diet (poultry),
feed, feed components, feed, layer, feed,
mixed, feed (dog), feed (fish), feed (mink),
feed (pig), feed (poultry), feedstuff, forage
grass, grain, mixed feed, grains (no
specification), hay, maize, maize germ/bran,
maize gluten, maize meal, maize screenings,
maize stalks (pith), oat and barley
(hammer-milled), oats, peanuts, pig grower
diet, piglet diet, rye, sorghum, triticale, wheat

Diet (poultry)   may contain the following
mycotoxins:

OCHRATOXIN A
incidence: 4/7345, conc. range: nc, country:
Hungary[209]
see also alfalfa, barley, barley, oats, barley
(high moisture), barley-soybean diet, bird
food, domestic, bird food, wild, broilers feed,
cereal grains, citrus pulp, coconut, expeller,
corn cob mix silage, diet (dairy cow), diet
(starter), dog food, eat, egg production
mixed feed, feed, feed wheat, oat and barley,
feed, commercial mix, feed, mixed, feed,
mixed (pelleted), feed (broilers), feed (cat),
feed (cattle), feed (cereals), feed (pig), feed ec
(pig), feed (poultry), feed ec (poultry), feed
(poultry, pig), feed ec (rabbit), feed (trout),
grain, mixed feed, grains, mixed, grains
(heated), hay, horse bean, maize, maize feed,
milo, maize gluten, maize meal, maize, white,
Makhana (*Euryale ferox* Salisb) puffs, milk
production mixed feed, millet, oat and barley
(hammer-milled), oats, palm products,
peanut cake, peas, peas and beans, pet food,
pig feedstuffs, pig grower diet, pig meal,
piglet diet, poultry feedstuffs, rice bran, rice
germ, rice germ cake, rye, sorghum, soybean

groats, sunflower, sunflower seeds, extracted,
tapioca, triticale, *Vicia faba*, wheat, wheat and
barley, wheat bran, wheat hay, wheat, oats

RUBRATOXIN B
incidence: 2/7345, conc. range: nc, country:
Hungary[209]
see also diet (grower), diet (starter), pig diet
(fattening), pig grower diet, soybean groats,
wheat

T-2 TOXIN
incidence: 1/7345, conc.: nc, country:
Hungary[209]
see also alfalfa, barley, bran, diet (grower),
feed, feed components, feed, layer, feed,
mixed, feed (dog), feed (fish), feed (mink),
feed (pig), feed (poultry), feedstuff, forage
grass, grain, mixed feed, grains (no specification), hay, maize, maize germ/bran, maize
gluten, maize meal, maize screenings, maize
stalks (pith), oat and barley (hammer-milled),
oats, peanuts, pig grower diet, piglet diet, rye,
sorghum, triticale, wheat

ZEARALENONE
incidence: 6/7345, conc. range: nc, country:
Hungary[209]
see also alfalfa, barley, barley, husked, barley,
unhusked (naked), barley and feed, bone meal,
bran, broilers feed, *Carthamus* cake, chick pea,
concentrate, mixed, corn cob mix silage,
cotton cake, cottonseed, cottonseed cake, diet
(dairy cow), diets (mixed), feed, feed
components, feed, mixed, feed, mixed
(primarily maize also maize, oats, wheat), feed
(bran), feed (broiler chicken), feed (cattle),
feed (chicken), feed (dairy), feed (developing
pig), feed (mill run, from wheat), feed
(miscellaneous), feed (pig), feed (poultry), feed
(poultry, pig), feed (starter chicken), feedstuff,
fish meal, forage grass, grain, bruised, grain,
mixed feed, grains (no specification), hay,
maize, maize ears, maize flakes, maize germ,
maize germ/bran, maize gluten, maize kernels,
maize meal, maize oil cake, maize screenings,
maize stalks (pith), maize, "Baby", maize,
hybrid, maize, shelled, maize, unshelled,
maize, white, maize grain, artificially dried,
maize grain, crib dried, maize grain, ensiled,

milk production mixed feed, oats, *Paspalum palidosum*, straw, peanut hulls/skins, rice bran, rice germ, rice germ cake, rye, silage, sorghum, soybeans, soybeans, extracted, sunflower cake, tapioca, triticale, wheat, wheat bran, wheat bran and chana testa, wheat soya meal

Diet (starter)   may contain the following mycotoxins:

OCHRATOXIN A
incidence: 4/7345, conc. range: nc, country: Hungary[209]
see also alfalfa, barley, barley, oats, barley (high moisture), barley-soybean diet, bird food, domestic, bird food, wild, broilers feed, cereal grains, citrus pulp, coconut, expeller, corn cob mix silage, diet (dairy cow), diet (poultry), dog food, eat, egg production mixed feed, feed, feed wheat, oat and barley, feed, commercial mix, feed, mixed, feed, mixed (pelleted), feed (broilers), feed (cat), feed (cattle), feed (cereals), feed (pig), feed ec (pig), feed (poultry), feed ec (poultry), feed (poultry, pig), feed ec (rabbit), feed (trout), grain, mixed feed, grains, mixed, grains (heated), hay, horse bean, maize, maize feed, milo, maize gluten, maize meal, maize, white, *Makhana* (*Euryale ferox* Salisb) puffs, milk production mixed feed, millet, oat and barley (hammer-milled), oats, palm products, peanut cake, peas, peas and beans, pet food, pig feedstuffs, pig grower diet, pig meal, piglet diet, poultry feedstuffs, rice bran, rice germ, rice germ cake, rye, sorghum, soybean groats, sunflower, sunflower seeds, extracted, tapioca, triticale, *Vicia faba*, wheat, wheat and barley, wheat bran, wheat hay, wheat, oats

RUBRATOXIN B
incidence: 2/7345, conc. range: nc, country: Hungary[209]
see also diet (grower), diet (poultry), pig diet (fattening), pig grower diet, soybean groats, wheat

Diets, mixed   may contain the following mycotoxins:

AFLATOXIN B$_1$
incidence: 32/111, conc. range: 1–20 µg/kg (20 sa), 21–100 µg/kg (8 sa), 101–300 µg/kg (4 sa), country: USA[370]
see also alfalfa, *Ambadi* cake, animal feedstuffs (dairy cake), bagasse, barley, bengalgram husk, bird food, bird food, wild, biri testa, blackgram, blackgram husk, bran, broiler mixed feed, calf fattening mixed feed, calf fattening mixed feed (containing 4–20 % peanut products), *Carthamus* cake, castor cake, cereals, cereal products, chick pea, coconut cake, cocos, concentrate, mixed, concentrates, cotton cake, cottonseed, cottonseed (dehulled), cottonseed cake, cottonseed extract, cottonseed meal, cottonseed meal (ammoniated), cottonseed meal (decorticated), cottonseed meats, cottonseed products, crumbles, crumbles, grower, cycad meal, dairy cattle feed, dairy cattle feed (containing 2–5 % peanut products), dairy cattle feed (containing 6–10 % peanut products), dairy cattle feed (containing 6–12 % peanut products), dairy cattle feed (containing more than 20 % peanut products), dog food, egg production mixed feed, feed, feed and ingredients, feed, compound, feed, layer, feed, mixed, feed (beef), feed (broilers), feed (calf), feed (cat), feed (cattle), feed (chicken), feed (dairy), feed (dog), feed (dug), feed (fish), feed (gluten), feed (horse), feed (miscellaneous), feed (pig), feed (poultry), feed (poultry, pig), feed (rabbit), feed (sheep), feed (50–60 % maize), fish meal, grain by-products, grains (no specification), greengram, hay/silage, horsegram, husk, *Jagni* cake, legume mixture, linseed, linseed cake, livol, *Mahua* cake, maize, maize germ, maize gluten, maize grits, maize husk, maize meal, maize oil cake, maize powder, maize screenings, maize, ground, maize, hybrid, maize, preharvest, maize, yellow, maize (dark grains), *Makhana* (*Euryale ferox* Salisb) puffs, manioc, milk production mixed feed, mung testa, murkool, mustard cake, neem cake, niger cake, oats, palm kernel expeller cake, palm kernels, palm products, peanut cake, peanut cake (deoiled),

peanut expeller, peanut hay, peanut meal, peanut, kernels, peanut, shells, peanuts, pellets, finisher, pig meal and pellets, pigeon pea, poultry feeds (peanut containing), rapeseed cake, redgram husk, rice, rice bran, rice bran (deoiled), rice chaff, rice crack, rice germ, rice germ cake, rice meal, rice straw, rice (damaged), rice (polish), safflower cake, sal seed cake, sesame, sesame cake, sorghum, soybean meal, soybeans, sunflower, sunflower cake, sunflower flour, tapioca, wheat, wheat bran, wheat bran and chana testa

Zearalenone
incidence: 2/10, conc. range: 70–150 µg/kg (2 sa), country: USA[370]
see also alfalfa, barley, barley, husked, barley, unhusked (naked), barley and feed, bone meal, bran, broilers feed, *Carthamus* cake, chick pea, concentrate, mixed, corn cob mix silage, cotton cake, cottonseed, cottonseed cake, diet (dairy cow), diet (poultry), feed, feed components, feed, mixed, feed, mixed (primarily maize also maize, oats, wheat), feed (bran), feed (broiler chicken), feed (cattle), feed (chicken), feed (dairy), feed (developing pig), feed (mill run, from wheat), feed (miscellaneous), feed (pig), feed (poultry), feed (poultry, pig), feed (starter chicken), feedstuff, fish meal, forage grass, grain, bruised, grain, mixed feed, grains (no specification), hay, maize, maize ears, maize flakes, maize germ, maize germ/bran, maize gluten, maize kernels, maize meal, maize oil cake, maize screenings, maize stalks (pith), maize, "Baby", maize, hybrid, maize, shelled, maize, unshelled, maize, white, maize grain, artificially dried, maize grain, crib dried, maize grain, ensiled, milk production mixed feed, oats, *Paspalum palidosum*, straw, peanut hulls/skins, rice bran, rice germ, rice germ cake, rye, silage, sorghum, soybeans, soybeans, extracted, sunflower cake, tapioca, triticale, wheat, wheat bran, wheat bran and chana testa, wheat soya meal

Dog Feed
see Feed (dog)

Dog food   may contain the following mycotoxins:

Aflatoxin B$_1$
incidence: 3/45, conc. range: 15–25 µg/kg, Ø conc.: 19 µg/kg, country: Brazil[4]
see also alfalfa, *Ambadi* cake, animal feedstuffs (dairy cake), bagasse, barley, bengalgram husk, bird food, bird food, wild, biri testa, blackgram, blackgram husk, bran, broiler mixed feed, calf fattening mixed feed, calf fattening mixed feed (containing 4–20 % peanut products), *Carthamus* cake, castor cake, cereals, cereal products, chick pea, coconut cake, concentrate, mixed, concentrates, cotton cake, cottonseed, cottonseed (dehulled), cottonseed cake, cottonseed extract, cottonseed meal, cottonseed meal (ammoniated), cottonseed meal (decorticated), cottonseed meats, cottonseed products, crumbles, crumbles, grower, cycad meal, dairy cattle feed, dairy cattle feed (containing 2–5 % peanut products), dairy cattle feed (containing 6–10 % peanut products), dairy cattle feed (containing 6–12 % peanut products), dairy cattle feed (containing more than 20 % peanut products), diets, mixed, egg production mixed feed, feed, feed and ingredients, feed, compound, feed, layer, feed, mixed, feed (beef), feed (broilers), feed (calf), feed (cat), feed (cattle), feed (chicken), feed (dairy), feed (dog), feed (dug), feed (fish), feed (gluten), feed (horse), feed (miscellaneous), feed (pig), feed (poultry), feed (poultry, pig), feed (rabbit), feed (sheep), feed (50–60 % maize), fish meal, grain by-products, grains (no specification), greengram, hay/silage, horsegram, husk, *Jagni* cake, cocos, legume mixture, linseed, linseed cake, livol, *Mahua* cake, maize, maize germ, maize gluten, maize grits, maize husk, maize meal, maize oil cake, maize powder, maize screenings, maize, ground, maize, hybrid, maize, preharvest, maize, yellow, maize (dark grains), *Makhana* (*Euryale ferox* Salisb) puffs, manioc, milk production mixed feed, mung testa, murkool, mustard cake, neem cake, niger cake, oats, palm kernel expeller cake, palm kernels, palm products, peanut cake, peanut cake (deoiled), peanut expeller,

peanut hay, peanut meal, peanut, kernels, peanut, shells, peanuts, pellets, finisher, pig meal and pellets, pigeon pea, poultry feeds (peanut containing), rapeseed cake, redgram husk, rice, rice bran, rice bran (deoiled), rice chaff, rice crack, rice germ, rice germ cake, rice meal, rice straw, rice (damaged), rice (polish), safflower cake, sal seed cake, sesame, sesame cake, sorghum, soybean meal, soybeans, sunflower, sunflower cake, sunflower flour, tapioca, wheat, wheat bran, wheat bran and chana testa

Aflatoxin B_2
incidence: 1/45, conc.: 12 μg/kg, country: Brazil[4]
see also animal feedstuffs (dairy cake), bird food, bird food, wild, biri testa, blackgram, blackgram husk, cottonseed, cottonseed cake, cottonseed extract, cottonseed meal, cottonseed meal (ammoniated), cottonseed meats, egg production mixed feed, feed, compound, feed (cat), feed (cattle), feed (dog), feed (pig), feed (poultry), feed (rabbit), feed (sheep), fish meal, horsegram, maize, maize gluten, maize husk, maize, ground, maize, preharvest, mung testa, mustard cake, niger cake, peanut cake, peanut cake (deoiled), peanut expeller, peanut hay, peanut meal, peanuts, peanut, kernels, redgram husk, rice bran, rice bran (deoiled), rice chaff, rice meal, rice (polish), sal seed cake, sesame cake, sorghum, soybean meal, soybeans, wheat, wheat bran

Fumonisin B_1
incidence: 1/35, conc.: 105 μg/kg, country: UK[42]
see also barley, bird food, wild, feed, feed, complete ration, feed, general, feed, layer, feed, maize-based, feed, mixed, feed, pelleted ration, feed, screenings, feed, sweet, feed (cat), feed (chicken), feed (dog), feed (gluten), feed (broilers), feed (horse), feed (maize), feed (pig), feed (poultry), feed (rat), feed (rodent), forage grass, maize, maize and maize screenings, maize bran, maize ears, maize fine fractions, maize flakes, maize germ, maize germ/bran, maize germ meal,

maize gluten, maize grits, maize kernels, maize meal, maize powder, maize screenings, maize, "Baby", maize, ground, maize, preharvest, maize, sweet feed, maize/oats mix, rat chow, silage, sorghum, soybeans, wheat

Fumonisin B_2
incidence: 1/35, conc.: 30 μg/kg, country: UK[42]
see also barley, bird food, wild, feed, feed, maize-based, feed, mixed, feed (cat), feed (dog), feed (gluten), feed (horse), feed (maize), feed (poultry), feed (rodent), maize, maize bran, maize fine fractions, maize flakes, maize germ, maize germ/bran, maize germ meal, maize gluten, maize grits, maize kernels, maize meal, maize powder, maize screenings, maize, "Baby", maize, ground, maize, preharvest, rat chow, wheat

Ochratoxin A
incidence: 2/35, conc. range: 1.1–1.3 μg/kg, Ø conc.: 1.2 μg/kg, country: UK[42]
see also alfalfa, barley, barley, oats, barley (high moisture), barley-soybean diet, bird food, domestic, bird food, wild, broilers feed, cereal grains, citrus pulp, coconut, expeller, corn cob mix silage, diet (dairy cow), diet (poultry), diet (starter), eat, egg production mixed feed, feed, feed wheat, oat and barley, feed, commercial mix, feed, mixed, feed, mixed (pelleted), feed (broilers), feed (cat), feed (cattle), feed (cereals), feed (pig), feed ec (pig), feed (poultry), feed ec (poultry), feed (poultry, pig), feed ec (rabbit), feed (trout), grain, mixed feed, grains, mixed, grains (heated), hay, horse bean, maize, maize feed, milo, maize gluten, maize meal, maize, white, *Makhana (Euryale ferox* Salisb) puffs, milk production mixed feed, millet, oat and barley (hammer-milled), oats, palm products, peanut cake, peas, peas and beans, pet food, pig feedstuffs, pig grower diet, pig meal, piglet diet, poultry feedstuffs, rice bran, rice germ, rice germ cake, rye, sorghum, soybean groats, sunflower, sunflower seeds, extracted, tapioca, triticale, *Vicia faba*, wheat, wheat and barley, wheat bran, wheat hay, wheat, oats

Eat may contain the following mycotoxins:

OCHRATOXIN A
incidence: 1/7345, conc. range: nc, country: Hungary[209]
see also alfalfa, barley, barley, oats, barley (high moisture), barley-soybean diet, bird food, domestic, bird food, wild, broilers feed, cereal grains, citrus pulp, coconut, expeller, corn cob mix silage, diet (dairy cow), diet (poultry), diet (starter), dog food, egg production mixed feed, feed, feed wheat, oat and barley, feed, commercial mix, feed, mixed, feed, mixed (pelleted), feed (broilers), feed (cat), feed (cattle), feed (cereals), feed (pig), feed ec (pig), feed (poultry), feed ec (poultry), feed (poultry, pig), feed ec (rabbit), feed (trout), grain, mixed feed, grains, mixed, grains (heated), hay, horse bean, maize, maize feed, milo, maize gluten, maize meal, maize, white, *Makhana* (*Euryale ferox* Salisb) puffs, milk production mixed feed, millet, oat and barley (hammer-milled), oats, palm products, peanut cake, peas, peas and beans, pet food, pig feedstuffs, pig grower diet, pig meal, piglet diet, poultry feedstuffs, rice bran, rice germ, rice germ cake, rye, sorghum, soybean groats, sunflower, sunflower seeds, extracted, tapioca, triticale, *Vicia faba*, wheat, wheat and barley, wheat bran, wheat hay, wheat, oats

Egg production mixed feed may contain the following mycotoxins:

AFLATOXIN B$_1$
incidence: 1/5, conc.: 10 µg/kg, country: Egypt[16]
see also alfalfa, *Ambadi* cake, animal feedstuffs (dairy cake), bagasse, barley, bengalgram husk, bird food, bird food, wild, biri testa, blackgram, blackgram husk, bran, broiler mixed feed, calf fattening mixed feed, calf fattening mixed feed (containing 4–20 % peanut products), *Carthamus* cake, castor cake, cereals, cereal products, chick pea, coconut cake, cocos, concentrate, mixed, concentrates, cotton cake, cottonseed, cottonseed (dehulled), cottonseed cake, cottonseed extract, cottonseed meal, cottonseed meal (ammoniated), cottonseed meal (decorticated), cottonseed meats, cottonseed products, crumbles, crumbles, grower, cycad meal, dairy cattle feed, dairy cattle feed (containing 2–5 % peanut products), dairy cattle feed (containing 6–10 % peanut products), dairy cattle feed (containing 6–12 % peanut products), dairy cattle feed (containing more than 20 % peanut products), diets, mixed, dog food, feed, feed and ingredients, feed, compound, feed, layer, feed, mixed, feed (beef), feed (broilers), feed (calf), feed (cat), feed (cattle), feed (chicken), feed (dairy), feed (dog), feed (dug), feed (fish), feed (gluten), feed (horse), feed (miscellaneous), feed (pig), feed (poultry), feed (poultry, pig), feed (rabbit), feed (sheep), feed (50–60 % maize), fish meal, grain by-products, grains (no specification), greengram, hay/silage, horsegram, husk, *Jagni* cake, legume mixture, linseed, linseed cake, livol, *Mahua* cake, maize, maize germ, maize gluten, maize grits, maize husk, maize meal, maize oil cake, maize powder, maize screenings, maize, ground, maize, hybrid, maize, preharvest, maize, yellow, maize (dark grains), *Makhana* (*Euryale ferox* Salisb) puffs, manioc, milk production mixed feed, mung testa, murkool, mustard cake, neem cake, niger cake, oats, palm kernel expeller cake, palm kernels, palm products, peanut cake, peanut cake (deoiled), peanut expeller, peanut hay, peanut meal, peanut, kernels, peanut, shells, peanuts, pellets, finisher, pig meal and pellets, pigeon pea, poultry feeds (peanut containing), rapeseed cake, redgram husk, rice, rice bran, rice bran (deoiled), rice chaff, rice crack, rice germ, rice germ cake, rice meal, rice straw, rice (damaged), rice (polish), safflower cake, sal seed cake, sesame, sesame cake, sorghum, soybean meal, soybeans, sunflower, sunflower cake, sunflower flour, tapioca, wheat, wheat bran, wheat bran and chana testa

AFLATOXIN B$_2$
incidence: 1/5, conc.: 20 µg/kg, country: Egypt[16]

see also animal feedstuffs (dairy cake), bird food, bird food, wild, biri testa, blackgram, blackgram husk, cottonseed, cottonseed cake, cottonseed extract, cottonseed meal, cottonseed meal (ammoniated), cottonseed meats, dog food, feed, compound, feed (cat), feed (cattle), feed (dog), feed (pig), feed (poultry), feed (rabbit), feed (sheep), fish meal, horsegram, maize, maize gluten, maize husk, maize, ground, maize, preharvest, mung testa, mustard cake, niger cake, peanut cake, peanut cake (deoiled), peanut expeller, peanut hay, peanut meal, peanuts, peanut, kernels, redgram husk, rice bran, rice bran (deoiled), rice chaff, rice meal, rice (polish), sal seed cake, sesame cake, sorghum, soybean meal, soybeans, wheat, wheat bran

Deoxynivalenol
incidence: 3/4, conc. range: 225–528 µg/kg, Ø conc.: 345 µg/kg, country: Egypt[16]
see also barley, barley, husked, barley, unhusked (naked), barley (pressed), bone meal, bran, broilers feed, calf fattening mixed feed, coconut, expeller, corn cob mix silage, cottonseed, cottonseed cake, dairy cattle feed, feed, feed components, feed, commercial mix, feed, mixed, feed, mixed (primarily maize), feed (barley), feed (cattle), feed (chicken), feed (dog), feed (fish), feed (mill run, from wheat), feed (mink), feed (pig), feed (poultry), feed (reindeer), feeds, grain, feeds, industrial, feedstuff, feedstuffs (rapeseed, turnip, fish meal, concentrates), fish meal, grain, mixed feed, grains, mixed, grains (no specification), maize, maize ears, maize fibre, maize germ, maize germ/bran, maize germ meal, maize gluten, maize kernels, maize meal, maize powder, maize screenings, maize stalks (pith), maize, "Baby", maize, hybrid, maize, white, oats, rice bran, rice germ cake, rye, silage, sorghum, soybeans, triticale, wheat, wheat and barley, wheat, red hard winter, wheat, soft white winter, wheat, spring, wheat, winter

Ochratoxin A
incidence: 1/3, conc.: 14 µg/kg, country: Egypt[16]

see also alfalfa, barley, barley, oats, barley (high moisture), barley-soybean diet, bird food, domestic, bird food, wild, broilers feed, cereal grains, citrus pulp, coconut, expeller, corn cob mix silage, diet (dairy cow), diet (poultry), diet (starter), dog food, eat, feed, feed wheat, oat and barley, feed, commercial mix, feed, mixed, feed, mixed (pelleted), feed (broilers), feed (cat), feed (cattle), feed (cereals), feed (pig), feed ec (pig), feed (poultry), feed ec (poultry), feed (poultry, pig), feed ec (rabbit), feed (trout), grain, mixed feed, grains, mixed, grains (heated), hay, horse bean, maize, maize feed, milo, maize gluten, maize meal, maize, white, *Makhana* (*Euryale ferox* Salisb) puffs, milk production mixed feed, millet, oat and barley (hammer-milled), oats, palm products, peanut cake, peas, peas and beans, pet food, pig feedstuffs, pig grower diet, pig meal, piglet diet, poultry feedstuffs, rice bran, rice germ, rice germ cake, rye, sorghum, soybean groats, sunflower, sunflower seeds, extracted, tapioca, triticale, *Vicia faba*, wheat, wheat and barley, wheat bran, wheat hay, wheat, oats

Feed   may contain the following mycotoxins:

AAL-toxin
incidence: 61/63, conc. range: 90–1470 µg/kg, Ø conc.: 560 µg/kg, country: USA[334]

Aflatoxin B_1
incidence: 12/42, conc. range: <5–40 µg/kg, country: South Africa[79]
incidence: 68/194, conc. range: <5–400 µg/kg, country: South Africa[79]
incidence: 42/825, conc. range: ≤200 µg/kg, country: USA[165]
incidence: 61?/61, conc. range: <1–60 µg/kg, country: The Netherlands[283]
incidence: 14/104*, conc. range: tr–1920 µg/kg, country: Czechoslovakia[374], *partly imported
incidence: 45/105, conc. range: ≤300 µg/kg, country: Germany[376]
incidence: 41/226, conc. range: 0.1–1000 µg/kg, country: Spain[377]

incidence: ?/226, conc. range: <4–34 µg/kg, country: The Netherlands[379]

see also alfalfa, *Ambadi* cake, animal feedstuffs (dairy cake), bagasse, barley, bengalgram husk, bird food, bird food, wild, biri testa, blackgram, blackgram husk, bran, broiler mixed feed, calf fattening mixed feed, calf fattening mixed feed (containing 4–20% peanut products), *Carthamus* cake, castor cake, cereals, cereal products, chick pea, coconut cake, cocos, concentrate, mixed, concentrates, cotton cake, cottonseed, cottonseed (dehulled), cottonseed cake, cottonseed extract, cottonseed meal, cottonseed meal (ammoniated), cottonseed meal (decorticated), cottonseed meats, cottonseed products, crumbles, crumbles, grower, cycad meal, dairy cattle feed, dairy cattle feed (containing 2–5% peanut products), dairy cattle feed (containing 6–10% peanut products), dairy cattle feed (containing 6–12% peanut products), dairy cattle feed (containing more than 20% peanut products), diets, mixed, dog food, egg production mixed feed, feed and ingredients, feed, compound, feed, layer, feed, mixed, feed (beef), feed (calf), feed (cat), feed (cattle), feed (chicken), feed (dairy), feed (dog), feed (dug), feed (fish), feed (gluten), feed (horse), feed (miscellaneous), feed (pig), feed (poultry), feed (poultry, pig), feed (rabbit), feed (sheep), feed (50–60% maize), fish meal, grain by-products, grains (no specification), greengram, hay/silage, horsegram, husk, *Jagni* cake, legume mixture, linseed, linseed cake, livol, *Mahua* cake, maize, maize germ, maize gluten, maize grits, maize husk, maize meal, maize oil cake, maize powder, maize screenings, maize, ground, maize, hybrid, maize, preharvest, maize, yellow, maize (dark grains), *Makhana* (*Euryale ferox* Salisb) puffs, manioc, milk production mixed feed, mung testa, murkool, mustard cake, neem cake, niger cake, oats, palm kernel expeller cake, palm kernels, palm products, peanut cake, peanut cake (deoiled), peanut expeller, peanut hay, peanut meal, peanut, kernels, peanut, shells, peanuts, pellets, finisher, pig meal and pellets, pigeon pea, poultry feeds (peanut containing), rapeseed cake, redgram husk, rice, rice bran, rice bran (deoiled), rice chaff, rice crack, rice germ, rice germ cake, rice meal, rice straw, rice (damaged), rice (polish), safflower cake, sal seed cake, sesame, sesame cake, sorghum, soybean meal, soybeans, sunflower, sunflower cake, sunflower flour, tapioca, wheat, wheat bran, wheat bran and chana testa

AFLATOXIN B

incidence: 7/20, conc. range: ≥ 25–>100 µg/kg, country: France[46]

see also barley, cocoa (oil cake), cottonseed cake, feed (cattle), feed (maize, gluten), feed (pig), feed (poultry), flour (wheat), lucern (dried), maize, maize gluten, maize grains, oats, peanut (oil cake), rice, broken, sorghum, soybean (oil cake), sunflower (oil cake), wheat

AFLATOXIN

incidence: 18/71, conc. range: 2–20 µg/kg (7 sa), >20 µg/kg (11 sa), country: Uruguay[91]

incidence: 2/120, conc. range: 14–24 µg/kg, Ø conc.: 19 µg/kg, country: Germany[343]

incidence: 4/118, conc. range: 1.4–4 µg/kg, country: Germany[353]

incidence: ?/142*, conc. range: 0–2780 µg/kg, country: USA[406], *wild bird seed

see also blackgram husk, bread crumbs, broiler finisher, broiler starter, cotton cake, cottonseed, cottonseed cake, cottonseed extract, cottonseed meal, feed (cattle), feed (cow), feed (dog), feed (horse), feed (maize, gluten), feed (pig), feed (poultry), feed (rabbit), feed (rat/mice), feed (sheep), feeds, grain, fish meal, flour (wheat), groundnut cake, grower's mash, horsegram, layer's mash, maize, maize gluten, maize, white, milo, peanut cake, peanut cake (deoiled), peanut, kernels, peanut (oil cake), pearlmillet, pig breeder's mash, pig finisher, pig starter, pod with haulms, poultry breeder's mash, rabbit pellets, redgram husk, rice, rice bran (deoiled), rice, broken, rice (polish), sesame cake, silk worm pupae, sorghum, soybean cake, soybean meal, wheat, wheat bran

Aflatoxins
incidence: 13/427, conc. range: 2–39 μg/kg,
country: Slovenia[424]
incidence: 50/342, conc. range: 5–1906 μg/kg,
country: Brazil[425]
see also animal feed (maize), animal feed
(mixed), barley, copra meal, cottonseed,
cottonseed fines, cottonseed meats, cottonseed
meal, feed ingredients (miscellaneous), feed,
mixed, feed (excluding peanuts, suspect),
feed (goat), feed (pig), feed (poultry),
feedstuff, grain, grain, mixed feed, maize,
maize, shelled, millet, peanut cake, peanut
meal, peanut meal and by-products, peanuts,
protein concentrates, rice, rice bran,
sorghum, soybean meal, sunflower

Citrinin
incidence: 2/60, conc. range:
2000–3000 μg/kg, country: USA[32]
see also barley, barley, oats, barley-soybean
diet, cottonseed cake, feed, mixed, feed
(cattle), feed (pig), fish meal, hay, maize,
maize, white, *Makhana* (*Euryale ferox* Salisb)
puffs, oats, palm products, peas and beans,
rice bran, rice germ, wheat, wheat and other
grains (moldy), wheat bran

Cyclopiazonic Acid
incidence: 55/63, conc. range:
120–1820 μg/kg, Ø conc.: 340 μg/kg, country:
USA[334]
see also chick mash, groundnut cake, maize,
millet, little, peanuts, rice bran, wheat

Deoxynivalenol
incidence: 11/167, conc. range: 10–120 μg/kg,
country: Finland[10]
incidence: 37/43 (barley), conc. range:
≥20–<500 μg/kg, country: UK[36]
incidence: 1/1, conc.: 1400 μg/kg, country:
Canada[71]
incidence: 1/1, conc. range: 40–60 μg/kg,
country: ?[71]
incidence: 17/58, conc. range: 8–500 μg/kg
(6 sa), 500–1000 μg/kg (4 sa), >1000 μg/kg
(7 sa), country: Uruguay[91]
incidence: 4/4, conc. range: 210–623.3 μg/kg,
Ø conc.: 361.7 μg/kg, country: USA*[120],
*probably

incidence: 6/?, conc. range: 40–1800 μg/kg,
country: USA[174]
incidence: 24/55, conc. range: <100 μg/kg
(6 sa), 100–1000 μg/kg (14 sa), >1000 μg/kg
(4 sa), country: Austria[196]
incidence: 193/319, conc. range: <100 μg/kg
(35 sa), 100–1000 μg/kg (104 sa),
>1000 μg/kg (54 sa), country: Austria[196]
incidence: 124/297, conc. range: <100 μg/kg
(17 sa), 100–1000 μg/kg (76 sa), >1000 μg/kg
(31 sa), country: Austria[196]
incidence: 112/244, conc. range: <100 μg/kg
(11 sa), 100–1000 μg/kg (78 sa), >1000 μg/kg
(23 sa), country: Austria[196]
incidence: 269/409, conc. range: <100 μg/kg
(24 sa), 100–1000 μg/kg (164 sa),
>1000 μg/kg (81 sa), country: Austria[196]
incidence: 174/389, conc. range: <100 μg/kg
(25 sa), 100–1000 μg/kg (134 sa),
>1000 μg/kg (15 sa), country: Austria[196]
incidence: 157/200, conc. range: <100 μg/kg
(26 sa), 100–1000 μg/kg (113 sa),
>1000 μg/kg (18 sa), country: Austria[196]
incidence: 63/63, conc. range:
340–6020 μg/kg, Ø conc.: 730 μg/kg, country:
USA[334]
incidence: 77/296, conc. range:
1.8–10,000 μg/kg, country: Germany[353]
incidence: 8/77, conc. range: <2–300 μg/kg,
Ø conc.: 159.1 μg/kg, country: Saudi
Arabia[389]
see also barley, barley, husked, barley,
unhusked (naked), barley (pressed), bone
meal, bran, broilers feed, calf fattening mixed
feed, coconut, expeller, corn cob mix silage,
cottonseed, cottonseed cake, dairy cattle feed,
egg production mixed feed, feed components,
feed, commercial mix, feed, mixed, feed,
mixed (primarily maize), feed (barley), feed
(cattle), feed (chicken), feed (dog), feed
(fish), feed (mill run, from wheat), feed
(mink), feed (pig), feed (poultry), feed
(reindeer), feeds, grain, feeds, industrial,
feedstuff, feedstuffs (rapeseed, turnip, fish
meal, concentrates), fish meal, grain, mixed
feed, grains, mixed, grains (no specification),
maize, maize ears, maize fibre, maize germ,
maize germ/bran, maize germ meal, maize

gluten, maize kernels, maize meal, maize powder, maize screenings, maize stalks (pith), maize, "Baby", maize, hybrid, maize, white, oats, rice bran, rice germ cake, rye, silage, sorghum, soybeans, triticale, wheat, wheat and barley, wheat, red hard winter, wheat, soft white winter, wheat, spring, wheat, winter

DEOXYNIVALENOL AND METABOLITES
incidence: 345/505, conc. range:
20–8510 µg/kg, country: Slovenia[424]

DIACETOXYSCIRPENOL
incidence: 2/2, conc. range: 500–580 µg/kg,
Ø conc.: 540 µg/kg, country: USA[71]
incidence: 2/?, conc. range: 380–500 µg/kg,
Ø conc.: 490 µg/kg, country: USA[174]
see also alfalfa, barley, feed components, feed (dog), feed (fish), feed (mink), feed (pig), feed (poultry), feed (reindeer), feeds, grain, feedstuff, forage grass, grain, mixed feed, grains (no specification), maize, maize gluten, oat and barley (hammer-milled), oats, peanuts, soybeans, wheat

DIACETOXYSCIRPENOL AND METABOLITES
incidence: 19/215, conc. range: 20–160 µg/kg,
country: Slovenia[424]

ERGOT ALKALOID
incidence: 15/35, conc. range: 150–450 µg/kg
(6 sa), 450–750 µg/kg (6 sa), >750 µg/kg
(3 sa), country: Uruguay[91]
incidence: 2/2, conc. range:
8000–18,000 µg/kg*, Ø conc.: 13,000 µg/kg,
country: Australia[331], *>90 % dihydroergosine
see also feedstuff, grain, grain diets, tall
fescue grass (*Festuga arundinacea* Screb)

FUMONISIN B$_1$
incidence: 1/1, conc.: 4 µg/kg, country: USA[58]
incidence: 13/13, conc. range:
1300–16,800 µg/kg, Ø conc.: 6246.2 µg/kg,
country: USA[101]
incidence: 18/18, conc. range: 0–2200 µg/kg,
Ø conc.: 536 µg/kg, country: Spain[159]
incidence: 1/1, conc.: 756 µg/kg, country:
USA[161]

incidence: 7/50, conc. range:
278–637.2 µg/kg, Ø conc.: 366.4 µg/kg,
country: Germany, France[225]
incidence: 23/63, conc. range: 20–2120 µg/kg,
Ø conc.: 280 µg/kg, country: USA[334]
incidence: 11/13, conc. range:
89.1–2105 µg/kg, Ø conc.: 585 µg/kg,
country: Spain[347]
incidence: 9/11, conc. range:
2150–10,650 µg/kg, Ø conc.: 5718 µg/kg,
country: Brazil[414]
see also barley, bird food, wild, dog food, feed, complete ration, feed, general, feed, layer, feed, maize-based, feed, mixed, feed, pelleted ration, feed, screenings, feed, sweet, feed (broilers), feed (cat), feed (chicken), feed (dog), feed (gluten), feed (horse), feed (maize), feed (pig), feed (poultry), feed (rat), feed (rodent), forage grass, maize, maize and maize screenings, maize bran, maize ears, maize fine fractions, maize flakes, maize germ, maize germ/bran, maize germ meal, maize gluten, maize grits, maize kernels, maize meal, maize powder, maize screenings, maize, "Baby", maize, ground, maize, preharvest, maize, sweet feed, maize/oats mix, rat chow, silage, sorghum, soybeans, wheat

FUMONISIN B$_2$
incidence: 13/13, conc. range:
100–6500 µg/kg, Ø conc.: 2253.8 µg/kg,
country: USA[101]
incidence: 8/18, conc. range: 0–600 µg/kg,
Ø conc.: 308 µg/kg, country: Spain[159]
incidence: 1/1, conc.: 252 µg/kg, country:
USA[161]
incidence: 2/50, conc. range:
109.1–144.8 µg/kg, Ø conc.: 127 µg/kg,
country: Germany, France[225]
incidence: 9/11, conc. range: 740–4060 µg/kg,
Ø conc.: 1991 µg/kg, country: Brazil[414]
see also barley, bird food, wild, dog food, feed, maize-based, feed, mixed, feed (cat), feed (dog), feed (gluten), feed (horse), feed (maize), feed (poultry), feed (rodent), maize, maize bran, maize fine fractions, maize flakes, maize germ, maize germ/bran, maize germ meal, maize gluten, maize grits, maize kernels, maize meal, maize powder, maize

screenings, maize, "Baby", maize, ground, maize, preharvest, rat chow, wheat

FUMONISIN B₃
incidence: 12/13, conc. range: 50–2650 μg/kg, Ø conc.: 560 μg/kg, country: USA[101]
see also feed (maize), feed (rodent), maize, maize bran, maize fine fractions, maize germ meal, maize kernels, maize powder, maize screenings

FUSARIC ACID
incidence: 15/15, conc. range: 70–12,390 μg/kg, Ø conc.: 3367 μg/kg, country: USA[413]
see also feedstuff (barley), feedstuff (dry corn), feedstuff (high-moisture corn), feedstuff (wheat), feedstuff (whole seeds)

NEOSOLANIOL
incidence: 1/77, conc.: 200 μg/kg, country: Saudi Arabia[389]
see also feed components, feed, mixed, feed (poultry), peanut cake, wheat

NIVALENOL
incidence: 4/77, conc.: 25–600 μg/kg, Ø conc.: 337.5 μg/kg, country: Saudi Arabia[389]
see also barley, barley, husked, barley, unhusked (naked), barley (pressed), bran, feed components, feed (cattle), feed (poultry), feed (reindeer), feeds, industrial, maize, maize ears, maize germ, maize germ/bran, maize gluten, maize kernels, maize meal, maize powder, maize screenings, maize, "Baby", oats, rye, silage, triticale, wheat

OCHRATOXIN A
incidence: 30/170, conc. range: 100–1000 μg/kg, country: Austria[29]
incidence: 3/10, conc. range: 140–360 μg/kg, country: Tunisia[31]
incidence: 57/177, conc. range: ≤2389 μg/kg, country: Germany[73]
incidence: 5/28, conc. range: ≤670 μg/kg, country: Germany[73]
incidence: 38/608, conc. range: ≤206 μg/kg, country: Germany[73]
incidence: 21/150, conc. range: ≤40 μg/kg, country: Germany[73]
incidence: 54/100, conc. range: ≤1300 μg/kg, country: Norway[291]
incidence: 68/384, conc. range: 0.1–≥20 μg/kg, country: Germany[332]
incidence: 42/170, conc. range: 0.1–145 μg/kg, country: Germany[353]
incidence: 8/150, conc. range: 50–200 μg/kg, country: Poland[359]
incidence: 216/235, conc. range: tr–400 μg/kg, country: Yugoslavia[373]
incidence: 24/326, conc. range: 20–580 μg/kg, country: Slovenia[424]
incidence: 9/342, conc. range: 14–745 μg/kg, country: Brazil[425]
see also alfalfa, barley, barley, oats, barley (high moisture), barley-soybean diet, bird food, domestic, bird food, wild, broilers feed, cereal grains, citrus pulp, coconut, expeller, corn cob mix silage, diet (dairy cow), diet (poultry), diet (starter), dog food, eat, egg production mixed feed, feed wheat, oat and barley, feed, commercial mix, feed, mixed, feed, mixed (pelleted), feed (broilers), feed (cat), feed (cattle), feed (cereals), feed (pig), feed ec (pig), feed (poultry), feed ec (poultry), feed (poultry, pig), feed ec (rabbit), feed (trout), grain, mixed feed, grains, mixed, grains (heated), hay, horse bean, maize, maize feed, milo, maize gluten, maize meal, maize, white, *Makhana* (*Euryale ferox* Salisb) puffs, milk production mixed feed, millet, oat and barley (hammer-milled), oats, palm products, peanut cake, peas, peas and beans, pet food, pig feedstuffs, pig grower diet, pig meal, piglet diet, poultry feedstuffs, rice bran, rice germ, rice germ cake, rye, sorghum, soybean groats, sunflower, sunflower seeds, extracted, tapioca, triticale, *Vicia faba*, wheat, wheat and barley, wheat bran, wheat hay, wheat, oats

PR TOXIN
incidence: 48/63, conc. range: 50–260 μg/kg, Ø conc.: 130 μg/kg, country: USA[334]

TENUAZONIC ACID
incidence: 1/1, conc.: 10 μg/kg, country: Australia[188]
see also feed (GP No. 1), oilseed rape meal, sunflower seed meal, sunflower seeds, ensiling

Type-A-Trichothecene (HT-2 Toxin, T-2 Toxin)
incidence: 7/174, conc. range:
18.8–29.3 µg/kg, country: Germany[353]

T-2 Toxin
incidence: 2/2, conc. range: 76–300 µg/kg,
Ø conc.: 188 µg/kg, country: USA[71]
see also alfalfa, barley, bran, diet (grower),
diet (poultry), feed components, feed, layer,
feed, mixed, feed (dog), feed (fish), feed
(mink), feed (pig), feed (poultry), feedstuff,
forage grass, grain, mixed feed, grains (no
specification), hay, maize, maize germ/bran,
maize gluten, maize meal, maize screenings,
maize stalks (pith), oat and barley (hammer-
milled), oats, peanuts, pig grower diet, piglet
diet, rye, sorghum, triticale, wheat

T-2 Toxin and Metabolites
incidence: 42/232, conc. range: 20–490 µg/kg,
country: Slovenia[424]

Zearalenone
incidence: 12/69, conc. range: 100–200 µg/kg
(2 sa), >200 µg/kg (10 sa), country:
Uruguay[91]
incidence: 14/14, conc. range: 65–5600 µg/kg,
country: USA, Canada[174]
incidence: 7/9, conc. range: tr–3600 µg/kg,
country: USA[174]
incidence: 1/1, conc.: 25,000 µg/kg, country:
USA[175,176]
incidence: 36/100, conc. range: ≤137 µg/kg,
country: Norway[291]
incidence: 291/384, conc. range:
1–≥20.1 µg/kg, country: Germany[332]
incidence: 2/63, conc. range: 120–310 µg/kg,
Ø conc.: 215 µg/kg, country: USA[334]
incidence: 14/120, conc. range: 50–600 µg/kg,
217 µg/kg, country: Germany[343]
incidence: 129/345, conc. range:
0.1–769 µg/kg, country: Germany[353]
incidence: 40/372, conc. range: 20–460 µg/kg,
country: Slovenia[424]
incidence: 20/342, conc. range:
100–4982 µg/kg, country: Brazil[425]
see also alfalfa, barley, barley, husked, barley,
unhusked (naked), barley and feed, bone
meal, bran, broilers feed, *Carthamus* cake,
chick pea, concentrate, mixed, corn cob mix
silage, cotton cake, cottonseed, cottonseed
cake, diet (dairy cow), diet (poultry), diets
(mixed), feed components, feed, mixed, feed,
mixed (primarily maize also maize, oats,
wheat), feed (bran), feed (broiler chicken),
feed (cattle), feed (chicken), feed (dairy), feed
(developing pig), feed (mill run, from
wheat), feed (miscellaneous), feed (pig), feed
(poultry), feed (poultry, pig), feed (starter
chicken), feedstuff, fish meal, forage grass,
grain, bruised, grain, mixed feed, grains (no
specification), hay, maize, maize ears, maize
flakes, maize germ, maize germ/bran, maize
gluten, maize kernels, maize meal, maize oil
cake, maize screenings, maize stalks (pith),
maize, "Baby", maize, hybrid, maize, shelled,
maize, unshelled, maize, white, maize grain,
artificially dried, maize grain, crib dried,
maize grain, ensiled, milk production mixed
feed, oats, *Paspalum palidosum*, straw, peanut
hulls/skins, rice bran, rice germ, rice germ
cake, rye, silage, sorghum, soybeans,
soybeans, extracted, sunflower cake, tapioca,
triticale, wheat, wheat bran, wheat bran and
chana testa, wheat soya meal

Feed components may contain the following
mycotoxins:

3-Acetyldeoxynivalenol
incidence: 13/15, conc. range: 14–114 µg/kg,
Ø conc.: 50 µg/kg, country: Germany[366]
see also barley, feed (poultry), feeds, grain,
feeds, industrial, feedstuffs (rapeseed, turnip,
fish meal, concentrates), maize, maize ears,
maize gluten, maize kernels, maize meal,
maize screenings, oats, wheat

15-Acetyldeoxynivalenol
incidence: 14/15, conc. range:
165–1780 µg/kg, Ø conc.: 573 µg/kg, country:
Germany[366]
see also barley, feed, commercial mix, feed
(poultry), maize, maize ears, maize germ,
maize germ/bran, maize gluten, maize
kernels, maize meal, maize screenings, maize,
"Baby", oats, silage, wheat

DEOXYNIVALENOL
incidence: 15/15, conc. range:
634–6682 µg/kg, Ø conc.: 1990 µg/kg,
country: Germany[366]
see also barley, barley, husked, barley,
unhusked (naked), barley (pressed), bone
meal, bran, broilers feed, calf fattening mixed
feed, coconut, expeller, corn cob mix silage,
cottonseed, cottonseed cake, dairy cattle feed,
egg production mixed feed, feed, feed,
commercial mix, feed, mixed, feed, mixed
(primarily maize), feed (barley), feed (cattle),
feed (chicken), feed (dog), feed (fish), feed
(mill run, from wheat), feed (mink), feed
(pig), feed (poultry), feed (reindeer), feeds,
grain, feeds, industrial, feedstuff, feedstuffs
(rapeseed, turnip, fish meal, concentrates),
fish meal, grain, mixed feed, grains, mixed,
grains (no specification), maize, maize ears,
maize fibre, maize germ, maize germ/bran,
maize germ meal, maize gluten, maize
kernels, maize meal, maize powder, maize
screenings, maize stalks (pith), maize,
"Baby", maize, hybrid, maize, white, oats,
rice bran, rice germ cake, rye, silage,
sorghum, soybeans, triticale, wheat, wheat
and barley, wheat, red hard winter, wheat,
soft white winter, wheat, spring, wheat,
winter

DIACETOXYSCIRPENOL
incidence: 1/15, conc.: 21 µg/kg, country:
Germany[366]
see also alfalfa, barley, feed, feed (dog), feed
(fish), feed (mink), feed (pig), feed (poultry),
feed (reindeer), feeds, grain, feedstuff, forage
grass, grain, mixed feed, grains (no specification), maize, maize gluten, oat and barley
(hammer-milled), oats, peanuts, soybeans,
wheat

FUSARENONE-X
incidence: 7/15, conc. range: 29–494 µg/kg,
Ø conc.: 124 µg/kg, country: Germany[366]
see also barley, feed (poultry), maize, maize
ears, maize germ/bran, maize gluten, maize
kernels, maize meal, maize screenings, oats,
wheat

HT-2 TOXIN
incidence: 13/15, conc. range: 5–99 µg/kg,
Ø conc.: 34 µg/kg, country: Germany[366]
see also barley, feed (dog), feed (fish), feed
(pig), feed (poultry), feed (reindeer), grains,
mixed, grains, mixed feed, grains (no
specification), maize, maize germ/bran, maize
gluten, maize meal, maize screenings, oats,
rye, silage, wheat

MONOACETOXYSCIRPENOL
incidence: 10/15, conc. range: 5–39 µg/kg,
Ø conc.: 14 µg/kg, country: Germany[366]
see also maize, maize gluten, maize meal,
maize screenings, silage, wheat

NEOSOLANIOL
incidence: 1/15, conc.: 9 µg/kg, country:
Germany[366]
see also feed, feed, mixed, feed (poultry),
peanut cake, wheat

NIVALENOL
incidence: 15/15, conc. range: 21–2050 µg/kg,
Ø conc.: 374 µg/kg, country: Germany[366]
see also barley, barley, husked, barley,
unhusked (naked), barley (pressed), bran,
feed, feed (cattle), feed (poultry), feed
(reindeer), feeds, industrial, maize, maize
ears, maize germ, maize germ/bran, maize
gluten, maize kernels, maize meal, maize
powder, maize screenings, maize, "Baby",
oats, rye, silage, triticale, wheat

T-2 TETRAOL
incidence: 3/15, conc. range: 11–56 µg/kg,
Ø conc.: 39 µg/kg, country: Germany[366]
see also barley, maize, silage, wheat

T-2 TOXIN
incidence: 10/15, conc. range: 7–70 µg/kg,
Ø conc.: 21 µg/kg, country: Germany[366]
see also alfalfa, barley, bran, diet (grower),
diet (poultry), feed, feed, layer, feed, mixed,
feed (dog), feed (fish), feed (mink), feed
(pig), feed (poultry), feedstuff, forage grass,
grain, mixed feed, grains (no specification),
hay, maize, maize germ/bran, maize gluten,
maize meal, maize screenings, maize stalks
(pith), oat and barley (hammer-milled), oats,
peanuts, pig grower diet, piglet diet, rye,
sorghum, triticale, wheat

T-2 TRIOL
incidence: 1/15, conc.: 8 μg/kg, country:
Germany[366]
see also barley, grain, mixed feed, grains (no
specification), maize, oats, wheat

ZEARALENONE
incidence: 15/15, conc. range: 57–1362 μg/kg,
Ø conc.: 328 μg/kg, country: Germany[366]
see also alfalfa, barley, barley, husked, barley,
unhusked (naked), barley and feed, bone
meal, bran, broilers feed, *Carthamus* cake,
chick pea, concentrate, mixed, corn cob mix
silage, cotton cake, cottonseed, cottonseed
cake, diet (dairy cow), diet (poultry), diets
(mixed), feed, feed, mixed, feed, mixed
(primarily maize also maize, oats, wheat),
feed (bran), feed (broiler chicken), feed
(cattle), feed (chicken), feed (dairy), feed
(developing pig), feed (mill run, from
wheat), feed (miscellaneous), feed (pig), feed
(poultry), feed (poultry, pig), feed (starter
chicken), feedstuff, fish meal, forage grass,
grain, bruised, grain, mixed feed, grains (no
specification), hay, maize, maize ears, maize
flakes, maize germ, maize germ/bran, maize
gluten, maize kernels, maize meal, maize oil
cake, maize screenings, maize stalks (pith),
maize, "Baby", maize, hybrid, maize, shelled,
maize, unshelled, maize, white, maize grain,
artificially dried, maize grain, crib dried,
maize grain, ensiled, milk production mixed
feed, oats, *Paspalum palidosum*, straw, peanut
hulls/skins, rice bran, rice germ, rice germ
cake, rye, silage, sorghum, soybeans,
soybeans, extracted, sunflower cake, tapioca,
triticale, wheat, wheat bran, wheat bran and
chana testa, wheat soya meal

Feed ingredients (miscellaneous)   may contain
the following mycotoxins:

AFLATOXINS
incidence: 12/116, conc. range: ≤135 μg/kg,
Ø conc.: 42 μg/kg, country: USA[280]
see also animal feed (maize), animal feed
(mixed), barley, copra meal, cottonseed,
cottonseed fines, cottonseed meal, cottonseed
meats, feed, feed, mixed, feed (excluding

peanuts, suspect), feed (goat), feed (pig), feed
(poultry), feedstuff, grain, grain, mixed feed,
maize, maize, shelled, millet, peanut cake,
peanut meal, peanut meal and by-products,
peanuts, protein concentrates, rice, rice bran,
sorghum, soybean meal, sunflower

Feed wheat, oat and barley   may contain the
following mycotoxins:

OCHRATOXIN A
incidence: 1/51, conc.: 5900 μg/kg, country:
Canada[133]
see also alfalfa, barley, barley, oats, barley
(high moisture), barley-soybean diet, bird
food, domestic, bird food, wild, broilers feed,
cereal grains, citrus pulp, coconut, expeller,
corn cob mix silage, diet (dairy cow), diet
(poultry), diet (starter), dog food, eat, egg
production mixed feed, feed, feed,
commercial mix, feed, mixed, feed, mixed
(pelleted), feed (broilers), feed (cat), feed
(cattle), feed (cereals), feed (pig), feed ec
(pig), feed (poultry), feed ec (poultry), feed
(poultry, pig), feed ec (rabbit), feed (trout),
grain, mixed feed, grains, mixed, grains
(heated), hay, horse bean, maize, maize feed,
milo, maize gluten, maize meal, maize, white,
Makhana (*Euryale ferox* Salisb) puffs, milk
production mixed feed, millet, oat and barley
(hammer-milled), oats, palm products,
peanut cake, peas, peas and beans, pet food,
pig feedstuffs, pig grower diet, pig meal,
piglet diet, poultry feedstuffs, rice bran, rice
germ, rice germ cake, rye, sorghum, soybean
groats, sunflower, sunflower seeds, extracted,
tapioca, triticale, *Vicia faba*, wheat, wheat and
barley, wheat bran, wheat hay, wheat, oats

Feed and ingredients   may contain the
following mycotoxins:

AFLATOXIN B$_1$
incidence: 45/100, conc. range: 7–300 μg/kg,
country: Germany[169]
incidence: 39/306, conc. range: ≤2000 μg/kg,
country: Poland[170]
see also alfalfa, *Ambadi* cake, animal feedstuffs
(dairy cake), bagasse, barley, bengalgram

husk, bird food, bird food, wild, biri testa, blackgram, blackgram husk, bran, broiler mixed feed, calf fattening mixed feed, calf fattening mixed feed (containing 4–20 % peanut products), *Carthamus* cake, castor cake, cereals, cereal products, chick pea, coconut cake, cocos, concentrate, mixed, concentrates, cotton cake, cottonseed, cottonseed (dehulled), cottonseed cake, cottonseed extract, cottonseed meal, cottonseed meal (ammoniated), cottonseed meal (decorticated), cottonseed meats, cottonseed products, crumbles, crumbles, grower, cycad meal, dairy cattle feed, dairy cattle feed (containing 2–5 % peanut products), dairy cattle feed (containing 6–10 % peanut products), dairy cattle feed (containing 6–12 % peanut products), dairy cattle feed (containing more than 20 % peanut products), diets, mixed, dog food, egg production mixed feed, feed, feed, compound, feed, layer, feed, mixed, feed (beef), feed (calf), feed (cat), feed (cattle), feed (chicken), feed (dairy), feed (dog), feed (dug), feed (fish), feed (gluten), feed (horse), feed (miscellaneous), feed (pig), feed (poultry), feed (poultry, pig), feed (rabbit), feed (sheep), feed (50–60 % maize), fish meal, grain by-products, grains (no specification), greengram, hay/silage, horsegram, husk, *Jagni* cake, legume mixture, linseed, linseed cake, livol, *Mahua* cake, maize, maize germ, maize gluten, maize grits, maize husk, maize meal, maize oil cake, maize powder, maize screenings, maize, ground, maize, hybrid, maize, preharvest, maize, yellow, maize (dark grains), *Makhana* (*Euryale ferox* Salisb) puffs, manioc, milk production mixed feed, mung testa, murkool, mustard cake, neem cake, niger cake, oats, palm kernel expeller cake, palm kernels, palm products, peanut cake, peanut cake (deoiled), peanut expeller, peanut hay, peanut meal, peanut, kernels, peanut, shells, peanuts, pellets, finisher, pig meal and pellets, pigeon pea, poultry feeds (peanut containing), rapeseed cake, redgram husk, rice, rice bran, rice bran (deoiled), rice chaff, rice crack, rice germ, rice germ cake, rice meal, rice straw, rice (damaged), rice (polish), safflower cake, sal seed cake, sesame, sesame cake, sorghum, soybean meal, soybeans, sunflower, sunflower cake, sunflower flour, tapioca, wheat, wheat bran, wheat bran and chana testa

Feed, commercial mix may contain the following mycotoxins:

15-Acetyldeoxynivalenol
incidence: 1/94, conc.: 11 µg/kg, country: Canada[240]
see also barley, feed components, feed (poultry), maize, maize ears, maize germ, maize germ/bran, maize gluten, maize kernels, maize meal, maize screenings, maize, "Baby", oats, silage, wheat

Deoxynivalenol
incidence: 3/94, conc. range: 19–45 µg/kg, Ø conc.: 27.7 µg/kg, country: Canada[240]
see also barley, barley, husked, barley, unhusked (naked), barley (pressed), bone meal, bran, broilers feed, calf fattening mixed feed, coconut, expeller, corn cob mix silage, cottonseed, cottonseed cake, dairy cattle feed, egg production mixed feed, feed, feed components, feed, mixed, feed, mixed (primarily maize), feed (barley), feed (cattle), feed (chicken), feed (dog), feed (fish), feed (mill run, from wheat), feed (mink), feed (pig), feed (poultry), feed (reindeer), feeds, grain, feeds, industrial, feedstuff, feedstuffs (rapeseed, turnip, fish meal, concentrates), fish meal, grain, mixed feed, grains, mixed, grains (no specification), maize, maize ears, maize fibre, maize germ, maize germ/bran, maize germ meal, maize gluten, maize kernels, maize meal, maize powder, maize screenings, maize stalks (pith), maize, "Baby", maize, hybrid, maize, white, oats, rice bran, rice germ cake, rye, silage, sorghum, soybeans, triticale, wheat, wheat and barley, wheat, red hard winter, wheat, soft white winter, wheat, spring, wheat, winter

Ochratoxin A
incidence: 1/94, conc.: 5 µg/kg, country: Canada[240]

see also alfalfa, barley, barley, oats, barley (high moisture), barley-soybean diet, bird food, domestic, bird food, wild, broilers feed, cereal grains, citrus pulp, coconut, expeller, corn cob mix silage, diet (dairy cow), diet (poultry), diet (starter), dog food, eat, egg production mixed feed, feed, feed wheat, oat and barley, feed, mixed, feed, mixed (pelleted), feed (broilers), feed (cat), feed (cattle), feed (cereals), feed (pig), feed ec (pig), feed (poultry), feed ec (poultry), feed (poultry, pig), feed ec (rabbit), feed (trout), grain, mixed feed, grains, mixed, grains (heated), hay, horse bean, maize, maize feed, milo, maize gluten, maize meal, maize, white, *Makhana* (*Euryale ferox* Salisb) puffs, milk production mixed feed, millet, oat and barley (hammer-milled), oats, palm products, peanut cake, peas, peas and beans, pet food, pig feedstuffs, pig grower diet, pig meal, piglet diet, poultry feedstuffs, rice bran, rice germ, rice germ cake, rye, sorghum, soybean groats, sunflower, sunflower seeds, extracted, tapioca, triticale, *Vicia faba*, wheat, wheat and barley, wheat bran, wheat hay, wheat, oats

Feed, complete ration   may contain the following mycotoxins:

Fumonisin B_1
incidence: 9/10, conc. range: 4–83 µg/kg, Ø conc.: 25.6 µg/kg, country: USA[58]
incidence: 14/16, conc. range: 1–121 µg/kg, Ø conc.: 26.2 µg/kg, country: USA[58]
incidence: 1/1, conc.: 3 µg/kg, country: USA[58]
incidence: 2/2, conc. range: 3–4 µg/kg, Ø conc.: 3.5 µg/kg, country: USA[58]
incidence: 1/2, conc.: 6 µg/kg, country: USA[58]
incidence: 11/11, conc. range: 2–91 µg/kg, Ø conc.: 29.2 µg/kg, country: USA[58]
incidence: 7/8, conc. range: 4–19 µg/kg, Ø conc.: 8.4 µg/kg, country: USA[58]
incidence: 1/1, conc.: 12 µg/kg, country: USA[58]
see also barley, bird food, wild, dog food, feed, feed, general, feed, layer, feed, maize-based, feed, mixed, feed, pelleted ration, feed, screenings, feed, sweet, feed (broilers), feed (cat), feed (chicken), feed (dog), feed (gluten), feed (horse), feed (maize), feed (pig), feed (poultry), feed (rat), feed (rodent), forage grass, maize, maize and maize screenings, maize bran, maize ears, maize fine fractions, maize flakes, maize germ, maize germ/bran, maize germ meal, maize gluten, maize grits, maize kernels, maize meal, maize powder, maize screenings, maize, "Baby", maize, ground, maize, preharvest, maize, sweet feed, maize/oats mix, rat chow, silage, sorghum, soybeans, wheat

Feed, compound   may contain the following mycotoxins:

Aflatoxin B_1
incidence: 8/8, conc. range: 0.6–36.5 µg/kg, Ø conc.: 7.5 µg/kg, country: UK[122]
see also alfalfa, *Ambadi* cake, animal feedstuffs (dairy cake), bagasse, barley, bengalgram husk, bird food, bird food, wild, biri testa, blackgram, blackgram husk, bran, broiler mixed feed, calf fattening mixed feed, calf fattening mixed feed (containing 4–20 % peanut products), *Carthamus* cake, castor cake, cereals, cereal products, chick pea, coconut cake, cocos, concentrate, mixed, concentrates, cotton cake, cottonseed, cottonseed (dehulled), cottonseed cake, cottonseed extract, cottonseed meal, cottonseed meal (ammoniated), cottonseed meal (decorticated), cottonseed meats, cottonseed products, crumbles, crumbles, grower, cycad meal, dairy cattle feed, dairy cattle feed (containing 2–5 % peanut products), dairy cattle feed (containing 6–10 % peanut products), dairy cattle feed (containing 6–12 % peanut products), dairy cattle feed (containing more than 20 % peanut products), diets, mixed, dog food, egg production mixed feed, feed, feed and ingredients, feed, layer, feed, mixed, feed (beef), feed (calf), feed (cat), feed (cattle), feed (chicken), feed (dairy), feed (dog), feed (dug), feed (fish), feed (gluten), feed (horse), feed (miscellaneous), feed (pig), feed (poultry), feed (poultry, pig), feed (rabbit), feed (sheep), feed (50–60 % maize), fish meal, grain by-products, grains (no

specification), greengram, hay/silage, horsegram, husk, *Jagni* cake, legume mixture, linseed, linseed cake, livol, *Mahua* cake, maize, maize germ, maize gluten, maize grits, maize husk, maize meal, maize oil cake, maize powder, maize screenings, maize, ground, maize, hybrid, maize, preharvest, maize, yellow, maize (dark grains), *Makhana* (*Euryale ferox* Salisb) puffs, manioc, milk production mixed feed, mung testa, murkool, mustard cake, neem cake, niger cake, oats, palm kernel expeller cake, palm kernels, palm products, peanut cake, peanut cake (deoiled), peanut expeller, peanut hay, peanut meal, peanut, kernels, peanut, shells, peanuts, pellets, finisher, pig meal and pellets, pigeon pea, poultry feeds (peanut containing), rapeseed cake, redgram husk, rice, rice bran, rice bran (deoiled), rice chaff, rice crack, rice germ, rice germ cake, rice meal, rice straw, rice (damaged), rice (polish), safflower cake, sal seed cake, sesame, sesame cake, sorghum, soybean meal, soybeans, sunflower, sunflower cake, sunflower flour, tapioca, wheat, wheat bran, wheat bran and chana testa

Aflatoxin B_2
incidence: 7/8, conc. range: tr–7.4 μg/kg, country: UK[122]
see also animal feedstuffs (dairy cake), bird food, bird food, wild, biri testa, blackgram, blackgram husk, cottonseed, cottonseed cake, cottonseed extract, cottonseed meal, cottonseed meal (ammoniated), cottonseed meats, dog food, egg production mixed feed, feed (cat), feed (cattle), feed (dog), feed (pig), feed (poultry), feed (rabbit), feed (sheep), fish meal, horsegram, maize, maize gluten, maize husk, maize, ground, maize, preharvest, mung testa, mustard cake, niger cake, peanut cake, peanut cake (deoiled), peanut expeller, peanut hay, peanut meal, peanuts, peanut, kernels, redgram husk, rice bran, rice bran (deoiled), rice chaff, rice meal, rice (polish), sal seed cake, sesame cake, sorghum, soybean meal, soybeans, wheat, wheat bran

Feed, general may contain the following mycotoxins:

Fumonisin B_1
incidence: 5/5, conc. range: 200–<1000 μg/kg, country: South Africa[44]
see also barley, bird food, wild, dog food, feed, feed, complete ration, feed, layer, feed, maize-based, feed, mixed, feed, pelleted ration, feed, screenings, feed, sweet, feed (broilers), feed (cat), feed (chicken), feed (dog), feed (gluten), feed (horse), feed (maize), feed (pig), feed (poultry), feed (rat), feed (rodent), forage grass, maize, maize and maize screenings, maize bran, maize ears, maize fine fractions, maize flakes, maize germ, maize germ/bran, maize germ meal, maize gluten, maize grits, maize kernels, maize meal, maize powder, maize screenings, maize, "Baby", maize, ground, maize, preharvest, maize, sweet feed, maize/oats mix, rat chow, silage, sorghum, soybeans, wheat

Feed, layer may contain the following mycotoxins:

Aflatoxin B_1
incidence: 15/15, conc. range: 65–150 μg/kg, country: India[386]
see also alfalfa, *Ambadi* cake, animal feedstuffs (dairy cake), bagasse, barley, bengalgram husk, bird food, bird food, wild, biri testa, blackgram, blackgram husk, bran, broiler mixed feed, calf fattening mixed feed, calf fattening mixed feed (containing 4–20 % peanut products), *Carthamus* cake, castor cake, cereals, cereal products, chick pea, coconut cake, cocos, concentrate, mixed, concentrates, cotton cake, cottonseed, cottonseed (dehulled), cottonseed cake, cottonseed extract, cottonseed meal, cottonseed meal (ammoniated), cottonseed meal (decorticated), cottonseed meats, cottonseed products, crumbles, crumbles, grower, cycad meal, dairy cattle feed, dairy cattle feed (containing 2–5 % peanut products), dairy cattle feed (containing 6–10 % peanut products), dairy cattle feed (containing 6–12 % peanut products), dairy cattle feed (containing more than 20 % peanut products), diets, mixed, dog food, egg production mixed feed, feed, feed and

ingredients, feed, compound, feed, mixed, feed (beef), feed (calf), feed (cat), feed (cattle), feed (chicken), feed (compound), feed (dairy), feed (dog), feed (dug), feed (fish), feed (gluten), feed (horse), feed (miscellaneous), feed (pig), feed (poultry), feed (poultry, pig), feed (rabbit), feed (sheep), feed (50–60 % maize), fish meal, grain by-products, grains (no specification), greengram, hay/silage, horsegram, husk, *Jagni* cake, legume mixture, linseed, linseed cake, livol, *Mahua* cake, maize, maize germ, maize gluten, maize grits, maize husk, maize meal, maize oil cake, maize powder, maize screenings, maize, ground, maize, hybrid, maize, preharvest, maize, yellow, maize (dark grains), *Makhana* (*Euryale ferox* Salisb) puffs, manioc, milk production mixed feed, mung testa, murkool, mustard cake, neem cake, niger cake, oats, palm kernel expeller cake, palm kernels, palm products, peanut cake, peanut cake (deoiled), peanut expeller, peanut hay, peanut meal, peanut, kernels, peanut, shells, peanuts, pellets, finisher, pig meal and pellets, pigeon pea, poultry feeds (peanut containing), rapeseed cake, redgram husk, rice, rice bran, rice bran (deoiled), rice chaff, rice crack, rice germ, rice germ cake, rice meal, rice straw, rice (damaged), rice (polish), safflower cake, sal seed cake, sesame, sesame cake, sorghum, soybean meal, soybeans, sunflower, sunflower cake, sunflower flour, tapioca, wheat, wheat bran, wheat bran and chana testa

FUMONISIN B$_1$
incidence: 13/15, conc. range: 80–8500 µg/kg, country: India[386]
see also barley, bird food, wild, dog food, feed, feed, complete ration, feed, general, feed, maize-based, feed, mixed, feed, pelleted ration, feed, screenings, feed, sweet, feed (broilers), feed (cat), feed (chicken), feed (dog), feed (gluten), feed (horse), feed (maize), feed (pig), feed (poultry), feed (rat), feed (rodent), forage grass, maize, maize and maize screenings, maize bran, maize ears, maize fine fractions, maize flakes, maize germ, maize germ/bran, maize germ meal,

maize gluten, maize grits, maize kernels, maize meal, maize powder, maize screenings, maize, "Baby", maize, ground, maize, preharvest, maize, sweet feed, maize/oats mix, rat chow, silage, sorghum, soybeans, wheat

T-2 TOXIN
incidence: 5/7345, conc. range: nc, country: Hungary[209]
see also alfalfa, barley, bran, diet (grower), diet (poultry), feed, feed components, feed, mixed, feed (dog), feed (fish), feed (mink), feed (pig), feed (poultry), feedstuff, forage grass, grain, mixed feed, grains (no specification), hay, maize, maize germ/bran, maize gluten, maize meal, maize screenings, maize stalks (pith), oat and barley (hammer-milled), oats, peanuts, pig grower diet, piglet diet, rye, sorghum, triticale, wheat

Feed, maize-based   may contain the following mycotoxins:

FUMONISIN B$_1$
incidence: 13/13, conc. range: 256–6342 µg/kg, Ø conc.: 2573 µg/kg, country: Uruguay[7]
incidence: 61/61, conc. range: 1000–126,000 µg/kg, Ø conc.: 25,000 µg/kg, country: USA[140]
incidence: 13/13, conc. range: 1000–32,000 µg/kg, country: USA[140]
incidence: 20/21, conc. range: 200–38,500 µg/kg, country: Brazil[143]
see also barley, bird food, wild, dog food, feed, feed, complete ration, feed, general, feed, layer, feed, mixed, feed, pelleted ration, feed, screenings, feed, sweet, feed (broilers), feed (cat), feed (chicken), feed (dog), feed (gluten), feed (horse), feed (maize), feed (pig), feed (poultry), feed (rat), feed (rodent), forage grass, maize, maize and maize screenings, maize bran, maize ears, maize fine fractions, maize flakes, maize germ, maize germ/bran, maize germ meal, maize gluten, maize grits, maize kernels, maize meal, maize powder, maize screenings, maize, "Baby", maize, ground, maize, preharvest, maize, sweet feed, maize/oats mix, rat chow, silage, sorghum, soybeans, wheat

Fumonisin B$_2$
incidence: 18/21, conc. range: 100–12,000
μg/kg, country: Brazil[143]
see also barley, bird food, wild, dog food,
feed, feed, mixed, feed (cat), feed (dog), feed
(gluten), feed (horse), feed (maize), feed
(poultry), feed (rodent), maize, maize bran,
maize fine fractions, maize flakes, maize
germ, maize germ/bran, maize germ meal,
maize gluten, maize grits, maize kernels,
maize meal, maize powder, maize screenings,
maize, "Baby", maize, ground, maize,
preharvest, rat chow, wheat

Feed, mixed   may contain the following
mycotoxins:

Aflatoxin B$_1$
incidence: 12/102, conc. range:
40–3000 μg/kg, country: India[127]
incidence: ?/30, conc. range: 20–80 μg/kg,
country: India[183]
incidence: ?/30, conc. range: 40–60 μg/kg,
country: India[183]
incidence: 4/4, conc. range: 16.8–146 μg/kg,
Ø conc.: 63.1 μg/kg, country: USA[387]
see also alfalfa, *Ambadi* cake, animal feedstuffs
(dairy cake), bagasse, barley, bengalgram husk,
bird food, bird food, wild, biri testa,
blackgram, blackgram husk, bran, broiler
mixed feed, calf fattening mixed feed, calf
fattening mixed feed (containing 4–20 %
peanut products), *Carthamus* cake, castor cake,
cereals, cereal products, chick pea, coconut
cake, cocos, concentrate, mixed, concentrates,
cotton cake, cottonseed, cottonseed
(dehulled), cottonseed cake, cottonseed
extract, cottonseed meal, cottonseed meal
(ammoniated), cottonseed meal
(decorticated), cottonseed meats, cottonseed
products, crumbles, crumbles, grower, cycad
meal, dairy cattle feed, dairy cattle feed
(containing 2–5 % peanut products), dairy
cattle feed (containing 6–10 % peanut
products), dairy cattle feed (containing
6–12 % peanut products), dairy cattle feed
(containing more than 20 % peanut
products), diets, mixed, dog food, egg
production mixed feed, feed, feed and

ingredients, feed, compound, feed, layer, feed
(beef), feed (calf), feed (cat), feed (cattle),
feed (chicken), feed (dairy), feed (dog), feed
(dug), feed (fish), feed (gluten), feed (horse),
feed (miscellaneous), feed (pig), feed
(poultry), feed (poultry, pig), feed (rabbit),
feed (sheep), feed (50–60 % maize), fish meal,
grain by-products, grains (no specification),
greengram, hay/silage, horsegram, husk, *Jagni*
cake, legume mixture, linseed, linseed cake,
livol, *Mahua* cake, maize, maize germ, maize
gluten, maize grits, maize husk, maize meal,
maize oil cake, maize powder, maize
screenings, maize, ground, maize, hybrid,
maize, preharvest, maize, yellow, maize (dark
grains), *Makhana* (*Euryale ferox* Salisb) puffs,
manioc, milk production mixed feed, mung
testa, murkool, mustard cake, neem cake,
niger cake, oats, palm kernel expeller cake,
palm products, palm kernels, peanut cake,
peanut cake (deoiled), peanut expeller,
peanut hay, peanut meal, peanut, kernels,
peanut, shells, peanuts, pellets, finisher, pig
meal and pellets, pigeon pea, poultry feeds
(peanut containing), rapeseed cake, redgram
husk, rice, rice bran, rice bran (deoiled), rice
chaff, rice crack, rice germ, rice germ cake,
rice meal, rice straw, rice (damaged), rice
(polish), safflower cake, sal seed cake, sesame,
sesame cake, sorghum, soybean meal,
soybeans, sunflower, sunflower cake,
sunflower flour, tapioca, wheat, wheat bran,
wheat bran and chana testa

Aflatoxins
incidence: 2/18, conc. range: 51–500 μg/kg,
country: Australia[21]
incidence: 4/27, conc. range: <5 μg/kg,
country: Spain[128]
incidence: 18/30, conc. range: 10–1500 μg/kg,
country: India[349]
incidence: 44/49, conc. range: 2–10 μg/kg,
Ø conc.: 3.2 μg/kg, country: Egypt[357]
see also animal feed (maize), animal feed
(mixed), barley, copra meal, cottonseed,
cottonseed fines, cottonseed meal, cottonseed
meats, feed, feed ingredients (miscellaneous),
feed (excluding peanuts, suspect), feed (goat),
feed (pig), feed (poultry), feedstuff, grain,

grain, mixed feed, maize, maize, shelled, millet, peanut cake, peanut meal, peanut meal and by-products, peanuts, protein concentrates, rice, rice bran, sorghum, soybean meal, sunflower

CITRININ
incidence: 1/7345, conc.: nc, country: Hungary[209]
see also barley, barley, oats, barley-soybean diet, cottonseed cake, feed, feed (cattle), feed (pig), fish meal, hay, maize, maize, white, *Makhana* (*Euryale ferox* Salisb) puffs, oats, palm products, peas and beans, rice bran, rice germ, wheat, wheat and other grains (moldy), wheat bran

DEOXYNIVALENOL
incidence: 1/1, conc.: 1000 µg/kg, country: USA[71]
see also barley, barley, husked, barley, unhusked (naked), barley (pressed), bone meal, bran, broilers feed, calf fattening mixed feed, coconut, expeller, corn cob mix silage, cottonseed, cottonseed cake, dairy cattle feed, egg production mixed feed, feed, feed components, feed, commercial mix, feed, mixed (primarily maize), feed (barley), feed (cattle), feed (chicken), feed (dog), feed (fish), feed (mill run, from wheat), feed (mink), feed (pig), feed (poultry), feed (reindeer), feeds, grain, feeds, industrial, feedstuff, feedstuffs (rapeseed, turnip, fish meal, concentrates), fish meal, grain, mixed feed, grains, mixed, grains (no specification), maize, maize ears, maize fibre, maize germ, maize germ/bran, maize germ meal, maize gluten, maize kernels, maize meal, maize powder, maize screenings, maize stalks (pith), maize, "Baby", maize, hybrid, maize, white, oats, rice bran, rice germ cake, rye, silage, sorghum, soybeans, triticale, wheat, wheat and barley, wheat, red hard winter, wheat, soft white winter, wheat, spring, wheat, winter

FUMONISIN B$_1$
incidence: 1/1, conc.: 8850 µg/kg, country: South Africa[100]

see also barley, bird food, wild, dog food, feed, feed, complete ration, feed, general, feed, layer, feed, maize-based, feed, pelleted ration, feed, screenings, feed, sweet, feed (broilers), feed (cat), feed (chicken), feed (dog), feed (gluten), feed (horse), feed (maize), feed (pig), feed (poultry), feed (rat), feed (rodent), forage grass, maize, maize and maize screenings, maize bran, maize ears, maize fine fractions, maize flakes, maize germ, maize germ/bran, maize germ meal, maize gluten, maize grits, maize kernels, maize meal, maize powder, maize screenings, maize, "Baby", maize, ground, maize, preharvest, maize, sweet feed, maize/oats mix, rat chow, silage, sorghum, soybeans, wheat

FUMONISIN B$_2$
incidence: 1/1, conc.: 3000 µg/kg, country: South Africa[100]
see also barley, bird food, wild, dog food, feed, feed, maize-based, feed (cat), feed (dog), feed (gluten), feed (horse), feed (maize), feed (poultry), feed (rodent), maize, maize bran, maize fine fractions, maize flakes, maize germ, maize germ/bran, maize germ meal, maize gluten, maize grits, maize kernels, maize meal, maize powder, maize screenings, maize, "Baby", maize, ground, maize, preharvest, rat chow, wheat

MONILIFORMIN
incidence: 15/15, conc. range: <10 µg/kg (7 sa), 10–100 µg/kg (5 sa), 100–150 µg/kg (3 sa), country: Austria[182]
see also barley, feed (poultry), maize, maize flakes, maize germ, maize germ/bran, maize gluten, maize meal, maize screenings, maize, "Baby", oats, rice bran, triticale, wheat, wheat, summer, wheat, winter

NEOSOLANIOL
incidence: ?/30, conc.: 20 µg/kg, country: India[183]
see also feed, feed components, feed (poultry), peanut cake, wheat

OCHRATOXIN A
incidence: 89/630, conc. range: 0.2–12.5 µg/kg, Ø conc.: 2.3 µg/kg, country: Germany[13]

incidence: 5/474, conc. range: 30–100 µg/kg, country: Canada[19]
incidence: 4/51, conc. range: 48–5900 µg/kg, country: Canada[20]
incidence: 1/25, conc.: 70,000 µg/kg? country: Australia[21]
incidence: 19/1240, conc. range: 10–200 µg/kg, country: Poland[26]
incidence: 10/203, conc. range: 10–50 µg/kg, country: Poland[27]
incidence: 27/812, conc. range: 25–250 µg/kg, country: UK[30]
incidence: 2/13, conc. range: 20–530 µg/kg, country: Canada[92]
incidence: 1/7345, conc.: nc, country: Hungary[209]
see also alfalfa, barley, barley, oats, barley (high moisture), barley-soybean diet, bird food, domestic, bird food, wild, broilers feed, cereal grains, citrus pulp, coconut, expeller, corn cob mix silage, diet (dairy cow), diet (poultry), diet (starter), dog food, eat, egg production mixed feed, feed, feed wheat, oat and barley, feed, commercial mix, feed, mixed (pelleted), feed (broilers), feed (cat), feed (cattle), feed (cereals), feed (pig), feed ec (pig), feed (poultry), feed ec (poultry), feed (poultry, pig), feed ec (rabbit), feed (trout), grain, mixed feed, grains, mixed, grains (heated), hay, horse bean, maize, maize feed, milo, maize gluten, maize meal, maize, white, *Makhana* (*Euryale ferox* Salisb) puffs, milk production mixed feed, millet, oat and barley (hammer-milled), oats, palm products, peanut cake, peas, peas and beans, pet food, pig feedstuffs, pig grower diet, pig meal, piglet diet, poultry feedstuffs, rice bran, rice germ, rice germ cake, rye, sorghum, soybean groats, sunflower, sunflower seeds, extracted, tapioca, triticale, *Vicia faba*, wheat, wheat and barley, wheat bran, wheat hay, wheat, oats

T-2 Toxin
incidence: 3/7345, conc. range: nc, country: Hungary[209]
see also alfalfa, barley, bran, diet (grower), diet (poultry), feed, feed components, feed, layer, feed (dog), feed (fish), feed (mink), feed (pig), feed (poultry), feedstuff, forage grass, grain, mixed feed, grains (no specification), hay, maize, maize germ/bran, maize gluten, maize meal, maize screenings, maize stalks (pith), oat and barley (hammer-milled), oats, peanuts, pig grower diet, piglet diet, rye, sorghum, triticale, wheat

Zearalenone
incidence: 1/1, conc.: 89 µg/kg, country: China[70]
incidence: 20/31, conc. range: 3.6–55.8 µg/kg, Ø conc.: 11.6 µg/kg, country: Germany[107]
incidence: 1/1, conc.: ~100 µg/kg, country: Germany[124]
incidence: ?/30, conc. range: 40 µg/kg, country: India[183]
incidence: ?/30, conc. range: 30 µg/kg, country: India[183]
incidence: 1/7345, conc.: nc, country: Hungary[209]
see also alfalfa, barley, barley, husked, barley, unhusked (naked), barley and feed, bone meal, bran, broilers feed, *Carthamus* cake, chick pea, concentrate, mixed, corn cob mix silage, cotton cake, cottonseed, cottonseed cake, diet (dairy cow), diet (poultry), diets (mixed), feed, feed components, feed, mixed (primarily maize also maize, oats, wheat), feed (bran), feed (broiler chicken), feed (cattle), feed (chicken), feed (dairy), feed (developing pig), feed (mill run, from wheat), feed (miscellaneous), feed (pig), feed (poultry), feed (poultry, pig), feed (starter chicken), feedstuff, fish meal, forage grass, grain, bruised, grain, mixed feed, grains (no specification), hay, maize, maize ears, maize flakes, maize germ, maize germ/bran, maize gluten, maize kernels, maize meal, maize oil cake, maize screenings, maize stalks (pith), maize, "Baby", maize, hybrid, maize, shelled, maize, unshelled, maize, white, maize grain, artificially dried, maize grain, crib dried, maize grain, ensiled, milk production mixed feed, oats, *Paspalum palidosum*, straw, peanut hulls/skins, rice bran, rice germ, rice germ cake, rye, silage, sorghum, soybeans, soybeans, extracted, sunflower cake, tapioca, triticale, wheat, wheat bran, wheat bran and chana testa, wheat soya meal

Feed, mixed (crumbled)   may contain the following mycotoxins:

Sterigmatocystin
incidence: 1/51, conc.: 2300 μg/kg, country: Canada[133]
see also feed (excluding peanut meal, suspect), wheat

Feed, mixed (pelleted)   may contain the following mycotoxins:

Ochratoxin A
incidence: 1/51, conc.: 140 μg/kg, country: Canada[133]
see also alfalfa, barley, barley, oats, barley (high moisture), barley-soybean diet, bird food, domestic, bird food, wild, broilers feed, cereal grains, citrus pulp, coconut, expeller, corn cob mix silage, diet (dairy cow), diet (poultry), diet (starter), dog food, eat, egg production mixed feed, feed, feed wheat, oat and barley, feed, commercial mix, feed, mixed, feed (broilers), feed (cat), feed (cattle), feed (cereals), feed (pig), feed ec (pig), feed (poultry), feed ec (poultry), feed (poultry, pig), feed ec (rabbit), feed (trout), grain, mixed feed, grains, mixed, grains (heated), hay, horse bean, maize, maize feed, milo, maize gluten, maize meal, maize, white, *Makhana* (*Euryale ferox* Salisb) puffs, milk production mixed feed, millet, oat and barley (hammer-milled), oats, palm products, peanut cake, peas, peas and beans, pet food, pig feedstuffs, pig grower diet, pig meal, piglet diet, poultry feedstuffs, rice bran, rice germ, rice germ cake, rye, sorghum, soybean groats, sunflower, sunflower seeds, extracted, tapioca, triticale, *Vicia faba*, wheat, wheat and barley, wheat bran, wheat hay, wheat, oats

Feed, mixed (primarily maize)   may contain the following mycotoxins:

Deoxynivalenol
incidence: 134/342, conc. range: 100–22,000 μg/kg, Ø conc.: 2700 μg/kg, country: USA[237]
see also barley, barley, husked, barley, unhusked (naked), barley (pressed), bone

meal, bran, broilers feed, calf fattening mixed feed, coconut, expeller, corn cob mix silage, cottonseed, cottonseed cake, dairy cattle feed, egg production mixed feed, feed, feed components, feed, commercial mix, feed, mixed, feed (barley), feed (cattle), feed (chicken), feed (dog), feed (fish), feed (mill run, from wheat), feed (mink), feed (pig), feed (poultry), feed (reindeer), feeds, grain, feeds, industrial, feedstuff, feedstuffs (rapeseed, turnip, fish meal, concentrates), fish meal, grain, mixed feed, grains, mixed, grains (no specification), maize, maize ears, maize kernels, maize fibre, maize germ, maize germ/bran, maize germ meal, maize gluten, maize kernels, maize meal, maize powder, maize screenings, maize stalks (pith), maize, "Baby", maize, hybrid, maize, white, oats, rice bran, rice germ cake, rye, silage, sorghum, soybeans, triticale, wheat, wheat and barley, wheat, red hard winter, wheat, soft white winter, wheat, spring, wheat, winter

Feed, mixed (primarily maize also maize, oats, wheat)   may contain the following mycotoxins:

Zearalenone
incidence: 40/342, conc. range: 100–8000 μg/kg, Ø conc.: 660 μg/kg, country: USA[237]
see also alfalfa, barley, barley, husked, barley, unhusked (naked), barley and feed, bone meal, bran, broilers feed, *Carthamus* cake, chick pea, concentrate, mixed, corn cob mix silage, cotton cake, cottonseed, cottonseed cake, diet (dairy cow), diet (poultry), diets (mixed), feed, feed components, feed, mixed, feed (bran), feed (broiler chicken), feed (cattle), feed (chicken), feed (dairy), feed (developing pig), feed (mill run, from wheat), feed (miscellaneous), feed (pig), feed (poultry), feed (poultry, pig), feed (starter chicken), feedstuff, fish meal, forage grass, grain, bruised, grain, mixed feed, grains (no specification), hay, maize, maize ears, maize flakes, maize germ, maize germ/bran, maize gluten, maize kernels, maize meal, maize oil cake, maize screenings, maize stalks (pith),

maize, "Baby", maize, hybrid, maize, shelled, maize, unshelled, maize, white, maize grain, artificially dried, maize grain, crib dried, maize grain, ensiled, milk production mixed feed, oats, *Paspalum palidosum*, straw, peanut hulls/skins, rice bran, rice germ, rice germ cake, rye, silage, sorghum, soybeans, soybeans, extracted, sunflower cake, tapioca, triticale, wheat, wheat bran, wheat bran and chana testa, wheat soya meal

Feed, pelleted ration   may contain the following mycotoxins:

Fumonisin B$_1$
incidence: 6/6, conc. range: 5–22 μg/kg, Ø conc.: 11.2 μg/kg, country: USA[58]
see also barley, bird food, wild, dog food, feed, feed, complete ration, feed, general, feed, layer, feed, maize-based, feed, mixed, feed, screenings, feed, sweet, feed (broilers), feed (cat), feed (chicken), feed (dog), feed (gluten), feed (horse), feed (maize), feed (pig), feed (poultry), feed (rat), feed (rodent), forage grass, maize, maize and maize screenings, maize bran, maize ears, maize fine fractions, maize flakes, maize germ, maize germ/bran, maize germ meal, maize gluten, maize grits, maize kernels, maize meal, maize powder, maize screenings, maize, "Baby", maize, ground, maize, preharvest, maize, sweet feed, maize/oats mix, rat chow, silage, sorghum, soybeans, wheat

Feed, screenings   may contain the following mycotoxins:

Fumonisin B$_1$
incidence: 8/9, conc. range: 16–209 μg/kg, Ø conc.: 86.7 μg/kg, country: USA[58]
incidence: 9/9, conc. range: 16–330 μg/kg, Ø conc.: 143.7 μg/kg, country: USA[58]
incidence: 4/4, conc. range: 14–76 μg/kg, Ø conc.: 49.5 μg/kg, country: USA[58]
incidence: 3/4, conc. range: 9–61 μg/kg, Ø conc.: 28 μg/kg, country: USA[58]
incidence: 16/16, conc. range: 14–95 μg/kg, Ø conc.: 45.8 μg/kg, country: USA[58]

incidence: 4/4, conc. range: 37–126 μg/kg, Ø conc.: 85.8 μg/kg, country: USA[58]
see also barley, bird food, wild, dog food, feed, feed, complete ration, feed, general, feed, layer, feed, maize-based, feed, mixed, feed, pelleted ration, feed, sweet, feed (broilers), feed (cat), feed (chicken), feed (dog), feed (gluten), feed (horse), feed (maize), feed (pig), feed (poultry), feed (rat), feed (rodent), forage grass, maize, maize and maize screenings, maize bran, maize ears, maize fine fractions, maize flakes, maize germ, maize germ/bran, maize germ meal, maize gluten, maize grits, maize kernels, maize meal, maize powder, maize screenings, maize, "Baby", maize, ground, maize, preharvest, maize, sweet feed, maize/oats mix, rat chow, silage, sorghum, soybeans, wheat

Feed, sweet   may contain the following mycotoxins:

Fumonisin B$_1$
incidence: 1/1, conc.: 8 μg/kg, country: USA[58]
incidence: 1/1, conc.: 16 μg/kg, country: USA[58]
incidence: 1/1, conc.: 20 μg/kg, country: USA[58]
incidence: 1/1, conc.: 4 μg/kg, country: USA[58]
incidence: 1/2, conc.: 13 μg/kg, country: USA[58]
incidence: 1/1, conc.: 29 μg/kg, country: USA[58]
see also barley, bird food, wild, dog food, feed, feed, complete ration, feed, general, feed, layer, feed, maize-based, feed, mixed, feed, pelleted ration, feed, screenings, feed (broilers), feed (cat), feed (chicken), feed (dog), feed (gluten), feed (horse), feed (maize), feed (pig), feed (poultry), feed (rat), feed (rodent), forage grass, maize, maize and maize screenings, maize bran, maize ears, maize fine fractions, maize flakes, maize germ, maize germ/bran, maize germ meal, maize gluten, maize grits, maize kernels, maize meal, maize powder, maize screenings, maize, "Baby", maize, ground, maize, preharvest, maize, sweet feed, maize/oats mix, rat chow, silage, sorghum, soybeans, wheat

Feed (barley)   may contain the following mycotoxins:

DEOXYNIVALENOL
incidence: 22/43*, conc. range:
$\geq$10–>100 µg/kg, country: UK[55], *imported ?
see also barley, barley, husked, barley, unhusked (naked), barley (pressed), bone meal, bran, broilers feed, calf fattening mixed feed, coconut, expeller, corn cob mix silage, cottonseed, cottonseed cake, dairy cattle feed, egg production mixed feed, feed, feed components, feed, commercial mix, feed, mixed, feed, mixed (primarily maize), feed (cattle), feed (chicken), feed (dog), feed (fish), feed (mill run, from wheat), feed (mink), feed (pig), feed (poultry), feed (reindeer), feeds, grain, feeds, industrial, feedstuff, feedstuffs (rapeseed, turnip, fish meal, concentrates), fish meal, grain, mixed feed, grains, mixed, grains (no specification), maize, maize ears, maize fibre, maize germ, maize germ/bran, maize germ meal, maize gluten, maize kernels, maize meal, maize powder, maize screenings, maize stalks (pith), maize, "Baby", maize, hybrid, maize, white, oats, rice bran, rice germ cake, rye, silage, sorghum, soybeans, triticale, wheat, wheat and barley, wheat, red hard winter, wheat, soft white winter, wheat, spring, wheat, winter

Feed (beef)   may contain the following mycotoxins:

AFLATOXIN B_1
incidence: 4/10, conc. range: <5 µg/kg (2 sa), 6–20 µg/kg (1 sa), 51–100 µg/kg (1 sa), country: UK[267]
see also alfalfa, *Ambadi* cake, animal feedstuffs (dairy cake), bagasse, barley, bengalgram husk, bird food, bird food, wild, biri testa, blackgram, blackgram husk, bran, broiler mixed feed, calf fattening mixed feed, calf fattening mixed feed (containing 4–20% peanut products), *Carthamus* cake, castor cake, cereals, cereal products, chick pea, coconut cake, cocos, concentrate, mixed, concentrates, cotton cake, cottonseed,

cottonseed (dehulled), cottonseed cake, cottonseed extract, cottonseed meal, cottonseed meal (ammoniated), cottonseed meal (decorticated), cottonseed meats, cottonseed products, crumbles, crumbles, grower, cycad meal, dairy cattle feed, dairy cattle feed (containing 2–5% peanut products), dairy cattle feed (containing 6–10% peanut products), dairy cattle feed (containing 6–12% peanut products), dairy cattle feed (containing more than 20% peanut products), diets, mixed, dog food, egg production mixed feed, feed, feed and ingredients, feed, compound, feed, layer, feed, mixed, feed (broilers), feed (calf), feed (cat), feed (cattle), feed (chicken), feed (dairy), feed (dog), feed (dug), feed (fish), feed (gluten), feed (horse), feed (miscellaneous), feed (pig), feed (poultry), feed (poultry, pig), feed (rabbit), feed (sheep), feed (50–60% maize), fish meal, grain by-products, grains (no specification), greengram, hay/silage, horsegram, husk, *Jagni* cake, legume mixture, linseed, linseed cake, livol, *Mahua* cake, maize, maize germ, maize gluten, maize grits, maize husk, maize meal, maize oil cake, maize powder, maize screenings, maize, ground, maize, hybrid, maize, preharvest, maize, yellow, maize (dark grains), *Makhana* (*Euryale ferox* Salisb) puffs, manioc, milk production mixed feed, mung testa, murkool, mustard cake, neem cake, niger cake, oats, palm kernel expeller cake, palm kernels, palm products, peanut cake, peanut cake (deoiled), peanut expeller, peanut hay, peanut meal, peanut, kernels, peanut, shells, peanuts, pellets, finisher, pig meal and pellets, pigeon pea, poultry feeds (peanut containing), rapeseed cake, redgram husk, rice, rice bran, rice bran (deoiled), rice chaff, rice crack, rice germ, rice germ cake, rice meal, rice straw, rice (damaged), rice (polish), safflower cake, sal seed cake, sesame, sesame cake, sorghum, soybean meal, soybeans, sunflower, sunflower cake, sunflower flour, tapioca, wheat, wheat bran, wheat bran and chana testa

Feed (bran)   may contain the following mycotoxins:

Zearalenone
incidence: 1/1, conc.: 3 μg/kg, country:
China[70]
see also alfalfa, barley, barley, husked, barley,
unhusked (naked), barley and feed, bone
meal, bran, broilers feed, *Carthamus* cake,
chick pea, concentrate, mixed, corn cob mix
silage, cotton cake, cottonseed, cottonseed
cake, diet (dairy cow), diet (poultry), diets
(mixed), feed, feed components, feed, mixed,
feed, mixed (primarily maize also maize, oats,
wheat), feed (broiler chicken), feed (cattle),
feed (chicken), feed (dairy), feed (developing
pig), feed (mill run, from wheat), feed
(miscellaneous), feed (pig), feed (poultry),
feed (poultry, pig), feed (starter chicken),
feedstuff, fish meal, forage grass, grain,
bruised, grain, mixed feed, grains (no
specification), hay, maize, maize ears, maize
flakes, maize germ, maize germ/bran, maize
gluten, maize kernels, maize meal, maize oil
cake, maize screenings, maize stalks (pith),
maize, "Baby", maize, hybrid, maize, shelled,
maize, unshelled, maize, white, maize grain,
artificially dried, maize grain, crib dried,
maize grain, ensiled, milk production mixed
feed, oats, *Paspalum palidosum*, straw, peanut
hulls/skins, rice bran, rice germ, rice germ
cake, rye, silage, sorghum, soybeans,
soybeans, extracted, sunflower cake, tapioca,
triticale, wheat, wheat bran, wheat bran and
chana testa, wheat soya meal

Feed (broiler chicken)   may contain the
following mycotoxins:

Zearalenone
incidence: 2/19, conc.: 405–1238 μg/kg,
Ø conc.: 821.5 μg/kg, country: Taiwan[221]
see also alfalfa, barley, barley, husked, barley,
unhusked (naked), barley and feed, bone
meal, bran, broilers feed, *Carthamus* cake,
chick pea, concentrate, mixed, corn cob mix
silage, cotton cake, cottonseed, cottonseed
cake, diet (dairy cow), diet (poultry), diets
(mixed), feed, feed components, feed, mixed,
feed, mixed (primarily maize also maize, oats,
wheat), feed (bran), feed (cattle), feed
(chicken), feed (dairy), feed (developing pig),

feed (mill run, from wheat), feed
(miscellaneous), feed (pig), feed (poultry),
feed (poultry, pig), feed (starter chicken),
feedstuff, fish meal, forage grass, grain,
bruised, grain, mixed feed, grains (no
specification), hay, maize, maize ears, maize
flakes, maize germ, maize germ/bran, maize
gluten, maize kernels, maize meal, maize oil
cake, maize screenings, maize stalks (pith),
maize, "Baby", maize, hybrid, maize, shelled,
maize, unshelled, maize, white, milk
production mixed feed, oats, *Paspalum
palidosum*, straw, peanut hulls/skins, rice
bran, rice germ, rice germ cake, rye, silage,
sorghum, soybeans, soybeans, extracted,
sunflower cake, tapioca, triticale, wheat,
wheat bran, wheat bran and chana testa,
wheat soya meal

Feed (broilers)   may contain the following
mycotoxins:

Aflatoxin B$_1$
incidence: 14/14, conc. range: tr–108 μg/kg,
country: India[386]
see also alfalfa, *Ambadi* cake, animal feedstuffs
(dairy cake), bagasse, barley, bengalgram
husk, bird food, bird food, wild, biri testa,
blackgram, blackgram husk, bran, broiler
mixed feed, calf fattening mixed feed, calf
fattening mixed feed (containing 4–20 %
peanut products), *Carthamus* cake, castor
cake, cereals, cereal products, chick pea,
coconut cake, cocos, concentrate, mixed,
concentrates, cotton cake, cottonseed,
cottonseed (dehulled), cottonseed cake,
cottonseed extract, cottonseed meal,
cottonseed meal (ammoniated), cottonseed
meal (decorticated), cottonseed meats,
cottonseed products, crumbles, crumbles,
grower, cycad meal, dairy cattle feed, dairy
cattle feed (containing 2–5 % peanut
products), dairy cattle feed (containing
6–10 % peanut products), dairy cattle feed
(containing 6–12 % peanut products), dairy
cattle feed (containing more than 20 %
peanut products), diets, mixed, dog food, egg
production mixed feed, feed, feed and
ingredients, feed, compound, feed, layer, feed,

mixed, feed (beef), feed (calf), feed (cat), feed
(cattle), feed (chicken), feed (dairy), feed
(dog), feed (dug), feed (fish), feed (gluten),
feed (horse), feed (miscellaneous), feed (pig),
feed (poultry), feed (poultry, pig), feed
(rabbit), feed (sheep), feed (50–60 % maize),
fish meal, grain by-products, grains (no
specification), greengram, hay/silage,
horsegram, husk, *Jagni* cake, legume mixture,
linseed, linseed cake, livol, *Mahua* cake,
maize, maize germ, maize gluten, maize grits,
maize husk, maize meal, maize oil cake,
maize powder, maize screenings, maize,
ground, maize, hybrid, maize, preharvest,
maize, yellow, maize (dark grains), *Makhana*
(*Euryale ferox* Salisb) puffs, manioc, milk
production mixed feed, mung testa, murkool,
mustard cake, neem cake, niger cake, oats,
palm kernel expeller cake, palm kernels, palm
products, peanut cake, peanut cake (deoiled),
peanut expeller, peanut hay, peanut meal,
peanut, kernels, peanut, shells, peanuts,
pellets, finisher, pig meal and pellets, pigeon
pea, poultry feeds (peanut containing),
rapeseed cake, redgram husk, rice, rice bran,
rice bran (deoiled), rice chaff, rice crack, rice
germ, rice germ cake, rice meal, rice straw,
rice (damaged), rice (polish), safflower cake,
sal seed cake, sesame, sesame cake, sorghum,
soybean meal, soybeans, sunflower, sunflower
cake, sunflower flour, tapioca, wheat, wheat
bran, wheat bran and chana testa

FUMONISIN B$_1$
incidence: 5/14, conc. range: 20–200 µg/kg,
country: India[386]
see also barley, bird food, wild, dog food,
feed, feed, complete ration, feed, general,
feed, layer, feed, maize-based, feed, mixed,
feed, pelleted ration, feed, screenings, feed,
sweet, feed (cat), feed (chicken), feed (dog),
feed (gluten), feed (horse), feed (maize), feed
(pig), feed (poultry), feed (rat), feed (rodent),
forage grass, maize, maize and maize
screenings, maize bran, maize ears, maize fine
fractions, maize flakes, maize germ, maize
germ/bran, maize germ meal, maize gluten,
maize grits, maize kernels, maize meal, maize
powder, maize screenings, maize, "Baby",

maize, ground, maize, preharvest, maize,
sweet feed, maize/oats mix, rat chow, silage,
sorghum, soybeans, wheat

OCHRATOXIN A
incidence: 1/3, conc.: 13 µg/kg, country:
Egypt[16]
see also alfalfa, barley, barley, oats, barley
(high moisture), barley-soybean diet, bird
food, domestic, bird food, wild, broilers feed,
cereal grains, citrus pulp, coconut, expeller,
corn cob mix silage, diet (dairy cow), diet
(poultry), diet (starter), dog food, eat, egg
production mixed feed, feed, feed wheat, oat
and barley, feed, commercial mix, feed,
mixed, feed, mixed (pelleted), feed (cat), feed
(cattle), feed (cereals), feed (pig), feed ec
(pig), feed (poultry), feed ec (poultry), feed
(poultry, pig), feed ec (rabbit), feed (trout),
grain, mixed feed, grains, mixed, grains
(heated), hay, horse bean, maize, maize feed,
milo, maize gluten, maize meal, maize, white,
Makhana (*Euryale ferox* Salisb) puffs, milk
production mixed feed, millet, oat and barley
(hammer-milled), oats, palm products,
peanut cake, peas, peas and beans, pet food,
pig feedstuffs, pig grower diet, pig meal,
piglet diet, poultry feedstuffs, rice bran, rice
germ, rice germ cake, rye, sorghum, soybean
groats, sunflower, sunflower seeds, extracted,
tapioca, triticale, *Vicia faba*, wheat, wheat and
barley, wheat bran, wheat hay, wheat, oats

Feed (calf)   may contain the following
mycotoxins:

AFLATOXIN B$_1$
incidence: 2/3, conc. range: tr–1670 µg/kg,
Ø conc.: 690 µg/kg, country: India[321]
see also alfalfa, *Ambadi* cake, animal feedstuffs
(dairy cake), bagasse, barley, bengalgram
husk, bird food, bird food, wild, biri testa,
blackgram, blackgram husk, bran, broiler
mixed feed, calf fattening mixed feed, calf
fattening mixed feed (containing 4–20 %
peanut products), *Carthamus* cake, castor
cake, cereals, cereal products, chick pea,
coconut cake, cocos, concentrate, mixed,
concentrates, cotton cake, cottonseed,

cottonseed (dehulled), cottonseed cake, cottonseed extract, cottonseed meal, cottonseed meal (ammoniated), cottonseed meal (decorticated), cottonseed meats, cottonseed products, crumbles, crumbles, grower, cycad meal, dairy cattle feed, dairy cattle feed (containing 2–5 % peanut products), dairy cattle feed (containing 6–10 % peanut products), dairy cattle feed (containing 6–12 % peanut products), dairy cattle feed (containing more than 20 % peanut products), diets, mixed, dog food, egg production mixed feed, feed, feed and ingredients, feed, compound, feed, layer, feed, mixed, feed (beef), feed (broilers), feed (cat), feed (cattle), feed (chicken), feed (dairy), feed (dog), feed (dug), feed (fish), feed (gluten), feed (horse), feed (miscellaneous), feed (pig), feed (poultry), feed (poultry, pig), feed (rabbit), feed (sheep), feed (50–60 % maize), fish meal, grain by-products, grains (no specification), greengram, hay/silage, horsegram, husk, *Jagni* cake, legume mixture, linseed, linseed cake, livol, *Mahua* cake, maize, maize germ, maize gluten, maize grits, maize husk, maize meal, maize oil cake, maize powder, maize screenings, maize, ground, maize, hybrid, maize, preharvest, maize, yellow, maize (dark grains), *Makhana* (*Euryale ferox* Salisb) puffs, manioc, milk production mixed feed, mung testa, murkool, mustard cake, neem cake, niger cake, oats, palm kernel expeller cake, palm products, palm kernels, peanut cake, peanut cake (deoiled), peanut expeller, peanut hay, peanut meal, peanut, kernels, peanut, shells, peanuts, pellets, finisher, pig meal and pellets, pigeon pea, poultry feeds (peanut containing), rapeseed cake, redgram husk, rice, rice bran, rice bran (deoiled), rice chaff, rice crack, rice germ, rice germ cake, rice meal, rice straw, rice (damaged), rice (polish), safflower cake, sal seed cake, sesame, sesame cake, sorghum, soybean meal, soybeans, sunflower, sunflower cake, sunflower flour, tapioca, wheat, wheat bran, wheat bran and chana testa

Feed (cat)   may contain the following mycotoxins:

AFLATOXICOL
incidence: 14/16, Ø conc.: 0.01 μg/kg, country: Mexico[395]
see also feed (dog)

AFLATOXIN B$_1$
incidence: 1/25, conc.: 16 μg/kg, country: Brazil[4]
incidence: 1/35, conc.: 2.1 μg/kg, country: UK[42]
incidence: 16/16, conc. range: ≤46.1 μg/kg, Ø conc.: 8.02 μg/kg, country: Mexico[395]
see also alfalfa, *Ambadi* cake, animal feedstuffs (dairy cake), bagasse, barley, bengalgram husk, bird food, bird food, wild, biri testa, blackgram, blackgram husk, bran, broiler mixed feed, calf fattening mixed feed, calf fattening mixed feed (containing 4–20 % peanut products), *Carthamus* cake, castor cake, cereals, cereal products, chick pea, coconut cake, cocos, concentrate, mixed, concentrates, cotton cake, cottonseed, cottonseed (dehulled), cottonseed cake, cottonseed extract, cottonseed meal, cottonseed meal (ammoniated), cottonseed meal (decorticated), cottonseed meats, cottonseed products, crumbles, crumbles, grower, cycad meal, dairy cattle feed, dairy cattle feed (containing 2–5 % peanut products), dairy cattle feed (containing 6–10 % peanut products), dairy cattle feed (containing 6–12 % peanut products), dairy cattle feed (containing more than 20 % peanut products), diets, mixed, dog food, egg production mixed feed, feed, feed and ingredients, feed, compound, feed, layer, feed, mixed, feed (beef), feed (broilers), feed (calf), feed (cattle), feed (chicken), feed (dairy), feed (dog), feed (dug), feed (fish), feed (gluten), feed (horse), feed (miscellaneous), feed (pig), feed (poultry), feed (poultry, pig), feed (rabbit), feed (sheep), feed (50–60 % maize), fish meal, grain by-products, grains (no specification), greengram, hay/silage, horsegram, husk, *Jagni* cake, legume mixture, linseed, linseed cake, livol, *Mahua* cake, maize, maize germ, maize gluten, maize grits, maize husk, maize meal, maize oil cake, maize powder, maize screenings, maize,

ground, maize, hybrid, maize, preharvest, maize, yellow, maize (dark grains), *Makhana* (*Euryale ferox* Salisb) puffs, manioc, milk production mixed feed, mung testa, murkool, mustard cake, neem cake, niger cake, oats, palm kernel expeller cake, palm products, palm kernels, peanut cake, peanut cake (deoiled), peanut expeller, peanut hay, peanut meal, peanut, kernels, peanut, shells, peanuts, pellets, finisher, pig meal and pellets, pigeon pea, poultry feeds (peanut containing), rapeseed cake, redgram husk, rice, rice bran, rice bran (deoiled), rice chaff, rice crack, rice germ, rice germ cake, rice meal, rice straw, rice (damaged), rice (polish), safflower cake, sal seed cake, sesame, sesame cake, sorghum, soybean meal, soybeans, sunflower, sunflower cake, sunflower flour, tapioca, wheat, wheat bran, wheat bran and chana testa

AFLATOXIN B_2
incidence: 4/16, Ø conc.: 0.01 µg/kg, country: Mexico[395]
see also animal feedstuffs (dairy cake), bird food, bird food, wild, biri testa, blackgram, blackgram husk, cottonseed, cottonseed cake, cottonseed extract, cottonseed meal, cottonseed meal (ammoniated), cottonseed meats, dog food, egg production mixed feed, feed, compound, feed (cattle), feed (dog), feed (pig), feed (poultry), feed (rabbit), feed (sheep), fish meal, horsegram, maize, maize gluten, maize husk, maize, ground, maize, preharvest, mung testa, mustard cake, niger cake, peanut cake, peanut cake (deoiled), peanut expeller, peanut hay, peanut meal, peanuts, peanut, kernels, redgram husk, rice bran, rice bran (deoiled), rice chaff, rice meal, rice (polish), sal seed cake, sesame cake, sorghum, soybean meal, soybeans, wheat, wheat bran

AFLATOXIN G_1
incidence: 12/16, Ø conc.: 0.04 µg/kg, country: Mexico[395]
see also animal feedstuffs (dairy cake), bird food, bird food, wild, concentrate, mixed, cottonseed, cottonseed cake, feed (chicken), feed (dog), maize, maize, ground, maize,

preharvest, meat meal, milk production mixed feed, murkool, peanut cake, peanut expeller, peanut hay, peanut meal, peanuts, rice bran, rice germ, sorghum, soybeans, wheat, wheat bran, wheat bran and chana testa

AFLATOXIN G_2
incidence: 4/16, Ø conc.: 0.05 µg/kg, country: Mexico[395]
see also animal feedstuffs (dairy cake), bird food, feed (dog), maize, maize, preharvest, peanut expeller, peanut hay, peanut meal, peanuts, soybeans

AFLATOXIN M_1
incidence: 14/16, Ø conc.: 3 µg/kg, country: Mexico[395]
see also feed (dog), maize, maize, ground

AFLATOXIN M_2
incidence: 11/16, Ø conc.: 0.76 µg/kg, country: Mexico[395]
see also feed (dog)

AFLATOXIN P_1
incidence: 13/16, Ø conc.: 0.01 µg/kg, country: Mexico[395]
see also feed (dog)

FUMONISIN B_1
incidence: 3/35, conc. range: 90–690 µg/kg, Ø conc.: 340 µg/kg, country: UK[42]
incidence: 2/2, conc. range: 220–990 µg/kg, Ø conc.: 605 µg/kg, country: USA[154]
see also barley, bird food, wild, dog food, feed, feed, complete ration, feed, general, feed, layer, feed, maize-based, feed, mixed, feed, pelleted ration, feed, screenings, feed, sweet, feed (broilers), feed (chicken), feed (dog), feed (gluten), feed (horse), feed (maize), feed (pig), feed (poultry), feed (rat), feed (rodent), forage grass, maize, maize and maize screenings, maize bran, maize ears, maize fine fractions, maize flakes, maize germ, maize germ/bran, maize germ meal, maize gluten, maize grits, maize kernels, maize meal, maize powder, maize screenings, maize, "Baby", maize, ground, maize, preharvest, maize, sweet feed, maize/oats mix, rat chow, silage, sorghum, soybeans, wheat

Fumonisin B$_2$
incidence: 2/35, conc. range: 60–80 µg/kg,
Ø conc.: 70 µg/kg, country: UK[42]
incidence: 2/2, conc. range: 125–140 µg/kg,
Ø conc.: 132.5 µg/kg, country: USA[154]
see also barley, bird food, wild, dog food,
feed, feed, maize-based, feed, mixed, feed
(dog), feed (gluten), feed (horse), feed
(maize), feed (poultry), feed (rodent), maize,
maize bran, maize fine fractions, maize
flakes, maize germ, maize germ/bran, maize
germ meal, maize gluten, maize grits, maize
kernels, maize meal, maize powder, maize
screenings, maize, "Baby", maize, ground,
maize, preharvest, rat chow, wheat

Ochratoxin A
incidence: 3/35, conc. range: 1.2–2.3 µg/kg,
Ø conc.: 1.6 µg/kg, country: UK[42]
see also alfalfa, barley, barley, oats, barley
(high moisture), barley-soybean diet, bird
food, domestic, bird food, wild, broilers feed,
cereal grains, citrus pulp, coconut, expeller,
corn cob mix silage, diet (dairy cow), diet
(poultry), diet (starter), dog food, eat, egg
production mixed feed, feed, feed wheat, oat
and barley, feed, commercial mix, feed,
mixed, feed, mixed (pelleted), feed (broilers),
feed (cattle), feed (cereals), feed (pig), feed ec
(pig), feed (poultry), feed ec (poultry), feed
(poultry, pig), feed ec (rabbit), feed (trout),
grain, mixed feed, grains, mixed, grains
(heated), hay, horse bean, maize, maize feed,
milo, maize gluten, maize meal, maize, white,
Makhana (Euryale ferox Salisb) puffs, milk
production mixed feed, millet, oat and barley
(hammer-milled), oats, palm products,
peanut cake, peas, peas and beans, pet food,
pig feedstuffs, pig grower diet, pig meal,
piglet diet, poultry feedstuffs, rice bran, rice
germ, rice germ cake, rye, sorghum, soybean
groats, sunflower, sunflower seeds, extracted,
tapioca, triticale, Vicia faba, wheat, wheat and
barley, wheat bran, wheat hay, wheat,
oats

Feed (cattle)   may contain the following
mycotoxins:

Aflatoxin B$_1$
incidence: 26/318, conc. range:
<10–750 µg/kg, country: UK[30]
incidence: 2/2, conc. range: 100–200 µg/kg,
Ø conc.: 150 µg/kg, country: Australia[121]
incidence: 378/498, conc. range: ≤25 µg/kg
(131 sa), 26–50 µg/kg (87 sa), 51–100 µg/kg
(61 sa), 101–200 µg/kg (60 sa), 201–500 µg/kg
(25 sa), 501–1000 µg/kg (9 sa),
1001–1500 µg/kg (4 sa), 1501–2000 µg/kg
(1 sa), Ø conc.: 103 µg/kg, country: India[247]
incidence: 4/4, conc. range: 59.4–192.6 µg/kg,
Ø conc.: 116 µg/kg, country: India[253]
incidence: 22/25, conc. range: 0–6000 µg/kg,
Ø conc.: 1620 µg/kg, country: India[321]
incidence: 14/96, conc. range:
11.5–287 µg/kg, country: Brazil[409]
see also alfalfa, Ambadi cake, animal feedstuffs
(dairy cake), bagasse, barley, bengalgram
husk, bird food, bird food, wild, biri testa,
blackgram, blackgram husk, bran, broiler
mixed feed, calf fattening mixed feed, calf
fattening mixed feed (containing 4–20 %
peanut products), Carthamus cake, castor
cake, cereals, cereal products, chick pea,
coconut cake, cocos, concentrate, mixed,
concentrates, cotton cake, cottonseed,
cottonseed (dehulled), cottonseed cake,
cottonseed extract, cottonseed meal,
cottonseed meal (ammoniated), cottonseed
meal (decorticated), cottonseed meats,
cottonseed products, crumbles, crumbles,
grower, cycad meal, dairy cattle feed, dairy
cattle feed (containing 2–5 % peanut
products), dairy cattle feed (containing
6–10 % peanut products), dairy cattle feed
(containing 6–12 % peanut products), dairy
cattle feed (containing more than 20 %
peanut products), diets, mixed, dog food, egg
production mixed feed, feed, feed and
ingredients, feed, compound, feed, layer,
feed, mixed, feed (beef), feed (broilers), feed
(calf), feed (cat), feed (chicken), feed (dairy),
feed (dog), feed (dug), feed (fish), feed
(gluten), feed (horse), feed (miscellaneous),
feed (pig), feed (poultry), feed (poultry, pig),
feed (rabbit), feed (sheep), feed (50–60 %
maize), fish meal, grain by-products, grains

(no specification), greengram, hay/silage, horsegram, husk, *Jagni* cake, legume mixture, linseed, linseed cake, livol, *Mahua* cake, maize, maize germ, maize gluten, maize grits, maize husk, maize meal, maize oil cake, maize powder, maize screenings, maize, ground, maize, hybrid, maize, preharvest, maize, yellow, maize (dark grains), *Makhana* (*Euryale ferox* Salisb) puffs, manioc, milk production mixed feed, mung testa, murkool, mustard cake, neem cake, niger cake, oats, palm kernel expeller cake, palm products, palm kernels, peanut cake, peanut cake (deoiled), peanut expeller, peanut hay, peanut meal, peanut, kernels, peanut, shells, peanuts, pellets, finisher, pig meal and pellets, pigeon pea, poultry feeds (peanut containing), rapeseed cake, redgram husk, rice, rice bran, rice bran (deoiled), rice chaff, rice crack, rice germ, rice germ cake, rice meal, rice straw, rice (damaged), rice (polish), safflower cake, sal seed cake, sesame, sesame cake, sorghum, soybean meal, soybeans, sunflower, sunflower cake, sunflower flour, tapioca, wheat, wheat bran, wheat bran and chana testa

AFLATOXIN B$_2$
incidence: 234/498, conc. range: nc, Ø conc.: 19.3 μg/kg, country: India[247]
incidence: 14/96, conc. range: 19–40 μg/kg, country: Brazil[409]
see also animal feedstuffs (dairy cake), bird food, bird food, wild, biri testa, blackgram, blackgram husk, cottonseed, cottonseed cake, cottonseed extract, cottonseed meal, cottonseed meal (ammoniated), cottonseed meats, dog food, egg production mixed feed, feed, compound, feed (cat), feed (dog), feed (pig), feed (poultry), feed (rabbit), feed (sheep), fish meal, horsegram, maize, maize gluten, maize husk, maize, ground, maize, preharvest, mung testa, mustard cake, niger cake, peanut cake, peanut cake (deoiled), peanut expeller, peanut hay, peanut meal, peanuts, peanut, kernels, redgram husk, rice bran, rice bran (deoiled), rice chaff, rice meal, rice (polish), sal seed cake, sesame cake, sorghum, soybean meal, soybeans, wheat, wheat bran

AFLATOXIN B
incidence: 7/20, conc. range: ≥25–>100 μg/kg, country: France[46]
see also barley, cocoa (oil cake), cottonseed cake, feed, feed (maize, gluten), feed (pig), feed (poultry), flour (wheat), lucern (dried), maize, maize gluten, maize grains, oats, peanut (oil cake), rice, broken, sorghum, soybean (oil cake), sunflower (oil cake), wheat

AFLATOXIN
incidence: 45/124, conc. range: 1–3000 μg/kg, country: India[229]
incidence: 378/498, conc. range: 3–1754 μg/kg, country: India[247]
incidence: 3/3, conc. range: <30 μg/kg (2 sa), >30 μg/kg (1 sa), country: India[380]
see also blackgram husk, bread crumbs, broiler finisher, broiler starter, cotton cake, cottonseed, cottonseed cake, cottonseed extract, cottonseed meal, feed, feed (cow), feed (dog), feed (horse), feed (maize, gluten), feed (pig), feed (poultry), feed (rabbit), feed (rat/mice), feed (sheep), feeds, grain, fish meal, flour (wheat), groundnut cake, grower's mash, horsegram, layer's mash, maize, maize gluten, maize, white, milo, peanut cake, peanut cake (deoiled), peanut, kernels, peanut (oil cake), pearlmillet, pig breeder's mash, pig finisher, pig starter, pod with haulms, poultry breeder's mash, rabbit pellets, redgram husk, rice, rice bran (deoiled), rice, broken, rice (polish), sesame cake, silk worm pupae, sorghum, soybean cake, soybean meal, wheat, wheat bran

AFLATOXIN B$_1$ + B$_2$ + G$_1$ + G$_2$
incidence: 11/59, conc. range: <50 μg/kg, country: Poland[84]
see also feed (pig), feed (poultry)

AFLATOXIN B+G
incidence: 5/23, conc. range: 5–2000 μg/kg, Ø conc.: 90 μg/kg, country: France[45]
see also feed (pig), feed (poultry), feed (rabbit), feed (rat), mice feed, peanut (oil cake)

CITRININ
incidence: 14/124, conc. range:
251–2000 µg/kg, country: India[229]
see also barley, barley, oats, barley-soybean
diet, cottonseed cake, feed, feed, mixed, feed
(pig), fish meal, hay, maize, maize, white,
Makhana (*Euryale ferox* Salisb) puffs, oats,
palm products, peas and beans, rice bran,
rice germ, wheat, wheat and other grains
(moldy), wheat bran

DEOXYNIVALENOL
incidence: 2/18, conc. range: 200 µg/kg,
Ø conc.: 200 µg/kg, country: Saudi Arabia[389]
see also barley, barley, husked, barley,
unhusked (naked), barley (pressed), bone
meal, bran, broilers feed, calf fattening mixed
feed, coconut, expeller, corn cob mix silage,
cottonseed, cottonseed cake, dairy cattle feed,
egg production mixed feed, feed, feed
components, feed, commercial mix, feed,
mixed, feed, mixed (primarily maize), feed
(barley), feed (chicken), feed (dog), feed
(fish), feed (mill run, from wheat), feed
(mink), feed (pig), feed (poultry), feed
(reindeer), feeds, grain, feeds, industrial,
feedstuff, feedstuffs (rapeseed, turnip, fish
meal, concentrates), fish meal, grain, mixed
feed, grains, mixed, grains (no specification),
maize, maize ears, maize fibre, maize germ,
maize germ/bran, maize germ meal, maize
gluten, maize kernels, maize meal, maize
powder, maize screenings, maize stalks (pith),
maize, "Baby", maize, hybrid, maize, white,
oats, rice bran, rice germ cake, rye, silage,
sorghum, soybeans, triticale, wheat, wheat
and barley, wheat, red hard winter, wheat,
soft white winter, wheat, spring, wheat,
winter

NIVALENOL
incidence: 1/18, conc.: 1250 µg/kg, country:
Saudi Arabia[389]
see also barley, barley, husked, barley,
unhusked (naked), barley (pressed), bran,
feed, feed components, feed (poultry), feed
(reindeer), feeds, industrial, maize, maize
ears, maize germ, maize germ/bran, maize
gluten, maize kernels, maize meal, maize

powder, maize screenings, maize, "Baby",
oats, rye, silage, triticale, wheat

OCHRATOXIN A
incidence: 1/59, conc. between 10 and
50 µg/kg, country: Poland[84]
incidence: 26/124, conc. range:
251–1500 µg/kg, country: India[229]
see also alfalfa, barley, barley, oats, barley
(high moisture), barley-soybean diet, bird
food, domestic, bird food, wild, broilers feed,
cereal grains, citrus pulp, coconut, expeller,
corn cob mix silage, diet (dairy cow), diet
(poultry), diet (starter), dog food, eat, egg
production mixed feed, feed, feed wheat, oat
and barley, feed, commercial mix, feed,
mixed, feed, mixed (pelleted), feed (broilers),
feed (cat), feed (cereals), feed (pig), feed ec
(pig), feed (poultry), feed ec (poultry), feed
(poultry, pig), feed ec (rabbit), feed (trout),
grain, mixed feed, grains, mixed, grains
(heated), hay, horse bean, maize, maize feed,
milo, maize gluten, maize meal, maize, white,
Makhana (*Euryale ferox* Salisb) puffs, milk
production mixed feed, millet, oat and barley
(hammer-milled), oats, palm products,
peanut cake, peas, peas and beans, pet food,
pig feedstuffs, pig grower diet, pig meal,
piglet diet, poultry feedstuffs, rice bran, rice
germ, rice germ cake, rye, sorghum, soybean
groats, sunflower, sunflower seeds, extracted,
tapioca, triticale, *Vicia faba*, wheat, wheat and
barley, wheat bran, wheat hay, wheat, oats

ZEARALENONE
incidence: 8/8, conc. range: 23–694 µg/kg,
Ø conc.: 168.6 µg/kg, country: USA[119]
incidence: 18/124, conc. range:
251–1500 µg/kg, country: India[229]
see also alfalfa, barley, barley, husked, barley,
unhusked (naked), barley and feed, bone
meal, bran, broilers feed, *Carthamus* cake,
chick pea, concentrate, mixed, corn cob mix
silage, cotton cake, cottonseed, cottonseed
cake, diet (dairy cow), diet (poultry), diets
(mixed), feed, feed components, feed, mixed,
feed, mixed (primarily maize also maize, oats,
wheat), feed (bran), feed (broiler chicken),
feed (chicken), feed (dairy), feed (developing

pig), feed (mill run, from wheat), feed (miscellaneous), feed (pig), feed (poultry), feed (poultry, pig), feed (starter chicken), feedstuff, fish meal, forage grass, grain, bruised, grain, mixed feed, grains (no specification), hay, maize, maize ears, maize flakes, maize germ, maize germ/bran, maize gluten, maize kernels, maize meal, maize oil cake, maize screenings, maize stalks (pith), maize, "Baby", maize, hybrid, maize, shelled, maize, unshelled, maize, white, maize grain, artificially dried, maize grain, crib dried, maize grain, ensiled, milk production mixed feed, oats, *Paspalum palidosum*, straw, peanut hulls/skins, rice bran, rice germ, rice germ cake, rye, silage, sorghum, soybeans, soybeans, extracted, sunflower cake, tapioca, triticale, wheat, wheat bran, wheat bran and chana testa, wheat soya meal

Feed (cattle and dairy)   may contain the following mycotoxins:

Fumonisins
incidence: 7/7, conc. range: $\leq$5000 μg/kg, country: USA[147]
see also feed (chicken), feed (dairy), feed (horse), maize, maize germ, maize gluten, maize meal, maize screenings

Feed (cereals)   may contain the following mycotoxins:

Ochratoxin A
incidence: 57/177, conc. range: $\leq$2389 μg/kg, Ø conc.: 103 μg/kg, country: Germany[184]
incidence: 38/608, conc. range: $\leq$206 μg/kg, country: Germany[185]
incidence: 21/150, conc. range: $\leq$40 μg/kg, country: Germany[186]
see also alfalfa, barley, barley, oats, barley (high moisture), barley-soybean diet, bird food, domestic, bird food, wild, broilers feed, cereal grains, citrus pulp, coconut, expeller, corn cob mix silage, diet (dairy cow), diet (poultry), diet (starter), dog food, eat, egg production mixed feed, feed, feed wheat, oat and barley, feed, commercial mix, feed, mixed, feed, mixed (pelleted), feed (broilers),

feed (cat), feed (cattle), feed (pig), feed ec (pig), feed (poultry), feed ec (poultry), feed (poultry, pig), feed ec (rabbit), feed (trout), grain, mixed feed, grains, mixed, grains (heated), hay, horse bean, maize, maize feed, milo, maize gluten, maize meal, maize, white, *Makhana* (*Euryale ferox* Salisb) puffs, milk production mixed feed, millet, oat and barley (hammer-milled), oats, palm products, peanut cake, peas, peas and beans, pet food, pig feedstuffs, pig grower diet, pig meal, piglet diet, poultry feedstuffs, rice bran, rice germ, rice germ cake, rye, sorghum, soybean groats, sunflower, sunflower seeds, extracted, tapioca, triticale, *Vicia faba*, wheat, wheat and barley, wheat bran, wheat hay, wheat, oats

Feed (chicken)   may contain the following mycotoxins:

Aflatoxin B$_1$
incidence: 4/4, conc. range: 0.2–0.5 μg/kg, country: Botswana[102]
incidence: 274/290, conc.range: 1–500 μg/kg, country: Indonesia[118]
see also alfalfa, *Ambadi* cake, animal feedstuffs (dairy cake), bagasse, barley, bengalgram husk, bird food, bird food, wild, biri testa, blackgram, blackgram husk, bran, broiler mixed feed, calf fattening mixed feed, calf fattening mixed feed (containing 4–20 % peanut products), *Carthamus* cake, castor cake, cereals, cereal products, chick pea, coconut cake, cocos, concentrate, mixed, concentrates, cotton cake, cottonseed, cottonseed (dehulled), cottonseed cake, cottonseed extract, cottonseed meal, cottonseed meal (ammoniated), cottonseed meal (decorticated), cottonseed meats, cottonseed products, crumbles, crumbles, grower, cycad meal, dairy cattle feed, dairy cattle feed (containing 2–5 % peanut products), dairy cattle feed (containing 6–10 % peanut products), dairy cattle feed (containing 6–12 % peanut products), dairy cattle feed (containing more than 20 % peanut products), diets, mixed, dog food, egg production mixed feed, feed, feed and ingredients, feed, compound, feed, layer,

feed, mixed, feed (beef), feed (broilers), feed (calf), feed (cat), feed (cattle), feed (dairy), feed (dog), feed (dug), feed (fish), feed (gluten), feed (horse), feed (miscellaneous), feed (pig), feed (poultry), feed (poultry, pig), feed (rabbit), feed (sheep), feed (50–60 % maize), fish meal, grain by-products, grains (no specification), greengram, hay/silage, horsegram, husk, *Jagni* cake, legume mixture, linseed, linseed cake, livol, *Mahua* cake, maize, maize germ, maize gluten, maize grits, maize husk, maize meal, maize oil cake, maize powder, maize screenings, maize, ground, maize, hybrid, maize, preharvest, maize, yellow, maize (dark grains), *Makhana* (*Euryale ferox* Salisb) puffs, manioc, milk production mixed feed, mung testa, murkool, mustard cake, neem cake, niger cake, oats, palm kernel expeller cake, palm products, palm kernels, peanut cake, peanut cake (deoiled), peanut expeller, peanut hay, peanut meal, peanut, kernels, peanut, shells, peanuts, pellets, finisher, pig meal and pellets, pigeon pea, poultry feeds (peanut containing), rapeseed cake, redgram husk, rice, rice bran, rice bran (deoiled), rice chaff, rice crack, rice germ, rice germ cake, rice meal, rice straw, rice (damaged), rice (polish), safflower cake, sal seed cake, sesame, sesame cake, sorghum, soybean meal, soybeans, sunflower, sunflower cake, sunflower flour, tapioca, wheat, wheat bran, wheat bran and chana testa

Aflatoxin G$_1$
incidence: 4/4, conc. range: 0.2–0.5 µg/kg, country: Botswana[102]
see also animal feedstuffs (dairy cake), bird food, bird food, wild, concentrate, mixed, cottonseed, cottonseed cake, feed (cat), feed (dog), maize, maize, ground, maize, preharvest, meat meal, milk production mixed feed, murkool, peanut cake, peanut expeller, peanut hay, peanut meal, peanuts, rice bran, rice germ, sorghum, soybeans, wheat, wheat bran, wheat bran and chana testa

Deoxynivalenol
incidence: 1/1, conc. range: <400 µg/kg, country: USA[71]

see also barley, barley, husked, barley, unhusked (naked), barley (pressed), bone meal, bran, broilers feed, calf fattening mixed feed, coconut, expeller, corn cob mix silage, cottonseed, cottonseed cake, dairy cattle feed, egg production mixed feed, feed, feed components, feed, commercial mix, feed, mixed, feed, mixed (primarily maize), feed (barley), feed (cattle), feed (dog), feed (fish), feed (mill run, from wheat), feed (mink), feed (pig), feed (poultry), feed (reindeer), feeds, grain, feeds, industrial, feedstuff, feedstuffs (rapeseed, turnip, fish meal, concentrates), fish meal, grain, mixed feed, grains, mixed, grains (no specification), maize, maize ears, maize fibre, maize germ, maize germ/bran, maize germ meal, maize gluten, maize kernels, maize meal, maize powder, maize screenings, maize stalks (pith), maize, "Baby", maize, hybrid, maize, white, oats, rice bran, rice germ cake, rye, silage, sorghum, soybeans, triticale, wheat, wheat and barley, wheat, red hard winter, wheat, soft white winter, wheat, spring, wheat, winter

Fumonisin B$_1$
incidence: ?/3, conc. range: 100–15,000 µg/kg, country: USA[158]
see also barley, bird food, wild, dog food, feed, feed, complete ration, feed, general, feed, layer, feed, maize-based, feed, mixed, feed, pelleted ration, feed, screenings, feed, sweet, feed (cat), feed (broilers), feed (dog), feed (gluten), feed (horse), feed (maize), feed (pig), feed (poultry), feed (rat), feed (rodent), forage grass, maize, maize and maize screenings, maize bran, maize ears, maize fine fractions, maize flakes, maize germ, maize germ/bran, maize germ meal, maize gluten, maize grits, maize kernels, maize meal, maize powder, maize screenings, maize, "Baby", maize, ground, maize, preharvest, maize, sweet feed, maize/oats mix, rat chow, silage, sorghum, soybeans, wheat

Fumonisins
incidence: 4/4, conc. range: 163–1050 µg/kg, Ø conc.: 572 µg/kg, country: Botswana[102]

see also feed (cattle and dairy), feed (dairy), feed (horse), maize, maize germ, maize gluten, maize meal, maize screenings

ZEARALENONE
incidence: 1/4, conc.: 40 µg/kg, country: Botswana[102]
see also alfalfa, barley, barley, husked, barley, unhusked (naked), barley and feed, bone meal, bran, broilers feed, *Carthamus* cake, chick pea, concentrate, mixed, corn cob mix silage, cotton cake, cottonseed, cottonseed cake, diet (dairy cow), diet (poultry), diets (mixed), feed, feed components, feed, mixed, feed, mixed (primarily maize also maize, oats, wheat), feed (bran), feed (broiler chicken), feed (cattle), feed (dairy), feed (developing pig), feed (mill run, from wheat), feed (miscellaneous), feed (pig), feed (poultry), feed (poultry, pig), feed (starter chicken), feedstuff, fish meal, forage grass, grain, bruised, grain, mixed feed, grains (no specification), hay, maize, maize ears, maize flakes, maize germ, maize germ/bran, maize gluten, maize kernels, maize meal, maize oil cake, maize screenings, maize stalks (pith), maize, "Baby", maize, hybrid, maize, shelled, maize, unshelled, maize, white, maize grain, artificially dried, maize grain, crib dried, maize grain, ensiled, milk production mixed feed, oats, *Paspalum palidosum*, straw, peanut hulls/skins, rice bran, rice germ, rice germ cake, rye, silage, sorghum, soybeans, soybeans, extracted, sunflower cake, tapioca, triticale, wheat, wheat bran, wheat bran and chana testa, wheat soya meal

Feed (cow)   may contain the following mycotoxins:

AFLATOXIN
incidence: 13/18, conc. range: 0–268 µg/kg, country: Nigeria[109]
see also blackgram husk, bread crumbs, broiler finisher, broiler starter, cotton cake, cottonseed, cottonseed cake, cottonseed extract, cottonseed meal, feed, feed (cattle), feed (dog), feed (horse), feed (maize, gluten), feed (pig), feed (poultry), feed (rabbit), feed (rat/mice), feed (sheep), feeds, grain, fish meal, flour (wheat), groundnut cake, grower's mash, horsegram, layer's mash, maize, maize gluten, maize, white, milo, peanut cake, peanut cake (deoiled), peanut, kernels, peanut (oil cake), pearlmillet, pig breeder's mash, pig finisher, pig starter, pod with haulms, poultry breeder's mash, rabbit pellets, redgram husk, rice, rice bran (deoiled), rice, broken, rice (polish), sesame cake, silk worm pupae, sorghum, soybean cake, soybean meal, wheat, wheat bran

Feed (dairy)   may contain the following mycotoxins:

AFLATOXIN B$_1$
incidence: 62/206, conc. range: <5 µg/kg (17 sa), 6–20 µg/kg (24 sa), 21–50 µg/kg (19 sa), >100 µg/kg (2 sa), country: UK[267]
see also alfalfa, *Ambadi* cake, animal feedstuffs (dairy cake), bagasse, barley, bengalgram husk, bird food, bird food, wild, biri testa, blackgram, blackgram husk, bran, broiler mixed feed, calf fattening mixed feed, calf fattening mixed feed (containing 4–20 % peanut products), *Carthamus* cake, castor cake, cereals, cereal products, chick pea, coconut cake, cocos, concentrate, mixed, concentrates, cotton cake, cottonseed, cottonseed (dehulled), cottonseed cake, cottonseed extract, cottonseed meal, cottonseed meal (ammoniated), cottonseed meal (decorticated), cottonseed meats, cottonseed products, crumbles, crumbles, grower, cycad meal, dairy cattle feed, dairy cattle feed (containing 2–5 % peanut products), dairy cattle feed (containing 6–10 % peanut products), dairy cattle feed (containing 6–12 % peanut products), dairy cattle feed (containing more than 20 % peanut products), diets, mixed, dog food, egg production mixed feed, feed, feed and ingredients, feed, compound, feed, layer, feed, mixed, feed (beef), feed (broilers), feed (calf), feed (cat), feed (cattle), feed (chicken), feed (dog), feed (dug), feed (fish), feed (gluten), feed (horse), feed (miscellaneous), feed (pig), feed (poultry), feed (poultry, pig),

feed (rabbit), feed (sheep), feed (50–60 %
maize), fish meal, grain by-products, grains
(no specification), greengram, hay/silage,
horsegram, husk, *Jagni* cake, legume mixture,
linseed, linseed cake, livol, *Mahua* cake,
maize, maize germ, maize gluten, maize grits,
maize husk, maize meal, maize oil cake,
maize powder, maize screenings, maize,
ground, maize, hybrid, maize, preharvest,
maize, yellow, maize (dark grains), *Makhana*
(*Euryale ferox* Salisb) puffs, manioc, milk
production mixed feed, mung testa, murkool,
mustard cake, neem cake, niger cake, oats,
palm kernel expeller cake, palm kernels, palm
products, peanut cake, peanut cake (deoiled),
peanut expeller, peanut hay, peanut meal,
peanut, kernels, peanut, shells, peanuts,
pellets, finisher, pig meal and pellets, pigeon
pea, poultry feeds (peanut containing),
rapeseed cake, redgram husk, rice, rice bran,
rice bran (deoiled), rice chaff, rice crack, rice
germ, rice germ cake, rice meal, rice straw,
rice (damaged), rice (polish), safflower cake,
sal seed cake, sesame, sesame cake, sorghum,
soybean meal, soybeans, sunflower, sunflower
cake, sunflower flour, tapioca, wheat, wheat
bran, wheat bran and chana testa

FUMONISINS
incidence: 1/1, conc. range: ≤5000 µg/kg,
country: USA[147]
see also feed (cattle and dairy), feed
(chicken), feed (horse), maize, maize germ,
maize gluten, maize meal, maize screenings

ZEARALENONE
incidence: 35/275, conc. range:
140–960 µg/kg, country: Hungary[344]
see also alfalfa, barley, barley, husked, barley,
unhusked (naked), barley and feed, bone
meal, bran, broilers feed, *Carthamus* cake,
chick pea, concentrate, mixed, corn cob mix
silage, cotton cake, cottonseed, cottonseed
cake, diet (dairy cow), diet (poultry), diets
(mixed), feed, feed components, feed, mixed,
feed, mixed (primarily maize also maize, oats,
wheat), feed (bran), feed (broiler chicken),
feed (cattle), feed (chicken), feed (mill run,
from wheat), feed (miscellaneous), feed (pig),

feed (poultry), feed (poultry, pig), feed
(starter chicken), feedstuff, fish meal, forage
grass, grain, bruised, grain, mixed feed, grains
(no specification), hay, maize, maize ears,
maize flakes, maize germ, maize germ/bran,
maize gluten, maize kernels, maize meal,
maize oil cake, maize screenings, maize stalks
(pith), maize, "Baby", maize, hybrid, maize,
shelled, maize, unshelled, maize, white, maize
grain, artificially dried, maize grain, crib
dried, maize grain, ensiled, milk production
mixed feed, oats, *Paspalum palidosum*, straw,
peanut hulls/skins, rice bran, rice germ, rice
germ cake, rye, silage, sorghum, soybeans,
soybeans, extracted, sunflower cake, tapioca,
triticale, wheat, wheat bran, wheat bran and
chana testa, wheat soya meal

Feed (developing pig)   may contain the
following mycotoxins:

ZEARALENONE
incidence: 1/7, conc.: 1203 µg/kg, country:
Taiwan[221]
see also alfalfa, barley, barley, husked, barley,
unhusked (naked), barley and feed, bone
meal, bran, broilers feed, *Carthamus* cake,
chick pea, concentrate, mixed, corn cob mix
silage, cotton cake, cottonseed, cottonseed
cake, diet (dairy cow), diet (poultry), diets
(mixed), feed, feed components, feed, mixed,
feed, mixed (primarily maize also maize, oats,
wheat), feed (bran), feed (broiler chicken),
feed (cattle), feed (chicken), feed (dairy), feed
(mill run, from wheat), feed (miscellaneous),
feed (pig), feed (poultry), feed (poultry, pig),
feed (starter chicken), feedstuff, fish meal,
forage grass, grain, bruised, grain, mixed
feed, grains (no specification), hay, maize,
maize ears, maize flakes, maize germ, maize
germ/bran, maize gluten, maize kernels,
maize meal, maize oil cake, maize screenings,
maize stalks (pith), maize, "Baby", maize,
hybrid, maize, shelled, maize, unshelled,
maize, white, maize grain, artificially dried,
maize grain, crib dried, maize grain, ensiled,
milk production mixed feed, oats, *Paspalum
palidosum*, straw, peanut hulls/skins, rice
bran, rice germ, rice germ cake, rye, silage,

sorghum, soybeans, soybeans, extracted, sunflower cake, tapioca, triticale, wheat, wheat bran, wheat bran and chana testa, wheat soya meal

Feed (dog)   may contain the following mycotoxins:

AFLATOXICOL
incidence: 9/19, Ø conc.: 0.3 µg/kg, country: Mexico[395]
see also feed (cat)

AFLATOXIN B$_1$
incidence: 15/19, conc. range: ≤39.7 µg/kg, Ø conc.: 5 µg/kg, country: Mexico[395]
see also alfalfa, *Ambadi* cake, animal feedstuffs (dairy cake), bagasse, barley, bengalgram husk, bird food, bird food, wild, biri testa, blackgram, blackgram husk, bran, broiler mixed feed, calf fattening mixed feed, calf fattening mixed feed (containing 4–20 % peanut products), *Carthamus* cake, castor cake, cereals, cereal products, chick pea, coconut cake, cocos, concentrate, mixed, concentrates, cotton cake, cottonseed, cottonseed (dehulled), cottonseed cake, cottonseed extract, cottonseed meal, cottonseed meal (ammoniated), cottonseed meal (decorticated), cottonseed meats, cottonseed products, crumbles, crumbles, grower, cycad meal, dairy cattle feed, dairy cattle feed (containing 2–5 % peanut products), dairy cattle feed (containing 6–10 % peanut products), dairy cattle feed (containing 6–12 % peanut products), dairy cattle feed (containing more than 20 % peanut products), diets, mixed, dog food, egg production mixed feed, feed, feed and ingredients, feed, compound, feed, layer, feed, mixed, feed (beef), feed (broilers), feed (calf), feed (cat), feed (cattle), feed (chicken), feed (dairy), feed (dug), feed (fish), feed (gluten), feed (horse), feed (miscellaneous), feed (pig), feed (poultry), feed (poultry, pig), feed (rabbit), feed (sheep), feed (50–60 % maize), fish meal, grain by-products, grains (no specification), greengram, hay/silage, horsegram, husk, *Jagni* cake, legume mixture,

linseed, linseed cake, livol, *Mahua* cake, maize, maize germ, maize gluten, maize grits, maize husk, maize meal, maize oil cake, maize powder, maize screenings, maize, ground, maize, hybrid, maize, preharvest, maize, yellow, maize (dark grains), *Makhana* (*Euryale ferox* Salisb) puffs, manioc, milk production mixed feed, mung testa, murkool, mustard cake, neem cake, niger cake, oats, palm kernel expeller cake, palm kernels, palm products, peanut cake, peanut cake (deoiled), peanut expeller, peanut hay, peanut meal, peanut, kernels, peanut, shells, peanuts, pellets, finisher, pig meal and pellets, pigeon pea, poultry feeds (peanut containing), rapeseed cake, redgram husk, rice, rice bran, rice bran (deoiled), rice chaff, rice crack, rice germ, rice germ cake, rice meal, rice straw, rice (damaged), rice (polish), safflower cake, sal seed cake, sesame, sesame cake, sorghum, soybean meal, soybeans, sunflower, sunflower cake, sunflower flour, tapioca, wheat, wheat bran, wheat bran and chana testa

AFLATOXIN B$_2$
incidence: 5/19, Ø conc.: 0.07 µg/kg, country: Mexico[395]
see also animal feedstuffs (dairy cake), bird food, bird food, wild, biri testa, blackgram, blackgram husk, cottonseed, cottonseed cake, cottonseed extract, cottonseed meal, cottonseed meal (ammoniated), cottonseed meats, dog food, egg production mixed feed, feed, compound, feed (cat), feed (cattle), feed (pig), feed (poultry), feed (rabbit), feed (sheep), fish meal, horsegram, maize, maize gluten, maize husk, maize, ground, maize, preharvest, mung testa, mustard cake, niger cake, peanut cake, peanut cake (deoiled), peanut expeller, peanut hay, peanut meal, peanuts, peanut, kernels, redgram husk, rice bran, rice bran (deoiled), rice chaff, rice meal, rice (polish), sal seed cake, sesame cake, sorghum, soybean meal, soybeans, wheat, wheat bran

AFLATOXIN G$_1$
incidence: 12/19, Ø conc.: 0.05 µg/kg, country: Mexico[395]

see also animal feedstuffs (dairy cake), bird food, bird food, wild, concentrate, mixed, cottonseed, cottonseed cake, feed (cat), feed (chicken), maize, maize, ground, maize, preharvest, meat meal, milk production mixed feed, murkool, peanut cake, peanut expeller, peanut hay, peanut meal, peanuts, rice bran, rice germ, sorghum, soybeans, wheat, wheat bran, wheat bran and chana testa

AFLATOXIN G_2
incidence: 4/19, Ø conc.: 0.03 µg/kg, country: Mexico[395]
see also animal feedstuffs (dairy cake), bird food, feed (cat), maize, maize, preharvest, peanut expeller, peanut hay, peanut meal, peanuts, soybeans

AFLATOXIN M_1
incidence: 12/19, Ø conc.: 2 µg/kg, country: Mexico[395]
see also feed (cat), maize, maize, ground

AFLATOXIN M_2
incidence: 17/19, Ø conc.: 0.14 µg/kg, country: Mexico[395]
see also feed (cat)

AFLATOXIN P_1
incidence: 11/19, conc. range: ≤12.5 µg/kg, Ø conc.: 1.16 µg/kg, country: Mexico[395]
see also feed (cat)

AFLATOXIN
incidence: 3/18, conc. range: <1.75–20 µg/kg, country: Turkey[340]
see also blackgram husk, bread crumbs, broiler finisher, broiler starter, cotton cake, cottonseed, cottonseed cake, cottonseed extract, cottonseed meal, feed, feed (cattle), feed (cow), feed (horse), feed (maize, gluten), feed (pig), feed (poultry), feed (rabbit), feed (rat/mice), feed (sheep), feeds, grain, fish meal, flour (wheat), groundnut cake, grower's mash, horsegram, layer's mash, maize, maize gluten, maize, white, milo, peanut cake, peanut cake (deoiled), peanut, kernels, peanut (oil cake), pearlmillet, pig breeder's mash, pig finisher, pig starter, pod with haulms, poultry breeder's mash, rabbit pellets, redgram husk, rice, rice bran

(deoiled), rice, broken, rice (polish), sesame cake, silk worm pupae, sorghum, soybean cake, soybean meal, wheat, wheat bran

DEOXYNIVALENOL
incidence: 2/2*, conc. range: 14–15 µg/kg, Ø conc.: 14.5 µg/kg, country: Finland[62], *imported ?
incidence: 1/1, conc.: 300 µg/kg, country: USA[71]
incidence: 2/2, conc. range: 70–1000 µg/kg, Ø conc.: 535 µg/kg, country: USA[71]
see also barley, barley, husked, barley, unhusked (naked), barley (pressed), bone meal, bran, broilers feed, calf fattening mixed feed, coconut, expeller, corn cob mix silage, cottonseed, cottonseed cake, dairy cattle feed, egg production mixed feed, feed, feed components, feed, commercial mix, feed, mixed, feed, mixed (primarily maize), feed (barley), feed (cattle), feed (chicken), feed (fish), feed (mill run, from wheat), feed (mink), feed (pig), feed (poultry), feed (reindeer), feeds, grain, feeds, industrial, feedstuff, feedstuffs (rapeseed, turnip, fish meal, concentrates), fish meal, grain, mixed feed, grains, mixed, grains (no specification), maize, maize ears, maize fibre, maize germ, maize germ/bran, maize germ meal, maize gluten, maize kernels, maize meal, maize powder, maize screenings, maize stalks (pith), maize, "Baby", maize, hybrid, maize, white, oats, rice bran, rice germ cake, rye, silage, sorghum, soybeans, triticale, wheat, wheat and barley, wheat, red hard winter, wheat, soft white winter, wheat, spring, wheat, winter

DIACETOXYSCIRPENOL
incidence: 1/2*, conc.: 60 µg/kg, country: Finland[62], *imported ?
see also alfalfa, barley, feed, feed components, feed (fish), feed (mink), feed (pig), feed (poultry), feed (reindeer), feeds, grain, feedstuff, forage grass, grain, mixed feed, grains (no specification), maize, maize gluten, oat and barley (hammer-milled), oats, peanuts, soybeans, wheat

Fumonisin B$_1$
incidence: 2/2, conc. range: 820–1410 µg/kg,
Ø conc.: 1115 µg/kg, country: USA[154]
see also barley, bird food, wild, dog food,
feed, feed, complete ration, feed, general,
feed, layer, feed, maize-based, feed, mixed,
feed, pelleted ration, feed, screenings, feed,
sweet, feed (broilers), feed (cat), feed
(chicken), feed (gluten), feed (horse), feed
(maize), feed (pig), feed (poultry), feed (rat),
feed (rodent), forage grass, maize, maize and
maize screenings, maize bran, maize ears,
maize fine fractions, maize flakes, maize
germ, maize germ/bran, maize germ meal,
maize gluten, maize grits, maize kernels,
maize meal, maize powder, maize screenings,
maize, "Baby", maize, ground, maize,
preharvest, maize, sweet feed, maize/oats mix,
rat chow, silage, sorghum, soybeans, wheat

Fumonisin B$_2$
incidence: 2/2, conc. range: 102–144 µg/kg,
Ø conc.: 123 µg/kg, country: USA[154]
see also barley, bird food, wild, dog food,
feed, feed, maize-based, feed, mixed, feed
(cat), feed (gluten), feed (horse), feed
(maize), feed (poultry), feed (rodent), maize,
maize bran, maize fine fractions, maize
flakes, maize germ, maize germ/bran, maize
germ meal, maize gluten, maize grits, maize
kernels, maize meal, maize powder, maize
screenings, maize, "Baby", maize, ground,
maize, preharvest, rat chow, wheat

HT-2 toxin
incidence: 1/2*, conc.: 12 µg/kg, country:
Finland[62], *imported ?
see also barley, feed components, feed (fish),
feed (pig), feed (poultry), feed (reindeer),
grain, mixed feed, grains, mixed, grains (no
specification), maize, maize germ/bran, maize
gluten, maize meal, maize screenings, oats,
rye, silage, wheat

T-2 Toxin
incidence: 1/2*, conc.: 37 µg/kg, country:
Finland[62], *imported ?
see also alfalfa, barley, bran, diet (grower),
diet (poultry), feed, feed components, feed,
layer, feed, mixed, feed (fish), feed (mink),
feed (pig), feed (poultry), feedstuff, forage
grass, grain, mixed feed, grains (no
specification), hay, maize, maize germ/bran,
maize gluten, maize meal, maize screenings,
maize stalks (pith), oat and barley
(hammer-milled), oats, peanuts, pig grower
diet, piglet diet, rye, sorghum, triticale, wheat
For further information see also dog food

Feed (dug)   may contain the following
mycotoxins:

Aflatoxin B$_1$
incidence: 2/3, conc. range: 20–50 µg/kg,
Ø conc.: 35 µg/kg, country: Australia[121]
see also alfalfa, *Ambadi* cake, animal feedstuffs
(dairy cake), bagasse, barley, bengalgram
husk, bird food, bird food, wild, biri testa,
blackgram, blackgram husk, bran, broiler
mixed feed, calf fattening mixed feed, calf
fattening mixed feed (containing 4–20 %
peanut products), *Carthamus* cake, castor
cake, cereals, cereal products, chick pea,
coconut cake, cocos, concentrate, mixed,
concentrates, cotton cake, cottonseed,
cottonseed (dehulled), cottonseed cake,
cottonseed extract, cottonseed meal,
cottonseed meal (ammoniated), cottonseed
meal (decorticated), cottonseed meats,
cottonseed products, crumbles, crumbles,
grower, cycad meal, dairy cattle feed, dairy
cattle feed (containing 2–5 % peanut
products), dairy cattle feed (containing
6–10 % peanut products), dairy cattle feed
(containing 6–12 % peanut products), dairy
cattle feed (containing more than 20 %
peanut products), diets, mixed, dog food, egg
production mixed feed, feed, feed and
ingredients, feed, compound, feed, layer,
feed, mixed, feed (beef), feed (broilers), feed
(calf), feed (cat), feed (cattle), feed (chicken),
feed (dairy), feed (dog), feed (fish), feed
(gluten), feed (horse), feed (miscellaneous),
feed (pig), feed (poultry), feed (poultry, pig),
feed (rabbit), feed (sheep), feed (50–60 %
maize), fish meal, grain by-products, grains
(no specification), greengram, hay/silage,
horsegram, husk, *Jagni* cake, legume mixture,
linseed, linseed cake, livol, *Mahua* cake,

maize, maize germ, maize gluten, maize grits, maize husk, maize meal, maize oil cake, maize powder, maize screenings, maize, ground, maize, hybrid, maize, preharvest, maize, yellow, maize (dark grains), *Makhana* (*Euryale ferox* Salisb) puffs, manioc, milk production mixed feed, mung testa, murkool, mustard cake, neem cake, niger cake, oats, palm kernel expeller cake, palm kernels, palm products, peanut cake, peanut cake (deoiled), peanut expeller, peanut hay, peanut meal, peanut, kernels, peanut, shells, peanuts, pellets, finisher, pig meal and pellets, pigeon pea, poultry feeds (peanut containing), rapeseed cake, redgram husk, rice, rice bran, rice bran (deoiled), rice chaff, rice crack, rice germ, rice germ cake, rice meal, rice straw, rice (damaged), rice (polish), safflower cake, sal seed cake, sesame, sesame cake, sorghum, soybean meal, soybeans, sunflower, sunflower cake, sunflower flour, tapioca, wheat, wheat bran, wheat bran and chana testa

Feed (excluding peanuts, suspect) may contain the following mycotoxins:

Aflatoxins
incidence: 4/123, conc. range: 40–270 μg/kg, country: UK[168]
see also animal feed (maize), animal feed (mixed), barley, copra meal, cottonseed, cottonseed fines, cottonseed meal, cottonseed meats, feed, feed ingredients (miscellaneous), feed, mixed, feed (goat), feed (pig), feed (poultry), feedstuff, grain, grain, mixed feed, maize, maize, shelled, millet, peanut cake, peanut meal, peanut meal and by-products, peanuts, protein concentrates, rice, rice bran, sorghum, soybean meal, sunflower

Feed (excluding peanut meal, suspect) may contain the following mycotoxins:

Sterigmatocystin
incidence: 3/188, conc. range: ≤3000 μg/kg, country: UK[168]
see also feed, mixed (crumbled), wheat

Feed (fish) may contain the following mycotoxins:

Aflatoxin B₁
incidence: 1/41, conc.: 20 μg/kg, country: South Africa[79]
see also alfalfa, *Ambadi* cake, animal feedstuffs (dairy cake), bagasse, barley, bengalgram husk, bird food, bird food, wild, biri testa, blackgram, blackgram husk, bran, broiler mixed feed, calf fattening mixed feed, calf fattening mixed feed (containing 4–20 % peanut products), *Carthamus* cake, castor cake, cereals, cereal products, chick pea, coconut cake, cocos, concentrate, mixed, concentrates, cotton cake, cottonseed, cottonseed (dehulled), cottonseed cake, cottonseed extract, cottonseed meal, cottonseed meal (ammoniated), cottonseed meal (decorticated), cottonseed meats, cottonseed products, crumbles, crumbles, grower, cycad meal, dairy cattle feed, dairy cattle feed (containing 2–5 % peanut products), dairy cattle feed (containing 6–10 % peanut products), dairy cattle feed (containing 6–12 % peanut products), dairy cattle feed (containing more than 20 % peanut products), diets, mixed, dog food, egg production mixed feed, feed, feed and ingredients, feed, compound, feed, layer, feed, mixed, feed (beef), feed (broilers), feed (calf), feed (cat), feed (cattle), feed (chicken), feed (dairy), feed (dog), feed (dug), feed (gluten), feed (horse), feed (miscellaneous), feed (pig), feed (poultry), feed (poultry, pig), feed (rabbit), feed (sheep), feed (50–60 % maize), fish meal, grain by-products, grains (no specification), greengram, hay/silage, horsegram, husk, *Jagni* cake, legume mixture, linseed, linseed cake, livol, *Mahua* cake, maize, maize germ, maize gluten, maize grits, maize husk, maize meal, maize oil cake, maize powder, maize screenings, maize, ground, maize, hybrid, maize, preharvest, maize, yellow, maize (dark grains), *Makhana* (*Euryale ferox* Salisb) puffs, manioc, milk production mixed feed, mung testa, murkool, mustard cake, neem cake, niger cake, oats, palm kernel expeller cake, palm kernels, palm products, peanut cake, peanut cake (deoiled), peanut expeller, peanut hay, peanut meal,

peanut, kernels, peanut, shells, peanuts, pellets, finisher, pig meal and pellets, pigeon pea, poultry feeds (peanut containing), rapeseed cake, redgram husk, rice, rice bran, rice bran (deoiled), rice chaff, rice crack, rice germ, rice germ cake, rice meal, rice straw, rice (damaged), rice (polish), safflower cake, sal seed cake, sesame, sesame cake, sorghum, soybean meal, soybeans, sunflower, sunflower cake, sunflower flour, tapioca, wheat, wheat bran, wheat bran and chana testa

DEOXYNIVALENOL
incidence: 3/3*, conc. range: 1–60 µg/kg, Ø conc.: 33 µg/kg, country: Finland[62], *imported ?
see also barley, barley, husked, barley, unhusked (naked), barley (pressed), bone meal, bran, broilers feed, calf fattening mixed feed, coconut, expeller, corn cob mix silage, cottonseed, cottonseed cake, dairy cattle feed, egg production mixed feed, feed, feed components, feed, commercial mix, feed, mixed, feed, mixed (primarily maize), feed (barley), feed (cattle), feed (chicken), feed (dog), feed (mill run, from wheat), feed (mink), feed (pig), feed (poultry), feed (reindeer), feeds, grain, feeds, industrial, feedstuff, feedstuffs (rapeseed, turnip, fish meal, concentrates), fish meal, grain, mixed feed, grains, mixed, grains (no specification), maize, maize ears, maize fibre, maize germ, maize germ/bran, maize germ meal, maize gluten, maize kernels, maize meal, maize powder, maize screenings, maize stalks (pith), maize, "Baby", maize, hybrid, maize, white, oats, rice bran, rice germ cake, rye, silage, sorghum, soybeans, triticale, wheat, wheat and barley, wheat, red hard winter, wheat, soft white winter, wheat, spring, wheat, winter

DIACETOXYSCIRPENOL
incidence: 1/3*, conc.: 41 µg/kg, country: Finland[62], *imported ?
see also alfalfa, barley, feed, feed components, feed (dog), feed (mink), feed (pig), feed (poultry), feed (reindeer), feeds, grain, feedstuff, forage grass, grain, mixed feed,

grains (no specification), maize, maize gluten, oat and barley (hammer-milled), oats, peanuts, soybeans, wheat

HT-2 TOXIN
incidence: 2/3*, conc. range: 10–90 µg/kg, Ø conc.: 50 µg/kg, country: Finland[62], *imported ?
see also barley, feed components, feed (dog), feed (pig), feed (poultry), feed (reindeer), grains, mixed, grains, mixed feed, grains (no specification), maize, maize germ/bran, maize gluten, maize meal, maize screenings, oats, rye, silage, wheat

T-2 TOXIN
incidence: 3/3*, conc. range: 206–705 µg/kg, Ø conc.: 440 µg/kg, country: Finland[62], *imported?
see also alfalfa, barley, bran, diet (grower), diet (poultry), feed, feed components, feed, layer, feed, mixed, feed (dog), feed (mink), feed (pig), feed (poultry), feedstuff, forage grass, grain, mixed feed, grains (no specification), hay, maize, maize germ/bran, maize gluten, maize meal, maize screenings, maize stalks (pith), oat and barley (hammer-milled), oats, peanuts, pig grower diet, piglet diet, rye, sorghum, triticale, wheat

Feed (gluten)   may contain the following mycotoxins:

AFLATOXIN B$_1$
incidence: 6/6, conc. range: 6.3–18.6 µg/kg, Ø conc.: 12 µg/kg, country: Japan[141]
see also alfalfa, *Ambadi* cake, animal feedstuffs (dairy cake), bagasse, barley, bengalgram husk, bird food, bird food, wild, biri testa, blackgram, blackgram husk, bran, broiler mixed feed, calf fattening mixed feed, calf fattening mixed feed (containing 4–20 % peanut products), *Carthamus* cake, castor cake, cereals, cereal products, chick pea, coconut cake, cocos, concentrate, mixed, concentrates, cotton cake, cottonseed, cottonseed (dehulled), cottonseed cake, cottonseed extract, cottonseed meal, cottonseed meal (ammoniated), cottonseed

meal (decorticated), cottonseed meats, cottonseed products, crumbles, crumbles, grower, cycad meal, dairy cattle feed, dairy cattle feed (containing 2–5 % peanut products), dairy cattle feed (containing 6–10 % peanut products), dairy cattle feed (containing 6–12 % peanut products), dairy cattle feed (containing more than 20 % peanut products), diets, mixed, dog food, egg production mixed feed, feed, feed and ingredients, feed, compound, feed, layer, feed, mixed, feed (beef), feed (broilers), feed (calf), feed (cat), feed (cattle), feed (chicken), feed (dairy), feed (dog), feed (dug), feed (fish), feed (horse), feed (miscellaneous), feed (pig), feed (poultry), feed (poultry, pig), feed (rabbit), feed (sheep), feed (50–60 % maize), fish meal, grain by-products, grains (no specification), greengram, hay/silage, horsegram, husk, *Jagni* cake, legume mixture, linseed, linseed cake, livol, *Mahua* cake, maize, maize germ, maize gluten, maize grits, maize husk, maize meal, maize oil cake, maize powder, maize screenings, maize, ground, maize, hybrid, maize, preharvest, maize, yellow, maize (dark grains), *Makhana* (*Euryale ferox* Salisb) puffs, manioc, milk production mixed feed, mung testa, murkool, mustard cake, neem cake, niger cake, oats, palm kernel expeller cake, palm kernels, palm products, peanut cake, peanut cake (deoiled), peanut expeller, peanut hay, peanut meal, peanut, kernels, peanut, shells, peanuts, pellets, finisher, pig meal and pellets, pigeon pea, poultry feeds (peanut containing), rapeseed cake, redgram husk, rice, rice bran, rice bran (deoiled), rice chaff, rice crack, rice germ, rice germ cake, rice meal, rice straw, rice (damaged), rice (polish), safflower cake, sal seed cake, sesame, sesame cake, sorghum, soybean meal, soybeans, sunflower, sunflower cake, sunflower flour, tapioca, wheat, wheat bran, wheat bran and chana testa

Fumonisin B$_1$
incidence: 6/6, conc. range: 300–2400 µg/kg, Ø conc.: 1100 µg/kg, country: Japan[141]
see also barley, bird food, wild, dog food, feed, feed, complete ration, feed, general, feed, layer, feed, maize-based, feed, mixed, feed, pelleted ration, feed, screenings, feed, sweet, feed (broilers), feed (cat), feed (chicken), feed (dog), feed (horse), feed (maize), feed (pig), feed (poultry), feed (rat), feed (rodent), forage grass, maize, maize and maize screenings, maize bran, maize ears, maize fine fractions, maize flakes, maize germ, maize germ/bran, maize germ meal, maize gluten, maize grits, maize kernels, maize meal, maize powder, maize screenings, maize, "Baby", maize, ground, maize, preharvest, maize, sweet feed, maize/oats mix, rat chow, silage, sorghum, soybeans, wheat

Fumonisin B$_2$
incidence: 6/6, conc. range: <100–8500 µg/kg, Ø conc.: 3700 µg/kg, country: Japan[141]
see also barley, bird food, wild, dog food, feed, feed, maize-based, feed, mixed, feed (cat), feed (dog), feed (horse), feed (maize), feed (poultry), feed (rodent), maize, maize bran, maize fine fractions, maize flakes, maize germ, maize germ/bran, maize germ meal, maize gluten, maize grits, maize kernels, maize meal, maize powder, maize screenings, maize, "Baby", maize, ground, maize, preharvest, rat chow, wheat

Feed (goat)   may contain the following mycotoxins:

Aflatoxins
incidence: 1/1, conc.: 510–2000 µg/kg, country: Australia[21]
see also animal feed (maize), animal feed (mixed), barley, copra meal, cottonseed, cottonseed fines, cottonseed meal, cottonseed meats, feed, feed ingredients (miscellaneous), feed, mixed, feed (excluding peanuts, suspect), feed (pig), feed (poultry), feedstuff, grain, grain, mixed feed, maize, maize, shelled, millet, peanut cake, peanut meal, peanut meal and by-products, peanuts, protein concentrates, rice, rice bran, sorghum, soybean meal, sunflower

Feed (GP No.1)   may contain the following mycotoxins:

TENUAZONIC ACID
incidence: 10/10, conc. range: <10–20 μg/kg,
country: Australia[188]
see also feed, oilseed rape meal, sunflower
seed meal, sunflower seeds, ensiling

Feed (horse)   may contain the following
mycotoxins:

AFLATOXIN B$_1$
incidence: 5/5, conc. range: 80–150 μg/kg,
country: India[386]
see also alfalfa, *Ambadi* cake, animal feedstuffs
(dairy cake), bagasse, barley, bengalgram
husk, bird food, bird food, wild, biri testa,
blackgram, blackgram husk, bran, broiler
mixed feed, calf fattening mixed feed, calf
fattening mixed feed (containing 4–20 %
peanut products), *Carthamus* cake, castor
cake, cereals, cereal products, chick pea,
coconut cake, cocos, concentrate, mixed,
concentrates, cotton cake, cottonseed,
cottonseed (dehulled), cottonseed cake,
cottonseed extract, cottonseed meal,
cottonseed meal (ammoniated), cottonseed
meal (decorticated), cottonseed meats,
cottonseed products, crumbles, crumbles,
grower, cycad meal, dairy cattle feed, dairy
cattle feed (containing 2–5 % peanut
products), dairy cattle feed (containing
6–10 % peanut products), dairy cattle feed
(containing 6–12 % peanut products), dairy
cattle feed (containing more than 20 %
peanut products), diets, mixed, dog food, egg
production mixed feed, feed, feed and
ingredients, feed, compound, feed, layer,
feed, mixed, feed (beef), feed (broilers), feed
(calf), feed (cat), feed (cattle), feed (chicken),
feed (dairy), feed (dog), feed (dug), feed
(fish), feed (gluten), feed (miscellaneous),
feed (pig), feed (poultry), feed (poultry, pig),
feed (rabbit), feed (sheep), feed (50–60 %
maize), fish meal, grain by-products, grains
(no specification), greengram, hay/silage,
horsegram, husk, *Jagni* cake, legume mixture,
linseed, linseed cake, livol, *Mahua* cake,
maize, maize germ, maize gluten, maize grits,
maize husk, maize meal, maize oil cake,
maize powder, maize screenings, maize,

ground, maize, hybrid, maize, preharvest,
maize, yellow, maize (dark grains), *Makhana*
(*Euryale ferox* Salisb) puffs, manioc, milk
production mixed feed, mung testa, murkool,
mustard cake, neem cake, niger cake, oats,
palm kernel expeller cake, palm kernels, palm
products, peanut cake, peanut cake (deoiled),
peanut expeller, peanut hay, peanut meal,
peanut, kernels, peanut, shells, peanuts,
pellets, finisher, pig meal and pellets, pigeon
pea, poultry feeds (peanut containing),
rapeseed cake, redgram husk, rice, rice bran,
rice bran (deoiled), rice chaff, rice crack, rice
germ, rice germ cake, rice meal, rice straw,
rice (damaged), rice (polish), safflower cake,
sal seed cake, sesame, sesame cake, sorghum,
soybean meal, soybeans, sunflower, sunflower
cake, sunflower flour, tapioca, wheat, wheat
bran, wheat bran and chana testa

AFLATOXIN
incidence: 2/20, conc. range: <1.75–14 μg/kg,
country: Turkey[340]
see also blackgram husk, bread crumbs,
broiler finisher, broiler starter, cotton cake,
cottonseed, cottonseed cake, cottonseed
extract, cottonseed meal, feed, feed (cattle),
feed (cow), feed (dog), feed (maize, gluten),
feed (pig), feed (poultry), feed (rabbit), feed
(rat/mice), feed (sheep), feeds, grain, fish
meal, flour (wheat), groundnut cake,
grower's mash, horsegram, layer's mash,
maize, maize gluten, maize, white, milo,
peanut cake, peanut cake (deoiled), peanut,
kernels, peanut (oil cake), pearlmillet, pig
breeder's mash, pig finisher, pig starter, pod
with haulms, poultry breeder's mash, rabbit
pellets, redgram husk, rice, rice bran
(deoiled), rice, broken, rice (polish), sesame
cake, silk worm pupae, sorghum, soybean
cake, soybean meal, wheat, wheat bran

FUMONISIN B$_1$
incidence: 14/20, conc. range:
400–23,600 μg/kg, Ø conc.: 3500 μg/kg,
country: Spain[155]
incidence: ?/5, conc. range:
<100–37,000 μg/kg, country: USA[158]
incidence: 1/5, conc. range: 560 μg/kg,
country: India[386]

see also barley, bird food, wild, dog food, feed, feed, complete ration, feed, general, feed, layer, feed, maize-based, feed, mixed, feed, pelleted ration, feed, screenings, feed, sweet, feed (broilers), feed (cat), feed (chicken), feed (dog), feed (gluten), feed (maize), feed (pig), feed (poultry), feed (rat), feed (rodent), forage grass, maize, maize and maize screenings, maize bran, maize ears, maize fine fractions, maize flakes, maize germ, maize germ/bran, maize germ meal, maize gluten, maize grits, maize kernels, maize meal, maize powder, maize screenings, maize, "Baby", maize, ground, maize, preharvest, maize, sweet feed, maize/oats mix, rat chow, silage, sorghum, soybeans, wheat

FUMONISIN B_2
incidence: 1/20, conc.: 300 µg/kg, country: Spain[155]
see also barley, bird food, wild, dog food, feed, feed, maize-based, feed, mixed, feed (cat), feed (dog), feed (gluten), feed (maize), feed (poultry), feed (rodent), maize, maize bran, maize fine fractions, maize flakes, maize germ, maize germ/bran, maize germ meal, maize gluten, maize grits, maize kernels, maize meal, maize powder, maize screenings, maize, "Baby", maize, ground, maize, preharvest, rat chow, wheat

FUMONISINS
incidence: 3/3, conc. range: ≤5000 µg/kg, country: USA[147]
see also feed (cattle and dairy), feed (chicken), feed (dairy), maize, maize germ, maize gluten, maize meal, maize screenings

Feed (maize)   may contain the following mycotoxins:

FUMONISIN B_1
incidence: ?/165, conc. range: 0–8550 µg/kg, Ø conc.: 570 µg/kg*, country: South Africa[147], *mean of all samples
see also barley, bird food, wild, dog food, feed, feed, complete ration, feed, general, feed, layer, feed, maize-based, feed, mixed, feed, pelleted ration, feed, screenings, feed,

sweet, feed (broilers), feed (cat), feed (chicken), feed (dog), feed (gluten), feed (horse), feed (pig), feed (poultry), feed (rat), feed (rodent), forage grass, maize, maize and maize screenings, maize bran, maize ears, maize fine fractions, maize flakes, maize germ, maize germ/bran, maize germ meal, maize gluten, maize grits, maize kernels, maize meal, maize powder, maize screenings, maize, "Baby", maize, ground, maize, preharvest, maize, sweet feed, maize/oats mix, rat chow, silage, sorghum, soybeans, wheat

FUMONISIN B_2
incidence: ?/165, conc. range: 0–1500 µg/kg, Ø conc.: 140 µg/kg*, country: South Africa[147], *mean of all samples
see also barley, bird food, wild, dog food, feed, feed, maize-based, feed, mixed, feed (cat), feed (dog), feed (gluten), feed (horse), feed (poultry), feed (rodent), maize, maize bran, maize fine fractions, maize flakes, maize germ, maize germ/bran, maize germ meal, maize gluten, maize grits, maize kernels, maize meal, maize powder, maize screenings, maize, "Baby", maize, ground, maize, preharvest, rat chow, wheat

FUMONISIN B_3
incidence: ?/165, conc. range: 0–740 µg/kg, Ø conc.: 50 µg/kg*, country: South Africa[147], *mean of all samples
see also feed, feed (rodent), maize, maize bran, maize fine fractions, maize germ meal, maize kernels, maize powder, maize screenings

Feed (maize, gluten)   may contain the following mycotoxins:

AFLATOXIN B
incidence: 7/20, conc. range: ≥25–>100 µg/kg, country: France[46]
incidence: ?/8, conc. range: 6–57 µg/kg, Ø conc.: 31.5 µg/kg, country: Pakistan[162]
see also barley, cocoa (oil cake), cottonseed cake, feed, feed (cattle), feed (pig), feed (poultry), flour (wheat), lucern (dried), maize, maize gluten, maize grains, oats,

peanut (oil cake), rice, broken, sorghum, soybean (oil cake), sunflower (oil cake), wheat

AFLATOXIN G
incidence: ?/8, conc.: 22 µg/kg, country: Pakistan[162]
see also cottonseed cake, maize gluten, maize grains, rice, broken

AFLATOXIN
incidence: 125/156, conc. range: 3–60 µg/kg, country: USA[86]
see also blackgram husk, bread crumbs, broiler finisher, broiler starter, cotton cake, cottonseed, cottonseed cake, cottonseed extract, cottonseed meal, feed, feed (cattle), feed (cow), feed (dog), feed (horse), feed (pig), feed (poultry), feed (rabbit), feed (rat/mice), feed (sheep), feeds, grain, fish meal, flour (wheat), groundnut cake, grower's mash, horsegram, layer's mash, maize, maize gluten, maize, white, milo, peanut cake, peanut cake (deoiled), peanut, kernels, peanut (oil cake), pearlmillet, pig breeder's mash, pig finisher, pig starter, pod with haulms, poultry breeder's mash, rabbit pellets, redgram husk, rice, rice bran (deoiled), rice, broken, rice (polish), sesame cake, silk worm pupae, sorghum, soybean cake, soybean meal, wheat, wheat bran

Feed (mill run, from wheat)   may contain the following mycotoxins:

DEOXYNIVALENOL
incidence: 1/1*, conc.: 550 µg/kg, country: Australia[72], *mainly
see also barley, barley, husked, barley, unhusked (naked), barley (pressed), bone meal, bran, broilers feed, calf fattening mixed feed, coconut, expeller, corn cob mix silage, cottonseed, cottonseed cake, dairy cattle feed, egg production mixed feed, feed, feed components, feed, commercial mix, feed, mixed, feed, mixed (primarily maize), feed (barley), feed (cattle), feed (chicken), feed (dog), feed (fish), feed (mink), feed (pig), feed (poultry), feed (reindeer), feeds, grain, feeds, industrial, feedstuff, feedstuffs (rapeseed, turnip, fish meal, concentrates), fish meal, grain, mixed feed, grains, mixed, grains (no specification), maize, maize ears, maize fibre, maize germ, maize germ/bran, maize germ meal, maize gluten, maize kernels, maize meal, maize powder, maize screenings, maize stalks (pith), maize, "Baby", maize, hybrid, maize, white, oats, rice bran, rice germ cake, rye, silage, sorghum, soybeans, triticale, wheat, wheat and barley, wheat, red hard winter, wheat, soft white winter, wheat, spring, wheat, winter

ZEARALENONE
incidence: 1/1*, conc.: 1020 µg/kg, country: Australia[72], *mainly
see also alfalfa, barley, barley, husked, barley, unhusked (naked), barley and feed, bone meal, bran, broilers feed, *Carthamus* cake, chick pea, concentrate, mixed, corn cob mix silage, cotton cake, cottonseed, cottonseed cake, diet (dairy cow), diet (poultry), diets (mixed), feed, feed components, feed, mixed, feed, mixed (primarily maize also maize, oats, wheat), feed (bran), feed (broiler chicken), feed (cattle), feed (chicken), feed (dairy), feed (developing pig), feed (miscellaneous), feed (pig), feed (poultry), feed (poultry, pig), feed (starter chicken), feedstuff, fish meal, forage grass, grain, bruised, grain, mixed feed, grains (no specification), hay, maize, maize ears, maize flakes, maize germ, maize germ/bran, maize gluten, maize kernels, maize meal, maize oil cake, maize screenings, maize stalks (pith), maize, "Baby", maize, hybrid, maize, shelled, maize, unshelled, maize, white, maize grain, artificially dried, maize grain, crib dried, maize grain, ensiled, milk production mixed feed, oats, *Paspalum palidosum*, straw, peanut hulls/skins, rice bran, rice germ, rice germ cake, rye, silage, sorghum, soybeans, soybeans, extracted, sunflower cake, tapioca, triticale, wheat, wheat bran, wheat bran and chana testa, wheat soya meal

Feed (mink)   may contain the following mycotoxins:

Deoxynivalenol
incidence: 2/2*, conc. range: 32–47 µg/kg,
Ø conc.: 39.5 µg/kg, country: Finland[62],
*imported?
see also barley, barley, husked, barley,
unhusked (naked), barley (pressed), bone
meal, bran, broilers feed, calf fattening mixed
feed, coconut, expeller, corn cob mix silage,
cottonseed, cottonseed cake, dairy cattle feed,
egg production mixed feed, feed, feed
components, feed, commercial mix, feed,
mixed, feed, mixed (primarily maize), feed
(barley), feed (cattle), feed (chicken), feed
(dog), feed (fish), feed (mill run, from
wheat), feed (pig), feed (poultry), feed
(reindeer), feeds, grain, feeds, industrial,
feedstuff, feedstuffs (rapeseed, turnip, fish
meal, concentrates), fish meal, grain, mixed
feed, grains, mixed, grains (no specification),
maize, maize ears, maize fibre, maize germ,
maize germ/bran, maize germ meal, maize
gluten, maize kernels, maize meal, maize
powder, maize screenings, maize stalks (pith),
maize, "Baby", maize, hybrid, maize, white,
oats, rice bran, rice germ cake, rye, silage,
sorghum, soybeans, triticale, wheat, wheat
and barley, wheat, red hard winter, wheat,
soft white winter, wheat, spring, wheat,
winter

Diacetoxyscirpenol
incidence: 2/2*, conc. range: 155–230 µg/kg,
Ø conc.: 192.5 µg/kg, country: Finland[62],
*imported?
see also alfalfa, barley, feed, feed components,
feed (dog), feed (fish), feed (pig), feed
(poultry), feed (reindeer), feeds, grain,
feedstuff, forage grass, grain, mixed feed,
grains (no specification), maize, maize
gluten, oat and barley (hammer-milled), oats,
peanuts, soybeans, wheat

T-2 Toxin
incidence: 2/2*, conc. range: 23–109 µg/kg, Ø
conc.: 66 µg/kg, country: Finland[62],
*imported?
see also alfalfa, barley, bran, diet (grower),
diet (poultry), feed, feed components, feed,
layer, feed, mixed, feed (dog), feed (fish),
feed (pig), feed (poultry), feedstuff, forage
grass, grain, mixed feed, grains (no
specification), hay, maize, maize germ/bran,
maize gluten, maize meal, maize screenings,
maize stalks (pith), oat and barley
(hammer-milled), oats, peanuts, pig grower
diet, piglet diet, rye, sorghum, triticale, wheat

Feed (miscellaneous) may contain the
following mycotoxins:

Aflatoxin B$_1$
incidence: 5/27, conc. range: 1–20 µg/kg
(4 sa), >300 µg/kg (1 sa), country: USA[370]
see also alfalfa, *Ambadi* cake, animal feedstuffs
(dairy cake), bagasse, barley, bengalgram
husk, bird food, bird food, wild, biri testa,
blackgram, blackgram husk, bran, broiler
mixed feed, calf fattening mixed feed, calf
fattening mixed feed (containing 4–20 %
peanut products), *Carthamus* cake, castor
cake, cereals, cereal products, chick pea,
coconut cake, cocos, concentrate, mixed,
concentrates, cotton cake, cottonseed,
cottonseed (dehulled), cottonseed cake,
cottonseed extract, cottonseed meal,
cottonseed meal (ammoniated), cottonseed
meal (decorticated), cottonseed meats,
cottonseed products, crumbles, crumbles,
grower, cycad meal, dairy cattle feed, dairy
cattle feed (containing 2–5 % peanut
products), dairy cattle feed (containing
6–10 % peanut products), dairy cattle feed
(containing 6–12 % peanut products), dairy
cattle feed (containing more than 20 %
peanut products), diets, mixed, dog food, egg
production mixed feed, feed, feed and
ingredients, feed, compound, feed, layer,
feed, mixed, feed (beef), feed (broilers), feed
(calf), feed (cat), feed (cattle), feed (chicken),
feed (dairy), feed (dog), feed (dug), feed
(fish), feed (gluten), feed (horse), feed (pig),
feed (poultry), feed (poultry, pig), feed
(rabbit), feed (sheep), feed (50–60 % maize),
fish meal, grain by-products, grains (no
specification), greengram, hay/silage,
horsegram, husk, *Jagni* cake, legume mixture,
linseed, linseed cake, livol, *Mahua* cake,
maize, maize germ, maize gluten, maize grits,

maize husk, maize meal, maize oil cake, maize powder, maize screenings, maize, ground, maize, hybrid, maize, preharvest, maize, yellow, maize (dark grains), *Makhana* (*Euryale ferox* Salisb) puffs, manioc, milk production mixed feed, mung testa, murkool, mustard cake, neem cake, niger cake, oats, palm kernel expeller cake, palm kernels, palm products, peanut cake, peanut cake (deoiled), peanut expeller, peanut hay, peanut meal, peanut, kernels, peanut, shells, peanuts, pellets, finisher, pig meal and pellets, pigeon pea, poultry feeds (peanut containing), rapeseed cake, redgram husk, rice, rice bran, rice bran (deoiled), rice chaff, rice crack, rice germ, rice germ cake, rice meal, rice straw, rice (damaged), rice (polish), safflower cake, sal seed cake, sesame, sesame cake, sorghum, soybean meal, soybeans, sunflower, sunflower cake, sunflower flour, tapioca, wheat, wheat bran, wheat bran and chana testa

Zearalenone
incidence: 1/21, conc.: 162 µg/kg, country: Taiwan[221]
incidence: 1/6, conc.: 203 µg/kg, country: Taiwan[221]
incidence: 2/9, conc. range: 126–152 µg/kg, Ø conc.: 139 µg/kg, country: Taiwan[221]
see also alfalfa, barley, barley, husked, barley, unhusked (naked), barley and feed, bone meal, bran, broilers feed, *Carthamus* cake, chick pea, concentrate, mixed, corn cob mix silage, cotton cake, cottonseed, cottonseed cake, diet (dairy cow), diet (poultry), diets (mixed), feed, feed components, feed, mixed, feed, mixed (primarily maize also maize, oats, wheat), feed (bran), feed (broiler chicken), feed (cattle), feed (chicken), feed (dairy), feed (developing pig), feed (mill run, from wheat), feed (pig), feed (poultry), feed (poultry, pig), feed (starter chicken), feedstuff, fish meal, forage grass, grain, bruised, grain, mixed feed, grains (no specification), hay, maize, maize ears, maize flakes, maize germ, maize germ/bran, maize gluten, maize kernels, maize meal, maize oil cake, maize screenings, maize stalks (pith), maize, "Baby", maize, hybrid, maize, shelled,

maize, unshelled, maize, white, maize grain, artificially dried, maize grain, crib dried, maize grain, ensiled, milk production mixed feed, oats, *Paspalum palidosum*, straw, peanut hulls/skins, rice bran, rice germ, rice germ cake, rye, silage, sorghum, soybeans, soybeans, extracted, sunflower cake, tapioca, triticale, wheat, wheat bran, wheat bran and chana testa, wheat soya meal

Feed (pig)   may contain the following mycotoxins:

Aflatoxin B$_1$
incidence: 1/1, conc.: 400 µg/kg, country: Australia[121]
incidence: 7/16, conc. range: 1.7–12.3 µg/kg, Ø conc.: 6.8 µg/kg, country: Colombia[125]
incidence: 146/163, conc. range: ≤25 µg/kg (36 sa), 26–50 µg/kg (45 sa), 51–100 µg/kg (35 sa), 101–200 µg/kg (15 sa), 201–500 µg/kg (11 sa), 501–1000 µg/kg (3 sa), 1001–1500 µg/kg (1 sa), Ø conc.: 91.3 µg/kg, country: India[247]
incidence: 7/70, conc. range: <5 µg/kg (2 sa), 6–20 µg/kg (2 sa), 51–100 µg/kg (1 sa), >100 µg/kg (2), country: UK[267]
incidence: 4/4, conc. range: 270–6000 µg/kg, Ø conc.: 2250 µg/kg, country: India[321]
see also alfalfa, *Ambadi* cake, animal feedstuffs (dairy cake), bagasse, barley, bengalgram husk, bird food, bird food, wild, biri testa, blackgram, blackgram husk, bran, broiler mixed feed, calf fattening mixed feed, calf fattening mixed feed (containing 4–20 % peanut products), *Carthamus* cake, castor cake, cereals, cereal products, chick pea, coconut cake, cocos, concentrate, mixed, concentrates, cotton cake, cottonseed, cottonseed (dehulled), cottonseed cake, cottonseed extract, cottonseed meal, cottonseed meal (ammoniated), cottonseed meal (decorticated), cottonseed meats, cottonseed products, crumbles, crumbles, grower, cycad meal, dairy cattle feed, dairy cattle feed (containing 2–5 % peanut products), dairy cattle feed (containing 6–10 % peanut products), dairy cattle feed (containing 6–12 % peanut products), dairy

cattle feed (containing more than 20 %
peanut products), diets, mixed, dog food, egg
production mixed feed, feed, feed and
ingredients, feed, compound, feed, layer,
feed, mixed, feed (beef), feed (broilers), feed
(calf), feed (cat), feed (cattle), feed (chicken),
feed (dairy), feed (dog), feed (dug), feed
(fish), feed (gluten), feed (horse), feed
(miscellaneous), feed (poultry), feed (poultry,
pig), feed (rabbit), feed (sheep), feed
(50–60 % maize), fish meal, grain
by-products, grains (no specification),
greengram, hay/silage, horsegram, husk, *Jagni*
cake, legume mixture, linseed, linseed cake,
livol, *Mahua* cake, maize, maize germ, maize
gluten, maize grits, maize husk, maize meal,
maize oil cake, maize powder, maize
screenings, maize, ground, maize, hybrid,
maize, preharvest, maize, yellow, maize (dark
grains), *Makhana* (*Euryale ferox* Salisb) puffs,
manioc, milk production mixed feed, mung
testa, murkool, mustard cake, neem cake,
niger cake, oats, palm kernel expeller cake,
palm kernels, palm products, peanut cake,
peanut cake (deoiled), peanut expeller,
peanut hay, peanut meal, peanut, kernels,
peanut, shells, peanuts, pellets, finisher, pig
meal and pellets, pigeon pea, poultry feeds
(peanut containing), rapeseed cake, redgram
husk, rice, rice bran, rice bran (deoiled), rice
chaff, rice crack, rice germ, rice germ cake,
rice meal, rice straw, rice (damaged), rice
(polish), safflower cake, sal seed cake, sesame,
sesame cake, sorghum, soybean meal, soybeans,
sunflower, sunflower cake, sunflower flour,
tapioca, wheat, wheat bran, wheat bran and
chana *testa*

Aflatoxin B$_2$
incidence: 1/16, conc.: 1.1 µg/kg, country:
Colombia[125]
incidence: 115/163, conc. range: nc, Ø conc.:
18.9 µg/kg, country: India[247]
see also animal feedstuffs (dairy cake), bird
food, bird food, wild, biri testa, blackgram,
blackgram husk, cottonseed, cottonseed cake,
cottonseed extract, cottonseed meal,
cottonseed meal (ammoniated), cottonseed
meats, dog food, egg production mixed feed,

feed, compound, feed (cat), feed (cattle), feed
(dog), feed (poultry), feed (rabbit), feed
(sheep), fish meal, horsegram, maize, maize
gluten, maize husk, maize, ground, maize,
preharvest, mung testa, mustard cake, niger
cake, peanut cake, peanut cake (deoiled),
peanut expeller, peanut hay, peanut meal,
peanuts, peanut, kernels, redgram husk, rice
bran, rice bran (deoiled), rice chaff, rice
meal, rice (polish), sal seed cake, sesame
cake, sorghum, soybean meal, soybeans,
wheat, wheat bran

Aflatoxin B
incidence: 19/30, conc. range:
≥25–>100 µg/kg, country: France[46]
see also barley, cocoa (oil cake), cottonseed
cake, feed, feed (cattle), feed (maize, gluten),
feed (poultry), flour (wheat), lucern (dried),
maize, maize gluten, maize grains, oats,
peanut (oil cake), rice, broken, sorghum,
soybean (oil cake), sunflower (oil cake),
wheat

Aflatoxin
incidence: 146/163, conc. range:
3–1231 µg/kg, country: India[247]
see also blackgram husk, bread crumbs,
broiler finisher, broiler starter, cotton cake,
cottonseed, cottonseed cake, cottonseed
extract, cottonseed meal, feed, feed (cattle),
feed (cow), feed (dog), feed (horse), feed
(maize, gluten), feed (poultry), feed (rabbit),
feed (rat/mice), feed (sheep), feeds, grain, fish
meal, flour (wheat), groundnut cake,
grower's mash, horsegram, layer's mash,
maize, maize gluten, maize, white, milo,
peanut cake, peanut cake (deoiled), peanut,
kernels, peanut (oil cake), pearlmillet, pig
breeder's mash, pig finisher, pig starter, pod
with haulms, poultry breeder's mash, rabbit
pellets, redgram husk, rice, rice bran
(deoiled), rice, broken, rice (polish), sesame
cake, silk worm pupae, sorghum, soybean
cake, soybean meal, wheat, wheat bran

Aflatoxin B$_1$ + B$_2$ + G$_1$ + G$_2$
incidence: 14/90, conc.: range: <50 µg/kg,
country: Poland[84]
see also feed (cattle), feed (poultry)

AFLATOXIN B + G
incidence: 3/12, conc. range: 10–50 µg/kg,
country: France[45]
see also feed (cattle), feed (poultry), feed
(rabbit), feed (rat), mice feed, peanut (oil
cake)

AFLATOXINS
incidence: 1/12, conc.: 51–500 µg/kg, country:
Australia[21]
see also animal feed (maize), animal feed
(mixed), barley, copra meal, cottonseed,
cottonseed fines, cottonseed meal, cottonseed
meats, feed, feed ingredients (miscellaneous),
feed, mixed, feed (excluding peanuts,
suspect), feed (goat), feed (poultry), feedstuff,
grain, grain, mixed feed, maize, maize,
shelled, millet, peanut cake, peanut meal,
peanut meal and by-products, peanuts,
protein concentrates, rice, rice bran,
sorghum, soybean meal, sunflower

CITRININ
incidence: 1/2, conc.: 8600 µg/kg, country:
UK[93]
see also barley, barley, oats, barley-soybean
diet, cottonseed cake, feed, feed, mixed, feed
(cattle), fish meal, hay, maize, maize, white,
Makhana (*Euryale ferox* Salisb) puffs, oats,
palm products, peas and beans, rice bran,
rice germ, wheat, wheat and other grains
(moldy), wheat bran

DEOXYNIVALENOL
incidence: 1/1*, conc.: 120 µg/kg, country:
Finland[62], *imported ?
incidence: 3/3, conc. range: 2000–9700 µg/kg,
Ø conc.: 5633 µg/kg, country: USA[71]
incidence: 3/3, conc. range: 1700–8000 µg/kg,
Ø conc.: 4100 µg/kg, country: Argentina[213]
see also barley, barley, husked, barley,
unhusked (naked), barley (pressed), bone
meal, bran, broilers feed, calf fattening mixed
feed, coconut, expeller, corn cob mix silage,
cottonseed, cottonseed cake, dairy cattle feed,
egg production mixed feed, feed, feed
components, feed, commercial mix, feed,
mixed, feed, mixed (primarily maize), feed
(barley), feed (cattle), feed (chicken), feed
(dog), feed (fish), feed (mill run, from

wheat), feed (mink), feed (poultry), feed
(reindeer), feeds, grain, feeds, industrial,
feedstuff, feedstuffs (rapeseed, turnip, fish
meal, concentrates), fish meal, grain, mixed
feed, grains, mixed, grains (no specification),
maize, maize ears, maize fibre, maize germ,
maize germ/bran, maize germ meal, maize
gluten, maize kernels, maize meal, maize
powder, maize screenings, maize stalks (pith),
maize, "Baby", maize, hybrid, maize, white,
oats, rice bran, rice germ cake, rye, silage,
sorghum, soybeans, triticale, wheat, wheat
and barley, wheat, red hard winter, wheat,
soft white winter, wheat, spring, wheat,
winter

DIACETOXYSCIRPENOL
incidence: 1/1*, conc.: 60 µg/kg, country:
Finland[62], *imported ?
see also alfalfa, barley, feed, feed components,
feed (dog), feed (fish), feed (mink), feed
(poultry), feed (reindeer), feeds, grain,
feedstuff, forage grass, grain, mixed feed,
grains (no specification), maize, maize
gluten, oat and barley (hammer-milled), oats,
peanuts, soybeans, wheat

FUMONISIN B$_1$
incidence: 42/47, conc. range:
400–11,600 µg/kg, Ø conc.: 2400 µg/kg,
country: Spain[155]
see also barley, bird food, wild, dog food,
feed, feed, complete ration, feed, general,
feed, layer, feed, maize-based, feed, mixed,
feed, pelleted ration, feed, screenings, feed,
sweet, feed (broilers), feed (cat), feed
(chicken), feed (dog), feed (gluten), feed
(horse), feed (maize), feed (poultry), feed
(rat), feed (rodent), forage grass, maize,
maize and maize screenings, maize bran,
maize ears, maize fine fractions, maize flakes,
maize germ, maize germ/bran, maize germ
meal, maize gluten, maize grits, maize
kernels, maize meal, maize powder, maize
screenings, maize, "Baby", maize, ground,
maize, preharvest, maize, sweet feed,
maize/oats mix, rat chow, silage, sorghum,
soybeans, wheat

Fumonisin B$_1$ + B$_2$
incidence: 12/12, conc. range:
7000–73,000 µg/kg, Ø conc.: 31,000 µg/kg,
country: USA[144]
incidence: 8/9, conc. range:
6000–33,000 µg/kg, Ø conc.: 14,750 µg/kg,
country: USA[144]
see also maize, maize and feed samples

HT-2 Toxin
incidence: 1/1*, conc.: 207 µg/kg, country:
Finland[62], *imported ?
see also barley, feed components, feed (dog),
feed (fish), feed (poultry), feed (reindeer),
grain, mixed feed, grains, mixed, grains (no
specification), maize, maize germ/bran, maize
gluten, maize meal, maize screenings, oats,
rye, silage, wheat

Ochratoxin A
incidence: 8/90, conc. range: 10–50 µg/kg,
country: Poland[84]
see also alfalfa, barley, barley, oats, barley
(high moisture), barley-soybean diet, bird
food, domestic, bird food, wild, broilers feed,
cereal grains, citrus pulp, coconut, expeller,
corn cob mix silage, diet (dairy cow), diet
(poultry), diet (starter), dog food, eat, egg
production mixed feed, feed, feed wheat, oat
and barley, feed, commercial mix, feed,
mixed, feed, mixed (pelleted), feed (broilers),
feed (cat), feed (cattle), feed (cereals), feed ec
(pig), feed (poultry), feed ec (poultry), feed
(poultry, pig), feed ec (rabbit), feed (trout),
grain, mixed feed, grains, mixed, grains
(heated), hay, horse bean, maize, maize feed,
milo, maize gluten, maize meal, maize, white,
Makhana (*Euryale ferox* Salisb) puffs, milk
production mixed feed, millet, oat and barley
(hammer-milled), oats, palm products,
peanut cake, peas, peas and beans, pet food,
pig feedstuffs, pig grower diet, pig meal,
piglet diet, poultry feedstuffs, rice bran, rice
germ, rice germ cake, rye, sorghum, soybean
groats, sunflower, sunflower seeds, extracted,
tapioca, triticale, *Vicia faba*, wheat, wheat and
barley, wheat bran, wheat hay, wheat, oats

T-2 Toxin
incidence: 2/2, conc. range: 4100–5800 µg/kg,
Ø conc.: 4950 µg/kg, country: Hungary[50]
incidence: 1/1*, conc.: 490 µg/kg, country:
Finland[62], *imported ?
see also alfalfa, barley, bran, diet (grower),
diet (poultry), feed, feed components, feed,
layer, feed, mixed, feed (dog), feed (fish),
feed (mink), feed (poultry), feedstuff, forage
grass, grain, mixed feed, grains (no
specification), hay, maize, maize germ/bran,
maize gluten, maize meal, maize screenings,
maize stalks (pith), oat and barley
(hammer-milled), oats, peanuts, pig grower
diet, piglet diet, rye, sorghum, triticale, wheat

Zearalenone
incidence: 15/15, conc. range: tr–109 µg/kg,
country: USA[119]
incidence: 2/4, conc. range: 243–300 µg/kg,
Ø conc.: 271.5 µg/kg, country: Taiwan[221]
see also alfalfa, barley, barley, husked, barley,
unhusked (naked), barley and feed, bone
meal, bran, broilers feed, *Carthamus* cake,
chick pea, concentrate, mixed, corn cob mix
silage, cotton cake, cottonseed, cottonseed
cake, diet (dairy cow), diet (poultry), diets
(mixed), feed, feed components, feed, mixed,
feed, mixed (primarily maize also maize, oats,
wheat), feed (bran), feed (broiler chicken),
feed (cattle), feed (chicken), feed (dairy), feed
(developing pig), feed (mill run, from
wheat), feed (miscellaneous), feed (poultry),
feed (poultry, pig), feed (starter chicken),
feedstuff, fish meal, forage grass, grain,
bruised, grain, mixed feed, grains (no
specification), hay, maize, maize ears, maize
flakes, maize germ, maize germ/bran, maize
gluten, maize kernels, maize meal, maize oil
cake, maize screenings, maize stalks (pith),
maize, "Baby", maize, hybrid, maize, shelled,
maize, unshelled, maize, white, maize grain,
artificially dried, maize grain, crib dried,
maize grain, ensiled, milk production mixed
feed, oats, *Paspalum palidosum*, straw, peanut
hulls/skins, rice bran, rice germ, rice germ
cake, rye, silage, sorghum, soybeans,
soybeans, extracted, sunflower cake, tapioca,

triticale, wheat, wheat bran, wheat bran and chana testa, wheat soya meal

Feed ec (pig)   composition: corn, sunflower pellets or soy pellets, vitamins or growth regulators, essential amino acids (lysine, methionine), coccidiostats
may contain the following mycotoxins:

OCHRATOXIN A
incidence: 5/40, conc. range: 34 µg/kg, country: Argentina[1]
see also alfalfa, barley, barley, oats, barley (high moisture), barley-soybean diet, bird food, domestic, bird food, wild, broilers feed, cereal grains, citrus pulp, coconut, expeller, corn cob mix silage, diet (dairy cow), diet (poultry), diet (starter), dog food, eat, egg production mixed feed, feed, feed wheat, oat and barley, feed, commercial mix, feed, mixed, feed, mixed (pelleted), feed (broilers), feed (cat), feed (cattle), feed (cereals), feed (pig), feed (poultry), feed ec (poultry), feed (poultry, pig), feed ec (rabbit), feed (trout), grain, mixed feed, grains, mixed, grains (heated), hay, horse bean, maize, maize feed, milo, maize gluten, maize meal, maize, white, *Makhana* (*Euryale ferox* Salisb) puffs, milk production mixed feed, millet, oat and barley (hammer-milled), oats, palm products, peanut cake, peas, peas and beans, pet food, pig feedstuffs, pig grower diet, pig meal, piglet diet, poultry feedstuffs, rice bran, rice germ, rice germ cake, rye, sorghum, soybean groats, sunflower, sunflower seeds, extracted, tapioca, triticale, *Vicia faba*, wheat, wheat and barley, wheat bran, wheat hay, wheat, oats

Feed (poultry)   may contain the following mycotoxins:

3-ACETYLDEOXYNIVALENOL
incidence: 3/50, conc. range: 207–1497 µg/kg, Ø conc.: 858.3 µg/kg, country: Slovak Republic[420]
see also barley, feed components, feeds, grain, feeds, industrial, feedstuffs (rapeseed, turnip, fish meal, concentrates), maize, maize ears, maize gluten, maize kernels, maize meal, maize screenings, oats, wheat

15-ACETYLDEOXYNIVALENOL
incidence: 2/50, conc. range: 130–229 µg/kg, Ø conc.: 179.5 µg/kg, country: Slovak Republic[420]
see also barley, feed components, feeds, grain, feeds, industrial, feedstuffs (rapeseed, turnip, fish meal, concentrates), maize, maize ears, maize gluten, maize kernels, maize meal, maize screenings, oats, wheat

AFLATOXIN B_1
incidence: 8/202, conc. range: <10–20 µg/kg, country: UK[30]
incidence: 63/130, conc. range: 10–123 µg/kg, Ø conc.: 27.3 µg/kg, country: Argentina[78]
incidence: 50/144, conc. range: <5–100 µg/kg, country: South Africa[79]
incidence: 14/14, conc. range: 0.38–108.61 µg/kg, Ø conc.: 37.88 µg/kg, country: India[98]
incidence: 12/30, conc. range: 1.5–23.2 µg/kg, Ø conc.: 6.9 µg/kg, country: Colombia[125]
incidence: 41/300, conc. range: 17–197 µg/kg, country: Argentina[243]
incidence: 1108/1368, conc. range: ≤25 µg/kg (371 sa), 26–50 µg/kg (212 sa), 51–100 µg/kg (224 sa), 101–200 µg/kg (134 sa), 201–500 µg/kg (112 sa), 501–1000 µg/kg (45 sa), 1001–1500 µg/kg (7 sa), 1501–2000 µg/kg (1 sa), 2001–3000 µg/kg (2 sa), Ø conc.: 110.4 µg/kg, country: India[247]
incidence: 14/20, conc. range: 46.1–259.1 µg/kg, Ø conc.: 118.3 µg/kg, country: India[253]
incidence: 8/127, conc. range: <5 µg/kg (1 sa), 6–20 µg/kg (6 sa), 21–50 µg/kg (1 sa), country: UK[267]
incidence: 1/3, conc.: 640 µg/kg, country: India[321]
incidence: 1/10, conc.: 1.4 µg/kg, country: Morocco[423]
incidence: 50/101, conc. range: ≤160 µg/kg, Ø conc.: 14.9 µg/kg, country: Bangladesh[428]
incidence: 8/216*, conc. range: 20–110 µg/kg, country: Morocco[429], *feed and ingredients
incidence: 17/99*, conc. range: 20–50 µg/kg (8 sa), 51–200 µg/kg (4 sa), ≤5625 µg/kg (4 sa), country: Morocco[429], *from bins, sacs and troughs

see also alfalfa, *Ambadi* cake, animal feedstuffs (dairy cake), bagasse, barley, bengalgram husk, bird food, bird food, wild, biri testa, blackgram, blackgram husk, bran, broiler mixed feed, calf fattening mixed feed, calf fattening mixed feed (containing 4–20 % peanut products), *Carthamus* cake, castor cake, cereals, cereal products, chick pea, coconut cake, cocos, concentrate, mixed, concentrates, cotton cake, cottonseed, cottonseed (dehulled), cottonseed cake, cottonseed extract, cottonseed meal, cottonseed meal (ammoniated), cottonseed meal (decorticated), cottonseed meats, cottonseed products, crumbles, crumbles, grower, cycad meal, dairy cattle feed, dairy cattle feed (containing 2–5 % peanut products), dairy cattle feed (containing 6–10 % peanut products), dairy cattle feed (containing 6–12 % peanut products), dairy cattle feed (containing more than 20 % peanut products), diets, mixed, dog food, egg production mixed feed, feed, feed and ingredients, feed, compound, feed, layer, feed, mixed, feed (beef), feed (broilers), feed (calf), feed (cat), feed (cattle), feed (chicken), feed (dairy), feed (dog), feed (dug), feed (fish), feed (gluten), feed (horse), feed (miscellaneous), feed (pig), feed (poultry, pig), feed (rabbit), feed (sheep), feed (50–60 % maize), fish meal, grain by-products, grains (no specification), greengram, hay/silage, horsegram, husk, *Jagni* cake, legume mixture, linseed, linseed cake, livol, *Mahua* cake, maize, maize germ, maize gluten, maize grits, maize husk, maize meal, maize oil cake, maize powder, maize screenings, maize, ground, maize, hybrid, maize, preharvest, maize, yellow, maize (dark grains), *Makhana* (*Euryale ferox* Salisb) puffs, manioc, milk production mixed feed, mung testa, murkool, mustard cake, neem cake, niger cake, oats, palm kernel expeller cake, palm kernels, palm products, peanut cake, peanut cake (deoiled), peanut expeller, peanut hay, peanut meal, peanut, kernels, peanut, shells, peanuts, pellets, finisher, pig meal and pellets, pigeon pea, poultry feeds (peanut containing),

rapeseed cake, redgram husk, rice, rice bran, rice bran (deoiled), rice chaff, rice crack, rice germ, rice germ cake, rice meal, rice straw, rice (damaged), rice (polish), safflower cake, sal seed cake, sesame, sesame cake, sorghum, soybean meal, soybeans, sunflower, sunflower cake, sunflower flour, tapioca, wheat, wheat bran, wheat bran and chana testa

AFLATOXIN B$_2$
incidence: 1/30, conc.: 2 µg/kg, country: Colombia[125]
incidence: 894/1368, conc. range: nc, Ø conc.: 16.6 µg/kg, country: India[247]
incidence: 5/20, conc. range: 46.1–259.1 µg/kg, Ø conc.: 99.3 µg/kg, country: India[253]
see also animal feedstuffs (dairy cake), bird food, bird food, wild, biri testa, blackgram, blackgram husk, cottonseed, cottonseed cake, cottonseed extract, cottonseed meal, cottonseed meal (ammoniated), cottonseed meats, dog food, egg production mixed feed, feed, compound, feed (cat), feed (cattle), feed (dog), feed (pig), feed (rabbit), feed (sheep), fish meal, horsegram, maize, maize gluten, maize husk, maize, ground, maize, preharvest, mung testa, mustard cake, niger cake, peanut cake, peanut cake (deoiled), peanut expeller, peanut hay, peanut meal, peanuts, peanut, kernels, redgram husk, rice bran, rice bran (deoiled), rice chaff, rice meal, rice (polish), sal seed cake, sesame cake, sorghum, soybean meal, soybeans, wheat, wheat bran

AFLATOXIN B
incidence: 34/60, conc. range: $\geq$25–>100 µg/kg, country: France[46]
see also barley, cocoa (oil cake), cottonseed cake, feed, feed (cattle), feed (maize, gluten), feed (pig), flour (wheat), lucern (dried), maize, maize gluten, maize grains, oats, peanut (oil cake), rice, broken, sorghum, soybean (oil cake), sunflower (oilcake), wheat

AFLATOXIN
incidence: 1108/1368, conc. range: 2–2410 µg/kg, country: India[247]
incidence: 4/10, conc. range: 11–30 µg/kg (3 sa), 31–100 µg/kg (1 sa), country: India[311]

incidence: 23/23, conc. range: <30 μg/kg
(4 sa), >30 μg/kg (3 sa), >200 μg/kg (10 sa),
>600 μg/kg (5 sa), >1750 μg/kg (1 sa),
country: India[380]
incidence: 361/1174, conc. range: 50 μg/kg
(197 sa), 51–200 μg/kg (96 sa),
201–2000 μg/kg (67 sa), country:
Egypt, USA[415]
incidence: 110/190, conc. range:
0–3.26 μg/kg, country: Kuwait[421]
see also blackgram husk, bread crumbs,
broiler finisher, broiler starter, cotton cake,
cottonseed, cottonseed cake, cottonseed
extract, cottonseed meal, feed, feed (cattle),
feed (cow), feed (dog), feed (horse), feed
(maize, gluten), feed (pig), feed (rabbit), feed
(rat/mice), feed (sheep), feeds, grain, fish
meal, flour (wheat), groundnut cake,
grower's mash, horsegram, layer's mash,
maize, maize gluten, maize, white, milo,
peanut cake, peanut cake (deoiled), peanut,
kernels, peanut (oil cake), pearlmillet, pig
breeder's mash, pig finisher, pig starter, pod
with haulms, poultry breeder's mash, rabbit
pellets, redgram husk, rice, rice bran
(deoiled), rice, broken, rice (polish), sesame
cake, silk worm pupae, sorghum, soybean
cake, soybean meal, wheat, wheat bran

Aflatoxin $B_1 + B_2 + G_1 + G_2$
incidence: 2/54, conc. range: <50 μg/kg,
country: Poland[84]
see also feed (cattle), feed (pig)

Aflatoxin $B_1 + B_2$
incidence: 3/3, conc. range: 6–20 μg/kg,
(1 sa), >50 μg/kg, (2 sa, with a maximum of
71 μg/kg), country: Guatemala[375]
see also cottonseed, maize, sorghum

Aflatoxin $G_1 + G_2$
incidence: 30/300, conc. range: nc, country:
Argentina[243]

Aflatoxin $B + G$
incidence: 4/15, conc. range: 2–10 μg/kg,
Ø conc.: 4 μg/kg, country: France[45]
see also feed (cattle), feed (pig), feed (rabbit),
feed (rat), mice feed, peanut (oil cake)

Aflatoxins
incidence: 8/93, conc. range: 5–≤2000 μg/kg,
country: Australia[21]
incidence: 2/4, conc. range: 210–310 μg/kg,
Ø conc.: 260 μg/kg, country: USA[208]
see also animal feed (maize), animal feed
(mixed), barley, copra meal, cottonseed,
cottonseed fines, cottonseed meal, cottonseed
meats, feed, feed ingredients (miscellaneous),
feed, mixed, feed (excluding peanuts, suspect),
feed (goat), feed (pig), feedstuff, grain, grain,
mixed feed, maize, maize, shelled, millet,
peanut cake, peanut meal, peanut meal and
by-products, peanuts, protein concentrates,
rice, rice bran, sorghum, soybean meal,
sunflower

Deoxynivalenol
incidence: 25/300, conc. range:
240–410 μg/kg, country: Argentina[243]
incidence: 49/277, conc. range:
<2–500 μg/kg, Ø conc.: 211.2 μg/kg, country:
Saudi Arabia[389]
incidence: 28/50, conc. range: 64–1230 μg/kg,
Ø conc.: 303 μg/kg, country: Slovak
Republic[420]
incidence: 105/120, conc. range:
0–1500 μg/kg, country: Kuwait[421]
see also barley, barley, husked, barley,
unhusked (naked), barley (pressed), bone
meal, bran, broilers feed, calf fattening mixed
feed, coconut, expeller, corn cob mix silage,
cottonseed, cottonseed cake, dairy cattle feed,
egg production mixed feed, feed, feed
components, feed, commercial mix, feed,
mixed, feed, mixed (primarily maize), feed
(barley), feed (cattle), feed (chicken), feed
(dog), feed (fish), feed (mill run, from
wheat), feed (mink), feed (pig), feed (reindeer),
feeds, grain, feeds, industrial, feedstuff,
feedstuffs (rapeseed, turnip, fish meal,
concentrates), fish meal, grain, mixed feed,
grains, mixed, grains (no specification),
maize, maize ears, maize fibre, maize germ,
maize germ/bran, maize germ meal, maize
gluten, maize kernels, maize meal, maize
powder, maize screenings, maize stalks (pith),
maize, "Baby", maize, hybrid, maize, white,
oats, rice bran, rice germ cake, rye, silage,

sorghum, soybeans, triticale, wheat, wheat and barley, wheat, red hard winter, wheat, soft white winter, wheat, spring, wheat, winter

DIACETOXYSCIRPENOL
incidence: 10/50, conc. range: 3–5 µg/kg, Ø conc.: 4 µg/kg, country: Slovak Republic[420]
see also alfalfa, barley, feed, feed components, feed (dog), feed (fish), feed (mink), feed (pig), feed (reindeer), feeds, grain, feedstuff, forage grass, grain, mixed feed, grains (no specification), maize, maize gluten, oat and barley (hammer-milled), oats, peanuts, soybeans, wheat

FUMONISIN B$_1$
incidence: 27/32, conc. range: 200–<1000 µg/kg, country: South Africa[44]
incidence: 6/22, conc. range: ≤500 µg/kg, country: Switzerland[89]
incidence: 5/14, conc. range: 20–260 µg/kg, Ø conc.: 100 µg/kg, country: India[98]
incidence: 6/22, conc. range: 0–480 µg/kg, Ø conc.: 240 µg/kg*, country: Switzerland[148]
incidence: 2/2, conc. range: 1100–1400 µg/kg, Ø conc.: 1300 µg/kg, country: Spain[155]
incidence: 42/50, conc. range: <1000 µg/kg (11 sa), 1000–10,000 µg/kg (18 sa), 11,000–20,000 µg/kg (10 sa), 21,000–30,000 µg/kg (3 sa, with a maximum of 28,000 µg/kg), country: India[391]
incidence: 49/50, conc. range: 43–798 µg/kg, Ø conc.: 235 µg/kg, country: Slovak Republic[430]
see also barley, bird food, wild, dog food, feed, feed, complete ration, feed, general, feed, layer, feed, maize-based, feed, mixed, feed, pelleted ration, feed, screenings, feed, sweet, feed (broilers), feed (cat), feed (chicken), feed (dog), feed (gluten), feed (horse), feed (maize), feed (pig), feed (rat), feed (rodent), forage grass, maize, maize and maize screenings, maize bran, maize ears, maize fine fractions, maize flakes, maize germ, maize germ/bran, maize germ meal, maize gluten, maize grits, maize kernels, maize meal, maize powder, maize screenings, maize, "Baby", maize, ground, maize,

preharvest, maize, sweet feed, maize/oats mix, rat chow, silage, sorghum, soybeans, wheat

FUMONISIN B$_2$
incidence: 2/22, conc. range: 0–115 µg/kg, Ø conc.: 90 µg/kg, country: Switzerland[148]
incidence: 42/50, conc. range: 26–362 µg/kg, Ø conc.: 87 µg/kg, country: Slovak Republic[430]
see also barley, bird food, wild, dog food, feed, feed, maize-based, feed, mixed, feed (cat), feed (dog), feed (gluten), feed (horse), feed (maize), feed (rodent), maize, maize bran, maize fine fractions, maize flakes, maize germ, maize germ/bran, maize germ meal, maize gluten, maize grits, maize kernels, maize meal, maize powder, maize screenings, maize, "Baby", maize, ground, maize, preharvest, rat chow, wheat

FUMONISIN
incidence: 120/127, conc. range: 0–6000 µg/kg, country: Kuwait[421]

FUSARENONE-X
incidence: 1/277, conc.: 500 µg/kg, country: Saudi Arabia[389]
see also barley, feed components, maize, maize ears, maize germ/bran, maize gluten, maize kernels, maize meal, maize screenings, oats, wheat

HT-2 TOXIN
incidence: 1/277, conc.: 18.8 µg/kg, country: Saudi Arabia[389]
incidence: 38/50, conc. range: 3–173 µg/kg, Ø conc.: 18 µg/kg, country: Slovak Republic[420]
see also barley, feed components, feed (dog), feed (fish), feed (pig), feed (reindeer), grains, mixed, grains, mixed feed, grains (no specification), maize, maize germ/bran, maize gluten, maize meal, maize screenings, oats, rye, silage, wheat

MONILIFORMIN
incidence: 26/50, conc. range: 42–1214 µg/kg, Ø conc.: 217 µg/kg, country: Slovak Republic[430]
see also barley, feed, mixed, maize, maize flakes, maize germ, maize germ/bran, maize

gluten, maize meal, maize screenings, maize, "Baby", oats, rice bran, triticale, wheat, wheat, summer, wheat, winter

NEOSOLANIOL
incidence: 1/277, conc.: 100 µg/kg, country: Saudi Arabia[389]
see also feed, feed components, feed, mixed, peanut cake, wheat

NIVALENOL
incidence: 1/277, conc.: 3.1 µg/kg, country: Saudi Arabia[389]
see also barley, barley, husked, barley, unhusked (naked), barley (pressed), bran, feed, feed components, feed (cattle), feed (reindeer), feeds, industrial, maize, maize ears, maize germ, maize germ/bran, maize gluten, maize kernels, maize meal, maize powder, maize screenings, maize, "Baby", oats, silage, rye, triticale, wheat

OCHRATOXIN A
incidence: 6/203, conc. range: <25–150 µg/kg, country: UK[30]
incidence: 1/54, conc.: 10–50 µg/kg, country: Poland[84]
incidence: 109/120, conc. range: 0–40 µg/kg, country: Kuwait[421]
see also alfalfa, barley, barley, oats, barley (high moisture), barley-soybean diet, bird food, domestic, bird food, wild, broilers feed, cereal grains, citrus pulp, coconut, expeller, corn cob mix silage, diet (dairy cow), diet (poultry), diet (starter), dog food, eat, egg production mixed feed, feed, feed wheat, oat and barley, feed, commercial mix, feed, mixed, feed, mixed (pelleted), feed (broilers), feed (cat), feed (cattle), feed (cereals), feed (pig), feed ec (pig), feed ec (poultry), feed (poultry, pig), feed (rabbit), feed (trout), grain, mixed feed, grains, mixed, grains (heated), hay, horse bean, maize, maize feed, milo, maize gluten, maize meal, maize, white, *Makhana* (*Euryale ferox* Salisb) puffs, milk production mixed feed, millet, oat and barley (hammer-milled), oats, palm products, peanut cake, peas, peas and beans, pet food, pig feedstuffs, pig grower diet, pig meal, piglet diet, poultry feedstuffs, rice bran, rice

germ, rice germ cake, rye, sorghum, soybean groats, sunflower, sunflower seeds, extracted, tapioca, triticale, *Vicia faba*, wheat, wheat and barley, wheat bran, wheat hay, wheat, oats

T-2 TOXIN
incidence: 45/50, conc. range: 1–130 µg/kg, Ø conc.: 13 µg/kg, country: Slovak Republic[420]
see also alfalfa, barley, bran, diet (grower), diet (poultry), feed, feed components, feed, layer, feed, mixed, feed (dog), feed (fish), feed (mink), feed (pig), feedstuff, forage grass, grain, mixed feed, grains (no specification), hay, maize, maize germ/bran, maize gluten, maize meal, maize screenings, maize stalks (pith), oat and barley (hammer-milled), oats, peanuts, pig grower diet, piglet diet, rye, sorghum, triticale, wheat

ZEARALENONE
incidence: 22/130, conc. range: 327–5850 µg/kg, Ø conc.: 2544.3 µg/kg, country: Argentina[78]
incidence: 3/300, conc. range: 30–280 µg/kg, Ø conc.: 143.3 µg/kg, country: Argentina[243]
incidence: 44/50, conc. range: 3–86 µg/kg, Ø conc.: 21 µg/kg, country: Slovak Republic[420]
incidence: 113/120, conc. range: 0–400 µg/kg, country: Kuwait[421]
see also alfalfa, barley, barley, husked, barley, unhusked (naked), barley and feed, bone meal, bran, broilers feed, *Carthamus* cake, chick pea, concentrate, mixed, corn cob mix silage, cotton cake, cottonseed, cottonseed cake, diet (dairy cow), diet (poultry), diets (mixed), feed, feed components, feed, mixed, feed, mixed (primarily maize also maize, oats, wheat), feed (bran), feed (broiler chicken), feed (cattle), feed (chicken), feed (dairy), feed (developing pig), feed (mill run, from wheat), feed (miscellaneous), feed (pig), feed (poultry, pig), feed (starter chicken), feedstuff, fish meal, forage grass, grain, bruised, grain, mixed feed, grains (no specification), hay, maize, maize ears, maize flakes, maize germ, maize germ/bran, maize gluten, maize kernels, maize meal, maize oil

cake, maize screenings, maize stalks (pith), maize, "Baby", maize, hybrid, maize, shelled, maize, unshelled, maize, white, maize grain, artificially dried, maize grain, crib dried, maize grain, ensiled, milk production mixed feed, oats, *Paspalum palidosum*, straw, peanut hulls/skins, rice bran, rice germ, rice germ cake, rye, silage, sorghum, soybeans, soybeans, extracted, sunflower cake, tapioca, triticale, wheat, wheat bran, wheat bran and chana testa, wheat soya meal

Feed ec (poultry)   composition: corn, sunflower pellets or soy pellets, vitamins or growth regulators, essential amino acids (lysine, methionine), coccidiostats
may contain the following mycotoxins:

OCHRATOXIN A
incidence: 15/40, conc. range: 25–30 μg/kg, country: Argentina[1]
see also alfalfa, barley, barley, oats, barley (high moisture), barley-soybean diet, bird food, domestic, bird food, wild, broilers feed, cereal grains, citrus pulp, coconut, expeller, corn cob mix silage, diet (dairy cow), diet (poultry), diet (starter), dog food, eat, egg production mixed feed, feed, feed wheat, oat and barley, feed, commercial mix, feed, mixed, feed, mixed (pelleted), feed (broilers), feed (cat), feed (cattle), feed (cereals), feed (pig), feed ec (pig), feed (poultry), feed (poultry, pig), feed ec (rabbit), feed (trout), grain, mixed feed, grains, mixed, grains (heated), hay, horse bean, maize, maize feed, milo, maize gluten, maize meal, maize, white, *Makhana* (*Euryale ferox* Salisb) puffs, milk production mixed feed, millet, oat and barley (hammer-milled), oats, palm products, peanut cake, peas, peas and beans, pet food, pig feedstuffs, pig grower diet, pig meal, piglet diet, poultry feedstuffs, rice bran, rice germ, rice germ cake, rye, sorghum, soybean groats, sunflower, sunflower seeds, extracted, tapioca, triticale, *Vicia faba*, wheat, wheat and barley, wheat bran, wheat hay, wheat, oats

Feed (poultry, pig)   may contain the following mycotoxins:

AFLATOXIN B₁

incidence: 2/75, conc. range: ≤10 μg/kg, country: France[77]
see also alfalfa, *Ambadi* cake, animal feedstuffs (dairy cake), bagasse, barley, bengalgram husk, bird food, bird food, wild, biri testa, blackgram, blackgram husk, bran, broiler mixed feed, calf fattening mixed feed, calf fattening mixed feed (containing 4–20 % peanut products), *Carthamus* cake, castor cake, cereals, cereal products, chick pea, coconut cake, cocos, concentrate, mixed, concentrates, cotton cake, cottonseed, cottonseed (dehulled), cottonseed cake, cottonseed extract, cottonseed meal, cottonseed meal (ammoniated), cottonseed meal (decorticated), cottonseed meats, cottonseed products, crumbles, crumbles, grower, cycad meal, dairy cattle feed, dairy cattle feed (containing 2–5 % peanut products), dairy cattle feed (containing 6–10 % peanut products), dairy cattle feed (containing 6–12 % peanut products), dairy cattle feed (containing more than 20 % peanut products), diets, mixed, dog food, egg production mixed feed, feed, feed and ingredients, feed, compound, feed, layer, feed, mixed, feed (beef), feed (broilers), feed (calf), feed (cat), feed (cattle), feed (chicken), feed (dairy), feed (dog), feed (dug), feed (fish), feed (gluten), feed (horse), feed (miscellaneous), feed (pig), feed (poultry), feed (rabbit), feed (sheep), feed (50–60 % maize), fish meal, grain by-products, grains (no specification), greengram, hay/silage, horsegram, husk, *Jagni* cake, legume mixture, linseed, linseed cake, livol, *Mahua* cake, maize, maize germ, maize gluten, maize grits, maize husk, maize meal, maize oil cake, maize powder, maize screenings, maize, ground, maize, hybrid, maize, preharvest, maize, yellow, maize (dark grains), *Makhana* (*Euryale ferox* Salisb) puffs, manioc, milk production mixed feed, mung testa, murkool, mustard cake, neem cake, niger cake, oats, palm kernel expeller cake, palm kernels, palm products, peanut cake, peanut cake (deoiled), peanut expeller, peanut hay, peanut meal,

peanut, kernels, peanut, shells, peanuts, pellets, finisher, pig meal and pellets, pigeon pea, poultry feeds (peanut containing), rapeseed cake, redgram husk, rice, rice bran, rice bran (deoiled), rice chaff, rice crack, rice germ, rice germ cake, rice meal, rice straw, rice (damaged), rice (polish), safflower cake, sal seed cake, sesame, sesame cake, sorghum, soybean meal, soybeans, sunflower, sunflower cake, sunflower flour, tapioca, wheat, wheat bran, wheat bran and chana testa

Ochratoxin A
incidence: 2/75, conc. range: $\leq 10\,\mu g/kg$, country: France[77]
see also alfalfa, barley, barley, oats, barley (high moisture), barley-soybean diet, bird food, domestic, bird food, wild, broilers feed, cereal grains, citrus pulp, coconut, expeller, corn cob mix silage, diet (dairy cow), diet (poultry), diet (starter), dog food, eat, egg production mixed feed, feed, feed wheat, oat and barley, feed, commercial mix, feed, mixed, feed, mixed (pelleted), feed (broilers), feed (cat), feed (cattle), feed (cereals), feed (pig), feed ec (pig), feed (poultry), feed ec (poultry), feed ec (rabbit), feed (trout), grain, mixed feed, grains, mixed, grains (heated), hay, horse bean, maize, maize feed, milo, maize gluten, maize meal, maize, white, *Makhana* (*Euryale ferox* Salisb) puffs, milk production mixed feed, millet, oat and barley (hammer-milled), oats, palm products, peanut cake, peas, peas and beans, pet food, pig feedstuffs, pig grower diet, pig meal, piglet diet, poultry feedstuffs, rice bran, rice germ, rice germ cake, rye, sorghum, soybean groats, sunflower, sunflower seeds, extracted, tapioca, triticale, *Vicia faba*, wheat, wheat and barley, wheat bran, wheat hay, wheat, oats

Zearalenone
incidence: 62/75, conc. range: $\leq 170,000\,\mu g/kg$, country: France[77]
see also alfalfa, barley, barley, husked, barley, unhusked (naked), barley and feed, bone meal, bran, broilers feed, *Carthamus* cake, chick pea, concentrate, mixed, corn cob mix silage, cotton cake, cottonseed, cottonseed

cake, diet (dairy cow), diet (poultry), diets (mixed), feed, feed components, feed, mixed, feed, mixed (primarily maize also maize, oats, wheat), feed (bran), feed (broiler chicken), feed (cattle), feed (chicken), feed (dairy), feed (developing pig), feed (mill run, from wheat), feed (miscellaneous), feed (pig), feed (poultry), feed (starter chicken), feedstuff, fish meal, forage grass, grain, bruised, grain, mixed feed, grains (no specification), hay, maize, maize ears, maize flakes, maize germ, maize germ/bran, maize gluten, maize kernels, maize meal, maize oil cake, maize screenings, maize stalks (pith), maize, "Baby", maize, hybrid, maize, shelled, maize, unshelled, maize, white, maize grain, artificially dried, maize grain, crib dried, maize grain, ensiled, milk production mixed feed, oats, *Paspalum palidosum*, straw, peanut hulls/skins, rice bran, rice germ, rice germ cake, rye, silage, sorghum, soybeans, soybeans, extracted, sunflower cake, tapioca, triticale, wheat, wheat bran, wheat bran and chana testa, wheat soya meal

Feed (rabbit)   may contain the following mycotoxins:

Aflatoxin B_1
incidence: 3/4, conc. range: $26{-}50\,\mu g/kg$ (2 sa), $201{-}500\,\mu g/kg$ (1 sa), $\varnothing$ conc.: $122.7\,\mu g/kg$, country: India[247]
see also alfalfa, *Ambadi* cake, animal feedstuffs (dairy cake), bagasse, barley, bengalgram husk, bird food, bird food, wild, biri testa, blackgram, blackgram husk, bran, broiler mixed feed, calf fattening mixed feed, calf fattening mixed feed (containing 4–20% peanut products), *Carthamus* cake, castor cake, cereals, cereal products, chick pea, coconut cake, cocos, concentrate, mixed, concentrates, cotton cake, cottonseed, cottonseed (dehulled), cottonseed cake, cottonseed extract, cottonseed meal, cottonseed meal (ammoniated), cottonseed meal (decorticated), cottonseed meats, cottonseed products, crumbles, crumbles, grower, cycad meal, dairy cattle feed, dairy cattle feed (containing 2–5% peanut

products), dairy cattle feed (containing 6–10 % peanut products), dairy cattle feed (containing 6–12 % peanut products), dairy cattle feed (containing more than 20 % peanut products), diets, mixed, dog food, egg production mixed feed, feed, feed and ingredients, feed, compound, feed, layer, feed, mixed, feed (beef), feed (broilers), feed (calf), feed (cat), feed (cattle), feed (chicken), feed (dairy), feed (dog), feed (dug), feed (fish), feed (gluten), feed (horse), feed (miscellaneous), feed (pig), feed (poultry), feed (poultry, pig), feed (sheep), feed (50–60 % maize), fish meal, grain by-products, grains (no specification), greengram, hay/silage, horsegram, husk, *Jagni* cake, legume mixture, linseed, linseed cake, livol, *Mahua* cake, maize, maize germ, maize gluten, maize grits, maize husk, maize meal, maize oil cake, maize powder, maize screenings, maize, ground, maize, hybrid, maize, preharvest, maize, yellow, maize (dark grains), *Makhana* (*Euryale ferox* Salisb) puffs, manioc, milk production mixed feed, mung testa, murkool, mustard cake, neem cake, niger cake, oats, palm kernel expeller cake, palm products, palm kernels, peanut cake, peanut cake (deoiled), peanut expeller, peanut hay, peanut meal, peanut, kernels, peanut, shells, peanuts, pellets, finisher, pig meal and pellets, pigeon pea, poultry feeds (peanut containing), rapeseed cake, redgram husk, rice, rice bran, rice bran (deoiled), rice chaff, rice crack, rice germ, rice germ cake, rice meal, rice straw, rice (damaged), rice (polish), safflower cake, sal seed cake, sesame, sesame cake, sorghum, soybean meal, soybeans, sunflower, sunflower cake, sunflower flour, tapioca, wheat, wheat bran, wheat bran and chana testa

Aflatoxin B$_2$
incidence: 3/4, Ø conc.: 32.4 µg/kg, country: India[247]
see also animal feedstuffs (dairy cake), bird food, bird food, wild, biri testa, blackgram, blackgram husk, cottonseed, cottonseed cake, cottonseed extract, cottonseed meal, cottonseed meal (ammoniated), cottonseed

meats, dog food, egg production mixed feed, feed, compound, feed (cat), feed (cattle), feed (dog), feed (pig), feed (poultry), feed (sheep), fish meal, horsegram, maize, maize gluten, maize husk, maize, ground, maize, preharvest, mung testa, mustard cake, niger cake, peanut cake, peanut cake (deoiled), peanut expeller, peanut hay, peanut meal, peanuts, peanut, kernels, redgram husk, rice bran, rice bran (deoiled), rice chaff, rice meal, rice (polish), sal seed cake, sesame cake, sorghum, soybean meal, soybeans, wheat, wheat bran

Aflatoxin
incidence: 3/4, conc. range: 37–381 µg/kg, country: India[247]
see also blackgram husk, bread crumbs, broiler finisher, broiler starter, cotton cake, cottonseed, cottonseed cake, cottonseed extract, cottonseed meal, feed, feed (cattle), feed (cow), feed (dog), feed (horse), feed (maize, gluten), feed (pig), feed (poultry), feed (rat/mice), feed (sheep), feeds, grain, fish meal, flour (wheat), groundnut cake, grower's mash, horsegram, layer's mash, maize, maize gluten, maize, white, milo, peanut cake, peanut cake (deoiled), peanut, kernels, peanut (oil cake), pearlmillet, pig breeder's mash, pig finisher, pig starter, pod with haulms, poultry breeder's mash, rabbit pellets, redgram husk, rice, rice bran (deoiled), rice, broken, rice (polish), sesame cake, silk worm pupae, sorghum, soybean cake, soybean meal, wheat, wheat bran

Aflatoxin B + G
incidence: 9/18, conc. range: 9–1500 µg/kg, Ø conc.: 220 µg/kg, country: France[45]
see also feed (cattle), feed (pig), feed (poultry), feed (rat), mice feed, peanut (oil cake)

Feed ec (rabbit)   composition: alfalfa, sunflower pellets or soy pellets, vitamins or growth regulators, essential amino acids (lysine, methionine), coccidiostats
may contain the following mycotoxins:

OCHRATOXIN A
incidence: 10/40, conc. range: 18.5–25 µg/kg,
country: Argentina[1]
see also alfalfa, barley, barley, oats, barley
(high moisture), barley-soybean diet, bird
food, domestic, bird food, wild, broilers feed,
cereal grains, citrus pulp, coconut, expeller,
corn cob mix silage, diet (dairy cow), diet
(poultry), diet (starter), dog food, eat, egg
production mixed feed, feed, feed wheat, oat
and barley, feed, commercial mix, feed,
mixed, feed, mixed (pelleted), feed (broilers),
feed (cat), feed (cattle), feed (cereals), feed
(pig), feed ec (pig), feed (poultry), feed ec
(poultry), feed (poultry, pig), feed (trout),
grain, mixed feed, grains, mixed, grains
(heated), hay, horse bean, maize, maize feed,
milo, maize gluten, maize meal, maize, white,
Makhana (*Euryale ferox* Salisb) puffs, milk
production mixed feed, millet, oat and barley
(hammer-milled), oats, palm products,
peanut cake, peas, peas and beans, pet food,
pig feedstuffs, pig grower diet, pig meal,
piglet diet, poultry feedstuffs, rice bran, rice
germ, rice germ cake, rye, sorghum, soybean
groats, sunflower, sunflower seeds, extracted,
tapioca, triticale, *Vicia faba*, wheat, wheat and
barley, wheat bran, wheat hay, wheat, oats

Feed (rat)   may contain the following
mycotoxins:

AFLATOXIN B + G
incidence: 7/20, conc. range: 5–650 µg/kg,
Ø conc.: 65 µg/kg, country: France[45]
see also feed (cattle), feed (pig), feed
(poultry), feed (rabbit), mice feed, peanut
(oil cake)

FUMONISIN B$_1$
incidence: ?/5, conc. range: 100–2000 µg/kg,
country: USA[158]
see also barley, bird food, wild, dog food,
feed, feed, complete ration, feed, general,
feed, layer, feed, maize-based, feed, mixed,
feed, pelleted ration, feed, screenings, feed,
sweet, feed (broilers), feed (cat), feed
(chicken), feed (dog), feed (gluten), feed
(horse), feed (maize), feed (pig), feed

(poultry), feed (rodent), forage grass, maize,
maize and maize screenings, maize bran,
maize ears, maize fine fractions, maize flakes,
maize germ, maize germ/bran, maize germ
meal, maize gluten, maize grits, maize
kernels, maize meal, maize powder, maize
screenings, maize, "Baby", maize, ground,
maize, preharvest, maize, sweet feed,
maize/oats mix, rat chow, silage, sorghum,
soybeans, wheat

Feed (rat/mice)   may contain the following
mycotoxins:

AFLATOXIN
incidence: 2/2, conc. range: >200 µg/kg
(2 sa), country: India[380]
see also blackgram husk, bread crumbs,
broiler finisher, broiler starter, cotton cake,
cottonseed, cottonseed cake, cottonseed
extract, cottonseed meal, feed, feed (cattle),
feed (cow), feed (dog), feed (horse), feed
(maize, gluten), feed (pig), feed (poultry),
feed (rabbit), feed (sheep), feeds, grain, fish
meal, flour (wheat), groundnut cake,
grower's mash, horsegram, layer's mash,
maize, maize gluten, maize, white, milo,
peanut cake, peanut cake (deoiled), peanut,
kernels, peanut (oil cake), pearlmillet, pig
breeder's mash, pig finisher, pig starter, pod
with haulms, poultry breeder's mash, rabbit
pellets, redgram husk, rice, rice bran
(deoiled), rice, broken, rice (polish), sesame
cake, silk worm pupae, sorghum, soybean
cake, soybean meal, wheat, wheat bran

Feed (reindeer)   may contain the following
mycotoxins:

DEOXYNIVALENOL
incidence: 1/1*, conc.: 59 µg/kg, country:
Finland[62], *imported ?
see also barley, barley, husked, barley,
unhusked (naked), barley (pressed), bone
meal, bran, broilers feed, calf fattening mixed
feed, coconut, expeller, corn cob mix silage,
cottonseed, cottonseed cake, dairy cattle feed,
egg production mixed feed, feed, feed
components, feed, commercial mix, feed,

mixed, feed, mixed (primarily maize), feed (barley), feed (cattle), feed (chicken), feed (dog), feed (fish), feed (mill run, from wheat), feed (mink), feed (pig), feed (poultry), feeds, grain, feeds, industrial, feedstuff, feedstuffs (rapeseed, turnip, fish meal, concentrates), fish meal, grain, mixed feed, grains, mixed, grains (no specification), maize, maize ears, maize fibre, maize germ, maize germ/bran, maize germ meal, maize gluten, maize kernels, maize meal, maize powder, maize screenings, maize stalks (pith), maize, "Baby", maize, hybrid, maize, white, oats, rice bran, rice germ cake, rye, silage, sorghum, soybeans, triticale, wheat, wheat and barley, wheat, red hard winter, wheat, soft white winter, wheat, spring, wheat, winter

Diacetoxyscirpenol
incidence: 1/1*, conc.: 766 µg/kg, country: Finland[62], *imported ?
see also alfalfa, barley, feed, feed components, feed (dog), feed (fish), feed (mink), feed (pig), feed (poultry), feeds, grain, feedstuff, forage grass, grain, mixed feed, grains (no specification), maize, maize gluten, oat and barley (hammer-milled), oats, peanuts, soybeans, wheat

HT-2 Toxin
incidence: 1/1*, conc.: 23 µg/kg, country: Finland[62], *imported ?
see also barley, feed components, feed (dog), feed (fish), feed (pig), feed (poultry), grain, mixed feed, grains, mixed, grains (no specification), maize, maize germ/bran, maize gluten, maize meal, maize screenings, oats, rye, silage, wheat

Nivalenol
incidence: 1/1*, conc.: 10 µg/kg, country: Finland[62], *imported ?
see also barley, barley, husked, barley, unhusked (naked), barley (pressed), bran, feed components, feed (cattle), feed (poultry), feeds, industrial, maize, maize ears, maize germ, maize germ/bran, maize gluten, maize kernels, maize meal, maize powder, maize screenings, maize, "Baby", oats, rye, silage, triticale, wheat

Feed (rodent)   may contain the following mycotoxins:

Fumonisin B$_1$
incidence: 12/12, conc. range: 32.3–663.7 µg/kg, Ø conc.: 221.6 µg/kg, country: USA[99]
see also barley, bird food, wild, dog food, feed, feed, complete ration, feed, general, feed, layer, feed, maize-based, feed, mixed, feed, pelleted ration, feed, screenings, feed, sweet, feed (broilers), feed (cat), feed (chicken), feed (dog), feed (gluten), feed (horse), feed (maize), feed (pig), feed (poultry), feed (rat), forage grass, maize, maize and maize screenings, maize bran, maize ears, maize fine fractions, maize flakes, maize germ, maize germ/bran, maize germ meal, maize gluten, maize grits, maize kernels, maize meal, maize powder, maize screenings, maize, "Baby", maize, ground, maize, preharvest, maize, sweet feed, maize/oats mix, rat chow, silage, sorghum, soybeans, wheat

Fumonisin B$_2$
incidence: 12/12, conc. range: 11.2–204.2 µg/kg, Ø conc.: 73.4 µg/kg, country: USA[99]
see also barley, bird food, wild, dog food, feed, feed, maize-based, feed, mixed, feed (cat), feed (dog), feed (gluten), feed (horse), feed (maize), feed (poultry), maize, maize bran, maize fine fractions, maize flakes, maize germ, maize germ/bran, maize germ meal, maize gluten, maize grits, maize kernels, maize meal, maize powder, maize screenings, maize, "Baby", maize, ground, maize, preharvest, rat chow, wheat

Fumonisin B$_3$
incidence: 11/12, conc. range: 2.6–54.7 µg/kg, Ø conc.: 20 µg/kg, country: USA[99]
see also feed, feed (maize), maize, maize bran, maize fine fractions, maize germ meal, maize kernels, maize powder, maize screenings

Feed (sheep)   may contain the following mycotoxins:

AFLATOXIN B$_1$
incidence: 2/3, conc. range: 101–200 µg/kg
(1 sa), 201–500 µg/kg (1 sa), Ø conc.:
222.7 µg/kg, country: India[247]
incidence: 1/1, conc.: 8400 µg/kg, country:
India[321]
see also alfalfa, *Ambadi* cake, animal feedstuffs
(dairy cake), bagasse, barley, bengalgram
husk, bird food, bird food, wild, biri testa,
blackgram, blackgram husk, bran, broiler
mixed feed, calf fattening mixed feed, calf
fattening mixed feed (containing 4–20 %
peanut products), *Carthamus* cake, castor
cake, cereals, cereal products, chick pea,
coconut cake, cocos, concentrate, mixed,
concentrates, cotton cake, cottonseed,
cottonseed (dehulled), cottonseed cake,
cottonseed extract, cottonseed meal,
cottonseed meal (ammoniated), cottonseed
meal (decorticated), cottonseed meats,
cottonseed products, crumbles, crumbles,
grower, cycad meal, dairy cattle feed, dairy
cattle feed (containing 2–5 % peanut
products), dairy cattle feed (containing
6–10 % peanut products), dairy cattle feed
(containing 6–12 % peanut products), dairy
cattle feed (containing more than 20 %
peanut products), diets, mixed, dog food, egg
production mixed feed, feed, feed and
ingredients, feed, compound, feed, layer,
feed, mixed, feed (beef), feed (broilers), feed
(calf), feed (cat), feed (cattle), feed (chicken),
feed (dairy), feed (dog), feed (dug), feed
(fish), feed (gluten), feed (horse), feed
(miscellaneous), feed (pig), feed (poultry),
feed (poultry, pig), feed (rabbit), feed
(50–60 % maize), fish meal, grain
by-products, grains (no specification),
greengram, hay/silage, horsegram, husk, *Jagni*
cake, legume mixture, linseed, linseed cake,
livol, *Mahua* cake, maize, maize germ, maize
gluten, maize grits, maize husk, maize meal,
maize oil cake, maize powder, maize
screenings, maize, ground, maize, hybrid,
maize, preharvest, maize, yellow, maize (dark
grains), *Makhana* (*Euryale ferox* Salisb) puffs,
manioc, milk production mixed feed, mung
testa, murkool, mustard cake, neem cake,

niger cake, oats, palm kernel expeller cake,
palm kernels, palm products, peanut cake,
peanut cake (deoiled), peanut expeller,
peanut hay, peanut meal, peanut, kernels,
peanut, shells, peanuts, pellets, finisher, pig
meal and pellets, pigeon pea, poultry feeds
(peanut containing), rapeseed cake, redgram
husk, rice, rice bran, rice bran (deoiled), rice
chaff, rice crack, rice germ, rice germ cake,
rice meal, rice straw, rice (damaged), rice
(polish), safflower cake, sal seed cake, sesame,
sesame cake, sorghum, soybean meal,
soybeans, sunflower, sunflower cake,
sunflower flour, tapioca, wheat, wheat bran,
wheat bran and chana testa

AFLATOXIN B$_2$
incidence: 2/3, Ø conc.: 19.6 µg/kg, country:
India[247]
see also animal feedstuffs (dairy cake), bird
food, bird food, wild, biri testa, blackgram,
blackgram husk, cottonseed, cottonseed cake,
cottonseed extract, cottonseed meal,
cottonseed meal (ammoniated), cottonseed
meats, dog food, egg production mixed feed,
feed, compound, feed (cat), feed (cattle), feed
(dog), feed (pig), feed (poultry), feed (rabbit),
fish meal, horsegram, maize, maize gluten,
maize husk, maize, ground, maize,
preharvest, mung testa, mustard cake, niger
cake, peanut cake, peanut cake (deoiled),
peanut expeller, peanut hay, peanut meal,
peanuts, peanut, kernels, redgram husk, rice
bran, rice bran (deoiled), rice chaff, rice
meal, rice (polish), sal seed cake, sesame
cake, sorghum, soybean meal, soybeans,
wheat, wheat bran

AFLATOXIN
incidence: 2/3, conc. range:
200.1–245.2 µg/kg, country: India[247]
see also blackgram husk, bread crumbs,
broiler finisher, broiler starter, cotton cake,
cottonseed, cottonseed cake, cottonseed
extract, cottonseed meal, feed, feed (cattle),
feed (cow), feed (dog), feed (horse), feed
(maize, gluten), feed (pig), feed (poultry),
feed (rabbit), feed (rat/mice), feeds, grain,
fish meal, flour (wheat), groundnut cake,

grower's mash, horsegram, layer's mash, maize, maize gluten, maize, white, milo, peanut cake, peanut cake (deoiled), peanut, kernels, peanut (oil cake), pearlmillet, pig breeder's mash, pig finisher, pig starter, pod with haulms, poultry breeder's mash, rabbit pellets, redgram husk, rice, rice bran (deoiled), rice, broken, rice (polish), sesame cake, silk worm pupae, sorghum, soybean cake, soybean meal, wheat, wheat bran

Feed (starter chicken)   may contain the following mycotoxins:

Zearalenone
incidence: 4/26, conc. range: 354–1973 µg/kg, country: Taiwan[221]
incidence: 14/96, conc. range:
11.5–287 µg/kg, country: Brazil[409]
see also alfalfa, barley, barley, husked, barley, unhusked (naked), barley and feed, bone meal, bran, broilers feed, *Carthamus* cake, chick pea, concentrate, mixed, corn cob mix silage, cotton cake, cottonseed, cottonseed cake, diet (dairy cow), diet (poultry), diets (mixed), feed, feed components, feed, mixed, feed, mixed (primarily maize also maize, oats, wheat), feed (bran), feed (broiler chicken), feed (cattle), feed (chicken), feed (dairy), feed (developing pig), feed (mill run, from wheat), feed (miscellaneous), feed (pig), feed (poultry), feed (poultry, pig), feedstuff, fish meal, forage grass, grain, bruised, grain, mixed feed, grains (no specification), hay, maize, maize ears, maize flakes, maize germ, maize germ/bran, maize gluten, maize kernels, maize meal, maize oil cake, maize screenings, maize stalks (pith), maize, "Baby", maize, hybrid, maize, shelled, maize, unshelled, maize, white, maize grain, artificially dried, maize grain, crib dried, maize grain, ensiled, milk production mixed feed, oats, *Paspalum palidosum*, straw, peanut hulls/skins, rice bran, rice germ, rice germ cake, rye, silage, sorghum, soybeans, soybeans, extracted, sunflower cake, tapioca, triticale, wheat, wheat bran, wheat bran and chana testa, wheat soya meal

Feed (trout)   may contain the following mycotoxins:

Ochratoxin A
incidence: 4/13, conc. range: 0.4–1.9 µg/kg, Ø conc.: 0.93 µg/kg, country: Italy[14]
see also alfalfa, barley, barley, oats, barley (high moisture), barley-soybean diet, bird food, domestic, bird food, wild, broilers feed, cereal grains, citrus pulp, coconut, expeller, corn cob mix silage, diet (dairy cow), diet (poultry), diet (starter), dog food, eat, egg production mixed feed, feed, feed wheat, oat and barley, feed, commercial mix, feed, mixed, feed, mixed (pelleted), feed (broilers), feed (cat), feed (cattle), feed (cereals), feed (pig), feed ec (pig), feed (poultry), feed ec (poultry), feed (poultry, pig), feed ec (rabbit), grain, mixed feed, grains, mixed, grains (heated), hay, horse bean, maize, maize feed, milo, maize gluten, maize meal, maize, white, *Makhana* (*Euryale ferox* Salisb) puffs, milk production mixed feed, millet, oat and barley (hammer-milled), oats, palm products, peanut cake, peas, peas and beans, pet food, pig feedstuffs, pig grower diet, pig meal, piglet diet, poultry feedstuffs, rice bran, rice germ, rice germ cake, rye, sorghum, soybean groats, sunflower, sunflower seeds, extracted, tapioca, triticale, *Vicia faba*, wheat, wheat and barley, wheat bran, wheat hay, wheat, oats

Feed (50–60 % maize)   may contain the following mycotoxins:

Aflatoxin B$_1$
incidence: 94/278, conc. range:
60–15,000 µg/kg, country: USA[166]
see also alfalfa, *Ambadi* cake, animal feedstuffs (dairy cake), bagasse, barley, bengalgram husk, bird food, bird food, wild, biri testa, blackgram, blackgram husk, bran, broiler mixed feed, calf fattening mixed feed, calf fattening mixed feed (containing 4–20 % peanut products), *Carthamus* cake, castor cake, cereals, cereal products, chick pea, coconut cake, cocos, concentrate, mixed, concentrates, cotton cake, cottonseed, cottonseed (dehulled), cottonseed cake,

cottonseed extract, cottonseed meal, cottonseed meal (ammoniated), cottonseed meal (decorticated), cottonseed meats, cottonseed products, crumbles, crumbles, grower, cycad meal, dairy cattle feed, dairy cattle feed (containing 2–5 % peanut products), dairy cattle feed (containing 6–10 % peanut products), dairy cattle feed (containing 6–12 % peanut products), dairy cattle feed (containing more than 20 % peanut products), diets, mixed, dog food, egg production mixed feed, feed, feed and ingredients, feed, compound, feed, layer, feed, mixed, feed (beef), feed (broilers), feed (calf), feed (cat), feed (cattle), feed (chicken), feed (dairy), feed (dog), feed (dug), feed (fish), feed (gluten), feed (horse), feed (miscellaneous), feed (pig), feed (poultry), feed (poultry, pig), feed (rabbit), feed (sheep), fish meal, grain by-products, grains (no specification), greengram, hay/silage, horsegram, husk, *Jagni* cake, legume mixture, linseed, linseed cake, livol, *Mahua* cake, maize, maize germ, maize gluten, maize grits, maize husk, maize meal, maize oil cake, maize powder, maize screenings, maize, ground, maize, hybrid, maize, preharvest, maize, yellow, maize (dark grains), *Makhana* (*Euryale ferox* Salisb) puffs, manioc, milk production mixed feed, mung testa, murkool, mustard cake, neem cake, niger cake, oats, palm kernel expeller cake, palm kernels, palm products, peanut cake, peanut cake (deoiled), peanut expeller, peanut hay, peanut meal, peanut, kernels, peanut, shells, peanuts, pellets, finisher, pig meal and pellets, pigeon pea, poultry feeds (peanut containing), rapeseed cake, redgram husk, rice, rice bran, rice bran (deoiled), rice chaff, rice crack, rice germ, rice germ cake, rice meal, rice straw, rice (damaged), rice (polish), safflower cake, sal seed cake, sesame, sesame cake, sorghum, soybean meal, soybeans, sunflower, sunflower cake, sunflower flour, tapioca, wheat, wheat bran, wheat bran and chana testa

Feeds, grain   may contain the following mycotoxins:

3-ACETYLDEOXYNIVALENOL
incidence: 169/281, max. conc.: 570 µg/kg, Ø conc.: 23 µg/kg, country: Finland[11]
see also barley, feed components, feed (poultry), feeds, industrial, feedstuffs (rapeseed, turnip, fish meal, concentrates), maize, maize ears, maize gluten, maize kernels, maize meal, maize screenings, oats, wheat

AFLATOXIN
incidence: 95/172, conc. range: 1–350 µg/kg, country: UK[37]
incidence: 45/105, conc. range: 1–300 µg/kg, country: UK[37], from Germany
see also blackgram husk, bread crumbs, broiler finisher, broiler starter, cotton cake, cottonseed, cottonseed cake, cottonseed extract, cottonseed meal, feed, feed (cattle), feed (cow), feed (dog), feed (horse), feed (maize, gluten), feed (pig), feed (poultry), feed (rabbit), feed (rat/mice), feed (sheep), fish meal, flour (wheat), groundnut cake, grower's mash, horsegram, layer's mash, maize, maize gluten, maize, white, milo, peanut cake, peanut cake (deoiled), peanut, kernels, peanut (oil cake), pearlmillet, pig breeder's mash, pig finisher, pig starter, pod with haulms, poultry breeder's mash, rabbit pellets, redgram husk, rice, rice bran (deoiled), rice, broken, rice (polish), sesame cake, silk worm pupae, sorghum, soybean cake, soybean meal, wheat, wheat bran

DEOXYNIVALENOL
incidence: 350/363, conc. range: ≤ 16,000 µg/kg, Ø conc.: 160 µg/kg, country: Finland[11]
see also barley, barley, husked, barley, unhusked (naked), barley (pressed), bone meal, bran, broilers feed, calf fattening mixed feed, coconut, expeller, corn cob mix silage, cottonseed, cottonseed cake, dairy cattle feed, egg production mixed feed, feed, feed components, feed, commercial mix, feed, mixed, feed, mixed (primarily maize), feed (barley), feed (cattle), feed (chicken), feed (dog), feed (fish), feed (mill run, from wheat), feed (mink), feed (pig), feed

(poultry), feed (reindeer), feeds, industrial, feedstuff, feedstuffs (rapeseed, turnip, fish meal, concentrates), fish meal, grain, mixed feed, grains, mixed, grains (no specification), maize, maize ears, maize fibre, maize germ, maize germ/bran, maize kernels, maize germ meal, maize gluten, maize meal, maize powder, maize screenings, maize stalks (pith), maize, "Baby", maize, hybrid, maize, white, oats, rice bran, rice germ cake, rye, silage, sorghum, soybeans, triticale, wheat, wheat and barley, wheat, red hard winter, wheat, soft white winter, wheat, spring, wheat, winter

DIACETOXYSCIRPENOL
incidence: 42/363, max. conc.: 1680 µg/kg, Ø conc.: 220 µg/kg, country: Finland[11]
see also alfalfa, barley, feed, feed components, feed (dog), feed (fish), feed (mink), feed (pig), feed (poultry), feed (reindeer), feedstuff, forage grass, grain, mixed feed, grains (no specification), maize, maize gluten, oat and barley (hammer-milled), oats, peanuts, soybeans, wheat

Feeds, industrial   may contain the following mycotoxins:

3-ACETYLDEOXYNIVALENOL
incidence: 21/68*, conc. range: 5–87 µg/kg, Ø conc.: 35 µg/kg, country: Finland[9], *imported ?
see also barley, feed components, feed (poultry), feeds, grain, feedstuffs (rapeseed, turnip, fish meal, concentrates), maize, maize ears, maize gluten, maize kernels, maize meal, maize screenings, oats, wheat

DEOXYNIVALENOL
incidence: 66/68*, conc. range: 14–1216 µg/kg, Ø conc.: 148 µg/kg, country: Finland[9], *imported ?
see also barley, barley, husked, barley, unhusked (naked), barley (pressed), bone meal, bran, broilers feed, calf fattening mixed feed, coconut, expeller, corn cob mix silage, cottonseed, cottonseed cake, dairy cattle feed, egg production mixed feed, feed, feed

components, feed, commercial mix, feed, mixed, feed, mixed (primarily maize), feed (barley), feed (cattle), feed (chicken), feed (dog), feed (fish), feed (mill run, from wheat), feed (mink), feed (pig), feed (poultry), feed (reindeer), feeds, grain, feedstuff, feedstuffs (rapeseed, turnip, fish meal, concentrates), fish meal, grain, mixed feed, grains, mixed, grains (no specification), maize, maize ears, maize fibre, maize germ, maize germ/bran, maize germ meal, maize gluten, maize kernels, maize meal, maize powder, maize screenings, maize stalks (pith), maize, "Baby", maize, hybrid, maize, white, oats, rice bran, rice germ cake, rye, silage, sorghum, soybeans, triticale, wheat, wheat and barley, wheat, red hard winter, wheat, soft white winter, wheat, spring, wheat, winter

NIVALENOL
incidence: 3/68*, conc. range: 15–67 µg/kg, Ø conc.: 38 µg/kg, country: Finland[9], *imported ?
see also barley, barley, husked, barley, unhusked (naked), barley (pressed), bran, feed, feed components, feed (cattle), feed (poultry), feed (reindeer), maize, maize ears, maize germ, maize germ/bran, maize gluten, maize kernels, maize meal, maize powder, maize screenings, maize, "Baby", oats, rye, silage, triticale, wheat

Feedstuff   may contain the following mycotoxins:

AFLATOXINS
incidence: 5/2022, conc. range: ≤800 µg/kg, country: Canada[135]
see also animal feed (maize), animal feed (mixed), barley, copra meal, cottonseed, cottonseed fines, cottonseed meal, cottonseed meats, feed, feed ingredients (miscellaneous), feed, mixed, feed (excluding peanuts, suspect), feed (goat), feed (pig), feed (poultry), grain, grain, mixed feed, maize, maize, shelled, millet, peanut cake, peanut meal, peanut meal and by-products, peanuts, protein concentrates, rice, rice bran, sorghum, soybean meal, sunflower

Deoxynivalenol

incidence: 13/over 200, conc. range: 25–7400 μg/kg, Ø conc.: 1216.9 μg/kg, country: USA, South Africa, Zambia[260]
see also barley, barley, husked, barley, unhusked (naked), barley (pressed), bone meal, bran, broilers feed, calf fattening mixed feed, coconut, expeller, corn cob mix silage, cottonseed, cottonseed cake, dairy cattle feed, egg production mixed feed, feed, feed components, feed, commercial mix, feed, mixed, feed, mixed (primarily maize), feed (barley), feed (cattle), feed (chicken), feed (dog), feed (fish), feed (mill run, from wheat), feed (mink), feed (pig), feed (poultry), feed (reindeer), feeds, grain, feeds, industrial, feedstuffs (rapeseed, turnip, fish meal, concentrates), fish meal, grain, mixed feed, grains, mixed, grains (no specification), maize, maize ears, maize fibre, maize germ, maize germ/bran, maize germ meal, maize gluten, maize kernels, maize meal, maize powder, maize screenings, maize stalks (pith), maize, "Baby", maize, hybrid, maize, white, oats, rice bran, rice germ cake, rye, silage, sorghum, soybeans, triticale, wheat, wheat and barley, wheat, red hard winter, wheat, soft white winter, wheat, spring, wheat, winter

Diacetoxyscirpenol

incidence: 2/over 200, conc. range: 380–500 μg/kg, Ø conc.: 440 μg/kg, country: USA[260]
see also alfalfa, barley, feed, feed components, feed (dog), feed (fish), feed (mink), feed (pig), feed (poultry), feed (reindeer), feeds, grain, forage grass, grain, mixed feed, grains (no specification), maize, maize gluten, oat and barley (hammer-milled), oats, peanuts, soybeans, wheat

Ergot alkaloids

incidence: 1/2022, conc.: nc, country: Canada[135]
see also feed, grain, grain diets, tall fescue grass (*Festuga arundinacea* Screb)

Ochratoxins

incidence: 3/2022, conc. range: nc, country: Canada[135]

T-2 Toxin

incidence: 1/2022, conc.: nc, country: Canada[135]
incidence: 1/over 200, conc.: 76 μg/kg, country: USA[260]
see also alfalfa, barley, bran, diet (grower), diet (poultry), feed, feed components, feed, layer, feed, mixed, feed (dog), feed (fish), feed (mink), feed (pig), feed (poultry), forage grass, grain, mixed feed, grains (no specification), hay, maize, maize germ/bran, maize gluten, maize meal, maize screenings, maize stalks (pith), oat and barley (hammer-milled), oats, peanuts, pig grower diet, piglet diet, rye, sorghum, triticale, wheat

Zearalenone

incidence: 28/65, conc. range: 100–2,909,000 μg/kg, country: USA[88]
incidence: 266/2022, conc. range: tr–141,000 μg/kg, Ø conc.: 3850 μg/kg, country: Canada[135]
see also alfalfa, barley, barley, husked, barley, unhusked (naked), barley and feed, bone meal, bran, broilers feed, *Carthamus* cake, chick pea, concentrate, mixed, corn cob mix silage, cotton cake, cottonseed, cottonseed cake, diet (dairy cow), diet (poultry), diets (mixed), feed, feed components, feed, mixed, feed, mixed (primarily maize also maize, oats, wheat), feed (bran), feed (broiler chicken), feed (cattle), feed (chicken), feed (dairy), feed (developing pig), feed (mill run, from wheat), feed (miscellaneous), feed (pig), feed (poultry), feed (poultry, pig), feed (starter chicken), fish meal, forage grass, grain, bruised, grain, mixed feed, grains (no specification), hay, maize, maize ears, maize flakes, maize germ, maize germ/bran, maize gluten, maize kernels, maize meal, maize oil cake, maize screenings, maize stalks (pith), maize, "Baby", maize, hybrid, maize, shelled, maize, unshelled, maize, white, maize grain, artificially dried, maize grain, crib dried, maize grain, ensiled, milk production mixed

feed, oats, *Paspalum palidosum*, straw, peanut hulls/skins, rice bran, rice germ, rice germ cake, rye, silage, sorghum, soybeans, soybeans, extracted, sunflower cake, tapioca, triticale, wheat, wheat bran, wheat bran and chana testa, wheat soya meal

Feedstuff (barley)   may contain the following mycotoxins:

Fusaric Acid
incidence: 2/2*, conc. range: 11,200–13,250 μg/kg, Ø conc.: 12,230 μg/kg, country: Canada[66], *mainly
see also feed, feedstuff (dry corn), feedstuff (high-moisture corn), feedstuff (wheat), feedstuff (whole feeds)

Feedstuff (dry corn)   may contain the following mycotoxins:

Fusaric Acid
incidence: 14/16*, conc. range: 3010–28,770 μg/kg, Ø conc.: 11,750 μg/kg, country: Canada[66], *mainly
see also feed, feedstuff (barley), feedstuff (high-moisture corn), feedstuff (wheat), feedstuff (whole feeds)

Feedstuff (high-moisture corn)   may contain the following mycotoxins:

Fusaric Acid
incidence: 11/14*, conc. range: 5470–135,640 μg/kg, Ø conc.: 26,370 μg/kg, country: Canada[66], *mainly
see also feed, feedstuff (barley), feedstuff (dry corn), feedstuff (wheat), feedstuff (whole feeds)

Feedstuff (wheat)   may contain the following mycotoxins:

Fusaric Acid
incidence: 7/8*, conc. range: 1400–30,400 μg/kg, Ø conc.: 11,610 μg/kg, country: Canada[66], *mainly
see also feed, feedstuff (barley), feedstuff (dry corn), feedstuff (high-moisture corn), feedstuff (whole feeds)

Feedstuff (whole seeds)   may contain the following mycotoxins:

Fusaric Acid
incidence: 7/8*, conc. range: 9740–125,740 μg/kg, Ø conc.: 35,760 μg/kg, country: Canada[66], *mainly
see also feed, feedstuff (barley), feedstuff (dry corn), feedstuff (high-moisture corn), feedstuff (wheat)

Feedstuffs (rapeseed, turnip, fish meal, concentrates)   may contain the following mycotoxins:

3-Acetyldeoxynivalenol
incidence: 2/20*, conc. range: 5–20 μg/kg, Ø conc.: 13 μg/kg, country: Finland[9], *imported ?
see also barley, feed components, feed (poultry), feeds, grain, feeds, industrial, maize, maize ears, maize gluten, maize kernels, maize meal, maize screenings, oats, wheat

Deoxynivalenol
incidence: 13/20*, conc. range: 11–151 μg/kg, Ø conc.: 37 μg/kg, country: Finland[9], *imported ?
see also barley, barley, husked, barley, unhusked (naked), barley (pressed), bone meal, bran, broilers feed, calf fattening mixed feed, coconut, expeller, corn cob mix silage, cottonseed, cottonseed cake, dairy cattle feed, egg production mixed feed, feed, feed components, feed, commercial mix, feed, mixed, feed, mixed (primarily maize), feed (barley), feed (cattle), feed (chicken), feed (dog), feed (fish), feed (mill run, from wheat), feed (mink), feed (pig), feed (poultry), feed (reindeer), feeds, grain, feeds, industrial, feedstuff, fish meal, grain, mixed feed, grains, mixed, grains (no specification), maize, maize ears, maize fibre, maize germ, maize germ/bran, maize germ meal, maize gluten, maize kernels, maize meal, maize powder, maize screenings, maize stalks (pith), maize, "Baby", maize, hybrid, maize, white, oats, rice bran, rice germ cake, rye, silage,

sorghum, soybeans, triticale, wheat, wheat and barley, wheat, red hard winter, wheat, soft white winter, wheat, spring, wheat, winter

Fish meal   may contain the following mycotoxins:

AFLATOXIN B_1
incidence: 18/257, conc. range: ≤25 μg/kg (6 sa), 26–50 μg/kg (4 sa), 51–100 μg/kg (4 sa), 101–200 μg/kg (3 sa), 201–500 μg/kg (1 sa), Ø conc.: 63.1 μg/kg, country: India[247]
see also alfalfa, *Ambadi* cake, animal feedstuffs (dairy cake), bagasse, barley, bengalgram husk, bird food, bird food, wild, biri testa, blackgram, blackgram husk, bran, broiler mixed feed, calf fattening mixed feed, calf fattening mixed feed (containing 4–20 % peanut products), *Carthamus* cake, castor cake, cereals, cereal products, chick pea, coconut cake, cocos, concentrate, mixed, concentrates, cotton cake, cottonseed, cottonseed (dehulled), cottonseed cake, cottonseed extract, cottonseed meal, cottonseed meal (ammoniated), cottonseed meal (decorticated), cottonseed meats, cottonseed products, crumbles, crumbles, grower, cycad meal, dairy cattle feed, dairy cattle feed (containing 2–5 % peanut products), dairy cattle feed (containing 6–10 % peanut products), dairy cattle feed (containing 6–12 % peanut products), dairy cattle feed (containing more than 20 % peanut products), diets, mixed, dog food, egg production mixed feed, feed, feed and ingredients, feed, compound, feed, layer, feed, mixed, feed (beef), feed (broilers), feed (calf), feed (cat), feed (cattle), feed (chicken), feed (dairy), feed (dog), feed (dug), feed (fish), feed (gluten), feed (horse), feed (miscellaneous), feed (pig), feed (poultry), feed (poultry, pig), feed (rabbit), feed (sheep), feed (50–60 % maize), grain by-products, grains (no specification), greengram, hay/silage, horsegram, husk, *Jagni* cake, legume mixture, linseed, linseed cake, livol, *Mahua* cake, maize, maize germ, maize gluten, maize grits, maize husk, maize meal, maize oil cake, maize powder, maize screenings, maize, ground, maize, hybrid, maize, preharvest, maize, yellow, maize (dark grains), *Makhana* (*Euryale ferox* Salisb) puffs, manioc, milk production mixed feed, mung testa, murkool, mustard cake, neem cake, niger cake, oats, palm kernel expeller cake, palm kernels, palm products, peanut cake, peanut cake (deoiled), peanut expeller, peanut hay, peanut meal, peanut, kernels, peanut, shells, peanuts, pellets, finisher, pig meal and pellets, pigeon pea, poultry feeds (peanut containing), rapeseed cake, redgram husk, rice, rice bran, rice bran (deoiled), rice chaff, rice crack, rice germ, rice germ cake, rice meal, rice straw, rice (damaged), rice (polish), safflower cake, sal seed cake, sesame, sesame cake, sorghum, soybean meal, soybeans, sunflower, sunflower cake, sunflower flour, tapioca, wheat, wheat bran, wheat bran and chana testa

AFLATOXIN B_2
incidence: 2/257, conc. range: nc, Ø conc.: 28 μg/kg, country: India[247]
see also animal feedstuffs (dairy cake), bird food, bird food, wild, biri testa, blackgram, blackgram husk, cottonseed, cottonseed cake, cottonseed extract, cottonseed meal, cottonseed meal (ammoniated), cottonseed meats, dog food, egg production mixed feed, feed, compound, feed (cat), feed (cattle), feed (dog), feed (pig), feed (poultry), feed (rabbit), feed (sheep), horsegram, maize, maize gluten, maize husk, maize, ground, maize, preharvest, mung testa, mustard cake, niger cake, peanut cake, peanut cake (deoiled), peanut expeller, peanut hay, peanut meal, peanuts, peanut, kernels, redgram husk, rice bran, rice bran (deoiled), rice chaff, rice meal, rice (polish), sal seed cake, sesame cake, sorghum, soybean meal, soybeans, wheat, wheat bran

AFLATOXIN
incidence: 4/15, conc. range: 0–137 μg/kg, country: Nigeria[109]
incidence: 18/257, conc. range: 6–276 μg/kg, country: India[247]

incidence: 5/5, conc. range: <30 μg/kg (3 sa), >30 μg/kg (2 sa), country: India[380]
see also blackgram husk, bread crumbs, broiler finisher, broiler starter, cotton cake, cottonseed, cottonseed cake, cottonseed extract, cottonseed meal, feed, feed (cattle), feed (cow), feed (dog), feed (horse), feed (maize, gluten), feed (pig), feed (poultry), feed (rabbit), feed (rat/mice), feed (sheep), feeds, grain, flour (wheat), groundnut cake, grower's mash, horsegram, layer's mash, maize, maize gluten, maize, white, milo, peanut cake, peanut cake (deoiled), peanut, kernels, peanut (oil cake), pearlmillet, pig breeder's mash, pig finisher, pig starter, pod with haulms, poultry breeder's mash, rabbit pellets, redgram husk, rice, rice bran (deoiled), rice, broken, rice (polish), sesame cake, silk worm pupae, sorghum, soybean cake, soybean meal, wheat, wheat bran

CITRININ
incidence: 2/4, conc. range: 40–70 μg/kg, Ø conc.: 55 μg/kg, country: Egypt[16]
see also barley, barley, oats, barley-soybean diet, cottonseed cake, feed, feed, mixed, feed (cattle), feed (pig), hay, maize, maize, white, *Makhana* (*Euryale ferox* Salisb) puffs, oats, palm products, peas and beans, rice bran, rice germ, wheat, wheat and other grains (moldy), wheat bran

DEOXYNIVALENOL
incidence: 4/4, conc. range: 521–3986 μg/kg, Ø conc.: 2124 μg/kg, country: Egypt[16]
see also barley, barley, husked, barley, unhusked (naked), barley (pressed), bone meal, bran, broilers feed, calf fattening mixed feed, coconut, expeller, corn cob mix silage, cottonseed, cottonseed cake, dairy cattle feed, egg production mixed feed, feed, feed components, feed, commercial mix, feed, mixed, feed, mixed (primarily maize), feed (barley), feed (cattle), feed (chicken), feed (dog), feed (fish), feed (mill run, from wheat), feed (mink), feed (pig), feed (poultry), feed (reindeer), feeds, grain, feeds, industrial, feedstuff, feedstuffs (rapeseed, turnip, fish meal, concentrates), grain, mixed feed, grains, mixed, grains (no specification), maize, maize ears, maize fibre, maize germ, maize germ/bran, maize germ meal, maize gluten, maize kernels, maize meal, maize powder, maize screenings, maize stalks (pith), maize, "Baby", maize, hybrid, maize, white, oats, rice bran, rice germ cake, rye, silage, sorghum, soybeans, triticale, wheat, wheat and barley, wheat, red hard winter, wheat, soft white winter, wheat, spring, wheat, winter

ZEARALENONE
incidence: 4/4, conc. range: 77–152 μg/kg, Ø conc.: 107 μg/kg, country: Egypt[16]
see also alfalfa, barley, barley, husked, barley, unhusked (naked), barley and feed, bone meal, bran, broilers feed, *Carthamus* cake, chick pea, concentrate, mixed, corn cob mix silage, cotton cake, cottonseed, cottonseed cake, diet (dairy cow), diet (poultry), diets (mixed), feed, feed components, feed, mixed, feed, mixed (primarily maize also maize, oats, wheat), feed (bran), feed (broiler chicken), feed (cattle), feed (chicken), feed (dairy), feed (developing pig), feed (mill run, from wheat), feed (miscellaneous), feed (pig), feed (poultry), feed (poultry, pig), feed (starter chicken), feedstuff, forage grass, grain, bruised, grain, mixed feed, grains (no specification), hay, maize, maize ears, maize flakes, maize germ, maize germ/bran, maize gluten, maize kernels, maize meal, maize oil cake, maize screenings, maize stalks (pith), maize, "Baby", maize, hybrid, maize, shelled, maize, unshelled, maize, white, maize grain, artificially dried, maize grain, crib dried, maize grain, ensiled, milk production mixed feed, oats, *Paspalum palidosum*, straw, peanut hulls/skins, rice bran, rice germ, rice germ cake, rye, silage, sorghum, soybeans, soybeans, extracted, sunflower cake, tapioca, triticale, wheat, wheat bran, wheat bran and chana testa, wheat soya meal

Flour (wheat) for feed may contain the following mycotoxins:

AFLATOXIN B
incidence: 18/25, conc. range: ≥25–>100 μg/kg, country: France[46]

see also barley, cocoa (oil cake), cottonseed
cake, feed, feed (cattle), feed (maize, gluten),
feed (pig), feed (poultry), lucern (dried),
maize, maize gluten, maize grains, oats,
peanut (oil cake), rice, broken, sorghum,
soybean (oil cake), sunflower (oil cake),
wheat

AFLATOXIN
incidence: 1/1, conc. range: <30 µg/kg (1 sa),
country: India[380]
see also blackgram husk, bread crumbs,
broiler finisher, broiler starter, cotton cake,
cottonseed, cottonseed cake, cottonseed
extract, cottonseed meal, feed, feed (cattle),
feed (cow), feed (dog), feed (horse), feed
(maize, gluten), feed (pig), feed (poultry),
feed (rabbit), feed (rat/mice), feed (sheep),
feeds, grain, fish meal, groundnut cake,
grower's mash, horsegram, layer's mash,
maize, maize gluten, maize, white, milo,
peanut cake, peanut cake (deoiled), peanut,
kernels, peanut (oil cake), pearlmillet, pig
breeder's mash, pig finisher, pig starter, pod
with haulms, poultry breeder's mash, rabbit
pellets, redgram husk, rice, rice bran
(deoiled), rice, broken, rice (polish), sesame
cake, silk worm pupae, sorghum, soybean
cake, soybean meal, wheat, wheat bran

Forage grass   may contain the following
mycotoxins:

DIACETOXYSCIRPENOL
incidence: 180/832, conc. range: 3–60 µg/kg,
country: Germany[372]
see also alfalfa, barley, feed, feed components,
feeds, grain, feed (dog), feed (fish), feed
(mink), feed (pig), feed (poultry), feed
(reindeer), feedstuff, grain, mixed feed, grains
(no specification), maize, maize gluten, oat
and barley (hammer-milled), oats, peanuts,
soybeans, wheat

FUMONISIN B$_1$
incidence: 4/40, conc. range:
1000–9000 µg/kg, Ø conc.: 5000 µg/kg,
country: New Zealand[156]
see also barley, bird food, wild, dog food,
feed, feed, complete ration, feed, general,

feed, layer, feed, maize-based, feed, mixed,
feed, pelleted ration, feed, screenings, feed,
sweet, feed (broilers), feed (cat), feed
(chicken), feed (dog), feed (gluten), feed
(horse), feed (maize), feed (pig), feed
(poultry), feed (rat), feed (rodent), maize,
maize and maize screenings, maize bran,
maize ears, maize fine fractions, maize flakes,
maize germ, maize germ/bran, maize germ
meal, maize gluten, maize grits, maize
kernels, maize meal, maize powder, maize
screenings, maize, "Baby", maize, ground,
maize, preharvest, maize, sweet feed,
maize/oats mix, rat chow, silage, sorghum,
soybeans, wheat

FUMONISIN B$_1$ METHYL ESTER
incidence: 4/40, conc. range: 500–4000 µg/kg,
Ø conc.: 1750 µg/kg, country: New Zealand[156]

T-2 TOXIN
incidence: 208/832, conc. range:
40–2780 µg/kg, country: Germany[372]
see also alfalfa, barley, bran, diet (grower),
diet (poultry), feed, feed components, feed,
layer, feed, mixed, feed (dog), feed (fish),
feed (mink), feed (pig), feed (poultry),
feedstuff, grain, mixed feed, grains (no
specification), hay, maize, maize germ/bran,
maize gluten, maize meal, maize screenings,
maize stalks (pith), oat and barley
(hammer-milled), oats, peanuts, pig grower
diet, piglet diet, rye, sorghum, triticale, wheat

ZEARALENONE
incidence: 557/832, conc. range:
10–4750 µg/kg, country: Germany[372]
see also alfalfa, barley, barley, husked, barley,
unhusked (naked), barley and feed, bone
meal, bran, broilers feed, *Carthamus* cake,
chick pea, concentrate, mixed, corn cob mix
silage, cotton cake, cottonseed, cottonseed
cake, diet (dairy cow), diet (poultry), diets
(mixed), feed, feed components, feed, mixed,
feed, mixed (primarily maize also maize, oats,
wheat), feed (bran), feed (broiler chicken),
feed (cattle), feed (chicken), feed (dairy), feed
(developing pig), feed (mill run, from
wheat), feed (miscellaneous), feed (pig), feed
(poultry), feed (poultry, pig), feed (starter

chicken), feedstuff, fish meal, grain, bruised, grain, mixed feed, grains (no specification), hay, maize, maize ears, maize flakes, maize germ, maize germ/bran, maize gluten, maize kernels, maize meal, maize oil cake, maize screenings, maize stalks (pith), maize, "Baby", maize, hybrid, maize, shelled, maize, unshelled, maize, white, milk production mixed feed, oats, *Paspalum palidosum*, straw, peanut hulls/skins, rice bran, maize grain, artificially dried, maize grain, crib dried, maize grain, ensiled, rice germ, rice germ cake, rye, silage, sorghum, soybeans, soybeans, extracted, sunflower cake, tapioca, triticale, wheat, wheat bran, wheat bran and chana testa, wheat soya meal

Grain   Grain for feed may contain the following mycotoxins:

AFLATOXINS
incidence: 22/55, conc. range:
0.5–>400 µg/kg*, country: Sweden[333],
*formic acid-treated (700 g litre^{-1})
incidence: 20/52, conc. range:
0.5–>400 µg/kg*, country: Sweden[333],
*formic acid-treated (850 g litre^{-1})
incidence: 2/37, conc. range: 5–400 µg/kg*,
country: Sweden[333], *propionic acid-treated
see also animal feed (maize), animal feed (mixed), barley, copra meal, cottonseed, cottonseed fines, cottonseed meal, cottonseed meats, feed, feed ingredients (miscellaneous), feed, mixed, feed (excluding peanuts, suspect), feed (goat), feed (pig), feed (poultry), feedstuff, grain, mixed feed, maize, maize, shelled, millet, peanut cake, peanut meal, peanut meal and by-products, peanuts, protein concentrates, rice, rice bran, sorghum, soybean meal, sunflower

ERGOT ALKALOID
incidence: 8/8, conc. range:
<100–50,000 µg/kg*, country: Australia[331]
incidence: 4/4, conc. range:
4000–30,000 µg/kg*, country: Australia[331],
*>90 % dihydroergosine
see also feed, feedstuff, grain diets, tall fescue grass (*Festuga arundinacea* Screb)

Grain by-products   may contain the following mycotoxins:

AFLATOXIN B$_1$
incidence: 44/85, conc. range: 1–20 µg/kg (39 sa), 21–100 µg/kg (4 sa), 101–300 µg/kg (1 sa), country: USA[370]
see also alfalfa, *Ambadi* cake, animal feedstuffs (dairy cake), bagasse, barley, bengalgram husk, bird food, bird food, wild, biri testa, blackgram, blackgram husk, bran, broiler mixed feed, calf fattening mixed feed, calf fattening mixed feed (containing 4–20 % peanut products), *Carthamus* cake, castor cake, cereals, cereal products, chick pea, coconut cake, cocos, concentrate, mixed, concentrates, cotton cake, cottonseed, cottonseed (dehulled), cottonseed cake, cottonseed extract, cottonseed meal, cottonseed meal (ammoniated), cottonseed meal (decorticated), cottonseed meats, cottonseed products, crumbles, crumbles, grower, cycad meal, dairy cattle feed, dairy cattle feed (containing 2–5 % peanut products), dairy cattle feed (containing 6–10 % peanut products), dairy cattle feed (containing 6–12 % peanut products), dairy cattle feed (containing more than 20 % peanut products), diets, mixed, dog food, egg production mixed feed, feed, feed and ingredients, feed, compound, feed, layer, feed, mixed, feed (beef), feed (broilers), feed (calf), feed (cat), feed (cattle), feed (chicken), feed (dairy), feed (dog), feed (dug), feed (fish), feed (gluten), feed (horse), feed (miscellaneous), feed (pig), feed (poultry), feed (poultry, pig), feed (rabbit), feed (sheep), feed (50–60 % maize), fish meal, grains (no specification), greengram, hay/silage, horsegram, husk, *Jagni* cake, legume mixture, linseed, linseed cake, livol, *Mahua* cake, maize, maize germ, maize gluten, maize grits, maize husk, maize meal, maize oil cake, maize powder, maize screenings, maize, ground, maize, hybrid, maize, preharvest, maize, yellow, maize (dark grains), *Makhana* (*Euryale ferox* Salisb) puffs, manioc, milk production mixed feed, mung testa, murkool, mustard cake, neem cake,

niger cake, oats, palm kernel expeller cake, palm kernels, palm products, peanut cake, peanut cake (deoiled), peanut expeller, peanut hay, peanut meal, peanut, kernels, peanut, shells, peanuts, pellets, finisher, pig meal and pellets, pigeon pea, poultry feeds (peanut containing), rapeseed cake, redgram husk, rice, rice bran, rice bran (deoiled), rice chaff, rice crack, rice germ, rice germ cake, rice meal, rice straw, rice (damaged), rice (polish), safflower cake, sal seed cake, sesame, sesame cake, sorghum,soybean meal, soybeans, sunflower, sunflower cake, sunflower flour, tapioca, wheat, wheat bran, wheat bran and chana testa

Grain diets   may contain the following mycotoxins:

Ergot Alkaloid
incidence: 2/2, conc. range: 40,000 µg/kg*, Ø conc.: 40,000 µg/kg, country: Australia[331], *>90 % dihydroergosine
see also feed, feedstuff, grain, tall fescue grass (*Festuga arundinacea* Screb)

Grain, bruised   may contain the following mycotoxins:

Zearalenone
incidence: 1/5, conc.: 26.9 µg/kg, country: Germany[107]
see also alfalfa, barley, barley, husked, barley, unhusked (naked), barley and feed, bone meal, bran, broilers feed, *Carthamus* cake, chick pea, concentrate, mixed, corn cob mix silage, cotton cake, cottonseed, cottonseed cake, diet (dairy cow), diet (poultry), diets (mixed), feed, feed components, feed, mixed, feed, mixed (primarily maize also maize, oats, wheat), feed (bran), feed (broiler chicken), feed (cattle), feed (chicken), feed (dairy), feed (developing pig), feed (mill run, from wheat), feed (miscellaneous), feed (pig), feed (poultry), feed (poultry, pig), feed (starter chicken), feedstuff, fish meal, forage grass, grain, mixed feed, grains (no specification), hay, maize, maize ears, maize flakes, maize germ, maize germ/bran, maize gluten, maize

kernels, maize meal, maize oil cake, maize screenings, maize stalks (pith), maize, "Baby", maize, hybrid, maize, shelled, maize, unshelled, maize, white, milk production mixed feed, oats, *Paspalum palidosum*, straw, peanut hulls/skins, rice bran, maize grain, artificially dried, maize grain, crib dried, maize grain, ensiled, rice germ, rice germ cake, rye, silage, sorghum, soybeans, soybeans, extracted, sunflower cake, tapioca, triticale, wheat, wheat bran, wheat bran and chana testa, wheat soya meal

Grain, mixed feed   may contain the following mycotoxins:

Aflatoxins
incidence: 1/65, conc.: 2600 µg/kg*, country: Sweden[25], *surface of stored grain
see also animal feed (maize), animal feed (mixed), barley, copra meal, cottonseed, cottonseed fines, cottonseed meal, cottonseed meats, feed, feed ingredients (miscellaneous), feed, mixed, feed (excluding peanuts, suspect), feed (goat), feed (pig), feed (poultry), feedstuff, grain, maize, maize, shelled, millet, peanut cake, peanut meal, peanut meal and by-products, peanuts, protein concentrates, rice, rice bran, sorghum, soybean meal, sunflower

Deoxynivalenol
incidence: 2/9, conc. range: 20–139 µg/kg, Ø conc.: 79.5 µg/kg, country: Germany[68]
see also barley, barley, husked, barley, unhusked (naked), barley (pressed), bone meal, bran, broilers feed, calf fattening mixed feed, coconut, expeller, corn cob mix silage, cottonseed, cottonseed cake, dairy cattle feed, egg production mixed feed, feed, feed components, feed, commercial mix, feed, mixed, feed, mixed (primarily maize), feed (barley), feed (cattle), feed (chicken), feed (dog), feed (fish), feed (mill run, from wheat), feed (mink), feed (pig), feed (poultry), feed (reindeer), feeds, grain, feeds, industrial, feedstuff, feedstuffs (rapeseed, turnip, fish meal, concentrates), fish meal, grains, mixed, grains (no specification),

maize, maize ears, maize fibre, maize germ, maize germ/bran, maize germ meal, maize gluten, maize kernels, maize meal, maize powder, maize screenings, maize stalks (pith), maize, "Baby", maize, hybrid, maize, white, oats, rice bran, rice germ cake, rye, silage, sorghum, soybeans, triticale, wheat, wheat and barley, wheat, red hard winter, wheat, soft white winter, wheat, spring, wheat, winter

DIACETOXYSCIRPENOL
incidence: 2/38, conc. range: 300–19,000 μg/kg, Ø conc.: 9650 μg/kg, country: Germany[68]
see also alfalfa, barley, feed, feed components, feed (dog), feed (fish), feed (mink), feed (pig), feed (poultry), feed (reindeer), feeds, grain, feedstuff, forage grass, grains (no specification), maize, maize gluten, oat and barley (hammer-milled), oats, peanuts, soybeans, wheat

HT-2 TOXIN
incidence: 3/9, conc. range: 200–400 μg/kg, country: Germany[68]
see also barley, feed components, feed (dog), feed (fish), feed (pig), feed (poultry), feed (reindeer), grains, mixed, grains (no specification), maize, maize germ/bran, maize gluten, maize meal, maize screenings, oats, rye, silage, wheat

OCHRATOXIN A
incidence: 60/88, conc. range: 40–1690 μg/kg, country: Sweden[25]
see also alfalfa, barley, barley, oats, barley (high moisture), barley-soybean diet, bird food, domestic, bird food, wild, broilers feed, cereal grains, citrus pulp, coconut, expeller, corn cob mix silage, diet (dairy cow), diet (poultry), diet (starter), dog food, eat, egg production mixed feed, feed, feed wheat, oat and barley, feed, commercial mix, feed, mixed, feed, mixed (pelleted), feed (broilers), feed (cat), feed (cattle), feed (cereals), feed (pig), feed ec (pig), feed (poultry), feed ec (poultry), feed (poultry, pig), feed ec (rabbit), feed (trout), grains, mixed, grains (heated), horse bean, maize, maize feed,

milo, maize gluten, maize meal, maize, white, *Makhana* (*Euryale ferox* Salisb) puffs, milk production mixed feed, millet, oat and barley (hammer-milled), oats, palm products, peanut cake, peas, peas and beans, pet food, pig feedstuffs, pig grower diet, pig meal, piglet diet, poultry feedstuffs, rice bran, rice germ, rice germ cake, rye, sorghum, soybean groats, sunflower, sunflower seeds, extracted, tapioca, triticale, *Vicia faba*, wheat, wheat and barley, wheat bran, wheat hay, wheat, oats

T-2 TOXIN
incidence: 1/10, conc.: 20 μg/kg, country: Germany[68]
see also alfalfa, barley, bran, diet (grower), diet (poultry), feed, feed components, feed, layer, feed, mixed, feed (dog), feed (fish), feed (mink), feed (pig), feed (poultry), feedstuff, forage grass, grains (no specification), hay, maize, maize germ/bran, maize gluten, maize meal, maize screenings, maize stalks (pith), oat and barley (hammer-milled), oats, peanuts, pig grower diet, piglet diet, rye, sorghum, triticale, wheat

T-2 TRIOL
incidence: 1/9, conc.: 400 μg/kg, country: Germany[68]
see also barley, feed components, grains (no specification), maize, oats, wheat

ZEARALENONE
incidence: 2/68, conc. range: 100–1200 μg/kg, Ø conc.: 650 μg/kg, country: Sweden[25]
incidence: 17/41, conc. range: 10–500 μg/kg, country: Germany[68]
incidence: 1/10, conc.: 300 μg/kg, country: Germany[68]
see also alfalfa, barley, barley, husked, barley, unhusked (naked), barley and feed, bone meal, bran, broilers feed, *Carthamus* cake, chick pea, concentrate, mixed, corn cob mix silage, cotton cake, cottonseed, cottonseed cake, diet (dairy cow), diet (poultry), diets (mixed), feed, feed components, feed, mixed, feed, mixed (primarily maize also maize, oats, wheat), feed (bran), feed (broiler chicken), feed (cattle), feed (chicken), feed (dairy), feed (developing pig), feed (mill run, from

wheat), feed (miscellaneous), feed (pig), feed (poultry), feed (poultry, pig), feed (starter chicken), feedstuff, fish meal, forage grass, grain, bruised, grains (no specification), hay, maize, maize ears, maize flakes, maize germ, maize germ/bran, maize gluten, maize kernels, maize meal, maize oil cake, maize screenings, maize stalks (pith), maize, "Baby", maize, hybrid, maize, shelled, maize, unshelled, maize, white, maize grain, artificially dried, maize grain, crib dried, maize grain, ensiled, milk production mixed feed, oats, *Paspalum palidosum*, straw, peanut hulls/skins, rice bran, rice germ, rice germ cake, rye, silage, sorghum, soybeans, soybeans, extracted, sunflower cake, tapioca, triticale, wheat, wheat bran, wheat bran and chana testa, wheat soya meal

Grains (no specification)   Grains for feed may contain the following mycotoxins:

Aflatoxin B_1
incidence: 4/47, conc. range: 1–20 µg/kg (3 sa), 101–300 µg/kg (1 sa), country: USA[370]
see also alfalfa, *Ambadi* cake, animal feedstuffs (dairy cake), bagasse, barley, bengalgram husk, bird food, bird food, wild, biri testa, blackgram, blackgram husk, bran, broiler mixed feed, calf fattening mixed feed, calf fattening mixed feed (containing 4–20 % peanut products), *Carthamus* cake, castor cake, cereals, cereal products, chick pea, coconut cake, cocos, concentrate, mixed, concentrates, cotton cake, cottonseed, cottonseed (dehulled), cottonseed cake, cottonseed extract, cottonseed meal, cottonseed meal (ammoniated), cottonseed meal (decorticated), cottonseed meats, cottonseed products, crumbles, crumbles, grower, cycad meal, dairy cattle feed, dairy cattle feed (containing 2–5 % peanut products), dairy cattle feed (containing 6–10 % peanut products), dairy cattle feed (containing 6–12 % peanut products), dairy cattle feed (containing more than 20 % peanut products), diets, mixed, dog food, egg production mixed feed, feed, feed and ingredients, feed, compound, feed, layer,

feed, mixed, feed (beef), feed (broilers), feed (calf), feed (cat), feed (cattle), feed (chicken), feed (dairy), feed (dog), feed (dug), feed (fish), feed (gluten), feed (horse), feed (miscellaneous), feed (pig), feed (poultry), feed (poultry, pig), feed (rabbit), feed (sheep), feed (50–60 % maize), fish meal, grain by-products, greengram, hay/silage, horsegram, husk, *Jagni* cake, legume mixture, linseed, linseed cake, livol, *Mahua* cake, maize, maize germ, maize gluten, maize grits, maize husk, maize meal, maize oil cake, maize powder, maize screenings, maize, ground, maize, hybrid, maize, preharvest, maize, yellow, maize (dark grains), *Makhana* (*Euryale ferox* Salisb) puffs, manioc, milk production mixed feed, mung testa, murkool, mustard cake, neem cake, niger cake, oats, palm kernel expeller cake, palm kernels, palm products, peanut cake, peanut cake (deoiled), peanut expeller, peanut hay, peanut meal, peanut, kernels, peanut, shells, peanuts, pellets, finisher, pig meal and pellets, pigeon pea, poultry feeds (peanut containing), rapeseed cake, redgram husk, rice, rice bran, rice bran (deoiled), rice chaff, rice crack, rice germ, rice germ cake, rice meal, rice straw, rice (damaged), rice (polish), safflower cake, sal seed cake, sesame, sesame cake, sorghum, soybean meal, soybeans, sunflower, sunflower cake, sunflower flour, tapioca, wheat, wheat bran, wheat bran and chana testa

Deoxynivalenol
incidence: 4/102, conc. range: 1000–2000 µg/kg, country: Germany[68]
incidence: 31/297, conc. range: 10–500 µg/kg, country: Germany[68]
see also barley, barley, husked, barley, unhusked (naked), barley (pressed), bone meal, bran, broilers feed, calf fattening mixed feed, coconut, expeller, corn cob mix silage, cottonseed, cottonseed cake, dairy cattle feed, egg production mixed feed, feed, feed components, feed, commercial mix, feed, mixed, feed, mixed (primarily maize), feed (barley), feed (cattle), feed (chicken), feed (dog), feed (fish), feed (mill run, from wheat), feed (mink), feed (pig), feed

(poultry), feed (reindeer), feeds, grain, feeds, industrial, feedstuff, feedstuffs (rapeseed, turnip, fish meal, concentrates), fish meal, grain, mixed feed, grains, mixed, maize, maize ears, maize fibre, maize germ, maize germ/bran, maize germ meal, maize gluten, maize kernels, maize meal, maize powder, maize screenings, maize stalks (pith), maize, "Baby", maize, hybrid, maize, white, oats, rice bran, rice germ cake, rye, silage, sorghum, soybeans, triticale, wheat, wheat and barley, wheat, red hard winter, wheat, soft white winter, wheat, spring, wheat, winter

DIACETOXYSCIRPENOL
incidence: 14/188, conc. range:
300–19,000 µg/kg, country: Germany[68]
incidence: 11/288, conc. range:
200–700 µg/kg, country: Germany[68]
incidence: 2/55, conc.: 110 µg/kg, country: Canada[432]
see also alfalfa, barley, feed, feed components, feed (dog), feed (fish), feed (mink), feed (pig), feed (poultry), feed (reindeer), feeds, grain, feedstuff, forage grass, grain, mixed feed, maize, maize gluten, oat and barley (hammer-milled), oats, peanuts, soybeans, wheat

HT-2 TOXIN
incidence: 2/89, conc. range:
600–10,000 µg/kg, Ø conc.: 5300 µg/kg, country: Germany[68]
incidence: 59/273, conc. range:
100–700 µg/kg, country: Germany[68]
incidence: 2/55, conc. range: 120–440 µg/kg, country: Canada[432]
see also barley, feed components, feed (dog), feed (fish), feed (pig), feed (poultry), feed (reindeer), grain, mixed feed, grains, mixed, maize, maize germ/bran, maize gluten, maize meal, maize screenings, oats, rye, silage, wheat

T-2 TOXIN
incidence: 4/198, conc. range:
300–14,000 µg/kg, country: Germany[68]
incidence: 37/298, conc. range:
200–700 µg/kg, country: Germany[68]
incidence: 5/55, conc. range: 160–310 µg/kg, country: Canada[432]

see also alfalfa, barley, bran, diet (grower), diet (poultry), feed, feed components, feed, layer, feed, mixed, feed (dog), feed (fish), feed (mink), feed (pig), feed (poultry), feedstuff, forage grass, grain, mixed feed, hay, maize, maize germ/bran, maize gluten, maize meal, maize screenings, maize stalks (pith), oat and barley (hammer-milled), oats, peanuts, pig grower diet, piglet diet, rye, sorghum, triticale, wheat

T-2 TRIOL
incidence: 2/88, conc. range: 300 µg/kg, country: Germany[68]
incidence: 7/281, conc. range: 100–700 µg/kg, country: Germany[68]
see also barley, feed components, grain, mixed feed, maize, oats, wheat

ZEARALENOL
incidence: 1/248, conc.: 150 µg/kg, country: Germany[68]
see also maize, shelled, maize, unshelled, oats

ZEARALENONE
incidence: 56/227, conc. range: 10–700 µg/kg, country: Germany[68]
incidence: 21/298, conc. range:
10–2000 µg/kg, country: Germany[68]
see also alfalfa, barley, barley, husked, barley, unhusked (naked), barley and feed, bone meal, bran, broilers feed, *Carthamus* cake, chick pea, concentrate, mixed, corn cob mix silage, cotton cake, cottonseed, cottonseed cake, diet (dairy cow), diet (poultry), diets (mixed), feed, feed components, feed, mixed, feed, mixed (primarily maize also maize, oats, wheat), feed (bran), feed (broiler chicken), feed (cattle), feed (chicken), feed (dairy), feed (developing pig), feed (mill run, from wheat), feed (miscellaneous), feed (pig), feed (poultry), feed (poultry, pig), feed (starter chicken), feedstuff, fish meal, forage grass, grain, bruised, grain, mixed feed, hay, maize, maize ears, maize flakes, maize germ, maize germ/bran, maize gluten, maize kernels, maize meal, maize oil cake, maize screenings, maize stalks (pith), maize, "Baby", maize, hybrid, maize, shelled, maize, unshelled, maize, white, maize grain, artificially dried, maize grain, crib dried, maize

grain, ensiled, milk production mixed feed, oats, *Paspalum palidosum*, straw, peanut hulls/skins, rice bran, rice germ, rice germ cake, rye, silage, sorghum, soybeans, soybeans, extracted, sunflower cake, tapioca, triticale, wheat, wheat bran, wheat bran and chana testa, wheat soya meal

Grains, mixed   for feed may contain the following mycotoxins:

DEOXYNIVALENOL
incidence: 4/11, conc. range: 90–200 µg/kg, Ø conc.: 145 µg/kg, country: Sweden[197]
see also barley, barley, husked, barley, unhusked (naked), barley (pressed), bone meal, bran, broilers feed, calf fattening mixed feed, coconut, expeller, corn cob mix silage, cottonseed, cottonseed cake, dairy cattle feed, egg production mixed feed, feed, feed components, feed, commercial mix, feed, mixed, feed, mixed (primarily maize), feed (barley), feed (cattle), feed (chicken), feed (dog), feed (fish), feed (mill run, from wheat), feed (mink), feed (pig), feed (poultry), feed (reindeer), feeds, grain, feeds, industrial, feedstuff, feedstuffs (rapeseed, turnip, fish meal, concentrates), fish meal, grain, mixed feed, grains (no specification), maize, maize ears, maize fibre, maize germ, maize germ/bran, maize germ meal, maize gluten, maize kernels, maize meal, maize powder, maize screenings, maize stalks (pith), maize, "Baby", maize, hybrid, maize, white, oats, rice bran, rice germ cake, rye, silage, sorghum, soybeans, triticale, wheat, wheat and barley, wheat, red hard winter, wheat, soft white winter, wheat, spring, wheat, winter

HT-2 TOXIN
incidence: 1/94, conc.: 66 µg/kg, country: Canada[240]
see also barley, feed components, feed (dog), feed (fish), feed (pig), feed (poultry), feed (reindeer), grain, mixed feed, grains (no specification), maize, maize germ/bran, maize gluten, maize meal, maize screenings, oats, rye, silage, wheat

OCHRATOXIN A
incidence: 1/94, conc.: 1600 µg/kg, country: Canada[240]
see also alfalfa, barley, barley, oats, barley (high moisture), barley-soybean diet, bird food, domestic, bird food, wild, broilers feed, cereal grains, citrus pulp, coconut, expeller, corn cob mix silage, diet (dairy cow), diet (poultry), diet (starter), dog food, eat, egg production mixed feed, feed, feed wheat, oat and barley, feed, commercial mix, feed, mixed, feed, mixed (pelleted), feed (broilers), feed (cat), feed (cattle), feed (cereals), feed (pig), feed ec (pig), feed (poultry), feed ec (poultry), feed (poultry, pig), feed ec (rabbit), feed (trout), grain, mixed feed, grains (heated), hay, horse bean, maize, maize feed, milo, maize gluten, maize meal, maize, white, *Makhana* (*Euryale ferox* Salisb) puffs, milk production mixed feed, millet, oat and barley (hammer-milled), oats, palm products, peanut cake, peas, peas and beans, pet food, pig feedstuffs, pig grower diet, pig meal, piglet diet, poultry feedstuffs, rice bran, rice germ, rice germ cake, rye, sorghum, soybean groats, sunflower, sunflower seeds, extracted, tapioca, triticale, *Vicia faba*, wheat, wheat and barley, wheat bran, wheat hay, wheat, oats

Grains (heated)   Grains for feed may contain the following mycotoxins:

OCHRATOXIN A
incidence: 4/4, conc. range: 20–100 µg/kg, country: Canada[172]
see also alfalfa, barley, barley, oats, barley (high moisture), barley-soybean diet, bird food, domestic, bird food, wild, broilers feed, cereal grains, citrus pulp, coconut, expeller, corn cob mix silage, diet (dairy cow), diet (poultry), diet (starter), dog food, eat, egg production mixed feed, feed, feed wheat, oat and barley, feed, commercial mix, feed, mixed, feed, mixed (pelleted), feed (broilers), feed (cat), feed (cattle), feed (cereals), feed (pig), feed ec (pig), feed (poultry), feed ec (poultry), feed (poultry, pig), feed ec (rabbit), feed (trout), grain, mixed feed, grains, mixed, hay, horse bean, maize, maize

feed, milo, maize gluten, maize meal, maize, white, *Makhana* (*Euryale ferox* Salisb) puffs, milk production mixed feed, millet, oat and barley (hammer-milled), oats, palm products, peanut cake, peas, peas and beans, pet food, pig feedstuffs, pig grower diet, pig meal, piglet diet, poultry feedstuffs, rice bran, rice germ, rice germ cake, rye, sorghum, soybean groats, sunflower, sunflower seeds, extracted, tapioca, triticale, *Vicia faba*, wheat, wheat and barley, wheat bran, wheat hay, wheat, oats

Greengram   Greengram for feed may contain the following mycotoxins:

AFLATOXIN B$_1$
incidence: 6/10, conc. range: tr–20 µg/kg, country: India[183]
see also alfalfa, *Ambadi* cake, animal feedstuffs (dairy cake), barley, bengalgram husk, bird food, bird food, wild, biri testa, blackgram, blackgram husk, bran, broiler mixed feed, calf fattening mixed feed, calf fattening mixed feed (containing 4–20 % peanut products), *Carthamus* cake, castor cake, cereals, cereal products, chick pea, coconut cake, cocos, concentrate, mixed, concentrates, cotton cake, cottonseed, cottonseed (dehulled), cottonseed cake, cottonseed extract, cottonseed meal, cottonseed meal (ammoniated), cottonseed meal (decorticated), cottonseed meats, cottonseed products, crumbles, crumbles, grower, cycad meal, dairy cattle feed, dairy cattle feed (containing 2–5 % peanut products), dairy cattle feed (containing 6–10 % peanut products), dairy cattle feed (containing 6–12 % peanut products), dairy cattle feed (containing more than 20 % peanut products), diets, mixed, dog food, egg production mixed feed, feed, feed and ingredients, feed, compound, feed, layer, feed, mixed, feed (beef), feed (broilers), feed (calf), feed (cat), feed (cattle), feed (chicken), feed (dairy), feed (dog), feed (dug), feed (fish), feed (gluten), feed (horse), feed (miscellaneous), feed (pig), feed (poultry), feed (poultry, pig), feed (rabbit), feed (sheep), feed (50–60 % maize), fish meal,

grain by-products, grains (no specification), hay/silage, horsegram, husk, *Jagni* cake, legume mixture, linseed, linseed cake, livol, *Mahua* cake, maize, maize germ, maize gluten, maize grits, maize husk, maize meal, maize oil cake, maize powder, maize screenings, maize, ground, maize, hybrid, maize, preharvest, maize, yellow, maize (dark grains), *Makhana* (*Euryale ferox* Salisb) puffs, manioc, milk production mixed feed, mung testa, murkool, mustard cake, neem cake, niger cake, oats, palm kernel expeller cake, palm kernels, palm products, peanut cake, peanut cake (deoiled), peanut expeller, peanut hay, peanut meal, peanut, kernels, peanut, shells, peanuts, pellets, finisher, pig meal and pellets, pigeon pea, poultry feeds (peanut containing), rapeseed cake, redgram husk, rice, rice bran, rice bran (deoiled), rice chaff, rice crack, rice germ, rice germ cake, rice meal, rice straw, rice (damaged), rice (polish), safflower cake, sal seed cake, sesame, sesame cake, sorghum, soybean meal, soybeans, sunflower, sunflower cake, sunflower flour, tapioca, wheat, wheat bran, wheat bran and chana testa

GROUNDNUT OIL CAKE
see Peanut cake and Peanut (oil cake)

Groundnut cake   may contain the following mycotoxins:

AFLATOXIN
incidence: 20/23, conc. range: 0–1862 µg/kg, country: Nigeria[109]
incidence: 17/17, conc. range: 11–30 µg/kg (2 sa), 31–100 µg/kg (2 sa), >100 µg/kg (≤1007 µg/kg) (13 sa), country: India[311]
incidence: 55/56, conc. range: 10–30 µg/kg (10 sa), 101–500 µg/kg (43 sa), >500 µg/kg (2 sa), country: India[311]
see also blackgram husk, bread crumbs, broiler finisher, broiler starter, cotton cake, cottonseed, cottonseed cake, cottonseed extract, cottonseed meal, feed, feed (cattle), feed (cow), feed (dog), feed (horse), feed (maize, gluten), feed (pig), feed (poultry), feed (rabbit), feed (rat/mice), feed (sheep),

feeds, grain, fish meal, flour (wheat), grower's mash, horsegram, layer's mash, maize, maize gluten, maize, white, milo, peanut cake, peanut cake (deoiled), peanut, kernels, peanut (oil cake), pearlmillet, pig breeder's mash, pig finisher, pig starter, pod with haulms, poultry breeder's mash, rabbit pellets, redgram husk, rice, rice bran (deoiled), rice, broken, rice (polish), sesame cake, silk worm pupae, sorghum, soybean cake, soybean meal, wheat, wheat bran

CYCLOPIAZONIC ACID
incidence: 2/8, conc. range: 5000–10,000 μg/kg (estimated), Ø conc.: 7500 μg/kg, country: India[33]
see also chick mash, feed, maize, millet, little, peanuts, rice bran, wheat

GROUNDNUT KERNELS
see Peanut, kernels

Grower's mash   may contain the following mycotoxins:

AFLATOXIN
incidence: 3/8, conc. range: 0–46 μg/kg, country: Nigeria[109]
see also blackgram husk, bread crumbs, broiler finisher, broiler starter, cotton cake, cottonseed, cottonseed cake, cottonseed extract, cottonseed meal, feed, feed (cattle), feed (cow), feed (dog), feed (horse), feed (maize, gluten), feed (pig), feed (poultry), feed (rabbit), feed (rat/mice), feed (sheep), feeds, grain, fish meal, flour (wheat), groundnut cake, horsegram, layer's mash, maize, maize gluten, maize, white, milo, peanut cake, peanut cake (deoiled), peanut, kernels, peanut (oil cake), pearlmillet, pig breeder's mash, pig finisher, pig starter, pod with haulms, poultry breeder's mash, rabbit pellets, redgram husk, rice, rice bran (deoiled), rice, broken, rice (polish), sesame cake, silk worm pupae, sorghum, soybean cake, soybean meal, wheat, wheat bran

Hay   may contain the following mycotoxins:

CITRININ
incidence: 1/7345, conc.: nc, country: Hungary[209]
see also barley, barley, oats, barley-soybean diet, cottonseed cake, feed, mixed, feed (cattle), feed (pig), fish meal, maize, maize, white, *Makhana* (*Euryale ferox* Salisb) puffs, oats, palm products, peas and beans, rice bran, rice germ, wheat, wheat and other grains (moldy), wheat bran

OCHRATOXIN A
incidence: 1/7345, conc.: nc, country: Hungary[209]
see also alfalfa, barley, barley, oats, barley (high moisture), barley-soybean diet, bird food, domestic, bird food, wild, broilers feed, cereal grains, citrus pulp, coconut, expeller, corn cob mix silage, diet (dairy cow), diet (poultry), diet (starter), dog food, eat, egg production mixed feed, feed, feed wheat, oat and barley, feed, commercial mix, feed, mixed, feed, mixed (pelleted), feed (broilers), feed (cat), feed (cattle), feed (cereals), feed (pig), feed ec (pig), feed (poultry), feed ec (poultry), feed (poultry, pig), feed ec (rabbit), feed (trout), grain, mixed feed, grains, mixed, grains (heated), horse bean, maize, maize feed, milo, maize gluten, maize meal, maize, white, *Makhana* (*Euryale ferox* Salisb) puffs, milk production mixed feed, millet, oat and barley (hammer-milled), oats, palm products, peanut cake, peas, peas and beans, pet food, pig feedstuffs, pig grower diet, pig meal, piglet diet, poultry feedstuffs, rice bran, rice germ, rice germ cake, rye, sorghum, soybean groats, sunflower, sunflower seeds, extracted, tapioca, triticale, *Vicia faba*, wheat, wheat and barley, wheat bran, wheat hay, wheat, oats

PATULIN
incidence: 1/7345, conc.: nc, country: Hungary[209]
see also diet (grower), maize meal, piglet diet, wheat

T-2 TOXIN
incidence: 2/7345, conc. range: nc, country: Hungary[209]

see also alfalfa, barley, bran, diet (grower), diet (poultry), feed, feed components, feed, layer, feed, mixed, feed (dog), feed (fish), feed (mink), feed (pig), feed (poultry), feedstuff, forage grass, grain, mixed feed, grains (no specification), maize, maize germ/bran, maize gluten, maize meal, maize screenings, maize stalks (pith), oat and barley (hammer-milled), oats, peanuts, pig grower diet, piglet diet, rye, sorghum, triticale, wheat

Zearalenone
incidence: 1/1, conc.: 14,000 μg/kg, country: USA[177]
incidence: 3/7345, conc. range: nc, country: Hungary[209]
see also alfalfa, barley, barley, husked, barley, unhusked (naked), barley and feed, bone meal, bran, broilers feed, *Carthamus* cake, chick pea, concentrate, mixed, corn cob mix silage, cotton cake, cottonseed, cottonseed cake, diet (dairy cow), diet (poultry), diets (mixed), feed, feed components, feed, mixed, feed, mixed (primarily maize also maize, oats, wheat), feed (bran), feed (broiler chicken), feed (cattle), feed (chicken), feed (dairy), feed (developing pig), feed (mill run, from wheat), feed (miscellaneous), feed (pig), feed (poultry), feed (poultry, pig), feed (starter chicken), feedstuff, fish meal, forage grass, grain, bruised, grain, mixed feed, grains (no specification), maize, maize ears, maize flakes, maize germ, maize germ/bran, maize gluten, maize kernels, maize meal, maize oil cake, maize screenings, maize stalks (pith), maize, "Baby", maize, hybrid, maize, shelled, maize, unshelled, maize, white, maize grain, artificially dried, maize grain, crib dried, maize grain, ensiled, milk production mixed feed, oats, *Paspalum palidosum*, straw, peanut hulls/skins, rice bran, rice germ, rice germ cake, rye, silage, sorghum, soybeans, soybeans, extracted, sunflower cake, tapioca, triticale, wheat, wheat bran, wheat bran and chana testa, wheat soya meal

Hay/silage   may contain the following mycotoxins:

Aflatoxin B$_1$
incidence: 3/47, conc. range: 1–20 μg/kg (3 sa), country: USA[370]
see also alfalfa, *Ambadi* cake, animal feedstuffs (dairy cake), bagasse barley, bengalgram husk, bird food, bird food, wild, biri testa, blackgram, blackgram husk, bran, broiler mixed feed, calf fattening mixed feed, calf fattening mixed feed (containing 4–20 % peanut products), *Carthamus* cake, castor cake, cereals, cereal products, chick pea, coconut cake, cocos, concentrate, mixed, concentrates, cotton cake, cottonseed, cottonseed (dehulled), cottonseed cake, cottonseed extract, cottonseed meal, cottonseed meal (ammoniated), cottonseed meal (decorticated), cottonseed meats, cottonseed products, crumbles, crumbles, grower, cycad meal, dairy cattle feed, dairy cattle feed (containing 2–5 % peanut products), dairy cattle feed (containing 6–10 % peanut products), dairy cattle feed (containing 6–12 % peanut products), dairy cattle feed (containing more than 20 % peanut products), diets, mixed, dog food, egg production mixed feed, feed, feed and ingredients, feed, compound, feed, layer, feed, mixed, feed (beef), feed (broilers), feed (calf), feed (cat), feed (cattle), feed (chicken), feed (dairy), feed (dog), feed (dug), feed (fish), feed (gluten), feed (horse), feed (miscellaneous), feed (pig), feed (poultry), feed (poultry, pig), feed (rabbit), feed (sheep), feed (50–60 % maize), fish meal, grain by-products, grains (no specification), greengram, horsegram, husk, *Jagni* cake, legume mixture, linseed, linseed cake, livol, *Mahua* cake, maize, maize germ, maize gluten, maize grits, maize husk, maize meal, maize oil cake, maize powder, maize screenings, maize, ground, maize, hybrid, maize, preharvest, maize, yellow, maize (dark grains), *Makhana* (*Euryale ferox* Salisb) puffs, manioc, milk production mixed feed, mung testa, murkool, mustard cake, neem cake, niger cake, oats, palm kernel expeller cake, palm kernels, palm products, peanut cake, peanut cake (deoiled), peanut expeller,

peanut hay, peanut meal, peanut, kernels, peanut, shells, peanuts, pellets, finisher, pig meal and pellets, pigeon pea, poultry feeds (peanut containing), rapeseed cake, redgram husk, rice, rice bran, rice bran (deoiled), rice chaff, rice crack, rice germ, rice germ cake, rice meal, rice straw, rice (damaged), rice (polish), safflower cake, sal seed cake, sesame, sesame cake, sorghum, soybean meal, soybeans, sunflower, sunflower cake, sunflower flour, tapioca, wheat, wheat bran, wheat bran and chana testa

Horse bean   Horse bean for feed may contain the following mycotoxins:

OCHRATOXIN A
incidence: 1/3, conc.: 12 µg/kg, country: Egypt[16]
see also alfalfa, barley, barley, oats, barley (high moisture), barley-soybean diet, bird food, domestic, bird food, wild, broilers feed, cereal grains, citrus pulp, coconut, expeller, corn cob mix silage, diet (dairy cow), diet (poultry), diet (starter), dog food, eat, egg production mixed feed, feed, feed wheat, oat and barley, feed, commercial mix, feed, mixed, feed, mixed (pelleted), feed (broilers), feed (cat), feed (cattle), feed (cereals), feed (pig), feed ec (pig), feed (poultry), feed ec (poultry), feed (poultry, pig), feed ec (rabbit), feed (trout), grain, mixed feed, grains, mixed, grains (heated), hay, maize, maize feed, milo, maize gluten, maize meal, maize, white, *Makhana* (*Euryale ferox* Salisb) puffs, milk production mixed feed, millet, oat and barley (hammer-milled), oats, palm products, peanut cake, peas, peas and beans, pet food, pig feedstuffs, pig grower diet, pig meal, piglet diet, poultry feedstuffs, rice bran, rice germ, rice germ cake, rye, sorghum, soybean groats, sunflower, sunflower seeds, extracted, tapioca, triticale, *Vicia faba*, wheat, wheat and barley, wheat bran, wheat hay, wheat, oats

Horsegram   Horsegram for feed may contain the following mycotoxins:

AFLATOXIN B$_1$
incidence: 16/35, conc. range: ≤25 µg/kg (8 sa), 26–50 µg/kg (2 sa), 51–100 µg/kg (5 sa), 101–200 µg/kg (1 sa), Ø conc.: 36.6 µg/kg, country: India[247]
see also alfalfa, *Ambadi* cake, animal feedstuffs (dairy cake), bagasse, barley, bengalgram husk, bird food, bird food, wild, biri testa, blackgram, blackgram husk, bran, broiler mixed feed, calf fattening mixed feed, calf fattening mixed feed (containing 4–20 % peanut products), *Carthamus* cake, castor cake, cereals, cereal products, chick pea, coconut cake, cocos, concentrate, mixed, concentrates, cotton cake, cottonseed, cottonseed (dehulled), cottonseed cake, cottonseed extract, cottonseed meal, cottonseed meal (ammoniated), cottonseed meal (decorticated), cottonseed meats, cottonseed products, crumbles, crumbles, grower, cycad meal, dairy cattle feed, dairy cattle feed (containing 2–5 % peanut products), dairy cattle feed (containing 6–10 % peanut products), dairy cattle feed (containing 6–12 % peanut products), dairy cattle feed (containing more than 20 % peanut products), diets, mixed, dog food, egg production mixed feed, feed, feed and ingredients, feed, compound, feed, layer, feed, mixed, feed (beef), feed (broilers), feed (calf), feed (cat), feed (cattle), feed (chicken), feed (dairy), feed (dog), feed (dug), feed (fish), feed (gluten), feed (horse), feed (miscellaneous), feed (pig), feed (poultry), feed (poultry, pig), feed (rabbit), feed (sheep), feed (50–60 % maize), fish meal, grain by-products, grains (no specification), greengram, hay/silage, husk, *Jagni* cake, legume mixture, linseed, linseed cake, livol, *Mahua* cake, maize, maize germ, maize gluten, maize grits, maize husk, maize meal, maize oil cake, maize powder, maize screenings, maize, ground, maize, hybrid, maize, preharvest, maize, yellow, maize (dark grains), *Makhana* (*Euryale ferox* Salisb) puffs, manioc, milk production mixed feed, mung testa, murkool, mustard cake, neem cake, niger cake, oats, palm kernel expeller cake,

palm kernels, palm products, peanut cake, peanut cake (deoiled), peanut expeller, peanut hay, peanut meal, peanut, kernels, peanut, shells, peanuts, pellets, finisher, pig meal and pellets, pigeon pea, poultry feeds (peanut containing), rapeseed cake, redgram husk, rice, rice bran, rice bran (deoiled), rice chaff, rice crack, rice germ, rice germ cake, rice meal, rice straw, rice (damaged), rice (polish), safflower cake, sal seed cake, sesame, sesame cake, sorghum, soybean meal, soybeans, sunflower, sunflower cake, sunflower flour, tapioca, wheat, wheat bran, wheat bran and chana testa

AFLATOXIN B$_2$
incidence: 1/35, conc.: 9 µg/kg, country: India[247]
see also animal feedstuffs (dairy cake), bird food, bird food, wild, biri testa, blackgram, blackgram husk, cottonseed, cottonseed cake, cottonseed extract, cottonseed meal, cottonseed meal (ammoniated), cottonseed meats, dog food, egg production mixed feed, feed, compound, feed (cat), feed (cattle), feed (dog), feed (pig), feed (poultry), feed (rabbit), feed (sheep), fish meal, maize, maize gluten, maize husk, maize, ground, maize, preharvest, mung testa, mustard cake, niger cake, peanut cake, peanut cake (deoiled), peanut expeller, peanut hay, peanut meal, peanuts, peanut, kernels, redgram husk, rice bran, rice bran (deoiled), rice chaff, rice meal, rice (polish), sal seed cake, sesame cake, sorghum, soybean meal, soybeans, wheat, wheat bran

AFLATOXIN
incidence: 16/35, conc. range: nd–107 µg/kg, country: India[247]
see also blackgram husk, bread crumbs, broiler finisher, broiler starter, cotton cake, cottonseed, cottonseed cake, cottonseed extract, cottonseed meal, feed, feed (cattle), feed (cow), feed (dog), feed (horse), feed (maize, gluten), feed (pig), feed (poultry), feed (rabbit), feed (rat/mice), feed (sheep), feeds, grain, fish meal, flour (wheat), groundnut cake, grower's mash, layer's mash,

maize, maize gluten, maize, white, milo, peanut cake, peanut cake (deoiled), peanut, kernels, peanut (oil cake), pearlmillet, pig breeder's mash, pig finisher, pig starter, pod with haulms, poultry breeder's mash, rabbit pellets, redgram husk, rice, rice bran (deoiled), rice, broken, rice (polish), sesame cake, silk worm pupae, sorghum, soybean cake, soybean meal, wheat, wheat bran

HORSE FEED
see Feed (horse)

Husk   may contain the following mycotoxins:

AFLATOXIN B$_1$
incidence: 2/10, conc. range: 10–20 µg/kg, Ø conc.: 15 µg/kg, country: India[183]
see also alfalfa, *Ambadi* cake, animal feedstuffs (dairy cake), bagasse, barley, bengalgram husk, bird food, bird food, wild, biri testa, blackgram, blackgram husk, bran, broiler mixed feed, calf fattening mixed feed, calf fattening mixed feed (containing 4–20 % peanut products), *Carthamus* cake, castor cake, cereals, cereal products, chick pea, coconut cake, cocos, concentrate, mixed, concentrates, cotton cake, cottonseed, cottonseed (dehulled), cottonseed cake, cottonseed extract, cottonseed meal, cottonseed meal (ammoniated), cottonseed meal (decorticated), cottonseed meats, cottonseed products, crumbles, crumbles, grower, cycad meal, dairy cattle feed, dairy cattle feed (containing 2–5 % peanut products), dairy cattle feed (containing 6–10 % peanut products), dairy cattle feed (containing 6–12 % peanut products), dairy cattle feed (containing more than 20 % peanut products), diets, mixed, dog food, egg production mixed feed, feed, feed and ingredients, feed, compound, feed, layer, feed, mixed, feed (beef), feed (broilers), feed (calf), feed (cat), feed (cattle), feed (chicken), feed (dairy), feed (dog), feed (dug), feed (fish), feed (gluten), feed (horse), feed (miscellaneous), feed (pig), feed (poultry), feed (poultry, pig), feed (rabbit), feed (sheep), feed (50–60 % maize), fish meal,

grain by-products, grains (no specification), greengram, hay/silage, horsegram, *Jagni* cake, legume mixture, linseed, linseed cake, livol, *Mahua* cake, maize, maize germ, maize gluten, maize grits, maize husk, maize meal, maize oil cake, maize powder, maize screenings, maize, ground, maize, hybrid, maize, preharvest, maize, yellow, maize (dark grains), *Makhana* (*Euryale ferox* Salisb) puffs, manioc, milk production mixed feed, mung testa, murkool, mustard cake, neem cake, niger cake, oats, palm kernel expeller cake, palm kernels, palm products, peanut cake, peanut cake (deoiled), peanut expeller, peanut hay, peanut meal, peanut, kernels, peanut, shells, peanuts, pellets, finisher, pig meal and pellets, pigeon pea, poultry feeds (peanut containing), rapeseed cake, redgram husk, rice, rice bran, rice bran (deoiled), rice chaff, rice crack, rice germ, rice germ cake, rice meal, rice straw, rice (damaged), rice (polish), safflower cake, sal seed cake, sesame, sesame cake, sorghum, soybean meal, soybeans, sunflower, sunflower cake, sunflower flour, tapioca, wheat, wheat bran, wheat bran and chana testa

Jagni cake   may contain the following mycotoxins:

Aflatoxin B$_1$
incidence: 5/6, conc. range: 72.7–140.6 μg/kg, Ø conc.: 110.7 μg/kg, country: India[253]
see also alfalfa *Ambadi* cake, animal feedstuffs (dairy cake), bagasse, barley, bengalgram husk, bird food, bird food, wild, biri testa, blackgram, blackgram husk, bran, broiler mixed feed, calf fattening mixed feed, calf fattening mixed feed (containing 4–20 % peanut products), *Carthamus* cake, castor cake, cereals, cereal products, chick pea, coconut cake, cocos, concentrate, mixed, concentrates, cotton cake, cottonseed, cottonseed (dehulled), cottonseed cake, cottonseed extract, cottonseed meal, cottonseed meal (ammoniated), cottonseed meal (decorticated), cottonseed meats, cottonseed products, crumbles, crumbles, grower, cycad meal, dairy cattle feed, dairy

cattle feed (containing 2–5 % peanut products), dairy cattle feed (containing 6–10 % peanut products), dairy cattle feed (containing 6–12 % peanut products), dairy cattle feed (containing more than 20 % peanut products), diets, mixed, dog food, egg production mixed feed, feed, feed and ingredients, feed, compound, feed, layer, feed, mixed, feed (beef), feed (broilers), feed (calf), feed (cat), feed (cattle), feed (chicken), feed (dairy), feed (dog), feed (dug), feed (fish), feed (gluten), feed (horse), feed (miscellaneous), feed (pig), feed (poultry), feed (poultry, pig), feed (rabbit), feed (sheep), feed (50–60 % maize), fish meal, grain by-products, grains (no specification), greengram, hay/silage, horsegram, husk, legume mixture, linseed, linseed cake, livol, *Mahua* cake, maize, maize germ, maize gluten, maize grits, maize husk, maize meal, maize oil cake, maize powder, maize screenings, maize, ground, maize, hybrid, maize, preharvest, maize, yellow, maize (dark grains), *Makhana* (*Euryale ferox* Salisb) puffs, manioc, milk production mixed feed, mung testa, murkool, mustard cake, neem cake, niger cake, oats, palm kernel expeller cake, palm kernels, palm products, peanut cake, peanut cake (deoiled), peanut expeller, peanut hay, peanut meal, peanut, kernels, peanut, shells, peanuts, pellets, finisher, pig meal and pellets, pigeon pea, poultry feeds (peanut containing), rapeseed cake, redgram husk, rice, rice bran, rice bran (deoiled), rice chaff, rice crack, rice germ, rice germ cake, rice meal, rice straw, rice (damaged), rice (polish), safflower cake, sal seed cake, sesame, sesame cake, sorghum, soybean meal, soybeans, sunflower, sunflower cake, sunflower flour, tapioca, wheat, wheat bran, wheat bran and chana testa

Layer's mash   may contain the following mycotoxins:

Aflatoxin
incidence: 7/20, conc. range: 0–260 μg/kg, country: Nigeria[109]

see also blackgram husk, bread crumbs, broiler finisher, broiler starter, cotton cake, cottonseed, cottonseed cake, cottonseed extract, cottonseed meal, feed, feed (cattle), feed (cow), feed (dog), feed (horse), feed (maize, gluten), feed (pig), feed (poultry), feed (rabbit), feed (rat/mice), feed (sheep), feeds, grain, fish meal, flour (wheat), groundnut cake, grower's mash, horsegram, maize, maize gluten, maize, white, milo, peanut cake, peanut cake (deoiled), peanut, kernels, peanut (oil cake), pearlmillet, pig breeder's mash, pig finisher, pig starter, pod with haulms, poultry breeder's mash, rabbit pellets, redgram husk, rice, rice bran (deoiled), rice, broken, rice (polish), sesame cake, silk worm pupae, sorghum, soybean cake, soybean meal, wheat, wheat bran

Legume mixture   may contain the following mycotoxins:

Aflatoxin B$_1$
incidence: ?/10, conc. range: 20–40 µg/kg, country: India[183]
see also alfalfa, *Ambadi* cake, animal feedstuffs (dairy cake), bagasse, barley, bengalgram husk, bird food, bird food, wild, biri testa, blackgram, blackgram husk, bran, broiler mixed feed, calf fattening mixed feed, calf fattening mixed feed (containing 4–20 % peanut products), *Carthamus* cake, castor cake, cereals, cereal products, chick pea, coconut cake, cocos, concentrate, mixed, concentrates, cotton cake, cottonseed, cottonseed (dehulled), cottonseed cake, cottonseed extract, cottonseed meal, cottonseed meal (ammoniated), cottonseed meal (decorticated), cottonseed meats, cottonseed products, crumbles, crumbles, grower, cycad meal, dairy cattle feed, dairy cattle feed (containing 2–5 % peanut products), dairy cattle feed (containing 6–10 % peanut products), dairy cattle feed (containing 6–12 % peanut products), dairy cattle feed (containing more than 20 % peanut products), diets, mixed, dog food, egg production mixed feed, feed, feed and ingredients, feed, compound, feed, layer,

feed, mixed, feed (beef), feed (broilers), feed (calf), feed (cat), feed (cattle), feed (chicken), feed (dairy), feed (dog), feed (dug), feed (fish), feed (gluten), feed (horse), feed (miscellaneous), feed (pig), feed (poultry), feed (poultry, pig), feed (rabbit), feed (sheep), feed (50–60 % maize), fish meal, grain by-products, grains (no specification), greengram, hay/silage, horsegram, husk, *Jagni* cake, linseed, linseed cake, livol, *Mahua* cake, maize, maize germ, maize gluten, maize grits, maize husk, maize meal, maize oil cake, maize powder, maize screenings, maize, ground, maize, hybrid, maize, preharvest, maize, yellow, maize (dark grains), *Makhana* (*Euryale ferox* Salisb) puffs, manioc, milk production mixed feed, mung testa, murkool, mustard cake, neem cake, niger cake, oats, palm kernel expeller cake, palm kernels, palm products, peanut cake, peanut cake (deoiled), peanut expeller, peanut hay, peanut meal, peanut, kernels, peanut, shells, peanuts, pellets, finisher, pig meal and pellets, pigeon pea, poultry feeds (peanut containing), rapeseed cake, redgram husk, rice, rice bran, rice bran (deoiled), rice chaff, rice crack, rice germ, rice germ cake, rice meal, rice straw, rice (damaged), rice (polish), safflower cake, sal seed cake, sesame, sesame cake, sorghum, soybean meal, soybeans, sunflower, sunflower cake, sunflower flour, tapioca, wheat, wheat bran, wheat bran and chana testa

Penicillic Acid
incidence: ?/10, conc.: 20 µg/kg, country: India[183]

Linseed   Linseed for feed may contain the following mycotoxins:

Aflatoxin B$_1$
incidence: 6/21, conc. range: 5–10 µg/kg, Ø conc.: 9 µg/kg, country: investigated in Germany[298]
see also alfalfa, *Ambadi* cake, animal feedstuffs (dairy cake), bagasse, barley, bengalgram husk, bird food, bird food, wild, biri testa, blackgram, blackgram husk, bran, broiler mixed feed, calf fattening mixed feed,

calf fattening mixed feed (containing 4–20 %
peanut products), *Carthamus* cake, castor
cake, cereals, cereal products, chick pea,
coconut cake, cocos, concentrate, mixed,
concentrates, cotton cake, cottonseed,
cottonseed (dehulled), cottonseed cake,
cottonseed extract, cottonseed meal, cottonseed meal (ammoniated), cottonseed meal
(decorticated), cottonseed meats, cottonseed
products, crumbles, crumbles, grower, cycad
meal, dairy cattle feed, dairy cattle feed
(containing 2–5 % peanut products), dairy
cattle feed (containing 6–10 % peanut
products), dairy cattle feed (containing
6–12 % peanut products), dairy cattle feed
(containing more than 20 % peanut
products), diets, mixed, dog food, egg
production mixed feed, feed, feed and
ingredients, feed, compound, feed, layer,
feed, mixed, feed (beef), feed (broilers), feed
(calf), feed (cat), feed (cattle), feed (chicken),
feed (dairy), feed (dog), feed (dug), feed
(fish), feed (gluten), feed (horse), feed
(miscellaneous), feed (pig), feed (poultry),
feed (poultry, pig), feed (rabbit), feed
(sheep), feed (50–60 % maize), fish meal,
grain by-products, grains (no specification),
greengram, hay/silage, horsegram, husk, *Jagni*
cake, legume mixture, linseed , linseed cake,
livol, *Mahua* cake, maize, maize germ, maize
gluten, maize grits, maize husk, maize meal,
maize oil cake, maize powder, maize
screenings, maize, ground, maize, hybrid,
maize, preharvest, maize, yellow, maize (dark
grains), *Makhana* (*Euryale ferox* Salisb) puffs,
manioc, milk production mixed feed, mung
testa, murkool, mustard cake, neem cake,
niger cake, oats, palm kernel expeller cake,
palm kernels, palm products, peanut cake,
peanut cake (deoiled), peanut expeller,
peanut hay, peanut meal, peanut, kernels,
peanut, shells, peanuts, pellets, finisher, pig
meal and pellets, pigeon pea, poultry feeds
(peanut containing), rapeseed cake, redgram
husk, rice, rice bran, rice bran (deoiled), rice
chaff, rice crack, rice germ, rice germ cake,
rice meal, rice straw, rice (damaged), rice
(polish), safflower cake, sal seed cake, sesame,

sesame cake, sorghum, soybean meal,
soybeans, sunflower, sunflower cake,
sunflower flour, tapioca, wheat, wheat bran,
wheat bran and chana testa

Linseed cake   may contain the following
mycotoxins:

AFLATOXIN B$_1$
incidence: 10/23, conc. range:
44.5–153.1 µg/kg, Ø conc.: 94.7 µg/kg,
country: India[253]
see also alfalfa *Ambadi* cake, animal feedstuffs
(dairy cake), bagasse, barley, bengalgram
husk, bird food, bird food, wild, biri testa,
blackgram, blackgram husk, bran, broiler
mixed feed, calf fattening mixed feed, calf
fattening mixed feed (containing 4–20 %
peanut products), *Carthamus* cake, castor
cake, cereals, cereal products, chick pea,
coconut cake, cocos, concentrate, mixed,
concentrates, cotton cake, cottonseed,
cottonseed (dehulled), cottonseed cake,
cottonseed extract, cottonseed meal,
cottonseed meal (ammoniated), cottonseed
meal (decorticated), cottonseed meats,
cottonseed products, crumbles, crumbles,
grower, cycad meal, dairy cattle feed, dairy
cattle feed (containing 2–5 % peanut
products), dairy cattle feed (containing
6–10 % peanut products), dairy cattle feed
(containing 6–12 % peanut products), dairy
cattle feed (containing more than 20 %
peanut products), diets, mixed, dog food, egg
production mixed feed, feed, feed and
ingredients, feed, compound, feed, layer,
feed, mixed, feed (beef), feed (broilers), feed
(calf), feed (cat), feed (cattle), feed (chicken),
feed (dairy), feed (dog), feed (dug), feed
(fish), feed (gluten), feed (horse), feed
(miscellaneous), feed (pig), feed (poultry),
feed (poultry, pig), feed (rabbit), feed
(sheep), feed (50–60 % maize), fish meal,
grain by-products, grains (no specification),
greengram, hay/silage, horsegram, husk, *Jagni*
cake, legume mixture, linseed, linseed cake,
livol, *Mahua* cake, maize, maize germ, maize
gluten, maize grits, maize husk, maize meal,
maize oil cake, maize powder, maize

screenings, maize, ground, maize, hybrid, maize, preharvest, maize, yellow, maize (dark grains), *Makhana* (*Euryale ferox* Salisb) puffs, manioc, milk production mixed feed, mung testa, murkool, mustard cake, neem cake, niger cake, oats, palm kernel expeller cake, palm kernels, palm products, peanut cake, peanut cake (deoiled), peanut expeller, peanut hay, peanut meal, peanut, kernels, peanut, shells, peanuts, pellets, finisher, pig meal and pellets, pigeon pea, poultry feeds (peanut containing), rapeseed cake, redgram husk, rice, rice bran, rice bran (deoiled), rice chaff, rice crack, rice germ, rice germ cake, rice meal, rice straw, rice (damaged), rice (polish), safflower cake, sal seed cake, sesame, sesame cake, sorghum, soybean meal, soybeans, sunflower, sunflower cake, sunflower flour, tapioca, wheat, wheat bran, wheat bran and chana testa

Livol   may contain the following mycotoxins:

Aflatoxin B$_1$
incidence: 1/1, conc.: 18 μg/kg, country: India[247]
see also alfalfa *Ambadi* cake, animal feedstuffs (dairy cake), bagasse, barley, bengalgram husk, bird food, bird food, wild, biri testa, blackgram, blackgram husk, bran, broiler mixed feed, calf fattening mixed feed, calf fattening mixed feed (containing 4–20 % peanut products), *Carthamus* cake, castor cake, cereals, cereal products, chick pea, coconut cake, cocos, concentrate, mixed, concentrates, cotton cake, cottonseed, cottonseed (dehulled), cottonseed cake, cottonseed extract, cottonseed meal, cottonseed meal (ammoniated), cottonseed meal (decorticated), cottonseed meats, cottonseed products, crumbles, crumbles, grower, cycad meal, dairy cattle feed, dairy cattle feed (containing 2–5 % peanut products), dairy cattle feed (containing 6–10 % peanut products), dairy cattle feed (containing 6–12 % peanut products), dairy cattle feed (containing more than 20 % peanut products), diets, mixed, dog food, egg production mixed feed, feed, feed and

ingredients, feed, compound, feed, layer, feed, mixed, feed (beef), feed (broilers), feed (calf), feed (cat), feed (cattle), feed (chicken), feed (dairy), feed (dog), feed (dug), feed (fish), feed (gluten), feed (horse), feed (miscellaneous), feed (pig), feed (poultry), feed (poultry, pig), feed (rabbit), feed (sheep), feed (50–60 % maize), fish meal, grain by-products, grains (no specification), greengram, hay/silage, horsegram, husk, *Jagni* cake, legume mixture, linseed, linseed cake, *Mahua* cake, maize, maize germ, maize gluten, maize grits, maize husk, maize meal, maize oil cake, maize powder, maize screenings, maize, ground, maize, hybrid, maize, preharvest, maize, yellow, maize (dark grains), *Makhana* (*Euryale ferox* Salisb) puffs, manioc, milk production mixed feed, mung testa, murkool, mustard cake, neem cake, niger cake, oats, palm kernel expeller cake, palm kernels, palm products, peanut cake, peanut cake (deoiled), peanut expeller, peanut hay, peanut meal, peanut, kernels, peanut, shells, peanuts, pellets, finisher, pig meal and pellets, pigeon pea, poultry feeds (peanut containing), rapeseed cake, redgram husk, rice, rice bran, rice bran (deoiled), rice chaff, rice crack, rice germ, rice germ cake, rice meal, rice straw, rice (damaged), rice (polish), safflower cake, sal seed cake, sesame, sesame cake, sorghum, soybean meal, soybeans, sunflower, sunflower cake, sunflower flour, tapioca, wheat, wheat bran, wheat bran and chana testa

Lucern (dried)   may contain the following mycotoxins:

Aflatoxin B
incidence: 11/19, conc. range: ≥25–>100 μg/kg, country: France[46]
see also barley, cocoa (oil cake), cottonseed cake, feed, feed (cattle), feed (maize, gluten), feed (pig), feed (poultry), flour (wheat), maize, maize gluten, maize grains, oats, peanut (oil cake), rice, broken, sorghum, soybean (oil cake), sunflower (oil cake), wheat

Lupin seeds may contain the following mycotoxins:

PHOMOPSIN A
incidence: ?/?, conc. range: ≤1,460,000 μg/kg?
Ø conc.: 236,000 μg/kg? country:
Australia[194]

Mahua **cake** may contain the following mycotoxins:

AFLATOXIN B$_1$
incidence: 4/16, conc. range:
72.7–343.17 μg/kg, Ø conc.: 200.9 μg/kg,
country: India[253]
see also alfalfa, *Ambadi* cake, animal
feedstuffs (dairy cake), bagasse, barley,
bengalgram husk, bird food, bird food, wild,
biri testa, blackgram, blackgram husk, bran,
broiler mixed feed, calf fattening mixed feed,
calf fattening mixed feed (containing 4–20 %
peanut products), *Carthamus* cake, castor
cake, cereals, cereal products, chick pea,
coconut cake, cocos, concentrate, mixed,
concentrates, cotton cake, cottonseed,
cottonseed (dehulled), cottonseed cake,
cottonseed extract, cottonseed meal,
cottonseed meal (ammoniated), cottonseed
meal (decorticated), cottonseed meats,
cottonseed products, crumbles, crumbles,
grower, cycad meal, dairy cattle feed, dairy
cattle feed (containing 2–5 % peanut
products), dairy cattle feed (containing
6–10 % peanut products), dairy cattle feed
(containing 6–12 % peanut products), dairy
cattle feed (containing more than 20 %
peanut products), diets, mixed, dog food, egg
production mixed feed, feed, feed and
ingredients, feed, compound, feed, layer,
feed, mixed, feed (beef), feed (broilers), feed
(calf), feed (cat), feed (cattle), feed (chicken),
feed (dairy), feed (dog), feed (dug), feed
(fish), feed (gluten), feed (horse), feed
(miscellaneous), feed (pig), feed (poultry),
feed (poultry, pig), feed (rabbit), feed
(sheep), feed (50–60 % maize), fish meal,
grain by-products, grains (no specification),
greengram, hay/silage, horsegram, husk, *Jagni*
cake, legume mixture, linseed, linseed cake,

livol, maize, maize germ, maize gluten, maize
grits, maize husk, maize meal, maize oil cake,
maize powder, maize screenings, maize,
ground, maize, hybrid, maize, preharvest,
maize, yellow, maize (dark grains), *Makhana*
(*Euryale ferox* Salisb) puffs, manioc, milk
production mixed feed, mung testa, murkool,
mustard cake, neem cake, niger cake, oats,
palm kernel expeller cake, palm kernels, palm
products, peanut cake, peanut cake (deoiled),
peanut expeller, peanut hay, peanut meal,
peanut, kernels, peanut, shells, peanuts,
pellets, finisher, pig meal and pellets, pigeon
pea, poultry feeds (peanut containing),
rapeseed cake, redgram husk, rice, rice bran,
rice bran (deoiled), rice chaff, rice crack, rice
germ, rice germ cake, rice meal, rice straw,
rice (damaged), rice (polish), safflower cake,
sal seed cake, sesame, sesame cake, sorghum,
soybean meal, soybeans, sunflower, sunflower
cake, sunflower flour, tapioca, wheat, wheat
bran, wheat bran and chana testa

Maize Maize for feed may contain the
following mycotoxins:

3-ACETYLDEOXYNIVALENOL
incidence: 8/30, conc. range: ≤7900 μg/kg,
Ø conc.: 1200 μg/kg, country: Romania[54]
incidence: 3/85, conc. range: ≤113 μg/kg,
Ø conc.: 107 μg/kg, country: Austria[239]
incidence: 2/60, conc. range: 140–180 μg/kg,
Ø conc.: 160 μg/kg, country: Canada[346]
incidence: 6/152, conc. range: 60–130 μg/kg,
Ø conc.: 90 μg/kg, country: Austria[363]
incidence: 11/24, conc. range: 14–137 μg/kg,
Ø conc.: 43 μg/kg, country: Germany[366]
incidence: 1/9*, conc.: 51 μg/kg, country:
Germany[366], *wp
see also barley, feed components, feed
(poultry), feeds, grain, feeds, industrial,
feedstuffs (rapeseed, turnip, fish meal,
concentrates), maize ears, maize gluten,
maize kernels, maize meal, maize screenings,
oats, wheat

15-ACETYLDEOXYNIVALENOL
incidence: 7/20, conc. range: 900–7900 μg/kg,
Ø conc.: 4100 μg/kg, country: USA[48]

incidence: 11/30, conc. range: ≤42,000 μg/kg,
Ø conc.: 4500 μg/kg, country: Romania[54]
incidence: 46/85, conc. range: ≤1112 μg/kg,
Ø conc.: 284 μg/kg, country: Austria[239]
incidence: 2/94, conc. range: 16–80 μg/kg,
Ø conc.: 48 μg/kg, country: Canada[240]
incidence: 15/60, conc. range: 100–320 μg/kg,
country: Canada[346]
incidence: 65/152, conc. range: 50–980 μg/kg,
Ø conc.: 188.3 μg/kg, country: Austria[363]
incidence: 24/24, conc. range: 28–800 μg/kg,
Ø conc.: 175 μg/kg, country: Germany[366]
incidence: 9/9*, conc. range: 40–484 μg/kg,
Ø conc.: 136 μg/kg, country: Germany[366],
*wp
see also barley, feed components, feed,
commercial mix, feed (poultry), maize ears,
maize germ, maize germ/bran, maize gluten,
maize kernels, maize meal, maize screenings,
maize, "Baby", oats, silage, wheat

Aflatoxin B$_1$
incidence: 187/238, conc. range: 1–19 μg/kg
(48 sa), 20–49 μg/kg (44 sa), 50–99 μg/kg
(36 sa), 100–249 μg/kg (37 sa), 250–499 μg/kg
(10 sa), 500–1000 μg/kg (9 sa), >1000 μg/kg
(3 sa), country: USA[38]
incidence: 3/3, conc. range: 30–1477 μg/kg,
Ø conc.: 545.7 μg/kg, country: Canada[39]
incidence: 19/111, conc. range: 5–5000 μg/kg,
country: USA[40]
incidence: 37/40*, conc. range:
0.1–203.3 μg/kg, Ø conc.: 36.8 μg/kg,
country: USA[47], *ncac
incidence: 4/111, conc. range: 19–24 μg/kg,
country: Poland[84]
incidence: 5/5, conc. range: 73–709 μg/kg,
Ø conc.: 424.2 μg/kg, country: USA[90]
incidence: 29/97*, conc. range: 0.1–5.8 μg/kg,
country: UK[97], *imported from France
incidence: 16/37*, conc. range:
0.1–16.4 μg/kg, country: UK[97], *imported
from Argentina
incidence: 3/5*, conc. range: 0.1–3.7 μg/kg,
country: UK[97], *imported
incidence: 14/15, conc. range: 9.4–96 μg/kg,
Ø conc.: 28.4 μg/kg, country: Vietnam[104]
incidence: 25/26, conc. range: tr–9220 μg/kg,
Ø conc.: 119 μg/kg, country: Indonesia[105]

incidence: 3/3, conc. range: 11–606 μg/kg,
Ø conc.: 313.3 μg/kg, country: Thailand[117]
incidence: 1/2, conc.: 1 μg/kg, country:
Thailand[117]
incidence: 50/52, conc. range:
1–20,000 μg/kg, country: Indonesia[118]
incidence: 1/3, conc.: 70 μg/kg, country:
Australia[121]
incidence: 3/3*, conc. range: 9.25–41.5 μg/kg,
Ø conc.: 22.9 μg/kg, country: UK[122], *ncac
incidence: 4/33, conc. range: 3.9–66.1 μg/kg,
Ø conc.: 20.7 μg/kg, country: Colombia[125]
incidence: 10/10, conc. range: 7–422 μg/kg,
Ø conc.: 156.9 μg/kg, country: USA[136]
incidence: 1/2, conc.: 70 μg/kg, country:
Australia[181]
incidence: ?/10, conc. range: 10–20 μg/kg,
country: India[183]
incidence: 98/98*, conc. range: <2 μg/kg
(86 sa), 2–20 μg/kg (11 sa), 109 μg/kg (1 sa),
Ø conc.: 1.9 μg/kg, country: Italy[206], *ncac
incidence: 104/104*, conc. range: <2 μg/kg
(101 sa), 2–20 μg/kg (3 sa), Ø conc.:
0.3 μg/kg, country: Italy[206], *ncac
incidence: 94/94*, conc. range: <2 μg/kg
(77 sa), 2–20 μg/kg (15 sa), >20 μg/kg (2 sa),
Ø conc.: 1.5 μg/kg, country: Italy[206], *ncac
incidence: 114/114*, conc. range: <2 μg/kg
(89 sa), 2–20 μg/kg (24 sa), 28 μg/kg (1 sa),
Ø conc.: 1.5 μg/kg, country: Italy[206], *ncac
incidence: 93/93*, conc. range: <2 μg/kg
(81 sa), 2–20 μg/kg (9 sa), >20 μg/kg (3 sa),
Ø conc.: 1.5 μg/kg, country: Italy[206], *ncac
incidence: 563/862, conc. range: ≤25 μg/kg
(213 sa), 26–50 μg/kg (84 sa), 51–100 μg/kg
(83 sa), 101–200 μg/kg (63 sa), 201–500 μg/kg
(59 sa), 501–1000 μg/kg (39 sa),
1001–1500 μg/kg (3 sa), 1501–2000 μg/kg
(6 sa), 2001–3000 μg/kg (4 sa),
3001–4000 μg/kg (4 sa), 4001–5000 μg/kg
(4 sa), >5000 μg/kg (1 sa), Ø conc.:
226.3 μg/kg, country: India[247]
incidence: 1/4, conc.: 112.6 μg/kg, country:
India[253]
incidence: 1/461, conc.: 42 μg/kg, country:
France[262]
incidence: 4/30, conc. range: <20 μg/kg
(3 sa), 101–1000 μg/kg (1 sa), country: UK[267]

incidence: 30/165*, conc. range: 5–900 µg/kg,
Ø conc.: 78.5 µg/kg, country: Brazil[271], *ncac
incidence: 14/163*, conc. range: 5–148 µg/kg,
Ø conc.: 34.9 µg/kg, country: Brazil[271], *ncac
incidence: 37/214, conc. range: <5–56 µg/kg,
country: USA[277]
incidence: 152/297*, conc. range:
<0.6–25.6 µg/kg , country: USA[286], *ncac
incidence: 3/30*, conc. range: 22–50 µg/kg,
Ø conc.: 34 µg/kg, country: Argentina[322],
*ncac
incidence: 1/1, conc.: 114 µg/kg*, country:
USA[368], *e
incidence: 316/644, conc. range: 1–20 µg/kg
(224 sa), 21–100 µg/kg (69 sa), 101–300 µg/kg
(13 sa), >300 µg/kg (10 sa), country: USA[370]
incidence: 1/1*, conc.: 13.8 µg/kg, country:
USA[378], *ncac
incidence: 30/1311, conc. range: 3–19 µg/kg,
country: USA[381]
incidence: 34/35*, conc. range:
0.11–4030 µg/kg, country: India[386], *ncac
incidence: 19/19*, conc. range: 5–126 µg/kg,
country: India[386], *ncac, rain-affected
incidence: 3/3*, conc. range: 11.5–72.5 µg/kg,
Ø conc.: 34.2 µg/kg, country: USA[387], *ncac
incidence: 4/36*, conc. range: 6–27 µg/kg,
Ø conc.: 12.8 µg/kg, country: Brazil[394], *ncac
incidence: 20/36*, conc. range: 25–289 µg/kg,
Ø conc.: 152 µg/kg, country: Brazil[394], *ncac
incidence: 33/38*, conc. range:
18–1600 µg/kg, Ø conc.: 451.6 µg/kg,
country: Brazil[394], *ncac
incidence: 13/15, conc. range: 30–200 µg/kg,
Ø conc.: 107.7 µg/kg, country: Egypt[397]
incidence: 10/10, conc. range: 25–160 µg/kg,
Ø conc.: 76.5 µg/kg, country: Egypt[397]
incidence: 12/12, conc. range: 5–200 µg/kg,
Ø conc.: 81.7 µg/kg, country: Egypt[397]
incidence: 18/20, conc. range: 15–160 µg/kg,
Ø conc.: 63.6 µg/kg, country: Egypt[397]
incidence: 82/214*, conc. range:
0.2–129 µg/kg, Ø conc.: 9.4 µg/kg, country:
Brazil[403], *ncac
incidence: 8/23*, conc. range: 12–878 µg/kg,
Ø conc.: 166.5 µg/kg, country: Brazil[419], *ncac
incidence: 31/34, conc. range: 22–4074 µg/kg,
country: Indonesia[426]

incidence: 41/61*, conc. range: ≤245 µg/kg,
Ø conc.: 49.2 µg/kg, country: Bangladesh[428],
*ncac
see also alfalfa, *Ambadi* cake, animal
feedstuffs (dairy cake), bagasse, barley,
bengalgram husk, bird food, bird food, wild,
biri testa, blackgram, blackgram husk, bran,
broiler mixed feed, calf fattening mixed feed,
calf fattening mixed feed (containing 4–20 %
peanut products), *Carthamus* cake, castor
cake, cereals, cereal products, chick pea,
coconut cake, cocos, concentrate, mixed,
concentrates, cotton cake, cottonseed,
cottonseed (dehulled), cottonseed cake,
cottonseed extract, cottonseed meal,
cottonseed meal (ammoniated), cottonseed
meal (decorticated), cottonseed meats,
cottonseed products, crumbles, crumbles,
grower, cycad meal, dairy cattle feed, dairy
cattle feed (containing 2–5 % peanut
products), dairy cattle feed (containing
6–10 % peanut products), dairy cattle feed
(containing 6–12 % peanut products), dairy
cattle feed (containing more than 20 %
peanut products), diets, mixed, dog food, egg
production mixed feed, feed, feed and
ingredients, feed, compound, feed, layer,
feed, mixed, feed (beef), feed (broilers), feed
(calf), feed (cat), feed (cattle), feed (chicken),
feed (dairy), feed (dog), feed (dug), feed
(fish), feed (gluten), feed (horse), feed
(miscellaneous), feed (pig), feed (poultry),
feed (poultry, pig), feed (rabbit), feed
(sheep), feed (50–60 % maize), fish meal,
grain by-products, grains (no specification),
greengram, hay/silage, horsegram, husk, *Jagni*
cake, legume mixture, linseed, linseed cake,
livol, *Mahua* cake, maize germ, maize gluten,
maize grits, maize husk, maize meal, maize
oil cake, maize powder, maize screenings,
maize, ground, maize, hybrid, maize,
preharvest, maize, yellow, maize (dark
grains), *Makhana* (*Euryale ferox* Salisb) puffs,
manioc, milk production mixed feed, mung
testa, murkool, mustard cake, neem cake,
niger cake, oats, palm kernel expeller cake,
palm kernels, palm products, peanut cake,
peanut cake (deoiled), peanut expeller,

peanut hay, peanut meal, peanut, kernels, peanut, shells, peanuts, pellets, finisher, pig meal and pellets, pigeon pea, poultry feeds (peanut containing), rapeseed cake, redgram husk, rice, rice bran, rice bran (deoiled), rice chaff, rice crack, rice germ, rice germ cake, rice meal, rice straw, rice (damaged), rice (polish), safflower cake, sal seed cake, sesame, sesame cake, sorghum, soybean meal, soybeans, sunflower, sunflower cake, sunflower flour, tapioca, wheat, wheat bran, wheat bran and chana testa

AFLATOXIN B_2
incidence: 3/3, conc. range: 3.2–70 µg/kg, Ø conc.: 29.4 µg/kg, country: Canada[39]
incidence: 4/111, conc. range: 1–2 µg/kg, country: Poland[84]
incidence: 5/5, conc. range: 24–120 µg/kg, Ø conc.: 65.2 µg/kg, country: USA[90]
incidence: 20/26, conc. range: tr–158 µg/kg, Ø conc.: 13.7 µg/kg, country: Indonesia[105]
incidence: 3/3, conc. range: 3–73 µg/kg, Ø conc.: 41.3 µg/kg, country: Thailand[117]
incidence: 3/3*, conc. range: 1.05–40 µg/kg, Ø conc.: 14.1 µg/kg, country: UK[122], *ncac
incidence: 1/33, conc.: 10.4 µg/kg, country: Colombia[125]
incidence: 245/862, conc. range: nc, Ø conc.: 37.4 µg/kg, country: India[247]
incidence: 1/461, conc.: 6 µg/kg, country: France[262]
incidence: 2/30*, conc. range: tr–3 µg/kg, country: Argentina[322], *ncac
incidence: 1/1, conc.: 10 µg/kg*, country: USA[368], *e
incidence: 3/36*, conc. range: 1.9–6 µg/kg, Ø conc.: 3.3 µg/kg, country: Brazil[394], *ncac
incidence: 20/36*, conc. range: 7–55 µg/kg, Ø conc.: 22.1 µg/kg, country: Brazil[394], *ncac
incidence: 32/38*, conc. range: 6–192 µg/kg, Ø conc.: 38.9 µg/kg, country: Brazil[394], *ncac
incidence: 43/214*, conc. range: 0.1–32 µg/kg, Ø conc.: 2.1 µg/kg, country: Brazil[403], *ncac
incidence: 3/23*, conc. range: 7–180 µg/kg, Ø conc.: 78.3 µg/kg, country: Brazil[419], *ncac
incidence: 24/34, conc. range: 11–3021 µg/kg, country: Indonesia[426]

see also animal feedstuffs (dairy cake), bird food, bird food, wild, biri testa, blackgram, blackgram husk, cottonseed, cottonseed cake, cottonseed extract, cottonseed meal, cottonseed meal (ammoniated), cottonseed meats, dog food, egg production mixed feed, feed, compound, feed (cat), feed (cattle), feed (dog), feed (pig), feed (poultry), feed (rabbit), feed (sheep), fish meal, horsegram, maize gluten, maize husk, maize, ground, maize, preharvest, mung testa, mustard cake, niger cake, peanut cake, peanut cake (deoiled), peanut expeller, peanut hay, peanut meal, peanuts, peanut, kernels, redgram husk, rice bran, rice bran (deoiled), rice chaff, rice meal, rice (polish), sal seed cake, sesame cake, sorghum, soybean meal, soybeans, wheat, wheat bran

AFLATOXIN $B_1 + B_2$
incidence: 8/8*, conc. range: 7–360 µg/kg, Ø conc.: 147.4 µg/kg, country: USA[326]; *ncac
see also cottonseed, feed (poultry), sorghum

AFLATOXIN B
incidence: 25/34, conc. range: ≥25–>100 µg/kg, country: France[46]
see also barley, cocoa (oil cake), cottonseed cake, feed, feed (cattle), feed (maize, gluten), feed (pig), feed (poultry), flour (wheat), lucern (dried), maize gluten, maize grains, oats, peanut (oil cake), rice, broken, sorghum, soybean (oil cake), sunflower (oil cake), wheat

AFLATOXIN G_1
incidence: 32/238, conc. range: 1–19 µg/kg (14 sa), 20–49 µg/kg (4 sa), 50–99 µg/kg (5 sa), 100–249 µg/kg (5 sa), 250–499 µg/kg (2 sa), 500–1000 µg/kg (1 sa), >1000 µg/kg (1 sa), country: USA[38]
incidence: 1/3, conc.: 74 µg/kg, country: Canada[39]
incidence: 20/26, conc. range: tr–158 µg/kg, Ø conc.: 13.7 µg/kg, country: Indonesia[105]
incidence: 2/3*, conc. range: 0.5–96.5 µg/kg, Ø conc.: 48.5 µg/kg, country: UK[122], *ncac
incidence: 1/461, conc.: 4 µg/kg, country: France[262]

incidence: 5/1311, conc. range: 2–8 µg/kg,
country: USA[381]
incidence: 2/36*, conc. range: 39–254 µg/kg,
Ø conc.: 146.5 µg/kg, country: Brazil[394], *ncac
incidence: 9/36*, conc. range: 25–112 µg/kg,
Ø conc.: 53.8 µg/kg, country: Brazil[394], *ncac
incidence: 11/214*, conc. range: 0.2–12 µg/kg,
Ø conc.: 1.9 µg/kg, country: Brazil[403], *ncac
incidence: 2/23*, conc. range: 28 µg/kg,
Ø conc.: 28 µg/kg, country: Brazil[419], *ncac
incidence: 2/34, conc. range: 101–528 µg/kg,
Ø conc.: 314.5 µg/kg, country: Indonesia[426]
see also animal feedstuffs (dairy cake), bird
food, bird food, wild, concentrate, mixed,
cottonseed, cottonseed cake, feed (cat), feed
(chicken), feed (dog), maize, ground, maize,
preharvest, meat meal, milk production
mixed feed, murkool, peanut cake, peanut
expeller, peanut hay, peanut meal, peanuts,
rice bran, rice germ, sorghum, soybeans,
wheat, wheat bran, wheat bran and chana
testa

Aflatoxin G$_2$

incidence: 3/26, conc. range: tr–1 µg/kg,
country: Indonesia[105]
incidence: 1/3*, conc.: 38.5 µg/kg, country:
UK[122], *ncac
incidence: 2/36*, conc. range: 26–58 µg/kg,
Ø conc.: 42 µg/kg, country: Brazil[394], *ncac
incidence: 5/36*, conc. range: 7–26 µg/kg,
Ø conc.: 14.8 µg/kg, country: Brazil[394], *ncac
incidence: 2/214*, conc. range: 0.4–4 µg/kg,
Ø conc.: 2.2 µg/kg, country: Brazil[403], *ncac
incidence: 2/23*, conc. range: 6–11 µg/kg,
Ø conc.: 8.5 µg/kg, country: Brazil[419], *ncac
incidence: 1/34, conc.: 144 µg/kg, country:
Indonesia[426]
see also animal feedstuffs (dairy cake), bird
food, feed (cat), feed (dog), maize,
preharvest, peanut expeller, peanut hay,
peanut meal, peanuts, soybeans

Aflatoxin M$_1$

incidence: 1/1, conc.: 6 µg/kg*, country:
USA[368], *e
see also feed (cat), feed (dog), maize, ground

Aflatoxin

incidence: 563/862, conc. range:
3–8260 µg/kg, country: India[247]
incidence: ?/23, conc. range: ≤416 µg/kg,
Ø conc.: 71 µg/kg, country: India[250]
incidence: 8/8*, conc. range: 7.5–1100 µg/kg,
Ø conc.: 413 µg/kg, country: USA[296], *ncac
incidence: 45/45*, conc. range: 0–19 µg/kg
(23 sa), 20–99 µg/kg (15 sa), 100–399 µg/kg
(5 sa), 400–799 µg/kg (2 sa), country: USA[303],
*ncac
incidence: 45/45*, conc. range: 0–19 µg/kg
(33 sa), 20–99 µg/kg (7 sa), 100–399 µg/kg
(2 sa), 400–799 µg/kg (3 sa), country: USA[303],
*ncac
incidence: 33/76*, conc. range: 11–30 µg/kg
(14 sa), 31–100 µg/kg (12 sa), >100 µg/kg
(≤806 µg/kg) (7 sa), country: India[311], *ncac
incidence: 113/250*, conc. range:
46–8665 µg/kg, country: USA[320], *ncac
incidence: 90/7937*, conc. range: ≤396 µg/kg,
country: USA[328], *ncac
incidence: 235/17,245*, conc. range: nc
µg/kg, country: USA[328], *ncac
incidence: 7/7, conc. range: <30 µg/kg (3 sa),
>30 µg/kg (1 sa), >200 µg/kg (2 sa),
>600 µg/kg (1 sa), country: India[380]
incidence: 6/923*, conc. range: 13–151 µg/kg,
Ø conc.: 45 µg/kg, country: USA[385], *ncac
incidence: 3/31*, conc. range: 0–380 µg/kg,
country: USA[405], *wild turkey feed
incidence: 20/39, conc. range: tr–750 µg/kg,
country: USA[427]
see also blackgram husk, bread crumbs,
broiler finisher, broiler starter, cotton cake,
cottonseed, cottonseed cake, cottonseed
extract, cottonseed meal, feed, feed (cattle),
feed (cow), feed (dog), feed (horse), feed
(maize, gluten), feed (pig), feed (poultry),
feed (rabbit), feed (rat/mice), feed (sheep),
feeds, grain, fish meal, flour (wheat),
groundnut cake, grower's mash, horsegram,
layer's mash, maize gluten, maize, white,
milo, peanut cake, peanut cake (deoiled),
peanut, kernels, peanut (oil cake), pearlmillet,
pig breeder's mash, pig finisher, pig starter,
pod with haulms, poultry breeder's mash,
rabbit pellets, redgram husk, rice, rice bran

(deoiled), rice, broken, rice (polish), sesame cake, silk worm pupae, sorghum, soybean cake, soybean meal, wheat, wheat bran

AFLATOXINS
incidence: 4/13, conc. range: 5–500 µg/kg, country: Australia[21]
incidence: 16/97*, conc. range: 0.4–6 µg/kg, country: UK[97], *imported from France
incidence: 12/37*, conc. range: 0.4–29.1 µg/kg, country: UK[97], *imported from Argentina
incidence: 3/5*, conc. range: 0.4–3.8 µg/kg, country: UK[97], *imported
incidence: 17/150, conc. range: 38–460 µg/kg, Ø conc.: 191 µg/kg, country: Brazil[111]
incidence: 77/99*, conc. range: 1–19 µg/kg (40 sa), 20–49 µg/kg (15 sa), 50–99 µg/kg (11 sa), 100–499 µg/kg (10 sa), ≥500 µg/kg (1 sa), country: USA[302], *ncac
incidence: 215/253*, conc. range: 1–19 µg/kg (70 sa), 20–49 µg/kg (72 sa), 50–99 µg/kg (37 sa), 100–499 µg/kg (31 sa), ≥500 µg/kg (5 sa), country: USA[302], *ncac
incidence: 41/95, conc. range: 10–300 µg/kg, country: India[349]
see also animal feed (maize), animal feed (mixed), barley, copra meal, cottonseed, cottonseed fines, cottonseed meats, cottonseed meal, feed, feed ingredients (miscellaneous), feed, mixed, feed (excluding peanuts, suspect), feed (goat), feed (pig), feed (poultry), feedstuff, grain, grain, mixed feed, maize, shelled, millet, peanut cake, peanut meal, peanut meal and by-products, peanuts, protein concentrates, rice, rice bran, sorghum, soybean meal, sunflower

BEAUVERICIN
incidence: 4/6, conc. range: 5000–10,000 µg/kg, Ø conc.: 7500 µg/kg, country: Italy[52], ncac
incidence: 5/8, conc. range: 100–3000 µg/kg, country: USA[113]
incidence: 12/12, conc. range: 1800–36,890 µg/kg, Ø conc.: 15,197.5 µg/kg, country: Poland[259]
incidence: 8/9*, conc. range: nd–60,000 µg/kg, Ø conc.: 18,888 µg/kg, country: Poland[269], *ncac

incidence: 5/5*, conc. range: 5000–30,000 µg/kg, Ø conc.: 15,000 µg/kg, country: Poland[269], *ncac
incidence: 4*/42, conc. range: 4000–40,000 µg/kg, Ø conc.: 21,000 µg/kg, country: Italy[275], *s
incidence: 18/105*, conc. range: 13–1846 µg/kg, Ø conc.: 393 µg/kg, country: Croatia[318], *ncac
incidence: 1/104*, conc.: 696 µg/kg, country: Croatia[318], *ncac
see also barley, oats, wheat, wheat, summer, wheat, winter

CITREOVIRIDIN
incidence: 5/8*, conc. range: 19–2790 µg/kg, Ø conc.: 1230.6 µg/kg, country: USA[326]; *ncac

CITRININ
incidence: 1/30, conc.: ≤580 µg/kg, country: Romania[54]
incidence: 1/1, conc.: 450 µg/kg, country: UK[93]
see also barley, barley, oats, barley-soybean diet, feed, feed, mixed, feed (cattle), feed (pig), fish meal, hay, maize, white, *Makhana* (*Euryale ferox* Salisb) puffs, oats, palm products, peas and beans, rice bran, rice germ, wheat, wheat and other grains (moldy), wheat bran

CYCLOPIAZONIC ACID
incidence: 10/26, conc. range: 400–5000 µg/kg (estimated), country: India[33]
incidence: 21/26, conc. range: ≤9000 µg/kg, country: Indonesia[34]
incidence: 19/26, conc. range: 30–9220 µg/kg, Ø conc.: 2117 µg/kg, country: Indonesia[105]
see also chick mash, feed, groundnut cake, millet, little, peanuts, rice bran, wheat

DEOXYNIVALENOL
incidence: 83/760, conc. range: 50–870 µg/kg, Ø conc.: 191 µg/kg, country: Hungary[6]
incidence: 17/20, conc. range: 400–65,800 µg/kg, Ø conc.: 20,300 µg/kg, country: USA[48]
incidence: 1/11, conc.: 200 µg/kg, country: Hungary[50]

incidence: 3/6, conc. range:
550–50,500 μg/kg, Ø conc.: 17,350 μg/kg,
country: Austria[51], ncac
incidence: 8/8, conc. range: 18–668 μg/kg,
Ø conc.: 188.75 μg/kg, country: Italy[53]
incidence: 14/30, conc. range:
≤160,000 μg/kg, Ø conc.: 13,000 μg/kg,
country: Romania[54]
incidence: 2/3, conc. range: 140–600 μg/kg,
Ø conc.: 370 μg/kg, country: France[61]
incidence: 3/12, conc. range:
4000–18,800 μg/kg, Ø conc.: 10,100 μg/kg,
country: Nigeria[64]
incidence: 6/28, conc. range: 30–120 μg/kg,
country: Germany[68]
incidence: 3/6, conc. range: 100–190 μg/kg,
Ø conc.: 140 μg/kg, country: The
Netherlands[103]
incidence: 3/8, conc. range: 500–5000 μg/kg,
Ø conc.: 2000 μg/kg, country: Zambia[114]
incidence: 1/11, conc.: 4000 μg/kg, country:
Zambia[114]
incidence: 12/14, conc. range:
500–16,000 μg/kg, Ø conc.: 6600 μg/kg,
country: Zambia[114]
incidence: 3/6 * **, conc. range:
200–300 μg/kg, Ø conc.: 233.3 μg/kg,
country: New Zealand[131], *field maize, **ncac
incidence: 2/7 * **, conc. range: 30–90 μg/kg,
Ø conc.: 60 μg/kg, country: New Zealand[131],
*harvest maize, **ncac
incidence: 4/5, conc. range: 20–100 μg/kg,
Ø conc.: 62.5 μg/kg, country: New Zealand[131]
incidence: 926/926, conc. range: <100 μg/kg
(531 sa), 100–500 μg/kg (202 sa),
500–1000 μg/kg (103 sa), 1000–2000 μg/kg
(68 sa), 2000–5000 μg/kg (19 sa),
>5000 μg/kg (3 sa), country: Austria[182]
incidence: 98/98*, conc. range: <500 μg/kg
(95 sa), 500–1000 μg/kg (3 sa), Ø conc.:
194 μg/kg, country: Italy[206], *ncac
incidence: 104/104*, conc. range: <500 μg/kg
(8 sa), 500–1000 μg/kg (17 sa), >1000 μg/kg
(79 sa), Ø conc.: 2716 μg/kg, country: Italy[206],
*ncac
incidence: 94/94*, conc. range: <500 μg/kg
(44 sa), 500–1000 μg/kg (20 sa), >1000 μg/kg
(33 sa), Ø conc.: 802 μg/kg, country: Italy[206],
*ncac

incidence: 114/114*, conc. range: <500 μg/kg
(96 sa), 500–1000 μg/kg (13 sa), >1000 μg/kg
(5 sa), Ø conc.: 298 μg/kg, country: Italy[206],
*ncac
incidence: 93/93*, conc. range: <500 μg/kg
(74 sa), 500–1000 μg/kg (15 sa), >1000 μg/kg
(4 sa), Ø conc.: 290 μg/kg, country: Italy[206],
*ncac
incidence: 2/7345, conc. range: nc, country:
Hungary[209]
incidence: 116/144, conc. range:
60–9200 μg/kg, Ø conc.: 932 μg/kg, country:
Austria[223]
incidence: 2/2, conc. range: 2500–7400 μg/kg,
Ø conc.: 4950 μg/kg, country: Africa[228]
incidence: 3/3*, conc. range: 24–36 μg/kg,
Ø conc.: 30.7 μg/kg, country: unknown[232],
*ncac
incidence: 129/342, conc. range:
230–41,600 μg/kg, Ø conc.: 3100 μg/kg,
country: USA[237]
incidence: 81/85, conc. range: ≤2435 μg/kg,
Ø conc.: 728 μg/kg, country: Austria[239]
incidence: 3/94, conc. range: 20–105 μg/kg,
Ø conc.: 52.7 μg/kg, country: Canada[240]
incidence: 2/86*, conc. range:
410–2020 μg/kg, Ø conc.: 1215 μg/kg,
country: India[241], *ncac
incidence: 9/154, conc. range: 43–315 μg/kg,
country: Spain[242]
incidence: 9/223*, conc. range:
15,000–28,000 μg/kg, Ø conc.: 22,250 μg/kg,
country: USA[251], *is
incidence: 24/52, conc. range:
500–10,700 μg/kg, Ø conc.: 4991.7 μg/kg,
country: USA[255]
incidence: 1/7, conc.: 1900 μg/kg, country:
Poland[345]
incidence: 600/673, conc. range:
≤17,500 μg/kg, Ø conc.: 530 μg/kg, country:
Canada[346]
incidence: 15/45*, conc. range:
100–1500 μg/kg, Ø conc.: 140 μg/kg, country:
USA[352], *ncac
incidence: 115/152, conc. range:
50–2810 μg/kg, Ø conc.: 388.3 μg/kg,
country: Austria[363]
incidence: 24/24, conc. range: 14–4700 μg/kg,
Ø conc.: 1101 μg/kg, country: Germany[366]

incidence: 9/9*, conc. range: 215–728 μg/kg,
Ø conc.: 540 μg/kg, country:
Germany[366], *wp
incidence: 15/18*, conc. range: tr–2800 μg/kg,
Ø conc.: 400 μg/kg, country: France[404], *ncac
incidence: 14/21*, conc. range: tr–558 μg/kg,
Ø conc.: 70 μg/kg, country: France[404], *ncac
incidence: 5/9*, conc. range: 100–500 μg/kg,
Ø conc.: 200 μg/kg, country: Cameroon[416],
*ncac
incidence: 9/9*, conc. range: 100–1300 μg/kg,
Ø conc.: 433 μg/kg, country: Cameroon[416],
*ncac
incidence: 10/61*, conc. range: ≤337 μg/kg,
Ø conc.: 22 μg/kg, country: Bangladesh[428],
*ncac
see also barley, barley, husked, barley,
unhusked (naked), barley (pressed), bone
meal, bran, broilers feed, calf fattening mixed
feed, coconut, expeller, corn cob mix silage,
cottonseed, cottonseed cake, dairy cattle feed,
egg production mixed feed, feed, feed
components, feed, commercial mix, feed,
mixed, feed, mixed (primarily maize), feed
(barley), feed (cattle), feed (chicken), feed
(dog), feed (fish), feed (mill run, from
wheat), feed (mink), feed (pig), feed
(poultry), feed (reindeer), feeds, grain, feeds,
industrial, feedstuff, feedstuffs (rapeseed,
turnip, fish meal, concentrates), fish meal,
grain, mixed feed, grains, mixed, grains (no
specification), maize ears, maize fibre, maize
germ, maize germ/bran, maize germ meal,
maize gluten, maize kernels, maize meal,
maize powder, maize screenings, maize stalks
(pith), maize, "Baby", maize, hybrid, maize,
white, oats, rice bran, rice germ cake, rye,
silage, sorghum, soybeans, triticale, wheat,
wheat and barley, wheat, red hard winter,
wheat, soft white winter, wheat, spring,
wheat, winter

DIACETOXYSCIRPENOL
incidence: 6/11, conc. range: 500–2100 μg/kg,
Ø conc.: 1233 μg/kg, country: Hungary[50]
incidence: 1/6*, conc.: 400 μg/kg, country:
Austria[51], *ncac
incidence: 1/30, conc.: 2.6 μg/kg, country:
Romania[54]

incidence: 4/6 * **, conc. range:
10–900 μg/kg, Ø conc.: 477.5 μg/kg, country:
New Zealand[131], *field maize, **ncac
incidence: 1/7 * ***, conc.: 30 μg/kg, country:
New Zealand[131], *harvest maize, **ncac
incidence: 1/94, conc.: 19 μg/kg, country:
Canada[240]
incidence: 3/60, conc. range: 490–1000 μg/kg,
country: Canada[346]
incidence: 1/24, conc.: 210 μg/kg, country:
Germany[366]
see also alfalfa, barley, feed, feed components,
feed (dog), feed (fish), feed (mink), feed
(pig), feed (poultry), feed (reindeer), feeds,
grain, feedstuff, forage grass, grain, mixed
feed, grains (no specification), maize gluten,
oat and barley (hammer-milled), oats,
peanuts, soybeans, wheat

FUMONISIN B₁
incidence: 24/24, conc. range: 68–6555 μg/kg,
country: Honduras[3]
incidence: 11/11, conc. range: 1270–3980
μg/kg, Ø conc.: 2269 μg/kg, country: Iran[22]
incidence: 6/6, conc. range:
125,000–250,000 μg/kg, Ø conc.:
187,500 μg/kg, country: Italy[52], ncac
incidence: 1/30, conc.: 140 μg/kg, country:
Romania[54]
incidence: 3/3, conc. range:
37,000–122,000 μg/kg, Ø conc.: 72,000 μg/kg,
country: USA[56]
incidence: 14/14, conc. range:
1300–27,000 μg/kg, country: USA[57]
incidence: 15/16, conc. range:
2000–195,000 μg/kg, Ø conc.: 45,500 μg/kg,
country: USA[58]
incidence: 1/2, conc.: 4000 μg/kg, country:
USA[58]
incidence: 5/5, conc. range:
2000–27,000 μg/kg, Ø conc.: 15,400 μg/kg,
country: USA[58]
incidence: 1/1, conc.: 3000 μg/kg, country:
USA[58]
incidence: 3/3, conc. range:
4000–33,000 μg/kg, Ø conc.: 22,000 μg/kg,
country: USA[58]
incidence: 1/1, conc.: 4000 μg/kg, country:
USA[58]

incidence: 1/1, conc.: 19,000 μg/kg, country: USA[58]

incidence: 2/2, conc. range: 20,000–150,000 μg/kg, Ø conc.: 85,000 μg/kg, country: USA[59]

incidence: 23/25, conc. range: 10–8400 μg/kg, country: Italy[60]

incidence: 92/97*, conc. range: 30–1557 μg/kg, country: UK[97], *imported from France

incidence: 37/37*, conc. range: 501–3406 μg/kg, country: UK[97], *imported from Argentina

incidence: 5/5*, conc. range: 30–528 μg/kg, country: UK[97], *imported

incidence: 8/15, conc. range: 271–3447 μg/kg, Ø conc.: 1101 μg/kg, country: Vietnam[104]

incidence: 8/8, conc. range: 300–9500 μg/kg, country: USA[113]

incidence: 3/3, conc. range: 63–464 μg/kg, Ø conc.: 217.3 μg/kg, country: Thailand[117]

incidence: 1/2, conc.: 453 μg/kg, country: Thailand[117]

incidence: 1/1, conc.: 86,000 μg/kg, country: USA[139]

incidence: 5/12, conc. range: 53–1327 μg/kg, Ø conc.: 506 μg/kg, country: Korea[145]

incidence: 5/22, conc. range: <50–1590 μg/kg, Ø conc.: 810 μg/kg, country: Thailand[149]

incidence: 48/55, conc. range: 200–19,200 μg/kg, Ø conc.: 4800 μg/kg, country: Spain[155]

incidence: 214/214*, conc. range: ≥200–≤1000 μg/kg (42 sa), >1000–≤3000 μg/kg (124 sa), >3000–≤6000 μg/kg (48 sa), country: Brazil[157], *ncac

incidence: 12/17, conc. range: 0–2200 μg/kg, Ø conc.: 588 μg/kg, country: Spain[159]

incidence: 38/44, conc. range: 100–160,000 μg/kg, Ø conc.: 18,500 μg/kg, country: Korea[160]

incidence: 1/1, conc.: 1848 μg/kg, country: USA[161]

incidence: 98/98*, conc. range: <1000 μg/kg (34 sa), 1000–5000 μg/kg (47 sa), >5000 μg/kg (17 sa), Ø conc.: 3347 μg/kg, country: Italy[206], *ncac

incidence: 104/104*, conc. range: <1000 μg/kg (47 sa), 1000–5000 μg/kg (55 sa), >5000 μg/kg (2 sa), Ø conc.: 1324 μg/kg, country: Italy[206], *ncac

incidence: 94/94*, conc. range: <1000 μg/kg (25 sa), 1000–5000 μg/kg (60 sa), >5000 μg/kg (9 sa), Ø conc.: 3103 μg/kg, country: Italy[206], *ncac

incidence: 114/114*, conc. range: <1000 μg/kg (32 sa), 1000–5000 μg/kg (67 sa), >5000 μg/kg (15 sa), Ø conc.: 2655 μg/kg, country: Italy[206], *ncac

incidence: 93/93*, conc. range: <1000 μg/kg (15 sa), 1000–5000 μg/kg (36 sa), >5000 μg/kg (42 sa), Ø conc.: 2655 μg/kg, country: Italy[206], *ncac

incidence: 20/20*, conc. range: <50–4100 μg/kg, country: USA[220], *food and feed

incidence: 5/141, conc. range: 7–33 μg/kg, country: Germany[248]

incidence: 81/121, conc. range: 0–5420 μg/kg, Ø conc.: 480.6 μg/kg, country: South Africa[258], ncac

incidence: 106/128, conc. range: 0–5030 μg/kg, Ø conc.: 332.2 μg/kg, country: South Africa[258], ncac

incidence: 20/56, conc. range: 50–33,400 μg/kg, Ø conc.: 2175 μg/kg, country: Hungary[266]

incidence: 4*/42, conc. range: 150,000–250,000 μg/kg, Ø conc.: 202,500 μg/kg, country: Italy[275], *s

incidence: 56/92, conc. range: 50–75,100 μg/kg, country: Hungary[289]

incidence: 35/35*, conc. range: 4–16,000 μg/kg, Ø conc.: 2500 μg/kg, country: Costa Rica[308], *ncac

incidence: 8/8, conc. range: 4900–18,520 μg/kg, Ø conc.: 10,590 μg/kg, country: Brazil[309]

incidence: 1/1, conc.: 5830 μg/kg, country: Brazil[309]

incidence: 27/27, conc. range: 1660–12,550 μg/kg, Ø conc.: 4790 μg/kg, country: Brazil[309]

incidence: 8/8, conc. range: 600–6930 μg/kg, Ø conc.: 3250 μg/kg, country: Brazil[309]

incidence: 3/3, conc. range: 2960–3890 µg/kg, Ø conc.: 3300 µg/kg, country: Brazil[309]
incidence: 1/1, conc.: 5450 µg/kg, country: Brazil[309]
incidence: 176/195*, conc. range: 870–49,310 µg/kg, country: Brazil[335], *ncac
incidence: 4/78*, conc. range: ≤1614 µg/kg, Ø conc.: 619 µg/kg, country: Taiwan[337], *imported from USA
incidence: 2/20*, conc. range: 281.4–334 µg/kg, Ø conc.: 307.7 µg/kg, country: Taiwan[337], *imported from Thailand
incidence: 2/10*, conc. range: 187–477 µg/kg, Ø conc.: 332 µg/kg, country: Taiwan[337], *imported from Australia
incidence: 20/20, conc. range: 580–7660 µg/kg, Ø conc.: 3180 µg/kg, country: Iran[338]
incidence: 19/25, conc. range: 136.4–8757 µg/kg, Ø conc.: 1709.7 µg/kg, country: Spain[347]
incidence: 23/26, conc. range: 90–10,870 µg/kg, country: Brazil[351]
incidence: 29/31, conc. range: 110–17,690 µg/kg, country: Brazil[351]
incidence: 45/45*, conc. range: tr–30,000 µg/kg, Ø conc.: 9500 µg/kg, country: USA[352], *ncac
incidence: 10/152, conc. range: 60–1750 µg/kg, Ø conc.: 483.3 µg/kg, country: Austria[363]
incidence: 23/23, conc. range: 1630–25,690 µg/kg, Ø conc.: 5610 µg/kg, country: Brazil[365]
incidence: 50/50*, conc. range: 185–27,050 µg/kg, Ø conc.: 2229 µg/kg, country: Argentina[382], *ncac
incidence: 26/35*, conc. range: 10–4700 µg/kg, country: India[386], *ncac
incidence: 19/19*, conc. range: 40–64,700 µg/kg, country: India[386], *ncac, rain-affected
incidence: 26/27*, conc. range: 70–2520 µg/kg, Ø conc.: 580 µg/kg, country: Brazil[390], *ncac
incidence: 86/86*, conc. range: 90–10,690 µg/kg, Ø conc.: 2390 µg/kg, country: Brazil[390], *ncac

incidence: 37/37*, conc. range: 370–13,460 µg/kg, Ø conc.: 4560 µg/kg, country: Brazil[390], *ncac
incidence: 91/100, conc. range: <1000 µg/kg (14 sa), 1000–20,000 µg/kg (15 sa), 21,000–40,000 µg/kg (19 sa), 41,000–60,000 µg/kg (27 sa), 61,000–80,000 µg/kg (14 sa), 81,000–100,000 µg/kg (2 sa, with a maximum of 87,000 µg/kg), country: India[391]
incidence: 1/1*, conc.: 33,103 µg/kg, country: Denmark[402], *from Ghana
incidence: 212/214*, conc. range: 200–6100 µg/kg, Ø conc.: 2200 µg/kg, country: Brazil[403], *ncac
incidence: 9/9*, conc. range: 1900–26,000 µg/kg, Ø conc.: 7477 µg/kg, country: Cameroon[416], *ncac
incidence: 7/9*, conc. range: 300–2000 µg/kg, Ø conc.: 1214 µg/kg, country: Cameroon[416], *ncac
incidence: 49/49*, conc. range: 142.2–1377.6 µg/kg, Ø conc.: 459.8 µg/kg, country: Croatia[417], *ncac
incidence: 30/30*, conc. range: 460–9950 µg/kg, Ø conc.: 3939 µg/kg, country: Argentina[422], *ncac
incidence: 6/61*, conc. range: ≤600 µg/kg, Ø conc.: 35 µg/kg, country: Bangladesh[428], *ncac
incidence: 5/5, conc. range: 765–6202.9 µg/kg, Ø conc.: 2241.41 µg/kg, country: USA[431]
see also barley, bird food, wild, dog food, feed, feed, complete ration, feed, general, feed, layer, feed, maize-based, feed, mixed, feed, pelleted ration, feed, screenings, feed, sweet, feed (broilers), feed (cat), feed (chicken), feed (dog), feed (gluten), feed (horse), feed (maize), feed (pig), feed (poultry), feed (rat), feed (rodent), forage grass, maize and maize screenings, maize bran, maize ears, maize fine fractions, maize flakes, maize germ, maize germ/bran, maize germ meal, maize gluten, maize grits, maize kernels, maize meal, maize powder, maize screenings, maize, "Baby", maize, ground, maize, preharvest, maize, sweet feed, maize/oats mix, rat chow, silage, sorghum, soybeans, wheat

Fumonisin B$_2$
incidence: 11/11, conc. range:
190–1175 µg/kg, Ø conc.: 512 µg/kg, country:
Iran[22]
incidence: 3/3, conc. range:
2000–23,000 µg/kg, Ø conc.: 12,000 µg/kg,
country: USA[56]
incidence: 14/14, conc. range:
100–12,600 µg/kg, country: USA[57]
incidence: 13/25, conc. range: 10–1330 µg/kg,
country: Italy[60]
incidence: 4/15, conc. range: 178–560 µg/kg,
Ø conc.: 276 µg/kg, country: Vietnam[104]
incidence: 7/8, conc. range: 800–4000 µg/kg,
country: USA[113]
incidence: 1/1, conc.: 55,000 µg/kg, country:
USA[139]
incidence: 4/12, conc. range: 69–680 µg/kg,
Ø conc.: 288 µg/kg, country: Korea[145]
incidence: 5/22, conc. range:
<100–350 µg/kg, Ø conc.: 230 µg/kg,
country: Thailand[149]
incidence: 22/55, conc. range:
200–5900 µg/kg, Ø conc.: 1900 µg/kg,
country: Spain[155]
incidence: 5/17, conc. range: 0–700 µg/kg,
Ø conc.: 338 µg/kg, country: Spain[159]
incidence: 34/44, conc. range:
900–46,000 µg/kg, Ø conc.: 5600 µg/kg,
country: Korea[160]
incidence: 1/1, conc.: 1092 µg/kg, country:
USA[161]
incidence: 20/20*, conc. range:
<50–1050 µg/kg, country: USA[220], *food and
feed
incidence: 49/121*, conc. range:
0–1600 µg/kg, Ø conc.: 253.8 µg/kg, country:
South Africa[258], *ncac
incidence: 65/128*, conc. range:
0–1670 µg/kg, Ø conc.: 216.6 µg/kg, country:
South Africa[258], *ncac
incidence: 8/8, conc. range:
3620–19,130 µg/kg, Ø conc.: 10,310 µg/kg,
country: Brazil[309]
incidence: 1/1, conc.: 3620 µg/kg, country:
Brazil[309]
incidence: 27/27, conc. range:
1200–10,240 µg/kg, Ø conc.: 3950 µg/kg,
country: Brazil[309]

incidence: 8/8, conc. range: 0–4790 µg/kg,
Ø conc.: 2340 µg/kg, country: Brazil[309]
incidence: 3/3, conc. range: 2170–2780 µg/kg,
Ø conc.: 2520 µg/kg, country: Brazil[309]
incidence: 1/1, conc.: 5090 µg/kg, country:
Brazil[309]
incidence: 190/195*, conc. range:
1960–29,160 µg/kg, country: Brazil[335], *ncac
incidence: 14/26, conc. range: 50–520 µg/kg,
country: Brazil[351]
incidence: 19/31, conc. range: 50–5240 µg/kg,
country: Brazil[351]
incidence: 10/45*, conc. range:
tr–10,700 µg/kg, Ø conc.: 1700 µg/kg,
country: USA[352], *ncac
incidence: 3/152, conc. range: 250–390 µg/kg,
Ø conc.: 290 µg/kg, country: Austria[363]
incidence: 23/23, conc. range:
380–8600 µg/kg, Ø conc.: 1860 µg/kg,
country: Brazil[365]
incidence: 50/50*, conc. range:
40–9950 µg/kg, Ø conc.: 812 µg/kg, country:
Argentina[382], *ncac
incidence: 17/27*, conc. range:
110–1140 µg/kg, Ø conc.: 200 µg/kg, country:
Brazil[390], *ncac
incidence: 84/86*, conc. range:
80–5170 µg/kg, Ø conc.: 1090 µg/kg, country:
Brazil[390], *ncac
incidence: 37/37*, conc. range:
200–6920 µg/kg, Ø conc.: 2200 µg/kg,
country: Brazil[390], *ncac
incidence: 1/1, conc.: 12,318 µg/kg, country:
Ghana[402]
incidence: 3/49*, conc. range:
68.4–3084 µg/kg, Ø conc.: 1087.2 µg/kg,
country: Croatia[417], *ncac
incidence: 29/30*, conc. range:
140–3060 µg/kg, Ø conc.: 1144 µg/kg,
country: Argentina[422], *ncac
incidence: 2/5, conc. range:
2357.23–9953.15 µg/kg, Ø conc.:
6155.19 µg/kg, country: USA[431]
see also barley, bird food, wild, dog food,
feed, feed, maize-based, feed, mixed, feed
(cat), feed (dog), feed (gluten), feed (horse),
feed (maize), feed (poultry), feed (rodent),
maize bran, maize fine fractions, maize

flakes, maize germ, maize germ/bran, maize germ meal, maize gluten, maize grits, maize kernels, maize meal, maize powder, maize screenings, maize, "Baby", maize, ground, maize, preharvest, rat chow, wheat

FUMONISIN B$_3$
incidence: 11/11, conc. range: 155–960 μg/kg, Ø conc.: 361 μg/kg, country: Iran[22]
incidence: 3/15, conc. range: 118–432 μg/kg, Ø conc.: 232 μg/kg, country: Vietnam[104]
incidence: 2/3, conc. range: 75–224 μg/kg, Ø conc.: 149.5 μg/kg, country: Thailand[117]
incidence: 1/2, conc.: 50 μg/kg, country: Thailand[117]
incidence: 32/44, conc. range: 50–31,000 μg/kg, Ø conc.: 2500 μg/kg, country: Korea[160]
incidence: 20/20*, conc. range: <50–420 μg/kg, country: USA,[220], *food and feed
incidence: 47/128*, conc. range: 0–400 μg/kg, Ø conc.: 82.6 μg/kg, country: South Africa[258], *ncac
incidence: 1/1, conc.: 7249 μg/kg, country: Ghana[402]
incidence: 28/30*, conc. range: 90–1070 μg/kg, Ø conc.: 432 μg/kg, country: Argentina[422], *ncac
incidence: 2/5, conc. range: 2261.21–8000.84 μg/kg, Ø conc.: 5131.03 μg/kg, country: USA[431]
see also feed, feed (maize), feed (rodent), maize bran, maize fine fractions, maize germ meal, maize kernels, maize powder, maize screenings

FUMONISIN B$_4$
incidence: 23/44, conc. range: 80–11,000 μg/kg, Ø conc.: 1600 μg/kg, country: Korea[160]

FUMONISIN B$_1$ + B$_2$
incidence: 21*/21**, conc. range: 14–1036 μg/kg, country: Germany[248], *also contaminated with fumonisin B$_3$, **imported from Argentina
incidence: 102/105*, conc. range: 12–11,661 μg/kg, Ø conc.: 645 μg/kg, country: Croatia[318], *ncac

incidence: 97/104*, conc. range: 12–2524 μg/kg, Ø conc.: 134 μg/kg, country: Croatia[318], *ncac
see also feed (pig), maize and feed samples

FUMONISIN B$_1$ + B$_2$ + B$_3$
incidence: 85/317, conc. range: 6–7132 μg/kg, country: Germany[248]

FUMONISIN C$_1$
incidence: 31/44, conc. range: 60–11,000 μg/kg, Ø conc.: 1900 μg/kg, country: Korea[160]

FUMONISIN C$_3$
incidence: 5/44, conc. range: 100–12,000 μg/kg, Ø conc.: 1700 μg/kg, country: Korea[160]

FUMONISIN C$_4$
incidence: 18/44, conc. range: 50–3300 μg/kg, Ø conc.: 500 μg/kg, country: Korea[160]

FUMONISIN
incidence: 233/234, conc. range: ≤5000–25,000 μg/kg, country: USA[147]
incidence: 248/250, conc. range: ≤5000–>25,000 μg/kg, country: USA[147]
incidence: 134/461, conc. range: ≤1710 μg/kg, Ø conc.: 310 μg/kg, country: Canada[346]
incidence: 13/99, conc. range: 1200–3200 μg/kg, Ø conc.: 2400 μg/kg, country: USA[370]
see also maize screenings, maize, shelled

FUMONISINS
incidence: 92/97*, conc. range: 30–2123 μg/kg, country: UK[97], *imported from France
incidence: 37/37*, conc. range: 1001–5007 μg/kg, country: UK[97], *imported from Argentina
incidence: 5/5*, conc. range: 30–771 μg/kg, country: UK[97], *imported
incidence: 147/150, conc. range: 96–22,600 μg/kg, Ø conc.: 5356.7 μg/kg, country: Brazil[111]
incidence: 36/36, conc. range: 810–23,700 μg/kg, Ø conc.: 9900 μg/kg, country: Brazil[336]

see also feed (cattle and dairy), feed
(chicken), feed (dairy), feed (horse), maize
germ, maize gluten, maize meal, maize
screenings

FUSAPROLIFERIN
incidence: 4/8, conc. range:
100–30,000 µg/kg, country: USA[113]

FUSARENONE-X
incidence: 4/12, conc. range:
3000–15,000 µg/kg, Ø conc.: 7500 µg/kg,
country: Nigeria[64]
incidence: 4/24, conc. range: 16–94 µg/kg,
Ø conc.: 55 µg/kg, country: Germany[366]
see also barley, feed components, feed
(poultry), maize ears, maize germ/bran,
maize gluten, maize kernels, maize meal,
maize screenings, oats, wheat

GRISEOFULVIN
incidence: 2/7345, conc. range: nc, country:
Hungary[209]

HT-2 TOXIN
incidence: 2/11, conc. range: 500–700 µg/kg,
Ø conc.: 600 µg/kg, country: Hungary[50]
incidence: 1/12, conc.: 3000 µg/kg, country:
Nigeria[64]
incidence: 1/28, conc.: 600 µg/kg, country:
Germany[68]
incidence: 1/94, conc.: 1 µg/kg, country:
Canada[240]
incidence: 14/60, conc. range:
110–1000 µg/kg, country: Canada[346]
incidence: 3/152, conc. range: 80–120 µg/kg,
Ø conc.: 105 µg/kg, country: Austria[363]
incidence: 17/24, conc. range: 5–41 µg/kg,
Ø conc.: 18 µg/kg, country: Germany[366]
incidence: 8/9*, conc. range: 7–1307 µg/kg,
Ø conc.: 183 µg/kg, country: Germany[366],
*wp
see also barley, feed components, feed (dog),
feed (fish), feed (pig), feed (poultry), feed
(reindeer), grains, mixed, grains (no
specification), maize germ/bran, maize
gluten, maize meal, maize screenings, oats,
rye, silage, wheat

MONILIFORMIN
incidence: 6/12, conc. range: 450–8530 µg/kg,
Ø conc.: 3767.5 µg/kg, country: Poland[259]
incidence: 8/9*, conc. range:
17,000–425,000 µg/kg, Ø conc.:
133,750 µg/kg, country: Poland[269], *ncac
incidence: 1*/42, conc.: nd–200,000 µg/kg,
country: Italy[275], *s
incidence: 23/152, conc. range: 50–800 µg/kg,
Ø conc.: 216.6 µg/kg, country: Austria[363]
see also barley, feed, mixed, feed (poultry),
maize flakes, maize germ, maize germ/bran,
maize gluten, maize meal, maize screenings,
maize, "Baby", oats, rice bran, triticale,
wheat, wheat, summer, wheat, winter

MONOACETOXYSCIRPENOL
incidence: 5/24, conc. range: 5–51 µg/kg,
Ø conc.: 37 µg/kg, country: Germany[366]
incidence: 6/9*, conc. range: 9–78 µg/kg,
Ø conc.: 26 µg/kg, country: Germany[366], *wp
see also feed components, maize gluten,
maize meal, maize screenings, silage, wheat

NIVALENOL
incidence: 3/760, conc. range: 130–260 µg/kg,
Ø conc.: 186.7 µg/kg, country: Hungary[6]
incidence: 2/3, conc. range: 1180–4280 µg/kg,
Ø conc.: 2730 µg/kg, country: France[61]
incidence: 3/12, conc. range: 800–1000 µg/kg,
Ø conc.: 933.3 µg/kg, country: Nigeria[64]
incidence: 2/15, conc. range: 501–1251 µg/kg,
Ø conc.: 858 µg/kg, country: Vietnam[104]
incidence: 4/85, conc. range: ≤133 µg/kg,
Ø conc.: 89 µg/kg, country: Austria[239]
incidence: 1/94, conc.: 311 µg/kg, country:
Canada[240]
incidence: 6/7, conc. range:
1800–32,500 µg/kg, Ø conc.: 8900 µg/kg,
country: Poland[345]
incidence: 2/60, conc. range: 120 µg/kg,
Ø conc.: 120 µg/kg, country: Canada[346]
incidence: 11/152, conc. range: 60–330 µg/kg,
Ø conc.: 110 µg/kg, country: Austria[363]
incidence: 20/24, conc. range: 21–1388 µg/kg,
Ø conc.: 406 µg/kg, country: Germany[366]
incidence: 9/9*, conc. range: 116–5910 µg/kg,
Ø conc.: 1087 µg/kg, country:
Germany[366], *wp

see also barley, barley, husked, barley, unhusked (naked), barley (pressed), bran, feed, feed components, feed (cattle), feed (poultry), feed (reindeer), feeds, industrial, maize ears, maize germ, maize germ/bran, maize gluten, maize kernels, maize meal, maize powder, maize screenings, maize, "Baby", oats, rye, silage, triticale, wheat

OCHRATOXIN A
incidence: 71/760, conc. range: 60–1850 μg/kg, Ø conc.: 320 μg/kg, country: Hungary[6]
incidence: 3/40*, conc. range: 1.7–82 μg/kg, Ø conc.: 80.3 μg/kg, country: Germany[13], *ncac
incidence: 50/191, conc. range: 45–5125 μg/kg, Ø conc.: 490 μg/kg, country: Yugoslavia[75]
incidence: 12/97*, conc. range: 0.1–1.4 μg/kg, country: UK[97], *imported from France
incidence: 1/37*, conc.: 0.3 μg/kg, country: UK[97], *imported from Argentina
incidence: 1/5*, conc.: 1.5 μg/kg, country: UK[97], *imported
incidence: 1/6, conc.: 73 μg/kg, country: The Netherlands[103]
incidence: 1/26, conc.: 3 μg/kg, country: Indonesia[105]
incidence: ?/7, conc. range: ≤16,000 μg/kg, country: USA[171]
incidence: 3/7345, conc. range: nc, country: Hungary[209]
incidence: 1/283, conc.: 110–150 μg/kg, country: USA[217]
incidence: 3/293, conc. range: 83–166 μg/kg, country: USA[218]
incidence: 12/463, conc. range: 20–200 μg/kg, Ø conc.: 74.6 μg/kg, country: France[262]
incidence: 6/461, conc. range: 20–200 μg/kg, Ø conc.: 52.5 μg/kg, country: France[262]
incidence: 1/1*, conc. range: ≈110–150 μg/kg, country: USA[292], *ncac
incidence: 10/105*, conc. range: 0.36–224 μg/kg, Ø conc.: 37.87 μg/kg, country: Croatia[318], *ncac
incidence: 36/104*, conc. range: 0.26–614 μg/kg, Ø conc.: 57.13 μg/kg, country: Croatia[318], *ncac

incidence: 4/60, conc. range: 10–22 μg/kg, country: Canada[346]
incidence: 17/51*, conc. range: 0.02–40 μg/kg, country: Croatia[371], *ncac
incidence: 2/36*, conc. range: 128–206 μg/kg, Ø conc.: 167 μg/kg, country: Brazil[394], *ncac
incidence: 19/49*, conc. range: 0.9–2.54 μg/kg, Ø conc.: 1.47 μg/kg, country: Croatia[417], *ncac
incidence: 4/61*, conc. range: ≤114 μg/kg, Ø conc.: 4.5 μg/kg, country: Bangladesh[428], *ncac
see also alfalfa, barley, barley, oats, barley (high moisture), barley-soybean diet, bird food, domestic, bird food, wild, broilers feed, cereal grains, citrus pulp, coconut, expeller, corn cob mix silage, diet (dairy cow), diet (poultry), diet (starter), dog food, eat, egg production mixed feed, feed, feed wheat, oat and barley, feed, commercial mix, feed, mixed, feed, mixed (pelleted), feed (broilers), feed (cat), feed (cattle), feed (cereals), feed (pig), feed ec (pig), feed (poultry), feed ec (poultry), feed (poultry, pig), feed ec (rabbit), feed (trout), grain, mixed feed, grains, mixed, grains (heated), hay, horse bean, maize feed, milo, maize gluten, maize meal, maize, white, *Makhana* (*Euryale ferox* Salisb) puffs, milk production mixed feed, millet, oat and barley (hammer-milled), oats, palm products, peanut cake, peas, peas and beans, pet food, pig feedstuffs, pig grower diet, pig meal, piglet diet, poultry feedstuffs, rice bran, rice germ, rice germ cake, rye, sorghum, soybean groats, sunflower, sunflower seeds, extracted, tapioca, triticale, *Vicia faba*, wheat, wheat and barley, wheat bran, wheat hay, wheat, oats

T-2 TETRAOL
incidence: 3/9*, conc. range: 20–703 μg/kg, Ø conc.: 267 μg/kg, country: Germany[366], *wp
see also barley, feed components, silage, wheat

T-2 TOXIN
incidence: 214/760, conc. range: 50–980 μg/kg, Ø conc.: 225.2 μg/kg, country: Hungary[6]

incidence: 8/11, conc. range: 700–7500 µg/kg, Ø conc.: 2663 µg/kg, country: Hungary[50]
incidence: 1/30, conc.: 63 µg/kg, country: Romania[54]
incidence: 1/3, conc.: 20 µg/kg, country: France[61]
incidence: 4/28, conc. range: 100–200 µg/kg, country: Germany[68]
incidence: 4/6 * **, conc. range: 7–200 µg/kg, Ø conc.: 106.8 µg/kg, country: New Zealand[131], *field maize, **ncac
incidence: 5/7 * **, conc. range: 5–200 µg/kg, Ø conc.: 73 µg/kg, country: New Zealand[131], *harvest maize, **ncac
incidence: 4/5, conc. range: 14–60 µg/kg, Ø conc.: 43.5 µg/kg, country: New Zealand[131]
incidence: 1/7345, conc.: nc, country: Hungary[209]
incidence: 8/86*, conc. range: 550–2920 µg/kg, country: India[241], *ncac
incidence: 12/60, conc. range: 100–1000 µg/kg, country: Canada[346]
incidence: 3/152, conc. range: 130–260 µg/kg, Ø conc.: 200 µg/kg, country: Austria[363]
incidence: 6/24, conc. range: 6–8 µg/kg, Ø conc.: 7 µg/kg, country: Germany[366]
incidence: 6/9*, conc. range: 6–323 µg/kg, Ø conc.: 62 µg/kg, country: Germany[366], *wp
incidence: 9/61*, conc. range: ≤1093 µg/kg, Ø conc.: 96 µg/kg, country: Bangladesh[428], *ncac
see also alfalfa, barley, bran, diet (grower), diet (poultry), feed, feed components, feed, layer, feed, mixed, feed (dog), feed (fish), feed (mink), feed (pig), feed (poultry), feedstuff, forage grass, grain, mixed feed, grains (no specification), hay, maize germ/bran, maize gluten, maize meal, maize screenings, maize stalks (pith), oat and barley (hammer-milled), oats, peanuts, pig grower diet, piglet diet, rye, sorghum, triticale, wheat

T-2 TRIOL
incidence: 2/28, conc. range: 300 µg/kg, country: Germany[68]
incidence: 1/9*, conc.: 68 µg/kg, country: Germany[366], *wp
see also barley, feed components, grain, mixed feed, grains (no specification), oats, wheat

ZEARALENOLS (ALPHA- AND BETA-ZEARANELOL)
incidence: 8/8, conc. range: 7–86 µg/kg, Ø conc.: 36.5 µg/kg, country: Italy[53]

ZEARALENONE
incidence: 140/760, conc. range: 60–1350 µg/kg, Ø conc.: 228.9 µg/kg, country: Hungary[6]
incidence: 17/20, conc. range: 200–13,200 µg/kg, Ø conc.: 3100 µg/kg, country: USA[48]
incidence: 8/11, conc. range: 100–5800 µg/kg, Ø conc.: 3575 µg/kg, country: Hungary[50]
incidence: 3/6, conc. range: 420–1000 µg/kg, Ø conc.: 740 µg/kg, country: Austria[51], ncac
incidence: 8/8, conc. range: 156–7433 µg/kg, Ø conc.: 2667 µg/kg, country: Italy[53]
incidence: 4/30, conc. range: ≤1200 µg/kg, Ø conc.: 420 µg/kg, country: Romania[54]
incidence: 2/3, conc. range: 2500–10,000 µg/kg, Ø conc.: 6250 µg/kg, country: France[61]
incidence: 19/67, conc. range: 10–700 µg/kg, country: Germany[68]
incidence: 2/28, conc. range: 20–30 µg/kg, Ø conc.: 25 µg/kg, country: Germany[68]
incidence: 5/191, conc. range: 43–10,000 µg/kg, Ø conc.: 5100 µg/kg, country: Yugoslavia[75]
incidence: 93/97*, conc. range: 4–584 µg/kg, country: UK[97], *imported from France
incidence: 37/37*, conc. range: 21–357 µg/kg, country: UK[97], *imported from Argentina
incidence: 5/5*, conc. range: 4–34 µg/kg, country: UK[97], *imported
incidence: 2/6, conc. range: 17–150 µg/kg, Ø conc.: 84 µg/kg, country: The Netherlands[103]
incidence: 7/26, conc. range: 1–14 µg/kg, Ø conc.: 6.3 µg/kg, country: Indonesia[105]
incidence: 6/7, conc. range: 4.9–35 µg/kg, Ø conc.: 12.2 µg/kg, country: Germany[107]
incidence: 7/8, conc. range: 400–1600 µg/kg, Ø conc.: 730 µg/kg, country: Zambia[114]
incidence: 3/11, conc. range: 80–160 µg/kg, Ø conc.: 110 µg/kg, country: Zambia[114]
incidence: 9/14, conc. range: 80–6000 µg/kg, Ø conc.: 1710 µg/kg, country: Zambia[114]

incidence: 663/2311, conc. range: 5–17,350 µg/kg, country: Austria[115]

incidence: 249/293, conc. range: ≤1510 µg/kg, Ø conc.: 225 µg/kg, country: Australia[116]

incidence: 11/52, conc. range: 1–13,500 µg/kg, country: Indonesia[118]

incidence: 4/4, conc. range: ~100–2500 µg/kg, country: Germany[124]

incidence: 6/6 * **, conc. range: 300–2200 µg/kg, Ø conc.: 783 µg/kg, country: New Zealand[131], *field maize, **ncac

incidence: 3/7 * **, conc. range: 200–16,000 µg/kg, Ø conc.: 6366.7 µg/kg, country: New Zealand[131], *harvest maize, **ncac

incidence: 3/5, conc. range: 40–100 µg/kg, Ø conc.: 80 µg/kg, country: New Zealand[131]

incidence: 1/?, conc.: 2000 µg/kg, country: USA[179]

incidence: 937/937, conc. range: <10 µg/kg (574 sa), 10–50 µg/kg (195 sa), 50–200 µg/kg (124 sa), 200–500 µg/kg (29 sa), 500–1000 µg/kg (11 sa), >1000 µg/kg (4 sa) country: Austria[182]

incidence: ?/10, conc. range: 20 µg/kg, country: India[183]

incidence: 98/98*, conc. range: <20 µg/kg (89 sa), 20–100 µg/kg (9 sa), Ø conc.: 79 µg/kg, country: Italy[206], *ncac

incidence: 104/104*, conc. range: <20 µg/kg (48 sa), 20–100 µg/kg (41 sa), >100 µg/kg (15 sa), Ø conc.: 453 µg/kg, country: Italy[206], *ncac

incidence: 94/94*, conc. range: <20 µg/kg (91 sa), 20–100 µg/kg (3 sa), Ø conc.: 49 µg/kg, country: Italy[206], *ncac

incidence: 114/114*, conc. range: <20 µg/kg (113 sa), 20–100 µg/kg (1 sa), Ø conc.: 13 µg/kg, country: Italy[206], *ncac

incidence: 93/93*, conc. range: <20 µg/kg (92 sa), 20–100 µg/kg (1 sa), Ø conc.: 27 µg/kg, country: Italy[206], *ncac

incidence: 4/7345, conc. range: nc, country: Hungary[209]

incidence: 28/65, conc. range: 100–2909 µg/kg, country: USA[215]

incidence: 38/223, conc. range: 100–5000 µg/kg, country: USA[216]

incidence: 68/144, conc. range: 5–5670 µg/kg, Ø conc.: 310 µg/kg, country: Austria[223]

incidence: 2/2, conc. range: 6400–12,800 µg/kg, Ø conc.: 9600 µg/kg, country: Africa[228]

incidence: 60/85, conc. range: ≤751 µg/kg, Ø conc.: 126 µg/kg, country: Austria[239]

incidence: 9/86*, conc. range: 760–1500 µg/kg, country: India[241], *ncac

incidence: 18/154, conc. range: 800–9600 µg/kg, country: Spain[242]

incidence: 6/223*, conc. range: 200–5000 µg/kg, Ø conc.: 1750 µg/kg, country: USA[251], *is

incidence: 27/31, conc. range: 100–10,000 µg/kg, country: USA[254]

incidence: 5/6, conc. range: 60–32,000 µg/kg, Ø conc.: 8634 µg/kg, country: South Africa[263]

incidence: 9/165*, conc. range: 260–2937 µg/kg, Ø conc.: 1079 µg/kg, country: Brazil[271], *ncac

incidence: 7/163*, conc. range: 260–9830 µg/kg, Ø conc.: 2980 µg/kg, country: Brazil[271], *ncac

incidence: 167/671, conc. range: ≤1000 µg/kg, Ø conc.: 80 µg/kg, country: Canada[346]

incidence: 4/4*, conc. range: 7–2910 µg/kg, Ø conc.: 830.3 µg/kg, country: USA[354], *ncac

incidence: 55/152, conc. range: 10–340 µg/kg, Ø conc.: 63.3 µg/kg, country: Austria[363]

incidence: 4/18*, conc. range: 3.4–5.8 µg/kg, Ø conc.: 5 µg/kg, country: Korea[364], *ncac

incidence: 23/24, conc. range: 2–310 µg/kg, Ø conc.: 42 µg/kg, country: Germany[366]

incidence: 9/9*, conc. range: 7–492 µg/kg, Ø conc.: 131 µg/kg, country: Germany[366], *wp

incidence: 38/70, conc. range: 4–2250 µg/kg, Ø conc.: 326 µg/kg, country: Romania[367]

incidence: 5/75, conc. range: 70–150 µg/kg (5 sa), country: USA[370]

incidence: 1/36*, conc.: 4640 µg/kg, country: Brazil[394], *ncac

incidence: 30/380*, conc. range: <54.2 µg/kg (1 sa), 54.2–100 µg/kg (4 sa), 100–500 µg/kg (23 sa), >500 µg/kg (2 sa), country: Brazil[398], *ncac

incidence: 65/214*, conc. range:
36.8–719 µg/kg, Ø conc.: 155 µg/kg, country:
Brazil[403], *ncac
incidence: 2/15*, conc. range: 4–140 µg/kg,
Ø conc.: 48.8 µg/kg, country: Italy[408], *ncac
incidence: 6/9*, conc. range: <50–50 µg/kg,
country: Cameroon[416], *ncac
incidence: 9/9*, conc. range: <50–1100 µg/kg,
country: Cameroon[416], *ncac
incidence: 41/49*, conc. range:
0.43–39.12 µg/kg, Ø conc.: 3.84 µg/kg,
country: Croatia[417], *ncac
incidence: 30/30*, conc. range:
3000–7000 µg/kg, Ø conc.: 5480 µg/kg,
country: Argentina[422], *ncac
incidence: 6/61*, conc. range: ≤30 µg/kg,
Ø conc.: 1.5 µg/kg, country: Bangladesh[428],
*ncac
see also alfalfa, barley, barley, husked, barley,
unhusked (naked), barley and feed, bone
meal, bran, broilers feed, *Carthamus* cake,
chick pea, concentrate, mixed, corn cob mix
silage, cotton cake, cottonseed, cottonseed
cake, diet (dairy cow), diet (poultry), diets
(mixed), feed, feed components, feed, mixed,
feed, mixed (primarily maize also maize, oats,
wheat), feed (bran), feed (broiler chicken),
feed (cattle), feed (chicken), feed (dairy), feed
(developing pig), feed (mill run, from
wheat), feed (miscellaneous), feed (pig), feed
(poultry), feed (poultry, pig), feed (starter
chicken), feedstuff, fish meal, forage grass,
grain, bruised, grain, mixed feed, grains (no
specification), hay, maize ears, maize flakes,
maize germ, maize germ/bran, maize gluten,
maize kernels, maize meal, maize oil cake,
maize screenings, maize stalks (pith), maize,
"Baby", maize, hybrid, maize, shelled, maize,
unshelled, maize, white, maize grain,
artificially dried, maize grain, crib dried,
maize grain, ensiled, milk production mixed
feed, oats, *Paspalum palidosum*, straw, peanut
hulls/skins, rice bran, rice germ, rice germ
cake, rye, silage, sorghum, soybeans,
soybeans, extracted, sunflower cake, tapioca,
triticale, wheat, wheat bran, wheat bran and
chana testa, wheat soya meal

Maize and feed samples may contain the
following mycotoxins:

FUMONISIN $B_1 + B_2$
incidence: 252/291, conc. range:
100–5000 µg/kg (156 sa), 5100–10,000 µg/kg
(48 sa), 10,100–50,000 µg/kg (41 sa),
>50,000 µg/kg (7 sa), country: USA[301]
see also feed (pig), maize

Maize and maize screenings may contain the
following mycotoxins:

FUMONISIN B_1
incidence: 29/29, conc. range:
2000–330,000 µg/kg, Ø conc.: 76,000 µg/kg,
country: USA[142]
see also barley, bird food, wild, dog food,
feed, feed, complete ration, feed, general,
feed, layer, feed, maize-based, feed, mixed,
feed, pelleted ration, feed, screenings, feed,
sweet, feed (broilers), feed (cat), feed
(chicken), feed (dog), feed (gluten), feed
(horse), feed (maize), feed (pig), feed
(poultry), feed (rat), feed (rodent), forage
grass, maize, maize bran, maize ears, maize
fine fractions, maize flakes, maize germ,
maize germ/bran, maize germ meal, maize
gluten, maize grits, maize kernels, maize
meal, maize powder, maize screenings, maize,
"Baby", maize, ground, maize, preharvest,
maize, sweet feed, maize/oats mix, rat chow,
silage, sorghum, soybeans, wheat

Maize bran may contain the following
mycotoxins:

FUMONISIN B_1
incidence: ?/23, conc. range: 0–4480 µg/kg,
Ø conc.: 900 µg/kg*, country: South Africa[146],
*mean of all samples
incidence: 4/4, conc. range: 230–600 µg/kg,
Ø conc.: 420 µg/kg*, country: South Africa[147],
*mean of all samples
incidence: ?/85, conc. range: 0–3540 µg/kg,
Ø conc.: 410 µg/kg*, country: South Africa[147],
*mean of all samples
see also barley, bird food, wild, dog food,
feed, feed, complete ration, feed, general,

feed, layer, feed, maize-based, feed, mixed, feed, pelleted ration, feed, screenings, feed, sweet, feed (broilers), feed (cat), feed (chicken), feed (dog), feed (gluten), feed (horse), feed (maize), feed (pig), feed (poultry), feed (rat), feed (rodent), forage grass, maize, maize and maize screenings, maize ears, maize fine fractions, maize flakes, maize germ, maize germ/bran, maize germ meal, maize gluten, maize grits, maize kernels, maize meal, maize powder, maize screenings, maize, "Baby", maize, ground, maize, preharvest, maize, sweet feed, maize/oats mix, rat chow, silage, sorghum, soybeans, wheat

Fumonisin B_2
incidence: ?/23, conc. range: 0–1790 µg/kg, Ø conc.: 260 µg/kg*, country: South Africa[146], *mean of all samples
incidence: 4/4, conc. range: 0–130 µg/kg, Ø conc.: 30 µg/kg*, country: South Africa[147], *mean of all samples
incidence: ?/85, conc. range: 0–1270 µg/kg, Ø conc.: 120 µg/kg*, country: South Africa[147], *mean of all samples
see also barley, bird food, wild, dog food, feed, feed, maize-based, feed, mixed, feed (cat), feed (dog), feed (gluten), feed (horse), feed (maize), feed (poultry), feed (rodent), maize, maize fine fractions, maize flakes, maize germ, maize germ/bran, maize germ meal, maize gluten, maize kernels, maize meal, maize powder, maize screenings, maize, "Baby", maize, ground, maize, preharvest, rat chow, wheat

Fumonisin B_3
incidence: 4/4, conc. range: 0–50 µg/kg, Ø conc.: 10 µg/kg*, country: South Africa[147], *mean of all samples
incidence: 85, conc. range: 0–1030 µg/kg, Ø conc.: 40 µg/kg*, country: South Africa[147], *mean of all samples
see also feed, feed (maize), feed (rodent), maize, maize fine fractions, maize germ meal, maize kernels, maize powder, maize screenings

Maize ears may contain the following mycotoxins:

3-Acetyldeoxynivalenol
incidence: 2/5*, conc. range: 900–5900 µg/kg, Ø conc.: 3400 µg/kg, country: Poland[348], *ncac
see also barley, feed components, feed (poultry), feeds, grain, feeds, industrial, feedstuffs (rapeseed, turnip, fish meal, concentrates), maize, maize gluten, maize kernels, maize meal, maize screenings, oats, wheat

15-Acetyldeoxynivalenol
incidence: 2/5*, conc. range: 5800–8800 µg/kg, Ø conc.: 7300 µg/kg, country: Poland[348] *ncac
see also barley, feed components, feed, commercial mix, feed (poultry), maize, maize germ, maize germ/bran, maize gluten, maize kernels, maize meal, maize screenings, maize, "Baby", oats, silage, wheat

Deoxynivalenol
incidence: 3/5*, conc. range: 26700–208,300 µg/kg, Ø conc.: 123,933 µg/kg, country: Poland[348] *ncac
see also barley, barley, husked, barley, unhusked (naked), barley (pressed), bone meal, bran, broilers feed, calf fattening mixed feed, coconut, expeller, corn cob mix silage, cottonseed, cottonseed cake, dairy cattle feed, egg production mixed feed, feed, feed components, feed, commercial mix, feed, mixed, feed, mixed (primarily maize), feed (barley), feed (cattle), feed (chicken), feed (dog), feed (fish), feed (mill run, from wheat), feed (mink), feed (pig), feed (poultry), feed (reindeer), feeds, grain, feeds, industrial, feedstuff, feedstuffs (rapeseed, turnip, fish meal, concentrates), fish meal, grain, mixed feed, grains, mixed, grains (no specification), maize, maize fibre, maize germ, maize germ/bran, maize germ meal, maize gluten, maize kernels, maize meal, maize powder, maize screenings, maize stalks (pith), maize, "Baby", maize, hybrid, maize, white, oats, rice bran, rice germ cake, rye, silage, sorghum, soybeans, triticale, wheat,

wheat and barley, wheat, red hard winter, wheat, soft white winter, wheat, spring, wheat, winter

Fumonisin B$_1$
incidence: 8/8*, conc. range: ≤300,000 µg/kg, country: Italy[369] *ncac
see also barley, bird food, wild, dog food, feed, feed, complete ration, feed, general, feed, layer, feed, maize-based, feed, mixed, feed, pelleted ration, feed, screenings, feed, sweet, feed (broilers), feed (cat), feed (chicken), feed (dog), feed (gluten), feed (horse), feed (maize), feed (pig), feed (poultry), feed (rat), feed (rodent), forage grass, maize, maize and maize screenings, maize bran, maize fine fractions, maize flakes, maize germ, maize germ/bran, maize germ meal, maize gluten, maize grits, maize kernels, maize meal, maize powder, maize screenings, maize, "Baby", maize, ground, maize, preharvest, maize, sweet feed, maize/oats mix, rat chow, silage, sorghum, soybeans, wheat

Fusarenone-X
incidence: 2/5*, conc. range: 3600–20,000 µg/kg, Ø conc.: 11,800 µg/kg, country: Poland[348] *ncac
see also barley, feed components, feed (poultry), maize, maize germ/bran, maize gluten, maize kernels, maize meal, maize screenings, oats, wheat

Nivalenol
incidence: 2/5*, conc. range: 54,300–56,200 µg/kg, Ø conc.: 55,250 µg/kg, country: Poland[348] *ncac
see also barley, barley, husked, barley, unhusked (naked), barley (pressed), bran, feed, feed components, feed (cattle), feed (poultry), feed (reindeer), feeds, industrial, maize, maize germ, maize germ/bran, maize gluten, maize kernels, maize meal, maize powder, maize screenings, maize, "Baby", oats, rye, silage, triticale, wheat

Zearalenone
incidence: 5/5*, conc. range: 1600–350,000 µg/kg, Ø conc.: 76,200 µg/kg, country: Poland[348], *ncac

see also alfalfa, barley, barley, husked, barley, unhusked (naked), barley and feed, bone meal, bran, broilers feed, *Carthamus* cake, chick pea, concentrate, mixed, corn cob mix silage, cotton cake, cottonseed, cottonseed cake, diet (dairy cow), diet (poultry), diets (mixed), feed, feed components, feed, mixed, feed, mixed (primarily maize also maize, oats, wheat), feed (bran), feed (broiler chicken), feed (cattle), feed (chicken), feed (dairy), feed (developing pig), feed (mill run, from wheat), feed (miscellaneous), feed (pig), feed (poultry), feed (poultry, pig), feed (starter chicken), feedstuff, fish meal, forage grass, grain, bruised, grain, mixed feed, grains (no specification), hay, maize, maize flakes, maize germ, maize germ/bran, maize gluten, maize kernels, maize meal, maize oil cake, maize screenings, maize stalks (pith), maize, "Baby", maize, hybrid, maize, shelled, maize, unshelled, maize grain, artificially dried, maize grain, crib dried, maize grain, ensiled, milk production mixed feed, oats, *Paspalum palidosum*, straw, peanut hulls/skins, rice bran, rice germ, rice germ cake, rye, silage, sorghum, soybeans, soybeans, extracted, sunflower cake, tapioca, triticale, wheat, wheat bran, wheat bran and chana testa, wheat soya meal

Maize feed, milo   may contain the following mycotoxins:

Ochratoxin A
incidence: 2/2, conc. range: 7–12 µg/kg, Ø conc.: 10 µg/kg, country: The Netherlands[103]
see also alfalfa, barley, barley, oats, barley (high moisture), barley-soybean diet, bird food, domestic, bird food, wild, broilers feed, cereal grains, citrus pulp, coconut, expeller, corn cob mix silage, diet (dairy cow), diet (poultry), diet (starter), dog food, eat, egg production mixed feed, feed, feed wheat, oat and barley, feed, commercial mix, feed, mixed, feed, mixed (pelleted), feed (broilers), feed (cat), feed (cattle), feed (cereals), feed (pig), feed ec (pig), feed (poultry), feed ec (poultry), feed (poultry, pig), feed ec

(rabbit), feed (trout), grains, mixed, grains, mixed feed, grains (heated), hay, horse bean, maize, maize gluten, maize meal, maize, white, *Makhana* (*Euryale ferox* Salisb) puffs, milk production mixed feed, millet, oat and barley (hammer-milled), oats, palm products, peas, peas and beans, pet food, pig feedstuffs, pig grower diet, pig meal, piglet diet, poultry feedstuffs, rice bran, rice germ, rice germ cake, rye, sorghum, soybean groats, sunflower, sunflower seeds, extracted, tapioca, triticale, *Vicia faba*, wheat, wheat and barley, wheat bran, wheat hay, wheat, oats

Maize fibre may contain the following mycotoxins:

DEOXYNIVALENOL
incidence: 2/2, conc. range: 50–110 μg/kg, Ø conc.: 80 μg/kg, country: UK[36]
see also barley, barley, husked, barley, unhusked (naked), barley (pressed), bone meal, bran, broilers feed, calf fattening mixed feed, coconut, expeller, corn cob mix silage, cottonseed, cottonseed cake, dairy cattle feed, egg production mixed feed, feed, feed components, feed, commercial mix, feed, mixed, feed, mixed (primarily maize), feed (barley), feed (cattle), feed (chicken), feed (dog), feed (fish), feed (mill run, from wheat), feed (mink), feed (pig), feed (poultry), feed (reindeer), feeds, grain, feeds, industrial, feedstuff, feedstuffs (rapeseed, turnip, fish meal, concentrates), fish meal, grain, mixed feed, grains, mixed, grains (no specification), maize, maize ears, maize germ, maize germ/bran, maize germ meal, maize gluten, maize kernels, maize powder, maize screenings, maize stalks (pith), maize, "Baby", maize, hybrid, maize, white, oats, rice bran, rice germ cake, rye, silage, sorghum, soybeans, triticale, wheat, wheat and barley, wheat, red hard winter, wheat, soft white winter, wheat, spring, wheat, winter

Maize fine fractions may contain the following mycotoxins:

FUMONISIN B$_1$
incidence: 10/10* **, conc. range: 8660–21,350 μg/kg, Ø conc.: 13,390 μg/kg, country: South Africa[190], *imported, **ncac
see also barley, bird food, wild, dog food, feed, feed, complete ration, feed, general, feed, layer, feed, maize-based, feed, mixed, feed, pelleted ration, feed, screenings, feed, sweet, feed (broilers), feed (cat), feed (chicken), feed (dog), feed (gluten), feed (horse), feed (maize), feed (pig), feed (poultry), feed (rat), feed (rodent), forage grass, maize, maize and maize screenings, maize bran, maize ears, maize flakes, maize germ, maize germ/bran, maize germ meal, maize gluten, maize grits, maize kernels, maize meal, maize powder, maize screenings, maize, "Baby", maize, ground, maize, preharvest, maize, sweet feed, maize/oats mix, rat chow, silage, sorghum, soybeans, wheat

FUMONISIN B$_2$
incidence: 10/10* **, conc. range: 2720–6320 μg/kg, Ø conc.: 3993 μg/kg, country: South Africa[190], *imported, **ncac
see also barley, bird food, wild, dog food, feed, feed, maize-based, feed, mixed, feed (cat), feed (dog), feed (gluten), feed (horse), feed (maize), feed (poultry), feed (rodent), maize, maize bran, maize flakes, maize germ, maize germ/bran, maize germ meal, maize gluten, maize kernels, maize meal, maize powder, maize screenings, maize, "Baby", maize, ground, maize, preharvest, rat chow, wheat

FUMONISIN B$_3$
incidence: 10/10* **, conc. range: 960–2450 μg/kg, Ø conc.: 1671 μg/kg, country: South Africa[190], *imported, **ncac
see also feed, feed (maize), feed (rodent), maize, maize bran, maize germ meal, maize kernels, maize powder, maize screenings

Maize flakes Maize flakes for feed may contain the following mycotoxins:

FUMONISIN B$_1$
incidence: 2/3, conc. range: 190–280 μg/kg, Ø conc.: 235 μg/kg, country: UK[95]

see also barley, bird food, wild, dog food, feed, feed, complete ration, feed, general, feed, maize-based, feed, mixed, feed, pelleted ration, feed, screenings, feed, sweet, feed (cat), feed (chicken), feed (dog), feed (gluten), feed (horse), feed (maize), feed (pig), feed (poultry), feed (rat), feed (rodent), forage grass, maize, maize and maize screenings, maize bran, maize ears, maize fine fractions, maize germ, maize germ/bran, maize germ meal, maize gluten, maize grits, maize kernels, maize meal, maize powder, maize screenings, maize, "Baby", maize, ground, maize, preharvest, maize, sweet feed, maize/oats mix, rat chow, silage, sorghum, soybeans, wheat

FUMONISIN B_2
incidence: 1/3, conc.: 50 µg/kg, country: UK[95]
see also barley, bird food, wild, dog food, feed, feed, maize-based, feed, mixed, feed (cat), feed (dog), feed (gluten), feed (horse), feed (maize), feed (poultry), feed (rodent), maize, maize bran, maize fine fractions, maize germ, maize germ/bran, maize germ meal, maize gluten, maize kernels, maize meal, maize powder, maize screenings, maize, "Baby", maize, ground, maize, preharvest, rat chow, wheat

MONILIFORMIN
incidence: 3/3, conc. range: 90–130 µg/kg, Ø conc.: 110 µg/kg, country: UK[95]
see also barley, feed, mixed, feed (poultry), maize, maize germ, maize germ/bran, maize gluten, maize meal, maize screenings, maize, "Baby", oats, rice bran, triticale, wheat, wheat, summer, wheat, winter

ZEARALENONE
incidence: 6/9*, conc. range: 5–150 µg/kg, country: UK[94], *imported?
incidence: 3/3, conc. range: 80–110 µg/kg, country: UK[95]
see also alfalfa, barley, barley, husked, barley, unhusked (naked), barley and feed, bone meal, bran, broilers feed, *Carthamus* cake, chick pea, concentrate, mixed, corn cob mix silage, cotton cake, cottonseed, cottonseed cake, diet (dairy cow), diet (poultry), diets (mixed), feed, feed components, feed, mixed,

feed, mixed (primarily maize also maize, oats, wheat), feed (bran), feed (broiler chicken), feed (cattle), feed (chicken), feed (dairy), feed (developing pig), feed (mill run, from wheat), feed (miscellaneous), feed (pig), feed (poultry), feed (poultry, pig), feed (starter chicken), feedstuff, fish meal, forage grass, grain, bruised, grain, mixed feed, grains (no specification), hay, maize, maize ears, maize germ, maize germ/bran, maize gluten, maize kernels, maize meal, maize oil cake, maize screenings, maize stalks (pith), maize, "Baby", maize, hybrid, maize, shelled, maize, unshelled, maize, white, maize grain, artificially dried, maize grain, crib dried, maize grain, ensiled, milk production mixed feed, oats, *Paspalum palidosum*, straw, peanut hulls/skins, rice bran, rice germ, rice germ cake, rye, silage, sorghum, soybeans, soybeans, extracted, sunflower cake, tapioca, triticale, wheat, wheat bran, wheat bran and chana testa, wheat soya meal

Maize germ   may contain the following mycotoxins:

15-ACETYLDEOXYNIVALENOL
incidence: 4/7, conc. range: 15–40 µg/kg, Ø conc.: 31.3 µg/kg, country: UK[95]
see also barley, feed components, feed, commercial mix, feed (poultry), maize, maize ears, maize germ/bran, maize gluten, maize kernels, maize meal, maize screenings, maize, "Baby", oats, silage, wheat

AFLATOXIN B_1
incidence: 4/5*, conc. range: 1–17 µg/kg, country: UK[94], *imported?
see also alfalfa, *Ambadi* cake, animal feedstuffs (dairy cake), bagasse, barley, bengalgram husk, bird food, bird food, wild, biri testa, blackgram, blackgram husk, bran, broiler mixed feed, calf fattening mixed feed, calf fattening mixed feed (containing 4–20 % peanut products), *Carthamus* cake, castor cake, cereals, cereal products, chick pea, coconut cake, cocos, concentrate, mixed, concentrates, cotton cake, cottonseed, cottonseed (dehulled), cottonseed cake,

cottonseed extract, cottonseed meal, cottonseed meal (ammoniated), cottonseed meal (decorticated), cottonseed meats, cottonseed products, crumbles, crumbles, grower, cycad meal, dairy cattle feed, dairy cattle feed (containing 2–5 % peanut products), dairy cattle feed (containing 6–10 % peanut products), dairy cattle feed (containing 6–12 % peanut products), dairy cattle feed (containing more than 20 % peanut products), diets, mixed, dog food, egg production mixed feed, feed, feed and ingredients, feed, compound, feed, layer, feed, mixed, feed (beef), feed (broilers), feed (calf), feed (cat), feed (cattle), feed (chicken), feed (dairy), feed (dog), feed (dug), feed (fish), feed (gluten), feed (horse), feed (miscell-aneous), feed (pig), feed (poultry), feed (poultry, pig), feed (rabbit), feed (sheep), feed (50–60 % maize), fish meal, grains (no specification), grain by-products, greengram, hay/silage, horsegram, husk, *Jagni* cake, legume mixture, linseed, linseed cake, livol, *Mahua* cake, maize, maize gluten, maize grits, maize husk, maize meal, maize oil cake, maize powder, maize screenings, maize, ground, maize, hybrid, maize, preharvest, maize, yellow, maize (dark grains), *Makhana* (*Euryale ferox* Salisb) puffs, manioc, milk production mixed feed, mung testa, murkool, mustard cake, neem cake, niger cake, oats, palm kernel expeller cake, palm kernels, palm products, peanut cake, peanut cake (deoiled), peanut expeller, peanut hay, peanut meal, peanut, kernels, peanut, shells, peanuts, pellets, finisher, pig meal and pellets, pigeon pea, poultry feeds (peanut containing), rapeseed cake, redgram husk, rice, rice bran, rice bran (deoiled), rice chaff, rice crack, rice germ, rice germ cake, rice meal, rice straw, rice (damaged), rice (polish), safflower cake, sal seed cake, sesame, sesame cake, sorghum, soybean meal, soybeans, sunflower, sunflower cake, sunflower flour, tapioca, wheat, wheat bran, wheat bran and chana testa

AFLATOXINS (TOTAL)
incidence: 4/5*, conc. range: 1–18 µg/kg, country: UK[94], *imported?

see also cottonseed, maize gluten, maize (dark grains), palm products, rice bran, soybeans, sunflower

DEOXYNIVALENOL
incidence: 7/7, conc. range: 25–360 µg/kg, Ø conc.: 145 µg/kg, country: UK[95]
see also barley, barley, husked, barley, unhusked (naked), barley (pressed), bone meal, bran, broilers feed, calf fattening mixed feed, coconut, expeller, corn cob mix silage, cottonseed, cottonseed cake, dairy cattle feed, egg production mixed feed, feed, feed components, feed, commercial mix, feed, mixed, feed, mixed (primarily maize), feed (barley), feed (cattle), feed (chicken), feed (dog), feed (fish), feed (mill run, from wheat), feed (mink), feed (pig), feed (poultry), feed (reindeer), feeds, grain, feeds, industrial, feedstuff, feedstuffs (rapeseed, turnip, fish meal, concentrates), fish meal, grain, mixed feed, grains, mixed, grains (no specification), maize, maize ears, maize fibre, maize germ/bran, maize germ meal, maize gluten, maize kernels, maize powder, maize screenings, maize stalks (pith), maize, "Baby", maize, hybrid, maize, white, oats, rice bran, rice germ cake, rye, silage, sorghum, soybeans, triticale, wheat, wheat and barley, wheat, red hard winter, wheat, soft white winter, wheat, spring, wheat, winter

FUMONISIN B$_1$
incidence: 7/7, conc. range: 4200–7400 µg/kg, Ø conc.: 5728.6 µg/kg, country: UK[95]
see also barley, bird food, wild, dog food, feed, feed, complete ration, feed, general, feed, layer, feed, maize-based, feed, mixed, feed, pelleted ration, feed, screenings, feed, sweet, feed (broilers), feed (cat), feed (chicken), feed (dog), feed (gluten), feed (horse), feed (maize), feed (pig), feed (poultry), feed (rat), feed (rodent), forage grass, maize, maize and maize screenings, maize bran, maize ears, maize fine fractions, maize flakes, maize germ/bran, maize germ meal, maize gluten, maize grits, maize kernels, maize meal, maize powder, maize screenings, maize, "Baby", maize, ground,

maize, preharvest, maize, sweet feed,
maize/oats mix, rat chow, silage, sorghum,
soybeans, wheat

FUMONISIN B_2
incidence: 7/7, conc. range: 580–1100 µg/kg,
Ø conc.: 768.6 µg/kg, country: UK[95]
see also barley, bird food, wild, dog food,
feed, feed, maize-based, feed, mixed, feed
(cat), feed (dog), feed (gluten), feed (horse),
feed (maize), feed (poultry), feed (rodent),
maize, maize bran, maize fine fractions,
maize flakes, maize germ/bran, maize germ
meal, maize gluten, maize kernels, maize
meal, maize powder, maize screenings, maize,
"Baby", maize, ground, maize, preharvest, rat
chow, wheat

FUMONISINS
incidence: 4/5*, conc. range: 100–1060 µg/kg,
country: UK[94], *imported?
see also feed (cattle and dairy), feed
(chicken), feed (dairy), feed (horse), maize,
maize gluten, maize meal, maize screenings

MONILIFORMIN
incidence: 7/7, conc. range: 660–1100 µg/kg,
Ø conc.: 778.6 µg/kg, country: UK[95]
see also barley, feed, mixed, feed (poultry),
maize, maize flakes, maize germ/bran, maize
gluten, maize meal, maize screenings, maize,
"Baby", oats, rice bran, triticale, wheat,
wheat, summer, wheat, winter

NIVALENOL
incidence: 1/7, conc.: 35 µg/kg, country: UK[95]
see also barley, barley, husked, barley,
unhusked (naked), barley (pressed), bran,
feed, feed components, feed (cattle), feed
(poultry), feed (reindeer), feeds, industrial,
maize, maize ears, maize germ/bran, maize
gluten, maize kernels, maize meal, maize
powder, maize screenings, maize, "Baby",
oats, rye, silage, triticale, wheat

ZEARALENONE
incidence: 1/5, conc.: 15 µg/kg, country:
UK[94], *imported?
incidence: 7/7, conc. range: 50–80 µg/kg,
Ø conc.: 67.1 µg/kg, country: UK[95]

see also alfalfa, barley, barley, husked, barley,
unhusked (naked), barley and feed, bone
meal, bran, broilers feed, *Carthamus* cake,
chick pea, concentrate, mixed, corn cob mix
silage, cotton cake, cottonseed, cottonseed
cake, diet (dairy cow), diet (poultry), diets
(mixed), feed, feed components, feed, mixed,
feed, mixed (primarily maize also maize, oats,
wheat), feed (bran), feed (broiler chicken),
feed (cattle), feed (chicken), feed (dairy), feed
(developing pig), feed (mill run, from
wheat), feed (miscellaneous), feed (pig), feed
(poultry), feed (poultry, pig), feed (starter
chicken), feedstuff, fish meal, forage grass,
grain, bruised, grain, mixed feed, grains (no
specification), hay, maize, maize ears, maize
flakes, maize germ/bran, maize gluten, maize
kernels, maize meal, maize oil cake, maize
screenings, maize stalks (pith), maize,
"Baby", maize, hybrid, maize, shelled, maize,
unshelled, maize, white, maize grain,
artificially dried, maize grain, crib dried,
maize grain, ensiled, milk production mixed
feed, oats, *Paspalum palidosum*, straw, peanut
hulls/skins, rice bran, rice germ, rice germ
cake, rye, silage, sorghum, soybeans,
soybeans, extracted, sunflower cake, tapioca,
triticale, wheat, wheat bran, wheat bran and
chana testa, wheat soya meal

Maize germ/bran   may contain the following
mycotoxins:

15-ACETYLDEOXYNIVALENOL
incidence: 6/6, conc. range: 50–170 µg/kg,
Ø conc.: 107.2 µg/kg, country: UK[95]
see also barley, feed components, feed,
commercial mix, feed (poultry), maize, maize
ears, maize germ, maize gluten, maize
kernels, maize meal, maize screenings, maize,
"Baby", oats, silage, wheat

DEOXYNIVALENOL
incidence: 6/6, conc. range: 230–1000 µg/kg,
Ø conc.: 731.7 µg/kg, country: UK[95]
see also barley, barley, husked, barley,
unhusked (naked), barley (pressed), bone
meal, bran, broilers feed, calf fattening mixed
feed, coconut, expeller, corn cob mix silage,

cottonseed, cottonseed cake, dairy cattle feed, egg production mixed feed, feed, feed components, feed, commercial mix, feed, mixed, feed, mixed (primarily maize), feed (barley), feed (cattle), feed (chicken), feed (dog), feed (fish), feed (mill run, from wheat), feed (mink), feed (pig), feed (poultry), feed (reindeer), feeds, grain, feeds, industrial, feedstuff, feedstuffs (rapeseed, turnip, fish meal, concentrates), fish meal, grain, mixed feed, grains, mixed, grains (no specification), maize, maize ears, maize fibre, maize germ, maize germ meal, maize gluten, maize kernels, maize powder, maize screenings, maize stalks (pith), maize, "Baby", maize, hybrid, maize, white, oats, rice bran, rice germ cake, rye, silage, sorghum, soybeans, triticale, wheat, wheat and barley, wheat, red hard winter, wheat, soft white winter, wheat, spring, wheat, winter

FUMONISIN B$_1$
incidence: 6/6, conc. range: 390–1500 µg/kg, Ø conc.: 1083.3 µg/kg, country: UK[95]
see also barley, bird food, wild, dog food, feed, feed, complete ration, feed, general, feed, layer, feed, maize-based, feed, mixed, feed, pelleted ration, feed, screenings, feed, sweet, feed (broilers), feed (cat), feed (chicken), feed (dog), feed (gluten), feed (horse), feed (maize), feed (pig), feed (poultry), feed (rat), feed (rodent), forage grass, maize, maize and maize screenings, maize bran, maize ears, maize fine fractions, maize flakes, maize germ, maize germ meal, maize gluten, maize grits, maize kernels, maize meal, maize powder, maize screenings, maize, "Baby", maize, ground, maize, preharvest, maize, sweet feed, maize/oats mix, rat chow, silage, sorghum, soybeans, wheat

FUMONISIN B$_2$
incidence: 6/6, conc. range: 110–250 µg/kg, Ø conc.: 190 µg/kg, country: UK[95]
see also barley, bird food, wild, dog food, feed, feed, maize-based, feed, mixed, feed (cat), feed (dog), feed (gluten), feed (horse), feed (maize), feed (poultry), feed (rodent), maize, maize bran, maize fine fractions, maize flakes, maize

germ, maize germ meal, maize gluten, maize kernels, maize meal, maize powder, maize screenings, maize, "Baby", maize, ground, maize, preharvest, rat chow, wheat

FUSARENONE-X
incidence: 4/6, conc. range: 15–30 µg/kg, Ø conc.: 20 µg/kg, country: UK[95]
see also barley, feed components, feed (poultry), maize, maize ears, maize gluten, maize kernels, maize meal, maize screenings, oats, wheat

HT-2 TOXIN
incidence: 6/6, conc. range: 10–35 µg/kg, Ø conc.: 20.8 µg/kg, country: UK[95]
see also barley, feed components, feed (dog), feed (fish), feed (pig), feed (poultry), feed (reindeer), grain, mixed feed, grains, mixed, grains (no specification), maize, maize gluten, maize meal, maize screenings, oats, rye, silage, wheat

MONILIFORMIN
incidence: 6/6, conc. range: 270–570 µg/kg, Ø conc.: 426.7 µg/kg, country: UK[95]
see also barley, feed, mixed, feed (poultry), maize, maize flakes, maize germ, maize gluten, maize meal, maize screenings, maize, "Baby", oats, rice bran, triticale, wheat, wheat, summer, wheat, winter

NIVALENOL
incidence: 6/6, conc. range: 80–400 µg/kg, Ø conc.: 196.7 µg/kg, country: UK[95]
see also barley, barley, husked, barley, unhusked (naked), barley (pressed), bran, feed, feed components, feed (cattle), feed (poultry), feed (reindeer), feeds, industrial, maize, maize ears, maize germ, maize gluten, maize kernels, maize meal, maize powder, maize screenings, maize, "Baby", oats, rye, silage, triticale, wheat

T-2 TOXIN
incidence: 4/6, conc. range: 10–20 µg/kg, Ø conc.: 15 µg/kg, country: UK[95]
see also alfalfa, barley, bran, diet (grower), diet (poultry), feed, feed components, feed, layer, feed, mixed, feed (dog), feed (fish), feed (mink), feed (pig), feed (poultry),

feedstuff, forage grass, grain, mixed feed,
grains (no specification), hay, maize, maize
gluten, maize meal, maize screenings, maize
stalks (pith), oat and barley (hammer-milled),
oats, peanuts, pig grower diet, piglet diet, rye,
sorghum, triticale, wheat

ZEARALENONE
incidence: 6/6, conc. range: 160–540 µg/kg,
Ø conc.: 330 µg/kg, country: UK[95]
see also alfalfa, barley, barley, husked, barley,
unhusked (naked), barley and feed, bone
meal, bran, broilers feed, *Carthamus* cake,
chick pea, concentrate, mixed, corn cob mix
silage, cotton cake, cottonseed, cottonseed
cake, diet (dairy cow), diet (poultry), diets
(mixed), feed, feed components, feed, mixed,
feed, mixed (primarily maize also maize, oats,
wheat), feed (bran), feed (broiler chicken),
feed (cattle), feed (chicken), feed (dairy), feed
(developing pig), feed (mill run, from
wheat), feed (miscellaneous), feed (pig), feed
(poultry), feed (poultry, pig), feed (starter
chicken), feedstuff, fish meal, forage grass,
grain, bruised, grain, mixed feed, grains (no
specification), hay, maize, maize ears, maize
flakes, maize germ, maize gluten, maize
kernels, maize meal, maize oil cake, maize
screenings, maize stalks (pith), maize,
"Baby", maize, hybrid, maize, shelled, maize,
unshelled, maize, white, maize grain,
artificially dried, maize grain, crib dried,
maize grain, ensiled, milk production mixed
feed, oats, *Paspalum palidosum*, straw, peanut
hulls/skins, rice bran, rice germ, rice germ
cake, rye, silage, sorghum, soybeans,
soybeans, extracted, sunflower cake, tapioca,
triticale, wheat, wheat bran, wheat bran and
chana testa, wheat soya meal

Maize germ meal   Maize germ meal for feed
may contain the following mycotoxins:

DEOXYNIVALENOL
incidence: 2/2, conc. range: 430–1300 µg/kg,
Ø conc.: 865 µg/kg, country: UK[36]
see also barley, barley, husked, barley,
unhusked (naked), barley (pressed), bone
meal, bran, broilers feed, calf fattening mixed

feed, coconut, expeller, corn cob mix silage,
cottonseed, cottonseed cake, dairy cattle feed,
egg production mixed feed, feed, feed
components, feed, commercial mix, feed,
mixed, feed, mixed (primarily maize), feed
(barley), feed (cattle), feed (chicken), feed
(dog), feed (fish), feed (mill run, from
wheat), feed (mink), feed (pig), feed
(poultry), feed (reindeer), feeds, grain, feeds,
industrial, feedstuff, feedstuffs (rapeseed,
turnip, fish meal, concentrates), fish meal,
grain, mixed feed, grains, mixed, grains (no
specification), maize, maize ears, maize fibre,
maize germ, maize germ/bran, maize gluten,
maize kernels, maize powder, maize
screenings, maize stalks (pith), maize, "Baby",
maize, hybrid, maize, white, oats, rice bran,
rice germ cake, rye, silage, sorghum,
soybeans, triticale, wheat, wheat and barley,
wheat, red hard winter, wheat, soft white
winter, wheat, spring, wheat, winter

FUMONISIN B$_1$
incidence: 2/2, conc. range: 480–2180 µg/kg,
Ø conc.: 1330 µg/kg*, country: South
Africa[147], *mean of all samples
incidence: 21/21, conc. range: 50–1010 µg/kg,
Ø conc.: 290 µg/kg*, country: South Africa[147],
*mean of all samples
see also barley, bird food, wild, dog food,
feed, feed, complete ration, feed, general,
feed, layer, feed, maize-based, feed, mixed,
feed, pelleted ration, feed, screenings, feed,
sweet, feed (broilers), feed (cat), feed
(chicken), feed (dog), feed (gluten), feed
(horse), feed (maize), feed (pig), feed
(poultry), feed (rat), feed (rodent), forage
grass, maize, maize and maize screenings,
maize bran, maize ears, maize fine fractions,
maize flakes, maize germ, maize germ/bran,
maize gluten, maize grits, maize kernels,
maize meal, maize powder, maize screenings,
maize, "Baby", maize, ground, maize,
preharvest, maize, sweet feed, maize/oats mix,
rat chow, silage, sorghum, soybeans, wheat

FUMONISIN B$_2$
incidence: 2/2, conc. range: 0–1550 µg/kg,
Ø conc.: 770 µg/kg*, country: South Africa[147],
*mean of all samples

incidence: 21/21, conc. range: 0–420 µg/kg,
Ø conc.: 80 µg/kg*, country: South Africa[147],
*mean of all samples
see also barley, bird food, wild, dog food,
feed, feed, maize-based, feed, mixed, feed
(cat), feed (dog), feed (gluten), feed (horse),
feed (maize), feed (poultry), feed (rodent),
maize, maize bran, maize fine fractions,
maize flakes, maize germ, maize germ/bran,
maize gluten, maize kernels, maize meal,
maize powder, maize screenings, maize,
"Baby", maize, ground, maize, preharvest, rat
chow, wheat

Fumonisin B$_3$
incidence: 2/2, conc. range: 50–900 µg/kg,
Ø conc.: 480 µg/kg*, country: South Africa[147],
*mean of all samples
incidence: 21/21, conc. range: 0–100 µg/kg,
Ø conc.: 10 µg/kg*, country: South Africa[147],
*mean of all samples
see also feed, feed (maize), feed (rodent),
maize, maize bran, maize fine fractions,
maize kernels, maize powder, maize
screenings

Maize gluten   may contain the following
mycotoxins:

3-Acetyldeoxynivalenol
incidence: 3/40, conc. range: 10–30 µg/kg,
Ø conc.: 20 µg/kg, country: UK[95]
incidence: 5/11, conc. range: 14–78 µg/kg,
Ø conc.: 40 µg/kg, country: Germany[366]
see also barley, feed components, feed
(poultry), feeds, grain, feeds, industrial,
feedstuffs (rapeseed, turnip, fish meal,
concentrates), maize, maize ears, maize
kernels, maize meal, maize screenings, oats,
wheat

15-Acetyldeoxynivalenol
incidence: 38/40, conc. range: 10–400 µg/kg,
Ø conc.: 135 µg/kg, country: UK[95]
incidence: 9/11, conc. range: 112–565 µg/kg,
Ø conc.: 325 µg/kg, country: Germany[366]
see also barley, feed components, feed,
commercial mix, feed (poultry), maize, maize
ears, maize germ, maize germ/bran, maize

kernels, maize meal, maize screenings, maize,
"Baby", oats, silage, wheat

Aflatoxin B$_1$
incidence: 18/32*, conc. range: 1–41 µg/kg,
country: UK[94], *imported?
incidence: 6/7, conc. range: 26–50 µg/kg
(3 sa), 51–100 µg/kg (3 sa), Ø conc.:
54.1 µg/kg, country: India[247]
incidence: 10/34, conc. range: 5–90 µg/kg,
Ø conc.: 50.5 µg/kg, country: Egypt[397]
incidence: 5/18, conc. range: 15–150 µg/kg,
Ø conc.: 63 µg/kg, country: Egypt[397]
incidence: 15/27, conc. range: 5–200 µg/kg,
Ø conc.: 63.4 µg/kg, country: Egypt[397]
incidence: 3/15, conc. range: 5–25 µg/kg,
Ø conc.: 18.3 µg/kg, country: Egypt[397]
see also alfalfa, *Ambadi* cake, animal feedstuffs
(dairy cake), bagasse, barley, bengalgram
husk, bird food, bird food, wild, biri testa,
blackgram, blackgram husk, bran, broiler
mixed feed, calf fattening mixed feed, calf
fattening mixed feed (containing 4–20 %
peanut products), *Carthamus* cake, castor
cake, cereals, cereal products, chick pea,
coconut cake, cocos, concentrate, mixed,
concentrates, cotton cake, cottonseed,
cottonseed (dehulled), cottonseed cake,
cottonseed extract, cottonseed meal,
cottonseed meal (ammoniated), cottonseed
meal (decorticated), cottonseed meats,
cottonseed products, crumbles, crumbles,
grower, cycad meal, dairy cattle feed, dairy
cattle feed (containing 2–5 % peanut
products), dairy cattle feed (containing
6–10 % peanut products), dairy cattle feed
(containing 6–12 % peanut products), dairy
cattle feed (containing more than 20 %
peanut products), diets, mixed, dog food, egg
production mixed feed, feed, feed and
ingredients, feed, compound, feed, layer, feed,
mixed, feed (beef), feed (broilers), feed (calf),
feed (cat), feed (cattle), feed (chicken), feed
(dairy), feed (dog), feed (dug), feed (fish),
feed (gluten), feed (horse), feed (miscellaneous), feed (pig), feed (poultry), feed
(poultry, pig), feed (rabbit), feed (sheep),
feed (50–60 % maize), fish meal, grain
by-products, grains (no specification),

greengram, hay/silage, horsegram, husk, *Jagni* cake, legume mixture, linseed, linseed cake, livol, maize, *Mahua* cake, maize germ, maize grits, maize husk, maize meal, maize oil cake, maize powder, maize screenings, maize, ground, maize, hybrid, maize, preharvest, maize, yellow, maize (dark grains), *Makhana* (*Euryale ferox* Salisb) puffs, manioc, milk production mixed feed, mung testa, murkool, mustard cake, neem cake, niger cake, oats, palm kernel expeller cake, palm kernels, palm products, peanut cake, peanut cake (deoiled), peanut expeller, peanut hay, peanut meal, peanut, kernels, peanut, shells, peanuts, pellets, finisher, pig meal and pellets, pigeon pea, poultry feeds (peanut containing), rapeseed cake, redgram husk, rice, rice bran, rice bran (deoiled), rice chaff, rice crack, rice germ, rice germ cake, rice meal, rice straw, rice (damaged), rice (polish), safflower cake, sal seed cake, sesame, sesame cake, sorghum, soybean meal, soybeans, sunflower, sunflower cake, sunflower flour, tapioca, wheat, wheat bran, wheat bran and chana testa

Aflatoxin B$_2$
incidence: 5/7, conc. range: nc, Ø conc.: 14 µg/kg, country: India[247]
see also animal feedstuffs (dairy cake), bird food, bird food, wild, biri testa, blackgram, blackgram husk, cottonseed, cottonseed cake, cottonseed extract, cottonseed meal, cottonseed meal (ammoniated), cottonseed meats, dog food, egg production mixed feed, feed, compound, feed (cat), feed (cattle), feed (dog), feed (pig), feed (poultry), feed (rabbit), feed (sheep), fish meal, horsegram, maize, maize husk, maize, ground, maize, preharvest, mung testa, mustard cake, niger cake, peanut cake, peanut cake (deoiled), peanut expeller, peanut hay, peanut meal, peanuts, peanut, kernels, redgram husk, rice bran, rice bran (deoiled), rice chaff, rice meal, rice (polish), sal seed cake, sesame cake, sorghum, soybean meal, soybeans, wheat, wheat bran

Aflatoxin B
incidence: 3/11, conc. range: 6–57 µg/kg, Ø conc.: 24.3 µg/kg, country: Pakistan[412]

see also barley, cocoa (oil cake), cottonseed cake, feed, feed (cattle), feed (maize, gluten), feed (pig), feed (poultry), flour (wheat), lucern (dried), maize, maize grains, oats, peanut (oil cake), rice, broken, sorghum, soybean (oil cake), sunflower (oil cake), wheat

Aflatoxin G
incidence: 3/11, conc. range: 0–22 µg/kg, Ø conc.: 7.3 µg/kg, country: Pakistan[412]
see also cottonseed cake, feed (maize, gluten), maize grains, rice, broken

Aflatoxin
incidence: 6/7, conc. range: 35.7–103 µg/kg, country: India[247]
see also blackgram husk, bread crumbs, broiler finisher, broiler starter, cotton cake, cottonseed, cottonseed cake, cottonseed extract, cottonseed meal, feed, feed (cattle), feed (cow), feed (dog), feed (horse), feed (maize, gluten), feed (pig), feed (poultry), feed (rabbit), feed (rat/mice), feed (sheep), feeds, grain, fish meal, flour (wheat), groundnut cake, grower's mash, horsegram, layer's mash, maize, maize, white, milo, peanut cake, peanut cake (deoiled), peanut, kernels, peanut (oil cake), pearlmillet, pig breeder's mash, pig finisher, pig starter, pod with haulms, poultry breeder's mash, rabbit pellets, redgram husk, rice, rice bran (deoiled), rice, broken, rice (polish), sesame cake, silk worm pupae, sorghum, soybean cake, soybean meal, wheat, wheat bran

Aflatoxins (Total)
incidence: 19/32*, conc. range: 1–47 µg/kg, country: UK[94], *imported?
see also cottonseed, maize germ, maize (dark grains), palm products, rice bran, soybeans, sunflower

Deoxynivalenol
incidence: 2/2, conc. range: 300–890 µg/kg, Ø conc.: 595 µg/kg, country: UK[36]
incidence: 39/40, conc. range: 50–5000 µg/kg, Ø conc.: 545.4 µg/kg, country: UK[95]
incidence: 6/6, conc. range: 220–1900 µg/kg, Ø conc.: 1000 µg/kg, country: The Netherlands[103]

incidence: 11/11, conc. range: 86–2455 µg/kg, Ø conc.: 1166 µg/kg, country: Germany[366]
see also barley, barley, husked, barley, unhusked (naked), barley (pressed), bone meal, bran, broilers feed, calf fattening mixed feed, coconut, expeller, corn cob mix silage, cottonseed, cottonseed cake, dairy cattle feed, egg production mixed feed, feed, feed components, feed, commercial mix, feed, mixed, feed, mixed (primarily maize), feed (barley), feed (cattle), feed (chicken), feed (dog), feed (fish), feed (mill run, from wheat), feed (mink), feed (pig), feed (poultry), feed (reindeer), feeds, grain, feeds, industrial, feedstuff, feedstuffs (rapeseed, turnip, fish meal, concentrates), fish meal, grain, mixed feed, grains, mixed, grains (no specification), maize, maize ears, maize fibre, maize germ, maize germ/bran, maize germ meal, maize kernels, maize powder, maize screenings, maize stalks (pith), maize, "Baby", maize, hybrid, maize, white, oats, rice bran, rice germ cake, rye, silage, sorghum, soybeans, triticale, wheat, wheat and barley, wheat, red hard winter, wheat, soft white winter, wheat, spring, wheat, winter

DIACETOXYSCIRPENOL
incidence: 2/40, conc. range: 70–200 µg/kg, Ø conc.: 135 µg/kg, country: UK[95]
see also alfalfa, barley, feed, feed components, feed (dog), feed (fish), feed (mink), feed (pig), feed (poultry), feed (reindeer), feeds, grain, feedstuff, forage grass, grain, mixed feed, grains (no specification), maize, oat and barley (hammer-milled), oats, peanuts, soybeans, wheat

FUMONISIN B$_1$
incidence: 39/40, conc. range: 40–4800 µg/kg, Ø conc.: 556.7 µg/kg, country: UK[95]
incidence: 24/29*, conc. range: <50–4550 µg/kg, Ø conc.: 1040 µg/kg, country: UK[150], *imported
see also barley, bird food, wild, dog food, feed, feed, complete ration, feed, general, feed, layer, feed, maize-based, feed, mixed, feed, pelleted ration, feed, screenings, feed, sweet, feed (broilers), feed (cat), feed (chicken), feed (dog), feed (gluten), feed (horse), feed (maize), feed (pig), feed (poultry), feed (rat), feed (rodent), forage grass, maize, maize and maize screenings, maize bran, maize ears, maize fine fractions, maize flakes, maize germ, maize germ/bran, maize germ meal, maize grits, maize kernels, maize meal, maize powder, maize screenings, maize, "Baby", maize, ground, maize, preharvest, maize, sweet feed, maize/oats mix, rat chow, silage, sorghum, soybeans, wheat

FUMONISIN B$_2$
incidence: 23/40, conc. range: 25–400 µg/kg, Ø conc.: 92 µg/kg, country: UK[95]
incidence: 24/29*, conc. range: <50–70 µg/kg, Ø conc.: 50 µg/kg, country: UK[150], *imported
see also barley, bird food, wild, dog food, feed, feed, maize-based, feed, mixed, feed (cat), feed (dog), feed (gluten), feed (horse), feed (maize), feed (poultry), feed (rodent), maize, maize bran, maize fine fractions, maize flakes, maize germ, maize germ/bran, maize germ meal, maize kernels, maize meal, maize powder, maize screenings, maize, "Baby", maize, ground, maize, preharvest, rat chow, wheat

FUMONISINS
incidence: 27/32*, conc. range: 100–4550 µg/kg, country: UK[94], *imported?
see also feed (cattle and dairy), feed (chicken), feed (dairy), feed (horse), maize, maize germ, maize meal, maize screenings

FUSARENONE-X
incidence: 3/40, conc. range: 30–80 µg/kg, Ø conc.: 53.3 µg/kg, country: UK[95]
see also barley, feed components, feed (poultry), maize, maize ears, maize germ/bran, maize kernels, maize meal, maize screenings, oats, wheat

HT-2 TOXIN
incidence: 3/40, conc. range: 10–30 µg/kg, Ø conc.: 20 µg/kg, country: UK[95]
incidence: 9/11, conc. range: 17–90 µg/kg, Ø conc.: 42 µg/kg, country: Germany[366]
see also barley, feed components, feed (dog), feed (fish), feed (pig), feed (poultry), feed

(reindeer), grain, mixed feed, grains, mixed, grains (no specification), maize, maize germ/bran, maize meal, maize screenings, oats, rye, silage, wheat

MONILIFORMIN
incidence: 17/40, conc. range: 50–320 µg/kg, Ø conc.: 113.5 µg/kg, country: UK[95]
see also barley, feed, mixed, feed (poultry), maize, maize flakes, maize germ, maize germ/bran, maize meal, maize screenings, maize, "Baby", oats, rice bran, triticale, wheat, wheat, summer, wheat, winter

MONOACETOXYSCIRPENOL
incidence: 1/40, conc.: 10 µg/kg, country: UK[95]
see also feed components, maize, maize meal, maize screenings, silage, wheat

NIVALENOL
incidence: 2/40, conc. range: 600–1400 µg/kg, Ø conc.: 1000 µg/kg, country: UK[95]
incidence: 4/11, conc. range: 82–268 µg/kg, Ø conc.: 144 µg/kg, country: Germany[366]
see also barley, barley, husked, barley, unhusked (naked), barley (pressed), bran, feed, feed components, feed (cattle), feed (poultry), feed (reindeer), feeds, industrial, maize, maize ears, maize germ, maize germ/bran, maize kernels, maize meal, maize powder, maize screenings, maize, "Baby", oats, rye, silage, triticale, wheat

OCHRATOXIN A
incidence: 2/40, conc. range: 2–22 µg/kg, Ø conc.: 12 µg/kg, country: UK[95]
see also alfalfa, barley, barley, oats, barley (high moisture), barley-soybean diet, bird food, domestic, bird food, wild, broilers feed, cereal grains, citrus pulp, coconut, expeller, corn cob mix silage, diet (dairy cow), diet (poultry), diet (starter), dog food, eat, egg production mixed feed, feed, feed wheat, oat and barley, feed, commercial mix, feed, mixed, feed, mixed (pelleted), feed (broilers), feed (cat), feed (cattle), feed (cereals), feed (pig), feed ec (pig), feed (poultry), feed ec (poultry), feed (poultry, pig), feed ec (rabbit), feed (trout), grain, mixed feed, grains, mixed,

grains (heated), hay, horse bean, maize, maize feed, milo, maize meal, maize, white, *Makhana* (*Euryale ferox* Salisb) puffs, milk production mixed feed, millet, oat and barley (hammer-milled), oats, palm products, peanut cake, peas, peas and beans, pet food, pig feedstuffs, pig grower diet, pig meal, piglet diet, poultry feedstuffs, rice bran, rice germ, rice germ cake, rye, sorghum, soybean groats, sunflower, sunflower seeds, extracted, tapioca, triticale, *Vicia faba*, wheat, wheat and barley, wheat bran, wheat hay, wheat, oats

T-2 TOXIN
incidence: 2/40, conc. range: 10–100 µg/kg, Ø conc.: 55 µg/kg, country: UK[95]
incidence: 7/11, conc. range: 6–40 µg/kg, Ø conc.: 17 µg/kg, country: Germany[366]
see also alfalfa, barley, bran, diet (grower), diet (poultry), feed, feed components, feed, layer, feed, mixed, feed (dog), feed (fish), feed (mink), feed (pig), feed (poultry), feedstuff, forage grass, grain, mixed feed, grains (no specification), hay, maize, maize germ/bran, maize meal, maize screenings, maize stalks (pith), oat and barley (hammer-milled), oats, peanuts, pig grower diet, piglet diet, rye, sorghum, triticale, wheat

ZEARALENONE
incidence: 3/32*, conc. range: 20–440 µg/kg, country: UK[94], *imported?
incidence: 8/40, conc. range: 80–480 µg/kg, Ø conc.: 270 µg/kg, country: UK[95]
incidence: 6/6, conc. range: 14–61 µg/kg, Ø conc.: 42 µg/kg, country: The Netherlands[103]
incidence: 9/11, conc. range: 3–350 µg/kg, Ø conc.: 60 µg/kg, country: Germany[366]
see also alfalfa, barley, barley, husked, barley, unhusked (naked), barley and feed, bone meal, bran, broilers feed, *Carthamus* cake, chick pea, concentrate, mixed, corn cob mix silage, cotton cake, cottonseed, cottonseed cake, diet (dairy cow), diet (poultry), diets (mixed), feed, feed components, feed, mixed, feed, mixed (primarily maize also maize, oats, wheat), feed (bran), feed (broiler chicken), feed (cattle), feed (chicken), feed (dairy), feed

(developing pig), feed (mill run, from wheat), feed (miscellaneous), feed (pig), feed (poultry), feed (poultry, pig), feed (starter chicken), feedstuff, fish meal, forage grass, grain, bruised, grain, mixed feed, grains (no specification), hay, maize, maize ears, maize flakes, maize germ, maize germ/bran, maize kernels, maize meal, maize oil cake, maize screenings, maize stalks (pith), maize, "Baby", maize, hybrid, maize, shelled, maize, unshelled, maize, white, maize grain, artificially dried, maize grain, crib dried, maize grain, ensiled, milk production mixed feed, oats, *Paspalum palidosum*, straw, peanut hulls/skins, rice bran, rice germ, rice germ cake, rye, silage, sorghum, soybeans, soybeans, extracted, sunflower cake, tapioca, triticale, wheat, wheat bran, wheat bran and chana testa, wheat soya meal

Maize grains   Maize grains for feed may contain the following mycotoxins:

Aflatoxin B
incidence: ?/8, conc. range: 6–11 µg/kg, Ø conc.: 8.5 µg/kg, country: Pakistan[162]
see also barley, cocoa (oilcake), cottonseed cake, feed, feed (cattle), feed (maize, gluten), feed (pig), feed (poultry), flour (wheat), lucern (dried), maize, maize gluten, oats, peanut (oil cake), rice, broken, sorghum, soybean (oil cake), sunflower (oilcake), wheat

Aflatoxin G
incidence: ?/8, conc. range: 4–5 µg/kg, Ø conc.: 4.5 µg/kg, country: Pakistan[162]
see also cottonseed cake, feed (maize, gluten), maize gluten, rice, broken

Maize grain, artificially dried   for feed may contain the following mycotoxins:

Zearalenone
incidence: 5/20, conc. range: 300–1200 µg/kg (4 sa), 1300–4000 µg/kg (1 sa), country: Yugoslavia[314]
see also alfalfa, barley, barley, husked, barley, unhusked (naked), barley and feed, bone meal, bran, broilers feed, *Carthamus* cake,

chick pea, concentrate, mixed, corn cob mix silage, cotton cake, cottonseed, cottonseed cake, diet (dairy cow), diet (poultry), diets (mixed), feed, feed components, feed, mixed, feed, mixed (primarily maize also maize, oats, wheat), feed (bran), feed (broiler chicken), feed (cattle), feed (chicken), feed (dairy), feed (developing pig), feed (mill run, from wheat), feed (miscellaneous), feed (pig), feed (poultry), feed (poultry, pig), feed (starter chicken), feedstuff, fish meal, forage grass, grain, bruised, grain, mixed feed, grains (no specification), hay, maize, maize ears, maize flakes, maize germ, maize germ/bran, maize gluten, maize kernels, maize meal, maize oil cake, maize screenings, maize stalks (pith), maize, "Baby", maize, hybrid, maize, shelled, maize, unshelled, maize grain, crib dried, maize grain, ensiled, milk production mixed feed, oats, *Paspalum palidosum*, straw, peanut hulls/skins, rice bran, rice germ, rice germ cake, rye, silage, sorghum, soybeans, soybeans, extracted, sunflower cake, tapioca, triticale, wheat, wheat bran, wheat bran and chana testa, wheat soya meal

Maize grain, crib dried   may contain the following mycotoxins:

Zearalenone
incidence: 10/20, conc. range: 300–1200 (1 sa) µg/kg, 1300–4000 (5 sa) µg/kg, 4100–8500 (4 sa) µg/kg, country: Yugoslavia[314]
see also alfalfa, barley, barley, husked, barley, unhusked (naked), barley and feed, bone meal, bran, broilers feed, *Carthamus* cake, chick pea, concentrate, mixed, corn cob mix silage, cotton cake, cottonseed, cottonseed cake, diet (dairy cow), diet (poultry), diets (mixed), feed, feed components, feed, mixed, feed, mixed (primarily maize also maize, oats, wheat), feed (bran), feed (broiler chicken), feed (cattle), feed (chicken), feed (dairy), feed (developing pig), feed (mill run, from wheat), feed (miscellaneous), feed (pig), feed (poultry), feed (poultry, pig), feed (starter chicken), feedstuff, fish meal, forage grass, grain, bruised, grain, mixed feed, grains (no specification), hay, maize, maize ears, maize

flakes, maize germ, maize germ/bran, maize gluten, maize kernels, maize meal, maize oil cake, maize screenings, maize stalks (pith), maize, "Baby", maize, hybrid, maize, shelled, maize, unshelled, maize grain, artificially dried, maize grain, ensiled, milk production mixed feed, oats, *Paspalum palidosum*, straw, peanut hulls/skins, rice bran, rice germ, rice germ cake, rye, silage, sorghum, soybeans, soybeans, extracted, sunflower cake, tapioca, triticale, wheat, wheat bran, wheat bran and chana testa, wheat soya meal

Maize grain, ensiled   may contain the following mycotoxins:

Zearalenone
incidence: 8/20, conc. range: 300–1200 µg/kg (2 sa), 1300–4000 µg/kg (4 sa), 4100–8500 µg/kg (2 sa), country: Yugoslavia[314]
see also alfalfa, barley, barley, husked, barley, unhusked (naked), barley and feed, bone meal, bran, broilers feed, *Carthamus* cake, chick pea, concentrate, mixed, corn cob mix silage, cotton cake, cottonseed, cottonseed cake, diet (dairy cow), diet (poultry), diets (mixed), feed, feed components, feed, mixed, feed, mixed (primarily maize also maize, oats, wheat), feed (bran), feed (broiler chicken), feed (cattle), feed (chicken), feed (dairy), feed (developing pig), feed (mill run, from wheat), feed (miscellaneous), feed (pig), feed (poultry), feed (poultry, pig), feed (starter chicken), feedstuff, fish meal, forage grass, grain, bruised, grain, mixed feed, grains (no specification), hay, maize, maize ears, maize flakes, maize germ, maize germ/bran, maize gluten, maize kernels, maize meal, maize oil cake, maize screenings, maize stalks (pith), maize, "Baby", maize, hybrid, maize, shelled, maize, unshelled, maize grain, artificially dried, maize grain, crib dried, milk production mixed feed, oats, *Paspalum palidosum*, straw, peanut hulls/skins, rice bran, rice germ, rice germ cake, rye, silage, sorghum, soybeans, soybeans, extracted, sunflower cake, tapioca, triticale, wheat, wheat bran, wheat bran and chana testa, wheat soya meal

Maize grits   Maize grits for feed may contain the following mycotoxins:

Aflatoxin B$_1$
incidence: 1/1, conc.: 2 µg/kg, country: Thailand[117]
see also alfalfa, *Ambadi* cake, animal feedstuffs (dairy cake), bagasse, barley, bengalgram husk, bird food, bird food, wild, biri testa, blackgram, blackgram husk, bran, broiler mixed feed, calf fattening mixed feed, calf fattening mixed feed (containing 4–20% peanut products), *Carthamus* cake, castor cake, cereals, cereal products, chick pea, coconut cake, cocos, concentrate, mixed, concentrates, cotton cake, cottonseed, cottonseed (dehulled), cottonseed cake, cottonseed extract, cottonseed meal, cottonseed meal (ammoniated), cottonseed meal (decorticated), cottonseed meats, cottonseed products, crumbles, crumbles, grower, cycad meal, dairy cattle feed, dairy cattle feed (containing 2–5% peanut products), dairy cattle feed (containing 6–10% peanut products), dairy cattle feed (containing 6–12% peanut products), dairy cattle feed (containing more than 20% peanut products), diets, mixed, dog food, egg production mixed feed, feed, feed and ingredients, feed, compound, feed, layer, feed, mixed, feed (beef), feed (broilers), feed (calf), feed (cat), feed (cattle), feed (chicken), feed (dairy), feed (dog), feed (dug), feed (fish), feed (gluten), feed (horse), feed (miscellaneous), feed (pig), feed (poultry), feed (poultry, pig), feed (rabbit), feed (sheep), feed (50–60% maize), fish meal, grain by-products, grains (no specification), greengram, hay/silage, horsegram, husk, *Jagni* cake, legume mixture, linseed, linseed cake, livol, maize, *Mahua* cake, maize germ, maize gluten, maize husk, maize meal, maize oil cake, maize powder, maize screenings, maize, ground, maize, hybrid, maize, preharvest, maize, yellow, maize (dark grains), *Makhana* (*Euryale ferox* Salisb) puffs, manioc, milk

production mixed feed, mung testa, murkool, mustard cake, neem cake, niger cake, oats, palm kernel expeller cake, palm kernels, palm products, peanut cake, peanut cake (deoiled), peanut expeller, peanut hay, peanut meal, peanut, kernels, peanut, shells, peanuts, pellets, finisher, pig meal and pellets, pigeon pea, poultry feeds (peanut containing), rapeseed cake, redgram husk, rice, rice bran, rice bran (deoiled), rice chaff, rice crack, rice germ, rice germ cake, rice meal, rice straw, rice (damaged), rice (polish), safflower cake, sal seed cake, sesame, sesame cake, sorghum, soybean meal, soybeans, sunflower, sunflower cake, sunflower flour, tapioca, wheat, wheat bran, wheat bran and chana testa

Fumonisin B_1
incidence: 1/1, conc.: 217 µg/kg, country: Thailand[117]
see also barley, bird food, wild, dog food, feed, feed, complete ration, feed, general, feed, layer, feed, maize-based, feed, mixed, feed, pelleted ration, feed, screenings, feed, sweet, feed (broilers), feed (cat), feed (chicken), feed (dog), feed (gluten), feed (horse), feed (maize), feed (pig), feed (poultry), feed (rat), feed (rodent), forage grass, maize, maize and maize screenings, maize bran, maize ears, maize fine fractions, maize flakes, maize germ, maize germ/bran, maize germ meal, maize gluten, maize kernels, maize meal, maize powder, maize screenings, maize, "Baby", maize, ground, maize, preharvest, maize, sweet feed, maize/oats mix, rat chow, silage, sorghum, soybeans, wheat

Maize husk   may contain the following mycotoxins:

Aflatoxin B_1
incidence: 1/1, conc.: 76 µg/kg, country: India[247]
see also alfalfa, *Ambadi* cake, animal feedstuffs (dairy cake), bagasse, barley, bengalgram husk, bird food, bird food, wild, biri testa, blackgram, blackgram husk, bran, broiler mixed feed, calf fattening mixed feed, calf

fattening mixed feed (containing 4–20 % peanut products), *Carthamus* cake, castor cake, cereals, cereal products, chick pea, coconut cake, cocos, concentrate, mixed, concentrates, cotton cake, cottonseed, cottonseed (dehulled), cottonseed cake, cottonseed extract, cottonseed meal, cottonseed meal (ammoniated), cottonseed meal (decorticated), cottonseed meats, cottonseed products, crumbles, crumbles, grower, cycad meal, dairy cattle feed, dairy cattle feed (containing 2–5 % peanut products), dairy cattle feed (containing 6–10 % peanut products), dairy cattle feed (containing 6–12 % peanut products), dairy cattle feed (containing more than 20 % peanut products), diets, mixed, dog food, egg production mixed feed, feed, feed and ingredients, feed, compound, feed, layer, feed, mixed, feed (beef), feed (broilers), feed (calf), feed (cat), feed (cattle), feed (chicken), feed (dairy), feed (dog), feed (dug), feed (fish), feed (gluten), feed (horse), feed (miscellaneous), feed (pig), feed (poultry), feed (poultry, pig), feed (rabbit), feed (sheep), feed (50–60 % maize), fish meal, grain by-products, grains (no specification), greengram, hay/silage, horsegram, husk, *Jagni* cake, legume mixture, linseed, linseed cake, livol, maize, *Mahua* cake, maize germ, maize gluten, maize grits, maize meal, maize oil cake, maize powder, maize screenings, maize, ground, maize, hybrid, maize, preharvest, maize, yellow, maize (dark grains), *Makhana* (*Euryale ferox* Salisb) puffs, manioc, milk production mixed feed, mung testa, murkool, mustard cake, neem cake, niger cake, oats, palm kernel expeller cake, palm kernels, palm products, peanut cake, peanut cake (deoiled), peanut expeller, peanut hay, peanut meal, peanut, kernels, peanut, shells, peanuts, pellets, finisher, pig meal and pellets, pigeon pea, poultry feeds (peanut containing), rapeseed cake, redgram husk, rice, rice bran, rice bran (deoiled), rice chaff, rice crack, rice germ, rice germ cake, rice meal, rice straw, rice (damaged), rice (polish), safflower cake, sal seed cake, sesame, sesame cake, sorghum, soybean meal, soybeans, sunflower, sunflower cake, sunflower flour, tapioca, wheat, wheat bran, wheat bran and chana testa

AFLATOXIN B$_2$
incidence: 1/1, conc.: 23 μg/kg, country: India[247]
see also animal feedstuffs (dairy cake), bird food, bird food, wild, biri testa, blackgram, blackgram husk, cottonseed, cottonseed cake, cottonseed extract, cottonseed meal, cottonseed meal (ammoniated), cottonseed meats, dog food, egg production mixed feed, feed, compound, feed (cat), feed (cattle), feed (dog), feed (pig), feed (poultry), feed (rabbit), feed (sheep), fish meal, horsegram, maize, maize gluten, maize, ground, maize, preharvest, mung testa, niger cake, peanut cake, peanut cake (deoiled), peanut expeller, peanut hay, peanut meal, peanuts, peanut, kernels, redgram husk, rice bran, rice bran (deoiled), rice chaff, rice meal, rice (polish), sal seed cake, sesame cake, sorghum, soybean meal, soybeans, wheat, wheat bran

Maize kernels   Maize kernels for feed may contain the following mycotoxins:

3-ACETYLDEOXYNIVALENOL
incidence: 2/5*, conc. range:
2400–7500 μg/kg, Ø conc.: 4950 μg/kg, country: Poland[348] *ncac
see also barley, feed components, feed (poultry), feeds, grain, feeds, industrial, feedstuffs (rapeseed, turnip, fish meal, concentrates), maize, maize ears, maize gluten, maize meal, maize screenings, oats, wheat

15-ACETYLDEOXYNIVALENOL
incidence: 2/5*, conc. range: 500–600 μg/kg, Ø conc.: 550 μg/kg, country: Poland[348] *ncac
see also barley, feed components, feed, commercial mix, feed (poultry), maize, maize ears, maize germ, maize germ/bran, maize gluten, maize meal, maize screenings, maize, "Baby", oats, silage, wheat

DEOXYNIVALENOL
incidence: 3/5*, conc. range:
12500–175,200 μg/kg, Ø conc.: 88,633 μg/kg, country: Poland[348] ncac
see also barley, barley, husked, barley, unhusked (naked), barley (pressed), bone meal, bran, broilers feed, calf fattening mixed feed, coconut, expeller, corn cob mix silage, cottonseed, cottonseed cake, dairy cattle feed, egg production mixed feed, feed, feed components, feed, commercial mix, feed, mixed, feed, mixed (primarily maize), feed (barley), feed (cattle), feed (chicken), feed (dog), feed (fish), feed (mill run, from wheat), feed (mink), feed (pig), feed (poultry), feed (reindeer), feeds, grain, feeds, industrial, feedstuff, feedstuffs (rapeseed, turnip, fish meal, concentrates), fish meal, grain, mixed feed, grains, mixed, grains (no specification), maize, maize ears, maize fibre, maize germ, maize germ/bran, maize germ meal, maize gluten, maize meal, maize powder, maize screenings, maize stalks (pith), maize, "Baby", maize, hybrid, maize, white, oats, rice bran, rice germ cake, rye, silage, sorghum, soybeans, triticale, wheat, wheat and barley, wheat, red hard winter, wheat, soft white winter, wheat, spring, wheat, winter

FUMONISIN B$_1$
incidence: 10/10* **, conc. range:
360–1270 μg/kg, Ø conc.: 807 μg/kg, country: South Africa[190], *imported, **ncac
see also barley, bird food, wild, dog food, feed, feed, complete ration, feed, general, feed, layer, feed, maize-based, feed, mixed, feed, pelleted ration, feed, screenings, feed, sweet, feed (broilers), feed (cat), feed (chicken), feed (dog), feed (gluten), feed (horse), feed (maize), feed (pig), feed (poultry), feed (rat), feed (rodent), forage grass, maize, maize and maize screenings, maize bran, maize ears, maize fine fractions, maize flakes, maize germ, maize germ/bran, maize germ meal, maize gluten, maize grits, maize meal, maize powder, maize screenings, maize, "Baby", maize, ground, maize,

preharvest, maize, sweet feed, maize/oats mix, rat chow, silage, sorghum, soybeans, wheat

Fumonisin B$_2$
incidence: 10/10* **, conc. range: 120–430 µg/kg, Ø conc.: 269 µg/kg, country: South Africa[190], *imported, **ncac
see also barley, bird food, wild, dog food, feed, feed, maize-based, feed, mixed, feed (cat), feed (dog), feed (gluten), feed (horse), feed (maize), feed (poultry), feed (rodent), maize, maize bran, maize fine fractions, maize flakes, maize germ, maize germ/bran, maize germ meal, maize gluten, maize meal, maize powder, maize screenings, maize, "Baby", maize, ground, maize, preharvest, rat chow, wheat

Fumonisin B$_3$
incidence: 10/10* **, conc. range: 50–210 µg/kg, Ø conc.: 102 µg/kg, country: South Africa[190], *imported, **ncac
see also feed, feed (maize), feed (rodent), maize, maize bran, maize fine fractions, maize germ meal, maize powder, maize screenings

Fusarenone-X
incidence: 2/5*, conc. range: 600–1800 µg/kg, Ø conc.: 1200 µg/kg, country: Poland[348] *ncac
see also barley, feed components, feed (poultry), maize, maize ears, maize germ/bran, maize gluten, maize meal, maize screenings, oats, wheat

Nivalenol
incidence: 2/5*, conc. range: 33200–42,500 µg/kg, Ø conc.: 37,850 µg/kg, country: Poland[348] *ncac
see also barley, barley, husked, barley, unhusked (naked), barley (pressed), bran, feed, feed components, feed (cattle), feed (poultry), feed (reindeer), feeds, industrial, maize, maize ears, maize germ, maize germ/bran, maize gluten, maize meal, maize powder, maize screenings, maize, "Baby", oats, rye, silage, triticale, wheat

Zearalenone
incidence: 5/5*, conc. range: 700–10,000 µg/kg, Ø conc.: 4140 µg/kg, country: Poland[348]; *ncac
see also alfalfa, barley, barley, husked, barley, unhusked (naked), barley and feed, bone meal, bran, broilers feed, *Carthamus* cake, chick pea, concentrate, mixed, corn cob mix silage, cotton cake, cottonseed, cottonseed cake, diet (dairy cow), diet (poultry), diets (mixed), feed, feed components, feed, mixed, feed, mixed (primarily maize also maize, oats, wheat), feed (bran), feed (broiler chicken), feed (cattle), feed (chicken), feed (dairy), feed (developing pig), feed (mill run, from wheat), feed (miscellaneous), feed (pig), feed (poultry), feed (poultry, pig), feed (starter chicken), feedstuff, fish meal, forage grass, grain, bruised, grain, mixed feed, grains (no specification), hay, maize, maize ears, maize flakes, maize germ, maize germ/bran, maize gluten, maize meal, maize oil cake, maize screenings, maize stalks (pith), maize, "Baby", maize, hybrid, maize, shelled, maize, unshelled, maize grain, artificially dried, maize grain, crib dried, maize grain, ensiled, milk production mixed feed, oats, *Paspalum palidosum*, straw, peanut hulls/skins, rice bran, rice germ, rice germ cake, rye, silage, sorghum, soybeans, soybeans, extracted, sunflower cake, tapioca, triticale, wheat, wheat bran, wheat bran and chana testa, wheat soya meal

Maize meal   Maize meal for feed may contain the following mycotoxins:

3-Acetyldeoxynivalenol
incidence: 3/3, conc. range: 60–90 µg/kg, Ø conc.: 73.3 µg/kg, country: UK[95]
see also barley, feed components, feed (poultry), feeds, grain, feeds, industrial, feedstuffs (rapeseed, turnip, fish meal, concentrates), maize, maize ears, maize gluten, maize kernels, maize screenings, oats, wheat

15-Acetyldeoxynivalenol
incidence: 3/3, conc. range: 470–680 µg/kg, Ø conc.: 576.7 µg/kg, country: UK[95]

see also barley, feed components, feed, commercial mix, feed (poultry), maize, maize ears, maize germ, maize germ/bran, maize gluten, maize kernels, maize screenings, maize, "Baby", oats, silage, wheat

Aflatoxin B_1
incidence: ?/10, conc. range: 10–20 µg/kg, country: India[183]
see also alfalfa, *Ambadi* cake, animal feedstuffs (dairy cake), bagasse, barley, bengalgram husk, bird food, bird food, wild, biri testa, blackgram, blackgram husk, bran, broiler mixed feed, calf fattening mixed feed, calf fattening mixed feed (containing 4–20 % peanut products), *Carthamus* cake, castor cake, cereals, cereal products, chick pea, coconut cake, cocos, concentrate, mixed, concentrates, cotton cake, cottonseed, cottonseed (dehulled), cottonseed cake, cottonseed extract, cottonseed meal, cottonseed meal (ammoniated), cottonseed meal (decorticated), cottonseed meats, cottonseed products, crumbles, crumbles, grower, cycad meal, dairy cattle feed, dairy cattle feed (containing 2–5 % peanut products), dairy cattle feed (containing 6–10 % peanut products), dairy cattle feed (containing 6–12 % peanut products), dairy cattle feed (containing more than 20 % peanut products), diets, mixed, dog food, egg production mixed feed, feed, feed and ingredients, feed, compound, feed, layer, feed, mixed, feed (beef), feed (broilers), feed (calf), feed (cat), feed (cattle), feed (chicken), feed (dairy), feed (dog), feed (dug), feed (fish), feed (gluten), feed (horse), feed (miscellaneous), feed (pig), feed (poultry), feed (poultry, pig), feed (rabbit), feed (sheep), feed (50–60 % maize), fish meal, grain by-products, grains (no specification), greengram, hay/silage, horsegram, husk, *Jagni* cake, legume mixture, linseed, linseed cake, livol, *Mahua* cake, maize, maize germ, maize gluten, maize grits, maize husk, maize oil cake, maize powder, maize screenings, maize, ground, maize, hybrid, maize, preharvest, maize, yellow, maize (dark grains), *Makhana* (*Euryale ferox* Salisb) puffs, manioc, milk

production mixed feed, mung testa, murkool, mustard cake, neem cake, niger cake, oats, palm kernel expeller cake, palm kernels, palm products, peanut cake, peanut cake (deoiled), peanut expeller, peanut hay, peanut meal, peanut, kernels, peanut, shells, peanuts, pellets, finisher, pig meal and pellets, pigeon pea, poultry feeds (peanut containing), rapeseed cake, redgram husk, rice, rice bran, rice bran (deoiled), rice chaff, rice crack, rice germ, rice germ cake, rice meal, rice straw, rice (damaged), rice (polish), safflower cake, sal seed cake, sesame, sesame cake, sorghum, soybean meal, soybeans, sunflower, sunflower cake, sunflower flour, tapioca, wheat, wheat bran, wheat bran and chana testa

Deoxynivalenol
incidence: 3/3, conc. range: 3500–4900 µg/kg, Ø conc.: 4433.3 µg/kg, country: UK[95]
incidence: 5/6, conc. range: 140–1000 µg/kg, Ø conc.: 670 µg/kg, country: The Netherlands[103]
see also barley, barley, husked, barley, unhusked (naked), barley (pressed), bone meal, bran, broilers feed, calf fattening mixed feed, coconut, expeller, corn cob mix silage, cottonseed, cottonseed cake, dairy cattle feed, egg production mixed feed, feed, feed components, feed, commercial mix, feed, mixed, feed, mixed (primarily maize), feed (barley), feed (cattle), feed (chicken), feed (dog), feed (fish), feed (mill run, from wheat), feed (mink), feed (pig), feed (poultry), feed (reindeer), feeds, grain, feeds, industrial, feedstuff, feedstuffs (rapeseed, turnip, fish meal, concentrates), fish meal, grain, mixed feed, grains, mixed, grains (no specification), maize, maize ears, maize fibre, maize germ, maize germ/bran, maize germ meal, maize gluten, maize kernels, maize powder, maize screenings, maize stalks (pith), maize, "Baby", maize, hybrid, maize, white, oats, rice bran, rice germ cake, rye, silage, sorghum, soybeans, triticale, wheat, wheat and barley, wheat, red hard winter, wheat, soft white winter, wheat, spring, wheat, winter

Fumonisin B_1
incidence: 3/3, conc. range:
4900–12,000 μg/kg, Ø conc.: 8133.3 μg/kg,
country: UK[95]
incidence: 10/10, conc. range:
380–2800 μg/kg, Ø conc.: 1220 μg/kg,
country: USA[257]
incidence: 8/8, conc. range: 1723–5964 μg/kg,
Ø conc.: 2899.5 μg/kg, country: Spain[347]
see also barley, bird food, wild, dog food,
feed, feed, complete ration, feed, general,
feed, layer, feed, maize-based, feed, mixed,
feed, pelleted ration, feed, screenings, feed,
sweet, feed (broilers), feed (cat), feed
(chicken), feed (dog), feed (gluten), feed
(horse), feed (maize), feed (pig), feed
(poultry), feed (rat), feed (rodent), forage
grass, maize, maize and maize screenings,
maize bran, maize ears, maize fine fractions,
maize flakes, maize germ, maize germ/bran,
maize germ meal, maize gluten, maize grits,
maize kernels, maize powder, maize
screenings, maize, "Baby", maize, ground,
maize, preharvest, maize, sweet feed,
maize/oats mix, rat chow, silage, sorghum,
soybeans, wheat

Fumonisin B_2
incidence: 3/3, conc. range: 700–1300 μg/kg,
Ø conc.: 1000 μg/kg, country: UK[95]
see also barley, bird food, wild, dog food, feed,
feed, maize-based, feed, mixed, feed (cat), feed
(dog), feed (gluten), feed (horse), feed (maize),
feed (poultry), feed (rodent), maize, maize
bran, maize fine fractions, maize flakes, maize
germ, maize germ/bran, maize germ meal,
maize gluten, maize kernels, maize powder,
maize screenings, maize, "Baby", maize,
ground, maize, preharvest, rat chow, wheat

Fumonisins
incidence: 5/9*, conc. range: 20–1030 μg/kg,
country: UK[94], *imported?
see also feed (cattle and dairy), feed
(chicken), feed (dairy), feed (horse), maize,
maize germ, maize gluten, maize screenings

Fusarenone-X
incidence: 3/3, conc. range: 35–100 μg/kg,
Ø conc.: 67 μg/kg, country: UK[95]

see also barley, feed components, feed
(poultry), maize, maize ears, maize
germ/bran, maize gluten, maize kernels,
maize screenings, oats, wheat

HT-2 Toxin
incidence: 3/3, conc. range: 125–210 μg/kg,
Ø conc.: 161.7 μg/kg, country: UK[95]
see also barley, feed components, feed (dog),
feed (fish), feed (pig), feed (poultry), feed
(reindeer), grain, mixed feed, grains, mixed,
grains (no specification), maize, maize
germ/bran, maize gluten, maize screenings,
oats, rye, silage, wheat

Moniliformin
incidence: 3/3, conc. range: 1200–3200 μg/kg,
Ø conc.: 1966.7 μg/kg, country: UK[95]
see also barley, feed, mixed, feed (poultry),
maize, maize flakes, maize germ, maize
germ/bran, maize gluten, maize screenings,
maize, "Baby", oats, rice bran, triticale,
wheat, wheat, summer, wheat, winter

Monoacetoxyscirpenol
incidence: 3/3, conc. range: 10–15 μg/kg,
Ø conc.: 11.7 μg/kg, country: UK[95]
see also feed components, maize, maize
gluten, maize screenings, silage, wheat

Nivalenol
incidence: 3/3, conc. range: 470–1600 μg/kg,
Ø conc.: 930 μg/kg, country: UK[95]
see also barley, barley, husked, barley,
unhusked (naked), barley (pressed), bran,
feed, feed components, feed (cattle), feed
(poultry), feed (reindeer), feeds, industrial,
maize, maize ears, maize germ, maize
germ/bran, maize gluten, maize kernels,
maize powder, maize screenings, maize,
"Baby", oats, rye, silage, triticale, wheat

Ochratoxin A
incidence: 1/6, conc.: 3 μg/kg, country: The
Netherlands[103]
see also alfalfa, barley, barley, oats, barley
(high moisture), barley-soybean diet, bird
food, domestic, bird food, wild, broilers feed,
cereal grains, citrus pulp, coconut, expeller,
corn cob mix silage, diet (dairy cow), diet
(poultry), diet (starter), dog food, eat, egg

production mixed feed, feed, feed wheat, oat and barley, feed, commercial mix, feed, mixed, feed, mixed (pelleted), feed (broilers), feed (cat), feed (cattle), feed (cereals), feed (pig), feed ec (pig), feed (poultry), feed ec (poultry), feed (poultry, pig), feed ec (rabbit), feed (trout), grain, mixed feed, grains, mixed, grains (heated), hay, horse bean, maize, maize feed, milo, maize gluten, maize, white, *Makhana* (*Euryale ferox* Salisb) puffs, milk production mixed feed, millet, oat and barley (hammer-milled), oats, palm products, peanut cake, peas, peas and beans, pet food, pig feedstuffs, pig grower diet, pig meal, piglet diet, poultry feedstuffs, rice bran, rice germ, rice germ cake, rye, sorghum, soybean groats, sunflower, sunflower seeds, extracted, tapioca, triticale, *Vicia faba*, wheat, wheat and barley, wheat bran, wheat hay, wheat, oats

PATULIN
incidence: ?/10, conc.: 20 µg/kg, country: India[183]
see also diet (grower), hay, piglet diet, wheat

T-2 TOXIN
incidence: 3/3, conc. range: 80–170 µg/kg, Ø conc.: 120 µg/kg, country: UK[95]
see also alfalfa, barley, bran, diet (grower), diet (poultry), feed, feed components, feed, layer, feed, mixed, feed (dog), feed (fish), feed (mink), feed (pig), feed (poultry), feedstuff, forage grass, grain, mixed feed, grains (no specification), hay, maize, maize germ/bran, maize gluten, maize screenings, maize stalks (pith), oat and barley (hammer-milled), oats, peanuts, pig grower diet, piglet diet, rye, sorghum, triticale, wheat

ZEARALENONE
incidence: 3/3, conc. range: 640–1500 µg/kg, Ø conc.: 1080 µg/kg, country: UK[95]
incidence: 5/6, conc. range: 20–240 µg/kg, Ø conc.: 150 µg/kg, country: The Netherlands[103]
incidence: ?/10, conc. range: 20 µg/kg, country: India[183]
incidence: 9/11, conc. range: 12–69 µg/kg, country: USA[293]

see also alfalfa, barley, barley, husked, barley, unhusked (naked), barley and feed, bone meal, bran, broilers feed, *Carthamus* cake, chick pea, concentrate, mixed, corn cob mix silage, cotton cake, cottonseed, cottonseed cake, diet (dairy cow), diet (poultry), diets (mixed), feed, feed components, feed, mixed, feed, mixed (primarily maize also maize, oats, wheat), feed (bran), feed (broiler chicken), feed (cattle), feed (chicken), feed (dairy), feed (developing pig), feed (mill run, from wheat), feed (miscellaneous), feed (pig), feed (poultry), feed (poultry, pig), feed (starter chicken), feedstuff, fish meal, forage grass, grain, bruised, grain, mixed feed, grains (no specification), hay, maize, maize ears, maize flakes, maize germ, maize germ/bran, maize gluten, maize kernels, maize oil cake, maize screenings, maize stalks (pith), maize, "Baby", maize, hybrid, maize, shelled, maize, unshelled, maize, white, maize grain, artificially dried, maize grain, crib dried, maize grain, ensiled, milk production mixed feed, oats, *Paspalum palidosum*, straw, peanut hulls/skins, rice bran, rice germ, rice germ cake, rye, silage, sorghum, soybeans, soybeans, extracted, sunflower cake, tapioca, triticale, wheat, wheat bran, wheat bran and chana testa, wheat soya meal

Maize oil cake   may contain the following mycotoxins:

AFLATOXIN B$_1$
incidence: ?/10, conc. range: 10–20 µg/kg, country: India[183]
see also alfalfa, *Ambadi* cake, animal feedstuffs (dairy cake), bagasse, barley, bengalgram husk, bird food, bird food, wild, biri testa, blackgram, blackgram husk, bran, broiler mixed feed, calf fattening mixed feed, calf fattening mixed feed (containing 4–20% peanut products), *Carthamus* cake, castor cake, cereals, cereal products, chick pea, coconut cake, cocos, concentrate, mixed, concentrates, cotton cake, cottonseed, cottonseed (dehulled), cottonseed cake, cottonseed extract, cottonseed meal, cottonseed meal (ammoniated), cottonseed

meal (decorticated), cottonseed meats, cottonseed products, crumbles, crumbles, grower, cycad meal, dairy cattle feed, dairy cattle feed (containing 2–5 % peanut products), dairy cattle feed (containing 6–10 % peanut products), dairy cattle feed (containing 6–12 % peanut products), dairy cattle feed (containing more than 20 % peanut products), diets, mixed, dog food, egg production mixed feed, feed, feed and ingredients, feed, compound, feed, layer, feed, mixed, feed (beef), feed (broilers), feed (calf), feed (cat), feed (cattle), feed (chicken), feed (dairy), feed (dog), feed (dug), feed (fish), feed (gluten), feed (horse), feed (miscell-aneous), feed (pig), feed (poultry), feed (poultry, pig), feed (rabbit), feed (sheep), feed (50–60 % maize), fish meal, grain by-products, grains (no specification), greengram, hay/silage, horsegram, husk, *Jagni* cake, legume mixture, linseed, linseed cake, livol, *Mahua* cake, maize, maize germ, maize gluten, maize grits, maize husk, maize meal, maize powder, maize screenings, maize, ground, maize, hybrid, maize, preharvest, maize, yellow, maize (dark grains), *Makhana* (*Euryale ferox* Salisb) puffs, manioc, milk production mixed feed, mung testa, murkool, mustard cake, neem cake, niger cake, oats, palm kernel expeller cake, palm kernels, palm products, peanut cake, peanut cake (deoiled), peanut expeller, peanut hay, peanut meal, peanut, kernels, peanut, shells, peanuts, pellets, finisher, pig meal and pellets, pigeon pea, poultry feeds (peanut containing), rapeseed cake, redgram husk, rice, rice bran, rice bran (deoiled), rice chaff, rice crack, rice germ, rice germ cake, rice meal, rice straw, rice (damaged), rice (polish), safflower cake, sal seed cake, sesame, sesame cake, sorghum, soybean meal, soybeans, sunflower, sunflower cake, sunflower flour, tapioca, wheat, wheat bran, wheat bran and chana testa

Zearalenone
incidence: ?/10, conc. range: 20 µg/kg, country: India[183]
see also alfalfa, barley, barley, husked, barley, unhusked (naked), barley and feed, bone

meal, bran, broilers feed, *Carthamus* cake, chick pea, concentrate, mixed, corn cob mix silage, cotton cake, cottonseed, cottonseed cake, diet (dairy cow), diet (poultry), diets (mixed), feed, feed components, feed, mixed, feed, mixed (primarily maize also maize, oats, wheat), feed (bran), feed (broiler chicken), feed (cattle), feed (chicken), feed (dairy), feed (developing pig), feed (mill run, from wheat), feed (miscellaneous), feed (pig), feed (poultry), feed (poultry, pig), feed (starter chicken), feedstuff, fish meal, forage grass, grain, bruised, grain, mixed feed, grains (no specification), hay, maize, maize ears, maize flakes, maize germ, maize germ/bran, maize gluten, maize kernels, maize meal, maize screenings, maize stalks (pith), maize, "Baby", maize, hybrid, maize, shelled, maize, unshelled, maize, white, maize grain, artificially dried, maize grain, crib dried, maize grain, ensiled, milk production mixed feed, oats, *Paspalum palidosum*, straw, peanut hulls/skins, rice bran, rice germ, rice germ cake, rye, silage, sorghum, soybeans, soybeans, extracted, sunflower cake, tapioca, triticale, wheat, wheat bran, wheat bran and chana testa, wheat soya meal

Maize powder   may contain the following mycotoxins:

Aflatoxin B$_1$
incidence: 13/17, conc. range: 8.6–75 µg/kg, Ø conc.: 30.1 µg/kg, country: Vietnam[104]
see also alfalfa, *Ambadi* cake, animal feedstuffs (dairy cake), bagasse, barley, bengalgram husk, bird food, bird food, wild, biri testa, blackgram, blackgram husk, bran, broiler mixed feed, calf fattening mixed feed, calf fattening mixed feed (containing 4–20 % peanut products), *Carthamus* cake, castor cake, cereals, cereal products, chick pea, coconut cake, cocos, concentrate, mixed, concentrates, cotton cake, cottonseed, cottonseed (dehulled), cottonseed cake, cottonseed extract, cottonseed meal, cottonseed meal (ammoniated), cottonseed meal (decorticated), cottonseed meats, cottonseed products, crumbles, crumbles,

grower, cycad meal, dairy cattle feed, dairy cattle feed (containing 2–5 % peanut products), dairy cattle feed (containing 6–10 % peanut products), dairy cattle feed (containing 6–12 % peanut products), dairy cattle feed (containing more than 20 % peanut products), diets, mixed, dog food, egg production mixed feed, feed, feed and ingredients, feed, compound, feed, layer, feed, mixed, feed (beef), feed (broilers), feed (calf), feed (cat), feed (cattle), feed (chicken), feed (dairy), feed (dog), feed (dug), feed (fish), feed (gluten), feed (horse), feed (miscell-aneous), feed (pig), feed (poultry), feed (poultry, pig), feed (rabbit), feed (sheep), feed (50–60 % maize), fish meal, grain by-products, grains (no specification), greengram, hay/silage, horsegram, husk, *Jagni* cake, legume mixture, linseed, linseed cake, livol, *Mahua* cake, maize, maize germ, maize gluten, maize grits, maize husk, maize meal, maize oil cake, maize screenings, maize, ground, maize, hybrid, maize, preharvest, maize, yellow, maize (dark grains), *Makhana* (*Euryale ferox* Salisb) puffs, manioc, milk production mixed feed, mung testa, murkool, mustard cake, neem cake, niger cake, oats, palm kernel expeller cake, palm kernels, palm products, peanut cake, peanut cake (deoiled), peanut expeller, peanut hay, peanut meal, peanut, kernels, peanut, shells, peanuts, pellets, finisher, pig meal and pellets, pigeon pea, poultry feeds (peanut containing), rapeseed cake, redgram husk, rice, rice bran, rice bran (deoiled), rice chaff, rice crack, rice germ, rice germ cake, rice meal, rice straw, rice (damaged), rice (polish), safflower cake, sal seed cake, sesame, sesame cake, sorghum, soybean meal, soybeans, sunflower, sunflower cake, sunflower flour, tapioca, wheat, wheat bran, wheat bran and chana testa

Deoxynivalenol
incidence: 4/17, conc. range:
1530–6510 μg/kg, Ø conc.: 3170 μg/kg, country: Vietnam[104]
see also barley, barley, husked, barley, unhusked (naked), barley (pressed), bone meal, bran, broilers feed, calf fattening mixed

feed, coconut, expeller, corn cob mix silage, cottonseed, cottonseed cake, dairy cattle feed, egg production mixed feed, feed, feed components, feed, commercial mix, feed, mixed, feed, mixed (primarily maize), feed (barley), feed (cattle), feed (chicken), feed (dog), feed (fish), feed (mill run, from wheat), feed (mink), feed (pig), feed (poultry), feed (reindeer), feeds, grain, feeds, industrial, feedstuff, feedstuffs (rapeseed, turnip, fish meal, concentrates), fish meal, grain, mixed feed, grains, mixed, grains (no specification), maize, maize ears, maize fibre, maize germ, maize germ/bran, maize germ meal, maize gluten, maize kernels, maize meal, maize screenings, maize stalks (pith), maize, "Baby", maize, hybrid, maize, white, oats, rice bran, rice germ cake, rye, silage, sorghum, soybeans, triticale, wheat, wheat and barley, wheat, red hard winter, wheat, soft white winter, wheat, spring, wheat, winter

Fumonisin B₁
incidence: 15/17, conc. range:
268–1516 μg/kg, Ø conc.: 780 μg/kg, country: Vietnam[104]
see also barley, bird food, wild, dog food, feed, feed, complete ration, feed, general, feed, layer, feed, maize-based, feed, mixed, feed, pelleted ration, feed, screenings, feed, sweet, feed (broilers), feed (cat), feed (chicken), feed (dog), feed (gluten), feed (horse), feed (maize), feed (pig), feed (poultry), feed (rat), feed (rodent), forage grass, maize, maize and maize screenings, maize bran, maize ears, maize fine fractions, maize flakes, maize germ, maize germ/bran, maize germ meal, maize gluten, maize grits, maize kernels, maize meal, maize screenings, maize, "Baby", maize, ground, maize, preharvest, maize, sweet feed, maize/oats mix, rat chow, silage, sorghum, soybeans, wheat

Fumonisin B₂
incidence: 12/17, conc. range: 155–401 μg/kg, Ø conc.: 289 μg/kg, country: Vietnam[104]
see also barley, bird food, wild, dog food, feed, feed, maize-based, feed, mixed, feed

(cat), feed (dog), feed (gluten), feed (horse), feed (maize), feed (poultry), feed (rodent), maize, maize bran, maize fine fractions, maize flakes, maize germ, maize germ/bran, maize germ meal, maize gluten, maize kernels, maize meal, maize screenings, maize, "Baby", maize, ground, maize, preharvest, rat chow, wheat

Fumonisin B$_3$
incidence: 10/17, conc. range: 101–268 µg/kg, Ø conc.: 176 µg/kg, country: Vietnam[104]
see also feed, feed (maize), feed (rodent), maize, maize bran, maize fine fractions, maize germ meal, maize kernels, maize screenings

Nivalenol
incidence: 2/17, conc. range: 780–1950 µg/kg, Ø conc.: 1365 µg/kg, country: Vietnam[104]
see also barley, barley, husked, barley, unhusked (naked), barley (pressed), bran, feed, feed components, feed (cattle), feed (poultry), feed (reindeer), feeds, industrial, maize, maize ears, maize germ, maize germ/bran, maize gluten, maize kernels, maize meal, maize screenings, maize, "Baby", oats, rye, silage, triticale, wheat

Maize screenings may contain the following mycotoxins:

3-Acetyldeoxynivalenol
incidence: 3/4, conc. range: 40–110 µg/kg, Ø conc.: 66.7 µg/kg, country: UK[95]
see also barley, feed components, feed (poultry), feeds, grain, feeds, industrial, feedstuffs (rapeseed, turnip, fish meal, concentrates), maize, maize ears, maize gluten, maize kernels, maize meal, oats, wheat

15-Acetyldeoxynivalenol
incidence: 4/4, conc. range: 250–590 µg/kg, Ø conc.: 375 µg/kg, country: UK[95]
see also barley, feed components, feed, commercial mix, feed (poultry), maize, maize ears, maize germ, maize germ/bran, maize gluten, maize kernels, maize meal, maize, "Baby", oats, silage, wheat

Aflatoxin B$_1$
incidence: 1/1, conc.: 20 µg/kg, country: USA[191]
see also alfalfa, *Ambadi* cake, animal feedstuffs (dairy cake), bagasse, barley, bengalgram husk, bird food, bird food, wild, biri testa, blackgram, blackgram husk, bran, broiler mixed feed, calf fattening mixed feed, calf fattening mixed feed (containing 4–20 % peanut products), *Carthamus* cake, castor cake, cereals, cereal products, chick pea, coconut cake, cocos, concentrate, mixed, concentrates, cotton cake, cottonseed, cottonseed (dehulled), cottonseed cake, cottonseed extract, cottonseed meal, cottonseed meal (ammoniated), cottonseed meal (decorticated), cottonseed meats, cottonseed products, crumbles, crumbles, grower, cycad meal, dairy cattle feed, dairy cattle feed (containing 2–5 % peanut products), dairy cattle feed (containing 6–10 % peanut products), dairy cattle feed (containing 6–12 % peanut products), dairy cattle feed (containing more than 20 % peanut products), diets, mixed, dog food, egg production mixed feed, feed, feed and ingredients, feed, compound, feed, layer, feed, mixed, feed (beef), feed (broilers), feed (calf), feed (cat), feed (cattle), feed (chicken), feed (dairy), feed (dog), feed (dug), feed (fish), feed (gluten), feed (horse), feed (miscellaneous), feed (pig), feed (poultry), feed (poultry, pig), feed (rabbit), feed (sheep), feed (50–60 % maize), fish meal, grain by-products, grains (no specification), greengram, hay/silage, horsegram, husk, *Jagni* cake, legume mixture, linseed, linseed cake, livol, *Mahua* cake, maize, maize germ, maize gluten, maize grits, maize husk, maize meal, maize oil cake, maize powder, maize, ground, maize, hybrid, maize, preharvest, maize, yellow, maize (dark grains), *Makhana* (*Euryale ferox* Salisb) puffs, manioc, milk production mixed feed, mung testa, murkool, mustard cake, neem cake, niger cake, oats, palm kernel expeller cake, palm kernels, palm products, peanut cake, peanut cake (deoiled), peanut expeller, peanut hay, peanut meal,

peanut, kernels, peanut, shells, peanuts, pellets, finisher, pig meal and pellets, pigeon pea, poultry feeds (peanut containing), rapeseed cake, redgram husk, rice, rice bran, rice bran (deoiled), rice chaff, rice crack, rice germ, rice germ cake, rice meal, rice straw, rice (damaged), rice (polish), safflower cake, sal seed cake, sesame, sesame cake, sorghum, soybean meal, soybeans, sunflower, sunflower cake, sunflower flour, tapioca, wheat, wheat bran, wheat bran and chana testa

DEOXYNIVALENOL
incidence: 4/4, conc. range: 2500–4400 µg/kg, Ø conc.: 3575 µg/kg, country: UK[95]
incidence: 1/1, conc.: 0.3 µg/kg, country: USA[191]
see also barley, barley, husked, barley, unhusked (naked), barley (pressed), bone meal, bran, broilers feed, calf fattening mixed feed, coconut, expeller, corn cob mix silage, cottonseed, cottonseed cake, dairy cattle feed, egg production mixed feed, feed, feed components, feed, commercial mix, feed, mixed, feed, mixed (primarily maize), feed (barley), feed (cattle), feed (chicken), feed (dog), feed (fish), feed (mill run, from wheat), feed (mink), feed (pig), feed (poultry), feed (reindeer), feeds, grain, feeds, industrial, feedstuff, feedstuffs (rapeseed, turnip, fish meal, concentrates), fish meal, grain, mixed feed, grains, mixed, grains (no specification), maize, maize ears, maize fibre, maize germ, maize germ/bran, maize germ meal, maize gluten, maize kernels, maize meal, maize powder, maize stalks (pith), maize, "Baby", maize, hybrid, maize, white, oats, rice bran, rice germ cake, rye, silage, sorghum, soybeans, triticale, wheat, wheat and barley, wheat, red hard winter, wheat, soft white winter, wheat, spring, wheat, winter

FUMONISIN B$_1$
incidence: 2/2, conc. range: 105,000–155,000 µg/kg, Ø conc.: 130,000 µg/kg, country: USA[58]
incidence: 5/7, conc. range: 3600–330,000 µg/kg, Ø conc.: 120,000 µg/kg, country: USA[58]

incidence: 1/1, conc.: 2000 µg/kg, country: USA[58]
incidence: 1/1, conc.: 119,000 µg/kg, country: USA[58]
incidence: 1/1, conc.: 125,000 µg/kg, country: USA[58]
incidence: 4/4, conc. range: 21,000–27,000 µg/kg, Ø conc.: 23,750 µg/kg, country: UK[95]
incidence: 1/1, conc.: 60,000 µg/kg, country: Italy[138]
incidence: 15/15, conc. range: 470–4340 µg/kg, Ø conc.: 2100 µg/kg*, country: South Africa[146], *mean of all samples
incidence: 24/24, conc. range: 230–3780 µg/kg, Ø conc.: 1150 µg/kg*, country: South Africa[147], *mean of all samples
incidence: 37/37, conc. range: 50–44,750 µg/kg, Ø conc.: 5010 µg/kg*, country: South Africa[147], *mean of all samples
incidence: ?/180, conc. range: 100–239,000 µg/kg, country: USA[151]
incidence: 6/6, conc. range: 1740–196,500 µg/kg, Ø conc.: 55,400 µg/kg, country: USA[153]
incidence: 14/15, conc. range: 0–1500 µg/kg, Ø conc.: 417 µg/kg, country: Spain[159]
incidence: 4/4, conc. range: 23,000–46,000 µg/kg, Ø conc.: 30,250 µg/kg, country: USA[189]
incidence: 1/1, conc.: 166,000 µg/kg, country: USA[191]
incidence: ?/17, conc. range: 61,000–160,000 µg/kg, country: USA[319]
incidence: 12/12, conc. range: 772.4–5462 µg/kg, Ø conc.: 2152.7 µg/kg, country: Spain[347]
see also barley, bird food, wild, dog food, feed, feed, complete ration, feed, general, feed, layer, feed, maize-based, feed, mixed, feed, pelleted ration, feed, screenings, feed, sweet, feed (broilers), feed (cat), feed (chicken), feed (dog), feed (gluten), feed (horse), feed (maize), feed (pig), feed (poultry), feed (rat), feed (rodent), forage grass, maize, maize and maize screenings, maize bran, maize ears, maize fine fractions, maize flakes, maize germ, maize germ/bran,

maize germ meal, maize gluten, maize grits, maize kernels, maize meal, maize powder, maize, "Baby", maize, ground, maize, preharvest, maize, sweet feed, maize/oats mix, rat chow, silage, sorghum, soybeans, wheat

FUMONISIN B_2
incidence: 4/4, conc. range: 2800–3500 µg/kg, Ø conc.: 3225 µg/kg, country: UK[95]
incidence: 1/1, conc.: 15,000 µg/kg, country: Italy[138]
incidence: 15/15, conc. range: 100–2600 µg/kg, Ø conc.: 970 µg/kg*, country: South Africa[146], *mean of all samples
incidence: 24/24, conc. range: 0–1420 µg/kg, Ø conc.: 400 µg/kg*, country: South Africa[147], *mean of all samples
incidence: 37/37, conc. range: 0–26,890 µg/kg, Ø conc.: 2360 µg/kg*, country: South Africa[147], *mean of all samples
incidence: 6/6, conc. range: 590–42,840 µg/kg, Ø conc.: 15,080 µg/kg, country: USA[153]
incidence: 7/15, conc. range: 0–500 µg/kg, Ø conc.: 236 µg/kg, country: Spain[159]
incidence: 1/1, conc.: 48,000 µg/kg, country: USA[191]
incidence: ?/17, conc. range: 19,000–49,000 µg/kg, country: USA[319]
see also barley, bird food, wild, dog food, feed, feed, maize-based, feed, mixed, feed (cat), feed (dog), feed (gluten), feed (horse), feed (maize), feed (poultry), feed (rodent), maize, maize bran, maize fine fractions, maize flakes, maize germ, maize germ/bran, maize germ meal, maize gluten, maize kernels, maize meal, maize powder, maize, "Baby", maize, ground, maize, preharvest, rat chow, wheat

FUMONISIN B_3
incidence: 24/24, conc. range: 0–550 µg/kg, Ø conc.: 160 µg/kg*, country: South Africa[147], *mean of all samples
incidence: 37/37, conc. range: 0–12,180 µg/kg, Ø conc.: 840 µg/kg*, country: South Africa[147], *mean of all samples
see also feed, feed (maize), feed (rodent), maize, maize bran, maize fine fractions, maize germ meal, maize kernels, maize powder

FUMONISIN
incidence: 85/85, conc. range: 2600–32,000 µg/kg, Ø conc.: 12,100 µg/kg, country: USA[370]
see also maize, maize, shelled

FUMONISINS
incidence: 4/4*, conc. range: 100–2700 µg/kg, country: UK[94], *imported?
incidence: 55/55, conc. range: ≤5000–>25,000 µg/kg, country: USA[147]
incidence: 9/9, conc. range: ≤5000–>25,000 µg/kg, country: USA[147]
see also feed (cattle and dairy), feed (chicken), feed (dairy), feed (horse), maize, maize germ, maize gluten, maize meal

FUSARENONE-X
incidence: 4/4, conc. range: 35–90 µg/kg, Ø conc.: 71.3 µg/kg, country: UK[95]
see also barley, feed components, feed (poultry), maize, maize ears, maize germ/bran, maize gluten, maize kernels, maize meal, oats, wheat

FUSARIN-C
incidence: 1/1, conc.: 390 µg/kg, country: USA[69]

HT-2 TOXIN
incidence: 4/4, conc. range: 60–250 µg/kg, Ø conc.: 150 µg/kg, country: UK[95]
see also barley, feed components, feed (dog), feed (fish), feed (pig), feed (poultry), feed (reindeer), grain, mixed feed, grains, mixed, grains (no specification), maize, maize germ/bran, maize gluten, maize meal, oats, rye, silage, wheat

MONILIFORMIN
incidence: 1/1, conc.: 2820 µg/kg, country: USA[69]
incidence: 4/4, conc. range: 2200–4600 µg/kg, Ø conc.: 3150 µg/kg, country: UK[95]
see also barley, feed, mixed, feed (poultry), maize, maize flakes, maize germ, maize germ/bran, maize gluten, maize meal, maize, "Baby", oats, rice bran, triticale, wheat, wheat, summer, wheat, winter

Monoacetoxyscirpenol
incidence: 3/4, conc. range: 10–20 μg/kg,
Ø conc.: 15 μg/kg, country: UK[95]
see also feed components, maize, maize
gluten, maize meal, silage, wheat

Nivalenol
incidence: 4/4, conc. range: 590–2300 μg/kg,
Ø conc.: 1322.5 μg/kg, country: UK[95]
see also barley, barley, husked, barley,
unhusked (naked), barley (pressed), bran,
feed, feed components, feed (cattle), feed
(poultry), feed (reindeer), feeds, industrial,
maize, maize ears, maize germ, maize
germ/bran, maize gluten, maize kernels,
maize meal, maize powder, maize, "Baby",
oats, rye, silage, triticale, wheat

T-2 Toxin
incidence: 4/4, conc. range: 40–130 μg/kg,
Ø conc.: 102.5 μg/kg, country: UK[95]
see also alfalfa, barley, bran, diet (grower),
diet (poultry), feed, feed components, feed,
layer, feed, mixed, feed (dog), feed (fish),
feed (mink), feed (pig), feed (poultry),
feedstuff, forage grass, grain, mixed feed,
grains (no specification), hay, maize, maize
germ/bran, maize gluten, maize meal, maize
stalks (pith), oat and barley
(hammer-milled), oats, peanuts, pig grower
diet, piglet diet, rye, sorghum, triticale, wheat

Zearalenone
incidence: 3/4*, conc. range: 20–510 μg/kg,
country: UK[94], *imported?
incidence: 4/4, conc. range: 1300–1800 μg/kg,
Ø conc.: 1450 μg/kg, country: UK[95]
see also alfalfa, barley, barley, husked, barley,
unhusked (naked), barley and feed, bone
meal, bran, broilers feed, *Carthamus* cake,
chick pea, concentrate, mixed, corn cob mix
silage, cotton cake, cottonseed, cottonseed
cake, diet (dairy cow), diet (poultry), diets
(mixed), feed, feed components, feed, mixed,
feed, mixed (primarily maize also maize, oats,
wheat), feed (bran), feed (broiler chicken),
feed (cattle), feed (chicken), feed (dairy), feed
(developing pig), feed (mill run, from
wheat), feed (miscellaneous), feed (pig), feed
(poultry), feed (poultry, pig), feed (starter

chicken), feedstuff, fish meal, forage grass,
grain, bruised, grain, mixed feed, grains (no
specification), hay, maize, maize ears, maize
flakes, maize germ, maize germ/bran, maize
gluten, maize kernels, maize meal, maize oil
cake, maize stalks (pith), maize, "Baby",
maize, hybrid, maize, shelled, maize,
unshelled, maize, white, maize grain,
artificially dried, maize grain, crib dried,
maize grain, ensiled, milk production mixed
feed, oats, *Paspalum palidosum*, straw, peanut
hulls/skins, rice bran, rice germ, rice germ
cake, rye, silage, sorghum, soybeans,
soybeans, extracted, sunflower cake, tapioca,
triticale, wheat, wheat bran, wheat bran and
chana testa, wheat soya meal

Maize stalks (pith)   may contain the following
mycotoxins:

Deoxynivalenol
incidence: ?/50, conc. range: 1500 μg/kg,
country: USA[220]
see also barley, barley, husked, barley,
unhusked (naked), barley (pressed), bone
meal, bran, broilers feed, calf fattening mixed
feed, coconut, expeller, corn cob mix silage,
cottonseed, cottonseed cake, dairy cattle feed,
egg production mixed feed, feed, feed
components, feed, commercial mix, feed,
mixed, feed, mixed (primarily maize), feed
(barley), feed (cattle), feed (chicken), feed
(dog), feed (fish), feed (mill run, from
wheat), feed (mink), feed (pig), feed
(poultry), feed (reindeer), feeds, grain, feeds,
industrial, feedstuff, feedstuffs (rapeseed,
turnip, fish meal, concentrates), fish meal,
grain, mixed feed, grains, mixed, grains (no
specification), maize, maize ears, maize fibre,
maize germ, maize germ/bran, maize germ
meal, maize gluten, maize kernels, maize
meal, maize powder, maize screenings, maize,
"Baby", maize, hybrid, maize, white, oats, rice
bran, rice germ cake, rye, silage, sorghum,
soybeans, triticale, wheat, wheat and barley,
wheat, red hard winter, wheat, soft white
winter, wheat, spring, wheat, winter

T-2 Toxin
incidence: ?/50, conc.: 110 µg/kg,
country: USA[220]
see also alfalfa, barley, bran, diet (grower),
diet (poultry), feed, feed components, feed,
layer, feed, mixed, feed (dog), feed (fish),
feed (mink), feed (pig), feed (poultry),
feedstuff, forage grass, grain, mixed feed,
grains (no specification), hay, maize, maize
germ/bran, maize gluten, maize meal, maize
screenings, oat and barley (hammer-milled),
oats, peanuts, pig grower diet, piglet diet, rye,
sorghum, triticale, wheat

Zearalenone
incidence: ?/50, conc.: 2800 µg/kg,
country: USA[220]
see also alfalfa, barley, barley, husked, barley,
unhusked (naked), barley and feed, bone
meal, bran, broilers feed, *Carthamus* cake,
chick pea, concentrate, mixed, corn cob mix
silage, cotton cake, cottonseed, cottonseed
cake, diet (dairy cow), diet (poultry), diets
(mixed), feed, feed components, feed, mixed,
feed, mixed (primarily maize also maize, oats,
wheat), feed (bran), feed (broiler chicken),
feed (cattle), feed (chicken), feed (dairy), feed
(developing pig), feed (mill run, from
wheat), feed (miscellaneous), feed (pig), feed
(poultry), feed (poultry, pig), feed (starter
chicken), feedstuff, fish meal, forage grass,
grain, bruised, grain, mixed feed, grains (no
specification), hay, maize, maize ears, maize
flakes, maize germ, maize germ/bran, maize
gluten, maize kernels, maize meal, maize oil
cake, maize screenings, maize, "Baby", maize,
hybrid, maize, shelled, maize, unshelled,
maize, white, maize grain, artificially dried,
maize grain, crib dried, maize grain, ensiled,
milk production mixed feed, oats, *Paspalum
palidosum*, straw, peanut hulls/skins, rice
bran, rice germ, rice germ cake, rye, silage,
sorghum, soybeans, soybeans, extracted,
sunflower cake, tapioca, triticale, wheat,
wheat bran, wheat bran and chana testa,
wheat soya meal

Maize, "Baby"   may contain the following
mycotoxins:

15-Acetyldeoxynivalenol
incidence: 4/4, conc. range: 30–90 µg/kg,
Ø conc.: 52.5 µg/kg, country: UK[95]
see also barley, feed components, feed,
commercial mix, feed (poultry), maize, maize
ears, maize germ, maize germ/bran, maize
gluten, maize kernels, maize meal, maize
screenings, oats, silage, wheat

Deoxynivalenol
incidence: 4/4, conc. range: 100–340 µg/kg,
Ø conc.: 192.5 µg/kg, country: UK[95]
see also barley, barley, husked, barley,
unhusked (naked), barley (pressed), bone
meal, bran, broilers feed, calf fattening mixed
feed, coconut, expeller, corn cob mix silage,
cottonseed, cottonseed cake, dairy cattle feed,
egg production mixed feed, feed, feed
components, feed, commercial mix, feed,
mixed, feed, mixed (primarily maize), feed
(barley), feed (cattle), feed (chicken), feed
(dog), feed (fish), feed (mill run, from
wheat), feed (mink), feed (pig), feed
(poultry), feed (reindeer), feeds, grain, feeds,
industrial, feedstuff, feedstuffs (rapeseed,
turnip, fish meal, concentrates), fish meal,
grain, mixed feed, grains, mixed, grains (no
specification), maize, maize ears, maize fibre,
maize germ, maize germ/bran, maize germ
meal, maize gluten, maize kernels, maize
meal, maize powder, maize screenings, maize
stalks (pith), maize, hybrid, maize, white,
oats, rice bran, rice germ cake, rye, silage,
sorghum, soybeans, triticale, wheat, wheat
and barley, wheat, red hard winter, wheat,
soft white winter, wheat, spring, wheat,
winter

Fumonisin B$_1$
incidence: 4/4, conc. range: 45–570 µg/kg,
Ø conc.: 296.3 µg/kg, country: UK[95]
see also barley, bird food, wild, dog food,
feed, feed, complete ration, feed, general,
feed, layer, feed, maize-based, feed, mixed,
feed, pelleted ration, feed, screenings, feed,
sweet, feed (broilers), feed (cat), feed
(chicken), feed (dog), feed (gluten), feed
(horse), feed (maize), feed (pig), feed
(poultry), feed (rat), feed (rodent), forage

grass, maize, maize and maize screenings, maize bran, maize ears, maize fine fractions, maize flakes, maize germ, maize germ/bran, maize germ meal, maize gluten, maize grits, maize kernels, maize meal, maize powder, maize screenings, maize, ground, maize, preharvest, maize, sweet feed, maize/oats mix, rat chow, silage, sorghum, soybeans, wheat

Fumonisin B_2
incidence: 3/4, conc. range: 40–85 µg/kg, Ø conc.: 70 µg/kg, country: UK[95]
see also barley, bird food, wild, dog food, feed, feed, maize-based, feed, mixed, feed (cat), feed (dog), feed (gluten), feed (horse), feed (maize), feed (poultry), feed (rodent), maize, maize bran, maize fine fractions, maize flakes, maize germ, maize germ/bran, maize germ meal, maize gluten, maize kernels, maize meal, maize powder, maize screenings, maize, ground, maize, preharvest, rat chow, wheat

Moniliformin
incidence: 1/4, conc.: 70 µg/kg, country: UK[95]
see also barley, feed, mixed, feed (poultry), maize, maize flakes, maize germ, maize germ/bran, maize gluten, maize meal, maize screenings, oats, rice bran, triticale, wheat, wheat, summer, wheat, winter

Nivalenol
incidence: 4/4, conc. range: 25–50 µg/kg, Ø conc.: 35 µg/kg, country: UK[95]
see also barley, barley, husked, barley, unhusked (naked), barley (pressed), bran, feed, feed components, feed (cattle), feed (poultry), feed (reindeer), feeds, industrial, maize, maize ears, maize germ, maize germ/bran, maize gluten, maize kernels, maize meal, maize powder, maize screenings, oats, rye, silage, triticale, wheat

Zearalenone
incidence: 4/4, conc. range: 40–80 µg/kg, Ø conc.: 55 µg/kg, country: UK[95]
see also alfalfa, barley, barley, husked, barley, unhusked (naked), barley and feed, bone meal, bran, broilers feed, *Carthamus* cake, chick pea, concentrate, mixed, corn cob mix silage,

cotton cake, cottonseed, cottonseed cake, diet (dairy cow), diet (poultry), diets (mixed), feed, feed components, feed, mixed, feed, mixed (primarily maize also maize, oats, wheat), feed (bran), feed (broiler chicken), feed (cattle), feed (chicken), feed (dairy), feed (developing pig), feed (mill run, from wheat), feed (miscellaneous), feed (pig), feed (poultry), feed (poultry, pig), feed (starter chicken), feedstuff, fish meal, forage grass, grain, bruised, grain, mixed feed, grains (no specification), hay, maize, maize ears, maize flakes, maize germ, maize germ/bran, maize gluten, maize kernels, maize meal, maize oil cake, maize screenings, maize stalks (pith), maize, hybrid, maize, shelled, maize, unshelled, maize, white, maize grain, artificially dried, maize grain, crib dried, maize grain, ensiled, milk production mixed feed, oats, *Paspalum palidosum*, straw, peanut hulls/skins, rice bran, rice germ, rice germ cake, rye, silage, sorghum, soybeans, soybeans, extracted, sunflower cake, tapioca, triticale, wheat, wheat bran, wheat bran and chana testa, wheat soya meal

Maize, ground   may contain the following mycotoxins:

Aflatoxin B_1
incidence: 7/7, conc. range: 1–84 µg/kg, Ø conc.: 18 µg/kg, country: Thailand[117]
incidence: 8/8*, conc. range: 210–3200 µg/kg, Ø conc.: 891.25 µg/kg, country: USA[299], *ncac
see also alfalfa, *Ambadi* cake, animal feedstuffs (dairy cake), bagasse, barley, bengalgram husk, bird food, bird food, wild, biri testa, blackgram, blackgram husk, bran, broiler mixed feed, calf fattening mixed feed, calf fattening mixed feed (containing 4–20 % peanut products), *Carthamus* cake, castor cake, cereals, cereal products, chick pea, coconut cake, cocos, concentrate, mixed, concentrates, cotton cake, cottonseed, cottonseed (dehulled), cottonseed cake, cottonseed extract, cottonseed meal, cottonseed meal (ammoniated), cottonseed meal (decorticated), cottonseed meats, cottonseed products, crumbles, crumbles, grower, cycad meal, dairy cattle feed, dairy

cattle feed (containing 2–5 % peanut products), dairy cattle feed (containing 6–10 % peanut products), dairy cattle feed (containing 6–12 % peanut products), dairy cattle feed (containing more than 20 % peanut products), diets, mixed, dog food, egg production mixed feed, feed, feed and ingredients, feed, compound, feed, layer, feed, mixed, feed (beef), feed (broilers), feed (calf), feed (cat), feed (cattle), feed (chicken), feed (dairy), feed (dog), feed (dug), feed (fish), feed (gluten), feed (horse), feed (miscell-aneous), feed (pig), feed (poultry), feed (poultry, pig), feed (rabbit), feed (sheep), feed (50–60 % maize), fish meal, grain by-products, grains (no specification), greengram, hay/silage, horsegram, husk, *Jagni* cake, legume mixture, linseed, linseed cake, livol, *Mahua* cake, maize, maize germ, maize gluten, maize grits, maize husk, maize meal, maize oil cake, maize powder, maize screenings, maize, hybrid, maize, preharvest, maize, yellow, maize (dark grains), *Makhana* (*Euryale ferox* Salisb) puffs, manioc, milk production mixed feed, mung testa, murkool, mustard cake, neem cake, niger cake, oats, palm kernel expeller cake, palm kernels, palm products, peanut cake, peanut cake (deoiled), peanut expeller, peanut hay, peanut meal, peanut, kernels, peanut, shells, peanuts, pellets, finisher, pig meal and pellets, pigeon pea, poultry feeds (peanut containing), rapeseed cake, redgram husk, rice, rice bran, rice bran (deoiled), rice chaff, rice crack, rice germ, rice germ cake, rice meal, rice straw, rice (damaged), rice (polish), safflower cake, sal seed cake, sesame, sesame cake, sorghum, soybean meal, soybeans, sunflower, sunflower cake, sunflower flour, tapioca, wheat, wheat bran, wheat bran and chana testa

Aflatoxin B_2
incidence: 6/7, conc. range: 1–15 µg/kg, Ø conc.: 4.7 µg/kg, country: Thailand[117]
incidence: 8/8*, conc. range: 15–290 µg/kg, Ø conc.: 100.25 µg/kg, country: USA[299], *ncac
see also animal feedstuffs (dairy cake), bird food, bird food, wild, biri testa, blackgram, blackgram husk, cottonseed, cottonseed cake, cottonseed extract, cottonseed meal, cottonseed meal (ammoniated), cottonseed meats, dog food, egg production mixed feed, feed, compound, feed (cat), feed (cattle), feed (dog), feed (pig), feed (poultry), feed (rabbit), feed (sheep), fish meal, horsegram, maize, maize gluten, maize husk, maize, preharvest, mung testa, mustard cake, niger cake, peanut cake, peanut cake (deoiled), peanut expeller, peanut hay, peanut meal, peanuts, peanut, kernels, redgram husk, rice bran, rice bran (deoiled), rice chaff, rice meal, rice (polish), sal seed cake, sesame cake, sorghum, soybean meal, soybeans, wheat, wheat bran

Aflatoxin G_1
incidence: 3/7, conc. range: 2–7 µg/kg, Ø conc.: 4.7 µg/kg, country: Thailand[117]
see also animal feedstuffs (dairy cake), bird food, bird food, wild, concentrate, mixed, cottonseed, cottonseed cake, feed (cat), feed (chicken), feed (dog), maize, maize, preharvest, meat meal, milk production mixed feed, murkool, peanut cake, peanut expeller, peanut hay, peanut meal, peanuts, rice bran, rice germ, sorghum, soybeans, wheat, wheat bran, wheat bran and chana testa

Aflatoxin M_1
incidence: 8/8*, conc. range: 1–35 µg/kg, Ø conc.: 6.75 µg/kg, country: USA[299], *ncac
see also feed (cat), feed (dog), maize

Fumonisin B_1
incidence: 7/7, conc. range: 65–1450 µg/kg, Ø conc.: 842.4 µg/kg, country: Thailand[117]
see also barley, bird food, wild, dog food, feed, feed, complete ration, feed, general, feed, layer, feed, maize-based, feed, mixed, feed, pelleted ration, feed, screenings, feed, sweet, feed (broilers), feed (cat), feed (chicken), feed (dog), feed (gluten), feed (horse), feed (maize), feed (pig), feed (poultry), feed (rat), feed (rodent), forage grass, maize, maize and maize screenings, maize bran, maize ears, maize fine fractions, maize flakes, maize germ, maize germ/bran, maize germ meal, maize gluten, maize grits,

maize kernels, maize meal, maize powder, maize screenings, maize, "Baby", maize, preharvest, maize, sweet feed, maize/oats mix, rat chow, silage, sorghum, soybeans, wheat

FUMONISIN B_2
incidence: 6/7, conc. range: 80–191 µg/kg, Ø conc.: 128.2 µg/kg, country: Thailand[117]
see also barley, bird food, wild, dog food, feed, feed, maize-based, feed, mixed, feed (cat), feed (dog), feed (gluten), feed (horse), feed (maize), feed (poultry), feed (rodent), maize, maize bran, maize fine fractions, maize flakes, maize germ, maize germ/bran, maize germ meal, maize gluten, maize kernels, maize meal, maize powder, maize screenings, maize, "Baby", maize, preharvest, rat chow, wheat

Maize, hybrid for feed may contain the following mycotoxins:

AFLATOXIN B_1
incidence: 3/5, Ø conc.: 30 µg/kg, country: Egypt[16]
see also alfalfa, *Ambadi* cake, animal feedstuffs (dairy cake), bagasse, barley, bengalgram husk, bird food, bird food, wild, biri testa, blackgram, blackgram husk, bran, broiler mixed feed, calf fattening mixed feed, calf fattening mixed feed (containing 4–20 % peanut products), *Carthamus* cake, castor cake, cereals, cereal products, chick pea, coconut cake, cocos, concentrate, mixed, concentrates, cotton cake, cottonseed, cottonseed (dehulled), cottonseed cake, cottonseed extract, cottonseed meal, cottonseed meal (ammoniated), cottonseed meal (decorticated), cottonseed meats, cottonseed products, crumbles, crumbles, grower, cycad meal, dairy cattle feed, dairy cattle feed (containing 2–5 % peanut products), dairy cattle feed (containing 6–10 % peanut products), dairy cattle feed (containing 6–12 % peanut products), dairy cattle feed (containing more than 20 % peanut products), diets, mixed, dog food, egg production mixed feed, feed, feed and ingredients, feed, compound, feed, layer, feed, mixed, feed (beef), feed (broilers), feed (calf), feed (cat), feed (cattle), feed (chicken), feed (dairy), feed (dog), feed (dug), feed (fish), feed (gluten), feed (horse), feed (miscellaneous), feed (pig), feed (poultry), feed (poultry, pig), feed (rabbit), feed (sheep), feed (50–60 % maize), fish meal, grain by-products, grains (no specification), greengram, hay/silage, horsegram, husk, *Jagni* cake, legume mixture, linseed, linseed cake, livol, *Mahua* cake, maize, maize germ, maize gluten, maize grits, maize husk, maize meal, maize oil cake, maize powder, maize screenings, maize, ground, maize, preharvest, maize, yellow, maize (dark grains), *Makhana* (*Euryale ferox* Salisb) puffs, manioc, milk production mixed feed, mung testa, murkool, mustard cake, neem cake, niger cake, oats, palm kernel expeller cake, palm kernels, palm products, peanut cake, peanut cake (deoiled), peanut expeller, peanut hay, peanut meal, peanut, kernels, peanut, shells, peanuts, pellets, finisher, pig meal and pellets, pigeon pea, poultry feeds (peanut containing), rapeseed cake, redgram husk, rice, rice bran, rice bran (deoiled), rice chaff, rice crack, rice germ, rice germ cake, rice meal, rice straw, rice (damaged), rice (polish), safflower cake, sal seed cake, sesame, sesame cake, sorghum, soybean meal, soybeans, sunflower, sunflower cake, sunflower flour, tapioca, wheat, wheat bran, wheat bran and chana testa

DEOXYNIVALENOL
incidence: 4/4, conc. range: 100–222 µg/kg, Ø conc.: 178 µg/kg, country: Egypt[16]
see also barley, barley, husked, barley, unhusked (naked), barley (pressed), bone meal, bran, broilers feed, calf fattening mixed feed, coconut, expeller, corn cob mix silage, cottonseed, cottonseed cake, dairy cattle feed, egg production mixed feed, feed, feed components, feed, commercial mix, feed, mixed, feed, mixed (primarily maize), feed (barley), feed (cattle), feed (chicken), feed (dog), feed (fish), feed (mill run, from wheat), feed (mink), feed (pig), feed (poultry), feed (reindeer), feeds, grain, feeds, industrial, feedstuff, feedstuffs (rapeseed,

turnip, fish meal, concentrates), fish meal, grain, mixed feed, grains, mixed, grains (no specification), maize, maize ears, maize fibre, maize germ, maize germ/bran, maize germ meal, maize gluten, maize kernels, maize meal, maize powder, maize screenings, maize stalks (pith), maize, "Baby", maize, white, oats, rice bran, rice germ cake, rye, silage, sorghum, soybeans, triticale, wheat, wheat and barley, wheat, red hard winter, wheat, soft white winter, wheat, spring, wheat, winter

ZEARALENONE
incidence: 4/4, conc. range: 2–79 μg/kg, Ø conc.: 42 μg/kg, country: Egypt[16]
see also alfalfa, barley, barley, husked, barley, unhusked (naked), barley and feed, bone meal, bran, broilers feed, *Carthamus* cake, chick pea, concentrate, mixed, corn cob mix silage, cotton cake, cottonseed, cottonseed cake, diet (dairy cow), diet (poultry), diets (mixed), feed, feed components, feed, mixed, feed, mixed (primarily maize also maize, oats, wheat), feed (bran), feed (broiler chicken), feed (cattle), feed (chicken), feed (dairy), feed (developing pig), feed (mill run, from wheat), feed (miscellaneous), feed (pig), feed (poultry), feed (poultry, pig), feed (starter chicken), feedstuff, fish meal, forage grass, grain, bruised, grain, mixed feed, grains (no specification), hay, maize, maize ears, maize flakes, maize germ, maize germ/bran, maize gluten, maize kernels, maize meal, maize oil cake, maize screenings, maize stalks (pith), maize, "Baby", maize, shelled, maize, unshelled, maize, white, maize grain, artificially dried, maize grain, crib dried, maize grain, ensiled, milk production mixed feed, oats, *Paspalum palidosum*, straw, peanut hulls/skins, rice bran, rice germ, rice germ cake, rye, silage, sorghum, soybeans, soybeans, extracted, sunflower cake, tapioca, triticale, wheat, wheat bran, wheat bran and chana testa, wheat soya meal

Maize, preharvest may contain the following mycotoxins:

AFLATOXIN B$_1$
incidence: 19/103*, conc. range: 3–130 μg/kg, Ø conc.: 22 μg/kg, country: Nigeria[207], *ncac
see also alfalfa, *Ambadi* cake, animal feedstuffs (dairy cake), bagasse, barley, bengalgram husk, bird food, bird food, wild, biri testa, blackgram, blackgram husk, bran, broiler mixed feed, calf fattening mixed feed, calf fattening mixed feed (containing 4–20 % peanut products), *Carthamus* cake, castor cake, cereals, cereal products, chick pea, coconut cake, cocos, concentrate, mixed, concentrates, cotton cake, cottonseed, cottonseed (dehulled), cottonseed cake, cottonseed extract, cottonseed meal, cottonseed meal (ammoniated), cottonseed meal (decorticated), cottonseed meats, cottonseed products, crumbles, crumbles, grower, cycad meal, dairy cattle feed, dairy cattle feed (containing 2–5 % peanut products), dairy cattle feed (containing 6–10 % peanut products), dairy cattle feed (containing 6–12 % peanut products), dairy cattle feed (containing more than 20 % peanut products), diets, mixed, dog food, egg production mixed feed, feed, feed and ingredients, feed, compound, feed, layer, feed, mixed, feed (beef), feed (broilers), feed (calf), feed (cat), feed (cattle), feed (chicken), feed (dairy), feed (dog), feed (dug), feed (fish), feed (gluten), feed (horse), feed (miscellaneous), feed (pig), feed (poultry), feed (poultry, pig), feed (rabbit), feed (sheep), feed (50–60 % maize), fish meal, grain by-products, grains (no specification), greengram, hay/silage, horsegram, husk, *Jagni* cake, legume mixture, linseed, linseed cake, livol, *Mahua* cake, maize, maize germ, maize gluten, maize grits, maize husk, maize meal, maize oil cake, maize powder, maize screenings, maize, ground, maize, hybrid, maize, yellow, maize (dark grains), *Makhana* (*Euryale ferox* Salisb) puffs, manioc, milk production mixed feed, mung testa, murkool, mustard cake, neem cake, niger cake, oats, palm kernel expeller cake, palm kernels, palm products, peanut cake, peanut cake (deoiled), peanut expeller, peanut hay, peanut meal,

peanut, kernels, peanut, shells, peanuts, pellets, finisher, pig meal and pellets, pigeon pea, poultry feeds (peanut containing), rapeseed cake, redgram husk, rice, rice bran, rice bran (deoiled), rice chaff, rice crack, rice germ, rice germ cake, rice meal, rice straw, rice (damaged), rice (polish), safflower cake, sal seed cake, sesame, sesame cake, sorghum, soybean meal, soybeans, sunflower, sunflower cake, sunflower flour, tapioca, wheat, wheat bran, wheat bran and chana testa

AFLATOXIN B$_2$
incidence: 8/103*, conc. range: 4–26 µg/kg, Ø conc.: 10 µg/kg, country: Nigeria[207], *ncac
see also animal feedstuffs (dairy cake), bird food, bird food, wild, biri testa, blackgram, blackgram husk, cottonseed, cottonseed cake, cottonseed extract, cottonseed meal, cottonseed meal (ammoniated), cottonseed meats, dog food, egg production mixed feed, feed, compound, feed (cat), feed (cattle), feed (dog), feed (pig), feed (poultry), feed (rabbit), feed (sheep), fish meal, horsegram, maize, maize gluten, maize husk, maize, ground, mung testa, mustard cake, niger cake, peanut cake, peanut cake (deoiled), peanut expeller, peanut hay, peanut meal, peanuts, peanut, kernels, redgram husk, rice bran, rice bran (deoiled), rice chaff, rice meal, rice (polish), sal seed cake, sesame cake, sorghum, soybean meal, soybeans, wheat, wheat bran

AFLATOXIN G$_1$
incidence: 3/103*, conc. range: 5–11 µg/kg, Ø conc.: 8 µg/kg, country: Nigeria[207], *ncac
see also animal feedstuffs (dairy cake), bird food, bird food, wild, concentrate, mixed, cottonseed, cottonseed cake, feed (cat), feed (chicken), feed (dog), maize, maize, ground, meat meal, milk production mixed feed, murkool, peanut cake, peanut expeller, peanut hay, peanut meal, peanuts, rice bran, rice germ, sorghum, soybeans, wheat, wheat bran, wheat bran and chana testa

AFLATOXIN G$_2$
incidence: 1/103*, conc.: 7 µg/kg, country: Nigeria[207], *ncac

see also animal feedstuffs (dairy cake), bird food, feed (cat), feed (dog), maize, peanut expeller, peanut hay, peanut meal, peanuts, soybeans

FUMONISIN B$_1$
incidence: 81/103*, conc. range: 70–1780 µg/kg, Ø conc.: 495 µg/kg, country: Nigeria[207], *ncac
see also barley, bird food, wild, dog food, feed, feed, complete ration, feed, general, feed, layer, feed, maize-based, feed, mixed, feed, pelleted ration, feed, screenings, feed, sweet, feed (broilers), feed (cat), feed (chicken), feed (dog), feed (gluten), feed (horse), feed (maize), feed (pig), feed (poultry), feed (rat), feed (rodent), forage grass, maize, maize and maize screenings, maize bran, maize ears, maize fine fractions, maize flakes, maize germ, maize germ/bran, maize germ meal, maize gluten, maize grits, maize kernels, maize meal, maize powder, maize screenings, maize, "Baby", maize, ground, maize, sweet feed, maize/oats mix, rat chow, silage, sorghum, soybeans, wheat

FUMONISIN B$_2$
incidence: 68/103*, conc. range: 53–230 µg/kg, Ø conc.: 114 µg/kg, country: Nigeria[207], *ncac
see also barley, bird food, wild, dog food, feed, feed, maize-based, feed, mixed, feed (cat), feed (dog), feed (gluten), feed (horse), feed (maize), feed (poultry), feed (rodent), maize, maize bran, maize fine fractions, maize flakes, maize germ, maize germ/bran, maize germ meal, maize gluten, maize kernels, maize meal, maize powder, maize screenings, maize, "Baby", maize, ground, rat chow, wheat

Maize, shelled   for feed may contain the following mycotoxins:

AFLATOXINS
incidence: 146/1012, conc. range: <20 µg/kg (100 sa), <100 µg/kg (32 sa), <200 µg/kg (10 sa), <300 µg/kg (2 sa), >300 µg/kg (2 sa), country: USA[195]
incidence: 70/173, conc. range: <20 µg/kg (61 sa), <100 µg/kg (5 sa), <200 µg/kg (1 sa),

<300 μg/kg (1 sa), >300 μg/kg (2 sa), country: USA[195]
incidence: 71/123, conc. range: <20 μg/kg (29 sa), <100 μg/kg (28 sa), <200 μg/kg (7 sa), <300 μg/kg (2 sa), >300 μg/kg (5 sa), country: USA[195]
see also animal feed (maize), animal feed (mixed), barley, copra meal, cottonseed, cottonseed fines, cottonseed meal, cottonseed meats, feed, feed ingredients (miscellaneous), feed, mixed, feed (excluding peanuts, suspect), feed (goat), feed (pig), feed (poultry), feedstuff, grain, grain, mixed feed, maize, millet, peanut cake, peanut meal, peanut meal and by-products, peanuts, protein concentrates, rice, rice bran, sorghum, soybean meal, sunflower

Fumonisin
incidence: 22/22, conc. range: ≤5000–10,000 μg/kg, country: USA[147]
incidence: 36/36, conc. range: ≤5000 μg/kg, country: USA[147]
see also maize, maize screenings

Zearalenol
incidence: 1/31, conc.: 19 μg/kg, country: USA[119]
see also grains (no specification), maize, unshelled, oats

Zearalenone
incidence: 31/31, conc. range: tr–480 μg/kg, country: USA[119]
see also alfalfa, barley, barley, husked, barley, unhusked (naked), barley and feed, bone meal, bran, broilers feed, *Carthamus* cake, chick pea, concentrate, mixed, corn cob mix silage, cotton cake, cottonseed, cottonseed cake, diet (dairy cow), diet (poultry), diets (mixed), feed, feed components, feed, mixed, feed, mixed (primarily maize also maize, oats, wheat), feed (bran), feed (broiler chicken), feed (cattle), feed (chicken), feed (dairy), feed (developing pig), feed (mill run, from wheat), feed (miscellaneous), feed (pig), feed (poultry), feed (poultry, pig), feed (starter chicken), feedstuff, fish meal, forage grass, grain, bruised, grain, mixed feed, grains (no specification), hay, maize, maize ears, maize

flakes, maize germ, maize germ/bran, maize gluten, maize kernels, maize meal, maize oil cake, maize screenings, maize stalks (pith), maize, "Baby", maize, hybrid, maize, unshelled, maize, white, maize grain, artificially dried, maize grain, crib dried, maize grain, ensiled, milk production mixed feed, oats, *Paspalum palidosum*, straw, peanut hulls/skins, rice bran, rice germ, rice germ cake, rye, silage, sorghum, soybeans, soybeans, extracted, sunflower cake, tapioca, triticale, wheat, wheat bran, wheat bran and chana testa, wheat soya meal

Maize, sweet feed   may contain the following mycotoxins:

Fumonisin B$_1$
incidence: 2/2, conc. range: 27,000–94,000 μg/kg, Ø conc.: 60,500 μg/kg, country: USA[58]
incidence: 1/1, conc.: 18,000 μg/kg, country: USA[58]
see also barley, bird food, wild, dog food, feed, feed, complete ration, feed, general, feed, layer, feed, maize-based, feed, mixed, feed, pelleted ration, feed, screenings, feed, sweet, feed (broilers), feed (cat), feed (chicken), feed (dog), feed (gluten), feed (horse), feed (maize), feed (pig), feed (poultry), feed (rat), feed (rodent), forage grass, maize, maize and maize screenings, maize bran, maize ears, maize fine fractions, maize flakes, maize germ, maize germ/bran, maize germ meal, maize gluten, maize grits, maize kernels, maize meal, maize powder, maize screenings, maize, "Baby", maize, ground, maize, preharvest, maize/oats mix, rat chow, silage, sorghum, soybeans, wheat

Maize, unshelled   for feed may contain the following mycotoxins:

Zearalenol
incidence: 3/7, conc. range: 14–106 μg/kg, Ø conc.: 49.3 μg/kg, country: USA[119]
see also grains (no specification), maize, shelled, oats

Zearalenone
incidence: 7/7, conc. range: tr–3656 μg/kg,
country: USA[119]
see also alfalfa, barley, barley, husked, barley,
unhusked (naked), barley and feed, bone
meal, bran, broilers feed, *Carthamus* cake,
chick pea, concentrate, mixed, corn cob mix
silage, cotton cake, cottonseed, cottonseed
cake, diet (dairy cow), diet (poultry), diets
(mixed), feed, feed components, feed, mixed,
feed, mixed (primarily maize also maize, oats,
wheat), feed (bran), feed (broiler chicken),
feed (cattle), feed (chicken), feed (dairy), feed
(developing pig), feed (mill run, from
wheat), feed (miscellaneous), feed (pig), feed
(poultry), feed (poultry, pig), feed (starter
chicken), feedstuff, fish meal, forage grass,
grain, bruised, grain, mixed feed, grains (no
specification), hay, maize, maize ears, maize
flakes, maize germ, maize germ/bran, maize
gluten, maize kernels, maize meal, maize oil
cake, maize screenings, maize stalks (pith),
maize, "Baby", maize, hybrid, maize, shelled,
maize, white, maize grain, artificially dried,
maize grain, crib dried, maize grain, ensiled,
milk production mixed feed, oats, *Paspalum
palidosum*, straw, peanut hulls/skins, rice
bran, rice germ, rice germ cake, rye, silage,
sorghum, soybeans, soybeans, extracted,
sunflower cake, tapioca, triticale, wheat,
wheat bran, wheat bran and chana testa,
wheat soya meal

Maize, white   for feed may contain the
following mycotoxins:

Aflatoxin
incidence: 394/1283*, conc. range: <10 μg/kg
(136 sa), 10–19 μg/kg (93 sa), 20–29 μg/kg
(45 sa), 30–100 μg/kg (91 sa), ≤306 μg/kg
(29 sa), country: USA[278], *ncac
see also blackgram husk, bread crumbs,
broiler finisher, broiler starter, cotton cake,
cottonseed, cottonseed cake, cottonseed
extract, cottonseed meal, feed, feed (cattle),
feed (cow), feed (dog), feed (horse), feed
(maize, gluten), feed (pig), feed (poultry),
feed (rabbit), feed (rat/mice), feed (sheep),
feeds, grain, fish meal, flour (wheat),
groundnut cake, grower's mash, horsegram,
layer's mash, maize, maize gluten, milo,
peanut cake, peanut cake (deoiled), peanut,
kernels, peanut (oil cake), pearlmillet, pig
breeder's mash, pig finisher, pig starter, pod
with haulms, poultry breeder's mash, rabbit
pellets, redgram husk, rice, rice bran
(deoiled), rice, broken, rice (polish), sesame
cake, silk worm pupae, sorghum, soybean
cake, soybean meal, wheat, wheat bran

Citrinin
incidence: 2/4, conc. range: 10 μg/kg,
Ø conc.: 10 μg/kg, country: Egypt[16]
see also barley, barley, oats, barley-soybean
diet, feed, feed, mixed, feed (cattle), feed
(pig), fish meal, hay, maize, *Makhana*
(*Euryale ferox* Salisb) puffs, oats, palm
products, peas and beans, rice bran, rice
germ, wheat, wheat and other grains
(moldy), wheat bran

Deoxynivalenol
incidence: 4/4, conc. range: 70–700 μg/kg,
Ø conc.: 345 μg/kg, country: Egypt[16]
see also barley, barley, husked, barley,
unhusked (naked), barley (pressed), bone
meal, bran, broilers feed, calf fattening mixed
feed, coconut, expeller, corn cob mix silage,
cottonseed, cottonseed cake, dairy cattle feed,
egg production mixed feed, feed, feed
components, feed, commercial mix, feed,
mixed, feed, mixed (primarily maize), feed
(barley), feed (cattle), feed (chicken), feed
(dog), feed (fish), feed (mill run, from
wheat), feed (mink), feed (pig), feed (poultry),
feed (reindeer), feeds, grain, feeds, industrial,
feedstuff, feedstuffs (rapeseed, turnip, fish
meal, concentrates), fish meal, grain, mixed
feed, grains, mixed, grains (no specification),
maize, maize ears, maize fibre, maize germ,
maize germ/bran, maize germ meal, maize
gluten, maize kernels, maize meal, maize
powder, maize screenings, maize stalks (pith),
maize, "Baby", maize, hybrid, oats, rice bran,
rice germ cake, rye, silage, sorghum,
soybeans, triticale, wheat, wheat and barley,
wheat, red hard winter, wheat, soft white
winter, wheat, spring, wheat, winter

Ochratoxin A
incidence: 1/3, conc.: 12 µg/kg, country: Egypt[16]
see also alfalfa, barley, barley, oats, barley (high moisture), barley-soybean diet, bird food, domestic, bird food, wild, broilers feed, cereal grains, citrus pulp, coconut, expeller, corn cob mix silage, diet (dairy cow), diet (poultry), diet (starter), dog food, eat, egg production mixed feed, feed, feed wheat, oat and barley, feed, commercial mix, feed, mixed, feed, mixed (pelleted), feed (broilers), feed (cat), feed (cattle), feed (cereals), feed (pig), feed ec (pig), feed (poultry), feed ec (poultry), feed (poultry, pig), feed ec (rabbit), feed (trout), grain, mixed feed, grains, mixed, grains (heated), hay, horse bean, maize, maize feed, milo, maize gluten, maize meal, *Makhana* (*Euryale ferox* Salisb) puffs, milk production mixed feed, millet, oat and barley (hammer-milled), oats, palm products, peanut cake, peas, peas and beans, pet food, pig feedstuffs, pig grower diet, pig meal, piglet diet, poultry feedstuffs, rice bran, rice germ, rice germ cake, rye, sorghum, soybean groats, sunflower, sunflower seeds, extracted, tapioca, triticale, *Vicia faba*, wheat, wheat and barley, wheat bran, wheat hay, wheat, oats

Zearalenone
incidence: 4/4, conc. range: 4–30 µg/kg, Ø conc.: 16 µg/kg, country: Egypt[16]
see also alfalfa, barley, barley, husked, barley, unhusked (naked), barley and feed, bone meal, bran, broilers feed, *Carthamus* cake, chick pea, concentrate, mixed, corn cob mix silage, cotton cake, cottonseed, cottonseed cake, diet (dairy cow), diet (poultry), diets (mixed), feed, feed components, feed, mixed, feed, mixed (primarily maize also maize, oats, wheat), feed (bran), feed (broiler chicken), feed (cattle), feed (chicken), feed (dairy), feed (developing pig), feed (mill run, from wheat), feed (miscellaneous), feed (pig), feed (poultry), feed (poultry, pig), feed (starter chicken), feedstuff, fish meal, forage grass, grain, bruised, grain, mixed feed, grains (no specification), hay, maize, maize ears, maize

flakes, maize germ, maize germ/bran, maize gluten, maize kernels, maize meal, maize oil cake, maize screenings, maize stalks (pith), maize, "Baby", maize, hybrid, maize, shelled, maize, unshelled, maize grain, artificially dried, maize grain, crib dried, maize grain, ensiled, milk production mixed feed, oats, *Paspalum palidosum*, straw, peanut hulls/skins, rice bran, rice germ, rice germ cake, rye, silage, sorghum, soybeans, soybeans, extracted, sunflower cake, tapioca, triticale, wheat, wheat bran, wheat bran and chana testa, wheat soya meal

Maize, yellow   for feed may contain the following mycotoxins:

Aflatoxin B$_1$
incidence: 72/210, conc. range: 5–200 µg/kg, Ø conc.: 48.3 µg/kg, country: Egypt[397]
incidence: 44/205, conc. range: 5–190 µg/kg, Ø conc.: 54.6 µg/kg, country: Egypt[397]
incidence: 85/228, conc. range: 5–190 µg/kg, Ø conc.: 43.8 µg/kg, country: Egypt[397]
incidence: 66/190, conc. range: 15–190 µg/kg, Ø conc.: 47.6 µg/kg, country: Egypt[397]
see also alfalfa, *Ambadi* cake, animal feedstuffs (dairy cake), bagasse, barley, bengalgram husk, bird food, bird food, wild, biri testa, blackgram, blackgram husk, bran, broiler mixed feed, calf fattening mixed feed, calf fattening mixed feed (containing 4–20 % peanut products), *Carthamus* cake, castor cake, cereals, cereal products, chick pea, coconut cake, cocos, concentrate, mixed, concentrates, cotton cake, cottonseed, cottonseed (dehulled), cottonseed cake, cottonseed extract, cottonseed meal, cottonseed meal (ammoniated), cottonseed meal (decorticated), cottonseed meats, cottonseed products, crumbles, crumbles, grower, cycad meal, dairy cattle feed, dairy cattle feed (containing 2–5 % peanut products), dairy cattle feed (containing 6–10 % peanut products), dairy cattle feed (containing 6–12 % peanut products), dairy cattle feed (containing more than 20 % peanut products), diets, mixed, dog food, egg production mixed feed, feed, feed and

ingredients, feed, compound, feed, layer, feed, mixed, feed (beef), feed (broilers), feed (calf), feed (cat), feed (cattle), feed (chicken), feed (dairy), feed (dog), feed (dug), feed (fish), feed (gluten), feed (horse), feed (miscellaneous), feed (pig), feed (poultry), feed (poultry, pig), feed (rabbit), feed (sheep), feed (50–60 % maize), fish meal, grain by-products, grains (no specification), greengram, hay/silage, horsegram, husk, *Jagni* cake, legume mixture, linseed, linseed cake, livol, *Mahua* cake, maize, maize germ, maize gluten, maize grits, maize husk, maize meal, maize oil cake, maize powder, maize screenings, maize, ground, maize, hybrid, maize, preharvest, maize (dark grains), *Makhana* (*Euryale ferox* Salisb) puffs, manioc, milk production mixed feed, mung testa, murkool, mustard cake, neem cake, niger cake, oats, palm kernel expeller cake, palm kernels, palm products, peanut cake, peanut cake (deoiled), peanut expeller, peanut hay, peanut meal, peanut, kernels, peanut, shells, peanuts, pellets, finisher, pig meal and pellets, pigeon pea, poultry feeds (peanut containing), rapeseed cake, redgram husk, rice, rice bran, rice bran (deoiled), rice chaff, rice crack, rice germ, rice germ cake, rice meal, rice straw, rice (damaged), rice (polish), safflower cake, sal seed cake, sesame, sesame cake, sorghum, soybean meal, soybeans, sunflower, sunflower cake, sunflower flour, tapioca, wheat, wheat bran, wheat bran and chana testa

Maize (dark grains) may contain the following mycotoxins:

AFLATOXIN B$_1$
incidence: 3/9*, conc. range: 1–2.6 µg/kg, country: UK[94], imported?
see also alfalfa, *Ambadi* cake, animal feedstuffs (dairy cake), bagasse, barley, bengalgram husk, bird food, bird food, wild, biri testa, blackgram, blackgram husk, bran, broiler mixed feed, calf fattening mixed feed, calf fattening mixed feed (containing 4–20 % peanut products), *Carthamus* cake, castor cake, cereals, cereal products, chick pea, coconut cake, cocos, concentrate, mixed, concentrates, cotton cake, cottonseed, cottonseed (dehulled), cottonseed cake, cottonseed extract, cottonseed meal, cottonseed meal (ammoniated), cottonseed meal (decorticated), cottonseed meats, cottonseed products, crumbles, crumbles, grower, cycad meal, dairy cattle feed, dairy cattle feed (containing 2–5 % peanut products), dairy cattle feed (containing 6–10 % peanut products), dairy cattle feed (containing 6–12 % peanut products), dairy cattle feed (containing more than 20 % peanut products), diets, mixed, dog food, egg production mixed feed, feed, feed and ingredients, feed, compound, feed, layer, feed, mixed, feed (beef), feed (broilers), feed (calf), feed (cat), feed (cattle), feed (chicken), feed (dairy), feed (dog), feed (dug), feed (fish), feed (gluten), feed (horse), feed (miscellaneous), feed (pig), feed (poultry), feed (poultry, pig), feed (rabbit), feed (sheep), feed (50–60 % maize), fish meal, grain by-products, grains (no specification), greengram, hay/silage, horsegram, husk, *Jagni* cake, legume mixture, linseed, linseed cake, livol, *Mahua* cake, maize, maize germ, maize gluten, maize grits, maize husk, maize meal, maize oil cake, maize powder, maize screenings, maize, ground, maize, hybrid, maize, preharvest, maize, yellow, *Makhana* (*Euryale ferox* Salisb) puffs, manioc, milk production mixed feed, mung testa, murkool, mustard cake, neem cake, niger cake, oats, palm kernel expeller cake, palm kernels, palm products, peanut cake, peanut cake (deoiled), peanut expeller, peanut hay, peanut meal, peanut, kernels, peanut, shells, peanuts, pellets, finisher, pig meal and pellets, pigeon pea, poultry feeds (peanut containing), rapeseed cake, redgram husk, rice, rice bran, rice bran (deoiled), rice chaff, rice crack, rice germ, rice germ cake, rice meal, rice straw, rice (damaged), rice (polish), safflower cake, sal seed cake, sesame, sesame cake, sorghum, soybean meal, soybeans, sunflower, sunflower cake, sunflower flour, tapioca, wheat, wheat bran, wheat bran and chana testa

Aflatoxins (Total)
incidence: 3/9*, conc. range: 1–2.9 µg/kg,
country: UK[94], *imported?
see also cottonseed, maize gluten, maize
germ, palm products, rice bran, soybeans,
sunflower

Maize/oats mix   may contain the following
mycotoxins:

Fumonisin B$_1$
incidence: 1/1, conc.: 24 µg/kg, country:
USA[58]
see also barley, bird food, wild, dog food,
feed, feed, complete ration, feed, general,
feed, layer, feed, maize-based, feed, mixed,
feed, pelleted ration, feed, screenings, feed,
sweet, feed (broilers), feed (cat), feed
(chicken), feed (dog), feed (gluten), feed
(horse), feed (maize), feed (pig), feed
(poultry), feed (rat), feed (rodent), forage
grass, maize, maize and maize screenings,
maize bran, maize ears, maize fine fractions,
maize flakes, maize germ, maize germ/bran,
maize germ meal, maize gluten, maize grits,
maize kernels, maize meal, maize powder,
maize screenings, maize, "Baby", maize,
ground, maize, preharvest, maize, sweet feed,
rat chow, silage, sorghum, soybeans, wheat

Makhana (Euryale ferox Salisb) puffs   may
contain the following mycotoxins:

Aflatoxin B$_1$
incidence: 7/39, conc. range: 100–2140 µg/kg,
country: India[356]
see also alfalfa, *Ambadi* cake, animal feedstuffs
(dairy cake), bagasse, barley, bengalgram
husk, bird food, bird food, wild, biri testa,
blackgram, blackgram husk, bran, broiler
mixed feed, calf fattening mixed feed, calf
fattening mixed feed (containing 4–20 %
peanut products), *Carthamus* cake, castor
cake, cereals, cereal products, chick pea,
coconut cake, cocos, concentrate, mixed,
concentrates, cotton cake, cottonseed,
cottonseed (dehulled), cottonseed cake,
cottonseed extract, cottonseed meal,
cottonseed meal (ammoniated), cottonseed
meal (decorticated), cottonseed meats,
cottonseed products, crumbles, crumbles,
grower, cycad meal, dairy cattle feed, dairy
cattle feed (containing 2–5 % peanut
products), dairy cattle feed (containing
6–10 % peanut products), dairy cattle feed
(containing 6–12 % peanut products), dairy
cattle feed (containing more than 20 %
peanut products), diets, mixed, dog food, egg
production mixed feed, feed, feed and
ingredients, feed, compound, feed, layer,
feed, mixed, feed (beef), feed (broilers), feed
(calf), feed (cat), feed (cattle), feed (chicken),
feed (dairy), feed (dog), feed (dug), feed
(fish), feed (gluten), feed (horse), feed
(miscellaneous), feed (pig), feed (poultry),
feed (poultry, pig), feed (rabbit), feed
(sheep), feed (50–60 % maize), fish meal,
grain by-products, grains (no specification),
greengram, hay/silage, horsegram, husk, *Jagni*
cake, legume mixture, linseed, linseed cake,
livol, maize, maize germ, maize gluten, maize
grits, maize husk, maize meal, maize oil cake,
maize powder, maize screenings, maize,
ground, maize, hybrid, maize, preharvest,
maize, yellow, maize (dark grains), manioc,
milk production mixed feed, mung testa,
murkool, mustard cake, neem cake, niger
cake, oats, palm kernel expeller cake, palm
kernels, palm products, peanut cake, peanut
cake (deoiled), peanut expeller, peanut hay,
peanut meal, peanut, kernels, peanut, shells,
peanuts, pellets, finisher, pig meal and pellets,
pigeon pea, poultry feeds (peanut
containing), rapeseed cake, redgram husk,
rice, rice bran, rice bran (deoiled), rice chaff,
rice crack, rice germ, rice germ cake, rice
meal, rice straw, rice (damaged), rice
(polish), safflower cake, sal seed cake, sesame,
sesame cake, sorghum, soybean meal,
soybeans, sunflower, sunflower cake,
sunflower flour, tapioca, wheat, wheat bran,
wheat bran and chana testa

Citrinin
incidence: 2/39, conc. range: 100–500 µg/kg,
Ø conc.: 300 µg/kg, country: India[356]
see also barley, barley, oats, barley-soybean
diet, feed, feed, mixed, feed (cattle), feed

(pig), fish meal, hay, maize, maize, white, oats, palm products, peas and beans, rice bran, rice germ, wheat, wheat and other grains (moldy), wheat bran

OCHRATOXIN A
incidence: 3/39, conc. range: 100–600 μg/kg, country: India[356]
see also alfalfa, barley, barley, oats, barley (high moisture), barley-soybean diet, bird food, domestic, bird food, wild, broilers feed, cereal grains, citrus pulp, coconut, expeller, corn cob mix silage, diet (dairy cow), diet (poultry), diet (starter), dog food, eat, egg production mixed feed, feed, feed wheat, oat and barley, feed, commercial mix, feed, mixed, feed, mixed (pelleted), feed (broilers), feed (cat), feed (cattle), feed (cereals), feed (pig), feed ec (pig), feed (poultry), feed ec (poultry), feed (poultry, pig), feed ec (rabbit), feed (trout), grain, mixed feed, grains, mixed, grains (heated), hay, horse bean, maize, maize feed, milo, maize gluten, maize meal, maize, white, milk production mixed feed, millet, oat and barley (hammer-milled), oats, palm products, peanut cake, peas, peas and beans, pet food, pig feedstuffs, pig grower diet, pig meal, piglet diet, poultry feedstuffs, rice bran, rice germ, rice germ cake, rye, sorghum, soybean groats, sunflower, sunflower seeds, extracted, tapioca, triticale, *Vicia faba*, wheat, wheat and barley, wheat bran, wheat hay, wheat, oats

Manioc   Manioc for feed may contain the following mycotoxins:

AFLATOXIN B$_1$
incidence: 5/18* (containing aflatoxins), conc. range: tr μg/kg, country: UK[36], *imported?
see also alfalfa, *Ambadi* cake, animal feedstuffs (dairy cake), bagasse, barley, bengalgram husk, bird food, bird food, wild, biri testa, blackgram, blackgram husk, bran, broiler mixed feed, calf fattening mixed feed, calf fattening mixed feed (containing 4–20% peanut products), *Carthamus* cake, castor cake, cereals, cereal products, chick pea, coconut cake, cocos, concentrate, mixed, concentrates, cotton cake, cottonseed, cottonseed (dehulled), cottonseed cake, cottonseed extract, cottonseed meal, cottonseed meal (ammoniated), cottonseed meal (decorticated), cottonseed meats, cottonseed products, crumbles, crumbles, grower, cycad meal, dairy cattle feed, dairy cattle feed (containing 2–5% peanut products), dairy cattle feed (containing 6–10% peanut products), dairy cattle feed (containing 6–12% peanut products), dairy cattle feed (containing more than 20% peanut products), diets, mixed, dog food, egg production mixed feed, feed, feed and ingredients, feed, compound, feed, layer, feed, mixed, feed (beef), feed (broilers), feed (calf), feed (cat), feed (cattle), feed (chicken), feed (dairy), feed (dog), feed (dug), feed (fish), feed (gluten), feed (horse), feed (miscellaneous), feed (pig), feed (poultry), feed (poultry, pig), feed (rabbit), feed (sheep), feed (50–60% maize), fish meal, grain by-products, grains (no specification), greengram, hay/silage, horsegram, husk, *Jagni* cake, legume mixture, linseed, linseed cake, livol, *Mahua* cake, maize, maize germ, maize gluten, maize grits, maize husk, maize meal, maize oil cake, maize powder, maize screenings, maize, ground, maize, hybrid, maize, preharvest, maize, yellow, maize (dark grains), *Makhana* (*Euryale ferox* Salisb) puffs, milk production mixed feed, mung testa, murkool, mustard cake, neem cake, niger cake, oats, palm kernel expeller cake, palm kernels, palm products, peanut cake, peanut cake (deoiled), peanut expeller, peanut hay, peanut meal, peanut, kernels, peanut, shells, peanuts, pellets, finisher, pig meal and pellets, pigeon pea, poultry feeds (peanut containing), rapeseed cake, redgram husk, rice, rice bran, rice bran (deoiled), rice chaff, rice crack, rice germ, rice germ cake, rice meal, rice straw, rice (damaged), rice (polish), safflower cake, sal seed cake, sesame, sesame cake, sorghum, soybean meal, soybeans, sunflower, sunflower cake, sunflower flour, tapioca, wheat, wheat bran, wheat bran and chana testa

Meat meal   may contain the following mycotoxins:

Aflatoxin G_1
incidence: 1/4, conc. range: 6–20 μg/kg, (1 sa with a maximum of 13 μg/kg), country: Guatemala[375]
see also animal feedstuffs (dairy cake), bird food, bird food, wild, concentrate, mixed, cottonseed, cottonseed cake, feed (cat), feed (chicken), feed (dog), maize, maize, ground, maize, preharvest, milk production mixed feed, murkool, peanut cake, peanut expeller, peanut hay, peanut meal, peanuts, rice bran, rice germ, sorghum, soybeans, wheat, wheat bran, wheat bran and chana testa

Mice feed   may contain the following mycotoxins:

Aflatoxin $B + G$
incidence: 2/12, conc. range: 1–20 μg/kg, Ø conc.: 3 μg/kg, country: France[45]
see also feed (cattle), feed (pig), feed (poultry), feed (rabbit), feed (rat), peanut (oil cake)

Milk production mixed feed   may contain the following mycotoxins:

Aflatoxin B_1
incidence: 2/5, Ø conc.: 50 μg/kg, country: Egypt[16]
see also alfalfa, *Ambadi* cake, animal feedstuffs (dairy cake), bagasse, barley, bengalgram husk, bird food, bird food, wild, biri testa, blackgram, blackgram husk, bran, broiler mixed feed, calf fattening mixed feed, calf fattening mixed feed (containing 4–20 % peanut products), *Carthamus* cake, castor cake, cereals, cereal products, chick pea, coconut cake, cocos, concentrate, mixed, concentrates, cotton cake, cottonseed, cottonseed (dehulled), cottonseed cake, cottonseed extract, cottonseed meal, cottonseed meal (ammoniated), cottonseed meal (decorticated), cottonseed meats, cottonseed products, crumbles, crumbles, grower, cycad meal, dairy cattle feed, dairy cattle feed (containing 2–5 % peanut products), dairy cattle feed (containing 6–10 % peanut products), dairy cattle feed (containing 6–12 % peanut products), dairy cattle feed (containing more than 20 % peanut products), diets, mixed, dog food, egg production mixed feed, feed, feed and ingredients, feed, compound, feed, layer, feed, mixed, feed (beef), feed (broilers), feed (calf), feed (cat), feed (cattle), feed (chicken), feed (dairy), feed (dog), feed (dug), feed (fish), feed (gluten), feed (horse), feed (miscellaneous), feed (pig), feed (poultry), feed (poultry, pig), feed (rabbit), feed (sheep), feed (50–60 % maize), fish meal, grain by-products, grains (no specification), greengram, hay/silage, horsegram, husk, *Jagni* cake, legume mixture, linseed, linseed cake, livol, *Mahua* cake, maize, maize germ, maize gluten, maize grits, maize husk, maize meal, maize oil cake, maize powder, maize screenings, maize, ground, maize, hybrid, maize, preharvest, maize, yellow, maize (dark grains), *Makhana* (*Euryale ferox* Salisb) puffs, manioc, mung testa, murkool, mustard cake, neem cake, niger cake, oats, palm kernel expeller cake, palm kernels, palm products, peanut cake, peanut cake (deoiled), peanut expeller, peanut hay, peanut meal, peanut, kernels, peanut, shells, peanuts, pellets, finisher, pig meal and pellets, pigeon pea, poultry feeds (peanut containing), rapeseed cake, redgram husk, rice, rice bran, rice bran (deoiled), rice chaff, rice crack, rice germ, rice germ cake, rice meal, rice straw, rice (damaged), rice (polish), safflower cake, sal seed cake, sesame, sesame cake, sorghum, soybean meal, soybeans, sunflower, sunflower cake, sunflower flour, tapioca, wheat, wheat bran, wheat bran and chana testa

Aflatoxin G_1
incidence: 2/5, Ø conc.: 50 μg/kg, country: Egypt[16]
see also animal feedstuffs (dairy cake), bird food, bird food, wild, concentrate, mixed, cottonseed, cottonseed cake, feed (cat), feed (chicken), feed (dog), maize, maize, ground, maize, preharvest, meat meal, murkool,

peanut cake, peanut expeller, peanut hay, peanut meal, peanuts, rice bran, rice germ, sorghum, soybeans, wheat, wheat bran, wheat bran and chana testa

OCHRATOXIN A
incidence: 2/3, Ø conc.: 19 µg/kg, country: Egypt[16]
see also alfalfa, barley, barley (high moisture), barley, oats, barley-soybean diet, bird food, domestic, bird food, wild, broilers feed, cereal grains, citrus pulp, coconut, expeller, corn cob mix silage, diet (dairy cow), diet (poultry), diet (starter), dog food, eat, egg production mixed feed, feed, feed wheat, oat and barley, feed, commercial mix, feed, mixed, feed, mixed (pelleted), feed (broilers), feed (cat), feed (cattle), feed (cereals), feed (pig), feed ec (pig), feed (poultry), feed ec (poultry), feed (poultry, pig), feed ec (rabbit), feed (trout), grain, mixed feed, grains, mixed, grains (heated), hay, horse bean, maize, maize feed, milo, maize gluten, maize meal, maize, white, *Makhana (Euryale ferox* Salisb) puffs, millet, oat and barley (hammer-milled), oats, palm products, peanut cake, peas, peas and beans, pet food, pig feedstuffs, pig grower diet, pig meal, piglet diet, poultry feedstuffs, rice bran, rice germ, rice germ cake, rye, sorghum, soybean groats, sunflower, sunflower seeds, extracted, tapioca, triticale, *Vicia faba*, wheat, wheat and barley, wheat bran, wheat hay, wheat, oats

ZEARALENONE
incidence: 4/4, conc. range: 5–20 µg/kg, Ø conc.: 11 µg/kg, country: Egypt[16]
see also alfalfa, barley, barley, husked, barley, unhusked (naked), barley and feed, bone meal, bran, broilers feed, *Carthamus* cake, chick pea, concentrate, mixed, corn cob mix silage, cotton cake, cottonseed, cottonseed cake, diet (dairy cow), diet (poultry), diets (mixed), feed, feed components, feed, mixed, feed, mixed (primarily maize also maize, oats, wheat), feed (bran), feed (broiler chicken), feed (cattle), feed (chicken), feed (developing pig), feed (mill run, from wheat), feed (miscellaneous), feed (pig), feed (poultry),

feed (poultry, pig), feed (starter chicken), feedstuff, fish meal, forage grass, grain, bruised, grain, mixed feed, grains (no specification), hay, maize, maize ears, maize flakes, maize germ, maize germ/bran, maize gluten, maize kernels, maize meal, maize oil cake, maize screenings, maize stalks (pith), maize, "Baby", maize, hybrid, maize, shelled, maize, unshelled, maize, white, maize grain, artificially dried, maize grain, crib dried, maize grain, ensiled, oats, *Paspalum palidosum*, straw, peanut hulls/skins, rice bran, rice germ, rice germ cake, rye, silage, sorghum, soybeans, soybeans, extracted, sunflower cake, tapioca, triticale, wheat, wheat bran, wheat bran and chana testa, wheat soya meal

Millet   Millet for feed may contain the following mycotoxins:

AFLATOXINS
incidence: 1/4, conc.: 51–500 µg/kg, country: Australia[21]
incidence: 1/8, conc.: 10–29 µg/kg (1 sa), country: India[349]
see also animal feed (maize), animal feed (mixed), barley, copra meal, cottonseed, cottonseed fines, cottonseed meal, cottonseed meats, feed, feed ingredients (miscellaneous), feed, mixed, feed (excluding peanuts, suspect), feed (goat), feed (pig), feed (poultry), feedstuff, grain, grain, mixed feed, maize, maize, shelled, peanut cake, peanut meal, peanut meal and by-products, peanuts, protein concentrates, rice, rice bran, sorghum, soybean meal, sunflower

OCHRATOXIN A
incidence: 2/8, conc. range: ≤145 µg/kg, country: India[349]
see also alfalfa, barley, barley, oats, barley (high moisture), barley-soybean diet, bird food, domestic, bird food, wild, broilers feed, cereal grains, citrus pulp, coconut, expeller, corn cob mix silage, diet (dairy cow), diet (poultry), diet (starter), dog food, eat, egg production mixed feed, feed, feed wheat, oat and barley, feed, commercial mix, feed,

mixed, feed, mixed (pelleted), feed (broilers), feed (cat), feed (cattle), feed (cereals), feed (pig), feed ec (pig), feed (poultry), feed ec (poultry), feed (poultry, pig), feed ec (rabbit), feed (trout), grain, mixed feed, grains, mixed, grains (heated), hay, horse bean, maize, maize feed, milo, maize gluten, maize meal, maize, white, *Makhana* (*Euryale ferox* Salisb) puffs, milk production mixed feed, oat and barley (hammer-milled), oats, palm products, peanut cake, peas, peas and beans, pet food, pig feedstuffs, pig grower diet, pig meal, piglet diet, poultry feedstuffs, rice bran, rice germ, rice germ cake, rye, sorghum, soybean groats, sunflower, sunflower seeds, extracted, tapioca, triticale, *Vicia faba*, wheat, wheat and barley, wheat bran, wheat hay, wheat, oats

Millet, little   for feed may contain the following mycotoxins:

Cyclopiazonic Acid
incidence: 1/1, conc.: 10,000 µg/kg (estimated), country: India[33]
see also chick mash, feed, groundnut cake, maize, peanuts, rice bran, wheat

Milo   may contain the following mycotoxins:

Aflatoxin
incidence: 1/1, conc. range: >30 µg/kg (1 sa), country: India[380]
see also blackgram husk, bread crumbs, broiler finisher, broiler starter, cotton cake, cottonseed, cottonseed cake, cottonseed extract, cottonseed meal, feed, feed (cattle), feed (cow), feed (dog), feed (horse), feed (maize, gluten), feed (pig), feed (poultry), feed (rabbit), feed (rat/mice), feed (sheep), feeds, grain, fish meal, flour (wheat), groundnut cake, grower's mash, horsegram, layer's mash, maize, maize gluten, maize, white, peanut cake, peanut cake (deoiled), peanut, kernels, peanut (oil cake), pearlmillet, pig breeder's mash, pig finisher, pig starter, pod with haulms, poultry breeder's mash, rabbit pellets, redgram husk, rice, rice bran (deoiled), rice, broken, rice (polish), sesame cake, silk worm pupae, sorghum, soybean cake, soybean meal, wheat, wheat bran

Mung testa   may contain the following mycotoxins:

Aflatoxin B$_1$
incidence: 1/1, conc.: 4 µg/kg, country: India[129]
see also alfalfa, *Ambadi* cake, animal feedstuffs (dairy cake), bagasse, barley, bengalgram husk, bird food, bird food, wild, biri testa, blackgram, blackgram husk, bran, broiler mixed feed, calf fattening mixed feed, calf fattening mixed feed (containing 4–20 % peanut products), *Carthamus* cake, castor cake, cereals, cereal products, chick pea, coconut cake, cocos, concentrate, mixed, concentrates, cotton cake, cottonseed, cottonseed (dehulled), cottonseed cake, cottonseed extract, cottonseed meal, cottonseed meal (ammoniated), cottonseed meal (decorticated), cottonseed meats, cottonseed products, crumbles, crumbles, grower, cycad meal, dairy cattle feed, dairy cattle feed (containing 2–5 % peanut products), dairy cattle feed (containing 6–10 % peanut products), dairy cattle feed (containing 6–12 % peanut products), dairy cattle feed (containing more than 20 % peanut products), diets, mixed, dog food, egg production mixed feed, feed, feed and ingredients, feed, compound, feed, layer, feed, mixed, feed (beef), feed (broilers), feed (cat), feed (cattle), feed (chicken), feed (dairy), feed (dog), feed (dug), feed (fish), feed (gluten), feed (horse), feed (miscellaneous), feed (pig), feed (poultry), feed (poultry, pig), feed (rabbit), feed (sheep), feed (50–60 % maize), fish meal, grain by-products, grains (no specification), greengram, hay/silage, horsegram, husk, *Jagni* cake, legume mixture, linseed, linseed cake, livol, *Mahua* cake, maize, maize germ, maize gluten, maize grits, maize husk, maize meal, maize oil cake, maize powder, maize screenings, maize, ground, maize, hybrid, maize, preharvest, maize, yellow, maize (dark grains), *Makhana* (*Euryale ferox* Salisb) puffs, manioc, milk production mixed feed, murkool, mustard cake, neem cake, niger cake, oats, palm kernel expeller cake, palm kernels, palm

products, peanut cake, peanut cake (deoiled), peanut expeller, peanut hay, peanut meal, peanut, kernels, peanut, shells, peanuts, pellets, finisher, pig meal and pellets, pigeon pea, poultry feeds (peanut containing), rapeseed cake, redgram husk, rice, rice bran, rice bran (deoiled), rice chaff, rice crack, rice germ, rice germ cake, rice meal, rice straw, rice (damaged), rice (polish), safflower cake, sal seed cake, sesame, sesame cake, sorghum, soybean meal, soybeans, sunflower, sunflower cake, sunflower flour, tapioca, wheat, wheat bran, wheat bran and chana testa

Aflatoxin B$_2$
incidence: 1/1, conc.: 3 µg/kg, country: India[129]
see also animal feedstuffs (dairy cake), bird food, bird food, wild, biri testa, blackgram, blackgram husk, cottonseed, cottonseed cake, cottonseed extract, cottonseed meal, cottonseed meal (ammoniated), cottonseed meats, dog food, egg production mixed feed, feed, compound, feed (cat), feed (cattle), feed (dog), feed (pig), feed (poultry), feed (rabbit), feed (sheep), fish meal, horsegram, maize, maize gluten, maize husk, maize, ground, maize, preharvest, mustard cake, niger cake, peanut cake, peanut cake (deoiled), peanut expeller, peanut hay, peanut meal, peanuts, peanut, kernels, redgram husk, rice bran, rice bran (deoiled), rice chaff, rice meal, rice (polish), sal seed cake, sesame cake, sorghum, soybean meal, soybeans, wheat, wheat bran

Murkool   may contain the following mycotoxins:

Aflatoxin B$_1$
incidence: ?/10, conc. range: 10–40 µg/kg, country: India[183]
see also alfalfa, *Ambadi* cake, animal feedstuffs (dairy cake), bagasse, barley, bengalgram husk, bird food, bird food, wild, biri testa, blackgram, blackgram husk, bran, broiler mixed feed, calf fattening mixed feed, calf fattening mixed feed (containing 4–20% peanut products), *Carthamus* cake, castor cake, cereals, cereal products, chick pea,

coconut cake, cocos, concentrate, mixed, concentrates, cotton cake, cottonseed, cottonseed (dehulled), cottonseed cake, cottonseed extract, cottonseed meal, cottonseed meal (ammoniated), cottonseed meal (decorticated), cottonseed meats, cottonseed products, crumbles, crumbles, grower, cycad meal, dairy cattle feed, dairy cattle feed (containing 2–5% peanut products), dairy cattle feed (containing 6–10% peanut products), dairy cattle feed (containing 6–12% peanut products), dairy cattle feed (containing more than 20% peanut products), diets, mixed, dog food, egg production mixed feed, feed, feed and ingredients, feed, compound, feed, layer, feed, mixed, feed (beef), feed (broilers), feed (calf), feed (cat), feed (cattle), feed (chicken), feed (dairy), feed (dog), feed (dug), feed (fish), feed (gluten), feed (horse), feed (miscellaneous), feed (pig), feed (poultry), feed (poultry, pig), feed (rabbit), feed (sheep), feed (50–60% maize), fish meal, grain by-products, grains (no specification), greengram, hay/silage, horsegram, husk, *Jagni* cake, legume mixture, linseed, linseed cake, livol, *Mahua* cake, maize, maize germ, maize gluten, maize grits, maize husk, maize meal, maize oil cake, maize powder, maize screenings, maize, ground, maize, hybrid, maize, preharvest, maize, yellow, maize (dark grains), *Makhana* (*Euryale ferox* Salisb) puffs, manioc, milk production mixed feed, mung testa, mustard cake, neem cake, niger cake, oats, palm kernel expeller cake, palm kernels, palm products, peanut cake, peanut cake (deoiled), peanut expeller, peanut hay, peanut meal, peanut, kernels, peanut, shells, peanuts, pellets, finisher, pig meal and pellets, pigeon pea, poultry feeds (peanut containing), rapeseed cake, redgram husk, rice, rice bran, rice bran (deoiled), rice chaff, rice crack, rice germ, rice germ cake, rice meal, rice straw, rice (damaged), rice (polish), safflower cake, sal seed cake, sesame, sesame cake, sorghum, soybean meal, soybeans, sunflower, sunflower cake, sunflower flour, tapioca, wheat, wheat bran, wheat bran and chana testa

AFLATOXIN G_1
incidence: ?/10, conc. range: tr μg/kg,
country: India[183]
see also animal feedstuffs (dairy cake), bird
food, bird food, wild, concentrate, mixed,
cottonseed, cottonseed cake, feed (cat), feed
(chicken), feed (dog), maize, maize, ground,
maize, preharvest, meat meal, milk
production mixed feed, peanut cake, peanut
expeller, peanut hay, peanut meal, peanuts,
rice bran, rice germ, sorghum, soybeans,
wheat, wheat bran, wheat bran and chana
testa

Mustard cake   may contain the following
mycotoxins:

AFLATOXIN B_1
incidence: 9/25, conc. range:
71.7–275.3 μg/kg, Ø conc.: 138 μg/kg,
country: India[253]
see also alfalfa, *Ambadi* cake, animal feedstuffs
(dairy cake), bagasse, barley, bengalgram
husk, bird food, bird food, wild, biri testa,
blackgram, blackgram husk, bran, broiler
mixed feed, calf fattening mixed feed, calf
fattening mixed feed (containing 4–20 %
peanut products), *Carthamus* cake, castor
cake, cereals, cereal products, chick pea,
coconut cake, cocos, concentrate, mixed,
concentrates, cotton cake, cottonseed,
cottonseed (dehulled), cottonseed cake,
cottonseed extract, cottonseed meal,
cottonseed meal (ammoniated), cottonseed
meal (decorticated), cottonseed meats,
cottonseed products, crumbles, crumbles,
grower, cycad meal, dairy cattle feed, dairy
cattle feed (containing 2–5 % peanut
products), dairy cattle feed (containing
6–10 % peanut products), dairy cattle feed
(containing 6–12 % peanut products), dairy
cattle feed (containing more than 20 %
peanut products), diets, mixed, dog food, egg
production mixed feed, feed, feed and
ingredients, feed, compound, feed, layer,
feed, mixed, feed (beef), feed (broilers), feed
(calf), feed (cat), feed (cattle), feed (chicken),
feed (dairy), feed (dog), feed (dug), feed
(fish), feed (gluten), feed (horse), feed

(miscellaneous), feed (pig), feed (poultry),
feed (poultry, pig), feed (rabbit), feed
(sheep), feed (50–60 % maize), fish meal,
grain by-products, grains (no specification),
greengram, hay/silage, horsegram, husk, *Jagni*
cake, legume mixture, linseed, linseed cake,
livol, *Mahua* cake, maize, maize germ, maize
gluten, maize grits, maize husk, maize meal,
maize oil cake, maize powder, maize
screenings, maize, ground, maize, hybrid,
maize, preharvest, maize, yellow, maize (dark
grains), *Makhana* (*Euryale ferox* Salisb) puffs,
manioc, milk production mixed feed, mung
testa, murkool, neem cake, niger cake, oats,
palm kernel expeller cake, palm kernels, palm
products, peanut cake, peanut cake (deoiled),
peanut expeller, peanut hay, peanut meal,
peanut, kernels, peanut, shells, peanuts,
pellets, finisher, pig meal and pellets, pigeon
pea, poultry feeds (peanut containing),
rapeseed cake, redgram husk, rice, rice bran,
rice bran (deoiled), rice chaff, rice crack, rice
germ, rice germ cake, rice meal, rice straw,
rice (damaged), rice (polish), safflower cake,
sal seed cake, sesame, sesame cake, sorghum,
soybean meal, soybeans, sunflower, sunflower
cake, sunflower flour, tapioca, wheat, wheat
bran, wheat bran and chana testa

AFLATOXIN B_2
incidence: 4/25, conc. range: 44.5–85.2 μg/kg,
Ø conc.: 64.9 μg/kg, country: India[253]
see also animal feedstuffs (dairy cake), bird
food, bird food, wild, biri testa, blackgram,
blackgram husk, cottonseed, cottonseed cake,
cottonseed extract, cottonseed meal,
cottonseed meal (ammoniated), cottonseed
meats, dog food, egg production mixed feed,
feed, compound, feed (cat), feed (cattle), feed
(dog), feed (pig), feed (poultry), feed (rabbit),
feed (sheep), fish meal, horsegram, maize,
maize gluten, maize husk, maize, ground,
maize, preharvest, mung testa, niger cake,
peanut cake, peanut cake (deoiled), peanut
expeller, peanut hay, peanut meal, peanuts,
peanut, kernels, redgram husk, rice bran, rice
bran (deoiled), rice chaff, rice meal, rice
(polish), sal seed cake, sesame cake, sorghum,
soybean meal, soybeans, wheat, wheat bran

Neem cake may contain the following mycotoxins:

Aflatoxin B_1
incidence: 3/10, conc. range:
59.4–165.9 µg/kg, Ø conc.: 126 µg/kg,
country: India[253]
see also alfalfa, *Ambadi* cake, animal feedstuffs (dairy cake), bagasse, barley, bengalgram husk, bird food, bird food, wild, biri testa, blackgram, blackgram husk, bran, broiler mixed feed, calf fattening mixed feed, calf fattening mixed feed (containing 4–20 % peanut products), *Carthamus* cake, castor cake, cereals, cereal products, chick pea, coconut cake, cocos, concentrate, mixed, concentrates, cotton cake, cottonseed, cottonseed (dehulled), cottonseed cake, cottonseed extract, cottonseed meal, cottonseed meal (ammoniated), cottonseed meal (decorticated), cottonseed meats, cottonseed products, crumbles, crumbles, grower, cycad meal, dairy cattle feed, dairy cattle feed (containing 2–5 % peanut products), dairy cattle feed (containing 6–10 % peanut products), dairy cattle feed (containing 6–12 % peanut products), dairy cattle feed (containing more than 20 % peanut products), diets, mixed, dog food, egg production mixed feed, feed, feed and ingredients, feed, compound, feed, layer, feed, mixed, feed (beef), feed (broilers), feed (calf), feed (cat), feed (cattle), feed (chicken), feed (dairy), feed (dog), feed (dug), feed (fish), feed (gluten), feed (horse), feed (miscellaneous), feed (pig), feed (poultry), feed (poultry, pig), feed (rabbit), feed (sheep), feed (50–60 % maize), fish meal, grain by-products, grains (no specification), greengram, hay/silage, horsegram, husk, *Jagni* cake, legume mixture, linseed, linseed cake, livol, *Mahua* cake, maize, maize germ, maize gluten, maize grits, maize husk, maize meal, maize oil cake, maize powder, maize screenings, maize, ground, maize, hybrid, maize, preharvest, maize, yellow, maize (dark grains), *Makhana* (*Euryale ferox* Salisb) puffs, manioc, milk production mixed feed, mung testa, murkool, mustard cake, niger cake,

oats, palm kernel expeller cake, palm kernels, palm products, peanut cake, peanut cake (deoiled), peanut expeller, peanut hay, peanut meal, peanut, kernels, peanut, shells, peanuts, pellets, finisher, pig meal and pellets, pigeon pea, poultry feeds (peanut containing), rapeseed cake, redgram husk, rice, rice bran, rice bran (deoiled), rice chaff, rice crack, rice germ, rice germ cake, rice meal, rice straw, rice (damaged), rice (polish), safflower cake, sal seed cake, sesame, sesame cake, sorghum, soybean meal, soybeans, sunflower, sunflower cake, sunflower flour, tapioca, wheat, wheat bran, wheat bran and chana testa

Niger cake may contain the following mycotoxins:

Aflatoxin B_1
incidence: 15/20, conc. range:
58.1–641.7 µg/kg, Ø conc.: 184.3 µg/kg,
country: India[253]
see also alfalfa, *Ambadi* cake, animal feedstuffs (dairy cake), bagasse, barley, bengalgram husk, bird food, bird food, wild, biri testa, blackgram, blackgram husk, bran, broiler mixed feed, calf fattening mixed feed, calf fattening mixed feed (containing 4–20 % peanut products), *Carthamus* cake, castor cake, cereals, cereal products, chick pea, coconut cake, cocos, concentrate, mixed, concentrates, cotton cake, cottonseed, cottonseed (dehulled), cottonseed cake, cottonseed extract, cottonseed meal, cottonseed meal (ammoniated), cottonseed meal (decorticated), cottonseed meats, cottonseed products, crumbles, crumbles, grower, cycad meal, dairy cattle feed, dairy cattle feed (containing 2–5 % peanut products), dairy cattle feed (containing 6–10 % peanut products), dairy cattle feed (containing 6–12 % peanut products), dairy cattle feed (containing more than 20 % peanut products), diets, mixed, dog food, egg production mixed feed, feed, feed and ingredients, feed, compound, feed, layer, feed, mixed, feed (beef), feed (broilers), feed (calf), feed (cat), feed (cattle), feed (chicken),

feed (dairy), feed (dog), feed (dug), feed (fish), feed (gluten), feed (horse), feed (miscellaneous), feed (pig), feed (poultry), feed (poultry, pig), feed (rabbit), feed (sheep), feed (50–60 % maize), fish meal, grain by-products, grains (no specification), greengram, hay/silage, horsegram, husk, *Jagni* cake, legume mixture, linseed, linseed cake, livol, *Mahua* cake, maize, maize germ, maize gluten, maize grits, maize husk, maize meal, maize oil cake, maize powder, maize screenings, maize, ground, maize, hybrid, maize, preharvest, maize, yellow, maize (dark grains), *Makhana* (*Euryale ferox* Salisb) puffs, manioc, milk production mixed feed, mung testa, murkool, mustard cake, neem cake, oats, palm kernel expeller cake, palm kernels, palm products, peanut cake, peanut cake (deoiled), peanut expeller, peanut hay, peanut meal, peanut, kernels, peanut, shells, peanuts, pellets, finisher, pig meal and pellets, pigeon pea, poultry feeds (peanut containing), rapeseed cake, redgram husk, rice, rice bran, rice bran (deoiled), rice chaff, rice crack, rice germ, rice germ cake, rice meal, rice straw, rice (damaged), rice (polish), safflower cake, sal seed cake, sesame, sesame cake, sorghum, soybean meal, soybeans, sunflower, sunflower cake, sunflower flour, tapioca, wheat, wheat bran, wheat bran and chana testa

Aflatoxin B$_2$
incidence: 9/20, conc. range: 44.5–98.8 µg/kg, Ø conc.: 77.7 µg/kg, country: India[253]
see also animal feedstuffs (dairy cake), bird food, bird food, wild, biri testa, blackgram, blackgram husk, cottonseed, cottonseed cake, cottonseed extract, cottonseed meal, cottonseed meal (ammoniated), cottonseed meats, dog food, egg production mixed feed, feed, compound, feed (cat), feed (cattle), feed (dog), feed (pig), feed (poultry), feed (rabbit), feed (sheep), fish meal, horsegram, maize, maize gluten, maize husk, maize, ground, maize, preharvest, mung testa, mustard cake, peanut cake, peanut cake (deoiled), peanut expeller, peanut hay, peanut meal, peanuts, peanut, kernels, redgram husk, rice bran, rice bran (deoiled), rice chaff, rice meal, rice

(polish), sal seed cake, sesame cake, sorghum, soybean meal, soybeans, wheat, wheat bran

Oat and barley (hammer-milled)   Oat and barley for feed may contain the following mycotoxins:

Diacetoxyscirpenol
incidence: 1/51, conc.: 800 µg/kg, country: Canada[133]
see also alfalfa, barley, feed, feed components, feed (dog), feed (fish), feed (mink), feed (pig), feed (poultry), feed (reindeer), feeds, grain, feedstuff, forage grass, grain, mixed feed, grains (no specification), maize, maize gluten, oats, peanuts, soybeans, wheat

Ochratoxin A
incidence: 1/51, conc.: 880 µg/kg, country: Canada[133]
see also alfalfa, barley, barley, oats, barley (high moisture), barley-soybean diet, bird food, domestic, bird food, wild, broilers feed, cereal grains, citrus pulp, coconut, expeller, corn cob mix silage, diet (dairy cow), diet (poultry), diet (starter), dog food, eat, egg production mixed feed, feed, feed wheat, oat and barley, feed, commercial mix, feed, mixed, feed, mixed (pelleted), feed (broilers), feed (cat), feed (cattle), feed (cereals), feed (pig), feed ec (pig), feed (poultry), feed ec (poultry), feed (poultry, pig), feed ec (rabbit), feed (trout), grain, mixed feed, grains, mixed, grains (heated), hay, horse bean, maize, maize feed, milo, maize gluten, maize meal, maize, white, *Makhana* (*Euryale ferox* Salisb) puffs, milk production mixed feed, millet, oats, palm products, peanut cake, peas, peas and beans, pet food, pig feedstuffs, pig grower diet, pig meal, piglet diet, poultry feedstuffs, rice bran, rice germ, rice germ cake, rye, sorghum, soybean groats, sunflower, sunflower seeds, extracted, tapioca, triticale, *Vicia faba*, wheat, wheat and barley, wheat bran, wheat hay, wheat, oats

T-2 Toxin
incidence: 1/51, conc.: 2500 µg/kg, country: Canada[133]

see also alfalfa, barley, bran, diet (grower), diet (poultry), feed, feed components, feed, layer, feed, mixed, feed (dog), feed (fish), feed (mink), feed (pig), feed (poultry), feedstuff, forage grass, grain, mixed feed, grains (no specification), hay, maize, maize germ/bran, maize gluten, maize meal, maize screenings, maize stalks (pith), oats, peanuts, pig grower diet, piglet diet, rye, sorghum, triticale, wheat

Oat/Barley
see Barley, Oats

Oats   Oats for feed may contain the following mycotoxins:

3-Acetyldeoxynivalenol
incidence: 23/36*, conc. range: 20–553 µg/kg, Ø conc.: 119 µg/kg, country: Finland[9], *imported?
incidence: 24/272, conc. range: 3–115 µg/kg, Ø conc.: 25.1 µg/kg, country: Germany[41]
incidence: 1/1, conc.: 1500 µg/kg, country: Norway[276]
incidence: 1/7*, conc.: 700 µg/kg, country: Finland[281], *1 na
see also barley, feed components, feed (poultry), feeds, grain, feeds, industrial, feedstuffs (rapeseed, turnip, fish meal, concentrates), maize, maize ears, maize gluten, maize kernels, maize meal, maize screenings, wheat

15-Acetyldeoxynivalenol
incidence: 7/162, conc. range: 2–25 µg/kg, Ø conc.: 13.1 µg/kg, country: Germany[41]
see also barley, feed components, feed, commercial mix, feed (poultry), maize, maize ears, maize germ, maize germ/bran, maize gluten, maize kernels, maize meal, maize screenings, maize, "Baby", silage, wheat

Aflatoxin B₁
incidence: 2/5, conc. range: 20–40 µg/kg, Ø conc.: 30 µg/kg, country: Australia[121]
incidence: 2/3, conc. range: 20–40 µg/kg, Ø conc.: 30 µg/kg, country: Australia[181]
incidence: 1/1, conc.: 2600 µg/kg, country: Sweden[244]
incidence: 3/304, conc. range: 6 µg/kg, country: USA[384]

see also alfalfa, *Ambadi* cake, animal feedstuffs (dairy cake), bagasse, barley, bengalgram husk, bird food, bird food, wild, biri testa, blackgram, blackgram husk, bran, broiler mixed feed, calf fattening mixed feed, calf fattening mixed feed (containing 4–20 % peanut products), *Carthamus* cake, castor cake, cereals, cereal products, chick pea, coconut cake, cocos, concentrate, mixed, concentrates, cotton cake, cottonseed, cottonseed (dehulled), cottonseed cake, cottonseed extract, cottonseed meal, cottonseed meal (ammoniated), cottonseed meal (decorticated), cottonseed meats, cottonseed products, crumbles, crumbles, grower, cycad meal, dairy cattle feed, dairy cattle feed (containing 2–5 % peanut products), dairy cattle feed (containing 6–10 % peanut products), dairy cattle feed (containing 6–12 % peanut products), dairy cattle feed (containing more than 20 % peanut products), diets, mixed, dog food, egg production mixed feed, feed, feed and ingredients, feed, compound, feed, layer, feed, mixed, feed (beef), feed (broilers), feed (calf), feed (cat), feed (cattle), feed (chicken), feed (dairy), feed (dog), feed (dug), feed (fish), feed (gluten), feed (horse), feed (miscellaneous), feed (pig), feed (poultry), feed (poultry, pig), feed (rabbit), feed (sheep), feed (50–60 % maize), fish meal, grain by-products, grains (no specification), greengram, hay/silage, horsegram, husk, *Jagni* cake, legume mixture, linseed, linseed cake, livol, *Mahua* cake, maize, maize germ, maize gluten, maize grits, maize husk, maize meal, maize oil cake, maize powder, maize screenings, maize, ground, maize, hybrid, maize, preharvest, maize, yellow, maize (dark grains), *Makhana* (*Euryale ferox* Salisb) puffs, manioc, milk production mixed feed, mung testa, murkool, mustard cake, neem cake, niger cake, palm kernel expeller cake, palm kernels, palm products, peanut cake, peanut cake (deoiled), peanut expeller, peanut hay, peanut meal, peanut, kernels, peanut, shells, peanuts, pellets, finisher, pig meal and pellets, pigeon pea, poultry feeds (peanut containing), rapeseed cake, redgram husk, rice, rice bran, rice bran (deoiled), rice chaff, rice crack, rice germ, rice germ cake, rice meal, rice straw, rice (damaged), rice (polish), safflower cake, sal seed cake, sesame, sesame cake, sorghum, soybean meal, soybeans, sunflower, sunflower cake, sunflower flour, tapioca, wheat, wheat bran, wheat bran and chana testa

AFLATOXIN B
incidence: 11/19, conc. range:
≥25–>100 µg/kg, country: France[46]
see also barley, cocoa (oilcake), cottonseed cake, feed, feed (cattle), feed (maize, gluten), feed (pig), feed (poultry), flour (wheat), lucern (dried), maize, maize gluten, maize grains, peanut (oil cake), rice, broken, sorghum, soybean (oil cake), sunflower (oilcake), wheat

ALTERNARIOL METHYL ETHER
incidence: 4/11, conc. range: 10–42 µg/kg, country: Germany[294]
see also barley, barley, oats, cereals, mixture, sunflower, wheat, wheat, oats

BEAUVERICIN
incidence: 1/1*, conc.: tr µg/kg, country: Finland[205], *ncac
see also barley, maize, wheat, wheat, summer, wheat, winter

CITRININ
incidence: 7/9, conc. range: <5 µg/kg, country: Bulgaria[8]
see also barley, barley, oats, barley-soybean diet, cottonseed cake, feed, feed, mixed, feed (cattle), feed (pig), fish meal, hay, maize, maize, white, *Makhana* (*Euryale ferox* Salisb) puffs, palm products, peas and beans, rice bran, rice germ, wheat, wheat and other grains (moldy), wheat bran

DEOXYNIVALENOL
incidence: 28/45, conc. range: 50–870 µg/kg, Ø conc.: 188.6 µg/kg, country: Hungary[6]
incidence: 36/36*, conc. range:
22–2581 µg/kg, Ø conc.: 296 µg/kg, country: Finland[9], *imported?
incidence: 2/24, conc. range:
<100–2000 µg/kg, country: Sweden[25]

incidence: 181/272, conc. range:
3–1480 µg/kg, Ø conc.: 167.5 µg/kg, country: Germany[41]
incidence: 2/2, conc. range: 6–8 µg/kg, Ø conc.: 7 µg/kg, country: Finland[62]
incidence: 11/67, conc. range: 20–500 µg/kg, country: Germany[68]
incidence: 2/7, conc. range: 10–20 µg/kg, country: Finland[112]
incidence: 2/2, conc. range: 300–1500 µg/kg, Ø conc.: 900 µg/kg country: Germany[124]
incidence: 3/6, conc. range: 420–520 µg/kg, Ø conc.: 470 µg/kg, country: Sweden[197]
incidence: 11/32, conc. range: 40–260 µg/kg, Ø conc.: 140 µg/kg, country: Sweden[197]
incidence: 7/10, conc. range: 20–99 µg/kg (6 sa), 108 µg/kg (1 sa), country: UK[203]
incidence: 870/982*, conc. range:
30–62,050 µg/kg, country: Norway[224], *ncac
incidence: 4/4*, conc. range: 270–4200 µg/kg, Ø conc.: 1892.5 µg/kg, country: Norway[235], *ncac
incidence: 5/5*, conc. range: <30–30 µg/kg, country: Norway[235], *ncac
incidence: 7/342, conc. range:
220–13,500 µg/kg, Ø conc.: 3800 µg/kg, country: USA[237]
incidence: 1/1, conc.: 12,400 µg/kg, country: Norway[276]
incidence: 6/7*, conc. range: 28–11,000 µg/kg, Ø conc.: 2259 µg/kg, country: Finland[281], *1 na
incidence: 29/73?, Ø conc.: 330 µg/kg, country: Canada[346]
incidence: 34/73, conc. range: ≤1200 µg/kg, Ø conc.: 330 µg/kg, country: Canada[410]
see also barley, barley, husked, barley, unhusked (naked), barley (pressed), bone meal, bran, broilers feed, calf fattening mixed feed, coconut, expeller, corn cob mix silage, cottonseed, cottonseed cake, dairy cattle feed, egg production mixed feed, feed, feed components, feed, commercial mix, feed, mixed, feed, mixed (primarily maize), feed (barley), feed (cattle), feed (chicken), feed (dog), feed (fish), feed (mill run, from wheat), feed (mink), feed (pig), feed (poultry), feed (reindeer), feeds, grain, feeds, industrial, feedstuff, feedstuffs (rapeseed,

turnip, fish meal, concentrates), fish meal, grain, mixed feed, grains, mixed, grains (no specification), maize, maize ears, maize fibre, maize germ, maize germ/bran, maize germ meal, maize gluten, maize kernels, maize meal, maize powder, maize screenings, maize stalks (pith), maize, "Baby", maize, hybrid, maize, white, rice bran, rice germ cake, rye, silage, sorghum, soybeans, triticale, wheat, wheat and barley, wheat, red hard winter, wheat, soft white winter, wheat, spring, wheat, winter

DIACETOXYSCIRPENOL
incidence: 1/14, conc.: 9000 μg/kg, country: Germany[68]
incidence: 12/99*, conc. range: 10–118 μg/kg, Ø conc.: 23 μg/kg, country: Poland[342]; *ncac
see also alfalfa, barley, feed, feed components, feed (dog), feed (fish), feed (mink), feed (pig), feed (poultry), feed (reindeer), feeds, grain, feedstuff, forage grass, grain, mixed feed, grains (no specification), maize, maize gluten, oat and barley (hammer-milled), peanuts, soybeans, wheat

ENNIATIN A₁
incidence: 1/1*, conc.: tr μg/kg, country: Finland[205], *ncac
see also barley, rye, wheat, wheat, summer, wheat, winter

ENNIATIN B
incidence: 1/1*, conc.: 23 μg/kg, country: Finland[205], *ncac
see also barley, rye, wheat, wheat, summer, wheat, winter

ENNIATIN B₁
incidence: 1/1*, conc.: tr μg/kg, country: Finland[205], *ncac
see also barley, rye, wheat, wheat, summer, wheat, winter

FUSARENONE-X
incidence: 14/982*, conc. range: 100–700 μg/kg, Ø conc.: 300 μg/kg, country: Norway[224], *ncac
see also barley, feed components, feed (poultry), maize, maize ears, maize germ/bran, maize gluten, maize kernels, maize meal, maize screenings, wheat

HT-2 TOXIN
incidence: 20/272, conc. range: 10–2018 μg/kg, Ø conc.: 250.5 μg/kg, country: Germany[41]
incidence: 24/99*, conc. range: 10–47 μg/kg, Ø conc.: 21 μg/kg, country: Poland[342], *ncac
see also barley, feed components, feed (dog), feed (fish), feed (pig), feed (poultry), feed (reindeer), grain, mixed feed, grains, mixed, grains (no specification), maize, maize germ/bran, maize gluten, maize meal, maize screenings, rye, silage, wheat

MONILIFORMIN
incidence: 8/21*, conc. range: tr–70 μg/kg, Ø conc.: <40 μg/kg, country: Norway[204], *ncac
incidence: 6/26*, conc. range: tr–88 μg/kg, Ø conc.: <40 μg/kg, country: Norway[204], *ncac
incidence: 24/26*, conc. range: tr–210 μg/kg, Ø conc.: 73 μg/kg, country: Norway[204], *ncac
incidence: 1/1*, conc.: 84 μg/kg, country: Finland[205], *ncac
see also barley, feed, mixed, feed (poultry), maize, maize flakes, maize germ, maize germ/bran, maize gluten, maize meal, maize screenings, maize, "Baby", rice bran, triticale, wheat, wheat, summer, wheat, winter

NIVALENOL
incidence: 6/45, conc. range: 50–180 μg/kg, Ø conc.: 110 μg/kg, country: Hungary[6]
incidence: 2/36*, conc. range: 26–29 μg/kg, Ø conc.: 28 μg/kg, country: Finland[9], *imported?
incidence: 116/272, conc. range: 2–628 μg/kg, Ø conc.: 79.7 μg/kg, country: Germany[41]
incidence: 8/18, conc. range: ≥50–300 μg/kg, Ø conc.: 194 μg/kg, country: Sweden[65]
incidence: 2/52, conc. range: ≥50–150 μg/kg, Ø conc.: 52 μg/kg, country: Sweden[65]
incidence: 12/46, conc. range: ≥50–4700 μg/kg, Ø conc.: 1094 μg/kg, country: Sweden[65]
incidence: 45/71, conc. range: ≥50–360 μg/kg, Ø conc.: 147 μg/kg, country: Sweden[65]
incidence: 18/59, conc. range: 100–700 μg/kg, country: Germany[68]
incidence: 81/982*, conc. range: 50–667 μg/kg, Ø conc.: 130 μg/kg, country: Norway[224], *ncac

incidence: 1/7*, conc.: 28 µg/kg, country:
Finland[281], *1 na
incidence: 4/26, conc. range: 130–750 µg/kg,
country: Canada[410]
see also barley, barley, husked, barley,
unhusked (naked), barley (pressed), bran,
feed, feed components, feed (cattle), feed
(poultry), feed (reindeer), feeds, industrial,
maize, maize ears, maize germ, maize
germ/bran, maize gluten, maize kernels,
maize meal, maize powder, maize screenings,
maize, "Baby", silage, rye, triticale, wheat

OCHRATOXIN A
incidence: 5/60, conc. range: 80–200 µg/kg,
Ø conc.: 100 µg/kg, country: Hungary[6]
incidence: 7/9, conc. range: 1.44–112.5 µg/kg,
Ø conc.: 18.85 µg/kg, country: Bulgaria[8]
incidence: 12/93*, conc. range:
0.1–58.8 µg/kg, Ø conc.: 9.5 µg/kg, country:
Germany[13], *ncac
incidence: 6/21, conc. range: ≤2 µg/kg,
Ø conc.: 0.53 µg/kg, country: UK[15]
incidence: 1/1, conc.: 28 µg/kg, country:
Denmark[85]
incidence: 2/72, conc. range: 16–51 µg/kg,
Ø conc.: 33.5 µg/kg, country: Canada[410]
see also alfalfa, barley, barley (high moisture),
barley, oats, barley-soybean diet, bird food,
domestic, bird food, wild, broilers feed, cereal
grains, citrus pulp, coconut, expeller, corn
cob mix silage, diet (dairy cow), diet
(poultry), diet (starter), dog food, eat, egg
production mixed feed, feed, feed wheat, oat
and barley, feed, commercial mix, feed,
mixed, feed, mixed (pelleted), feed (broilers),
feed (cat), feed (cattle), feed (cereals), feed
(pig), feed ec (pig), feed (poultry), feed ec
(poultry), feed (poultry, pig), feed ec (rabbit),
feed (trout), grain, mixed feed, grains, mixed,
grains (heated), hay, horse bean, maize,
maize feed, milo, maize gluten, maize meal,
maize, white, Makhana (Euryale ferox Salisb)
puffs, milk production mixed feed, millet, oat
and barley (hammer-milled), palm products,
peanut cake, peas, peas and beans, pet food,
pig feedstuffs, pig grower diet, pig meal,
piglet diet, poultry feedstuffs, rice bran, rice
germ, rice germ cake, rye, sorghum, soybean

groats, sunflower, sunflower seeds, extracted,
tapioca, triticale, Vicia faba, wheat, wheat and
barley, wheat bran, wheat hay, wheat, oats

T-2 TOXIN
incidence: 4/45, conc. range: 120–360 µg/kg,
Ø conc.: 292.5 µg/kg, country: Hungary[6]
incidence: 107/272, conc. range:
2–1686 µg/kg, Ø conc.: 74.1 µg/kg, country:
Germany[41]
incidence: 6/21, conc. range: ≤2.2 µg/kg,
Ø conc.: 0.53 µg/kg, country: UK[43]
incidence: 1/2, conc.: 50 µg/kg, country:
Finland[62]
incidence: 1/14, conc.: 100 µg/kg, country:
Germany[68]
incidence: 14/68, conc. range: 100–700 µg/kg,
country: Germany[68]
incidence: 1/7*, conc.: 18 µg/kg, country:
Finland[281], *1 na
incidence: 15/99*, conc. range: 10–703 µg/kg,
Ø conc.: 60 µg/kg, country: Poland[342]; *ncac
incidence: 1/73, conc.: 1000 µg/kg, country:
Canada[410]
see also alfalfa, barley, bran, diet (grower),
diet (poultry), feed, feed components, feed,
layer, feed, mixed, feed (dog), feed (fish),
feed (mink), feed (pig), feed (poultry),
feedstuff, forage grass, grain, mixed feed,
grains (no specification), hay, maize, maize
germ/bran, maize gluten, maize meal, maize
screenings, maize stalks (pith), oat and barley
(hammer-milled), peanuts, pig grower diet,
piglet diet, rye, sorghum, triticale, wheat

T-2 TRIOL
incidence: 3/66, conc. range: 100–300 µg/kg,
country: Germany[68]
see also barley, feed components, grain,
mixed feed, grains (no specification), maize,
wheat

ZEARALENOL
incidence: 1/57, conc.: 150 µg/kg, country:
Germany[68]
see also grains (no specification), maize,
shelled, maize, unshelled

ZEARALENONE
incidence: 17/60, conc. range: 50–290 µg/kg,
Ø conc.: 101.5 µg/kg, country: Hungary[6]
incidence: 70/272, conc. range: 1–223 µg/kg,
Ø conc.: 15.7 µg/kg, country: Germany[41]
incidence: 6/12, conc. range: 20–400 µg/kg,
country: Germany[68]
incidence: 11/68, conc. range: 10–80 µg/kg,
country: Germany[68]
incidence: 2/7, conc. range: 8.2–11 µg/kg,
Ø conc.: 9.6 µg/kg, country: Germany[107]
incidence: 1/2, conc.: 100 µg/kg, country:
Germany[124]
incidence: 1/1, conc.: 750 µg/kg, country:
Norway[276]
incidence: 1/73, conc.: 300 µg/kg, country:
Canada[410]
see also alfalfa, barley, barley, husked, barley,
unhusked (naked), barley and feed, bone
meal, bran, broilers feed, *Carthamus* cake,
chick pea, concentrate, mixed, corn cob mix
silage, cotton cake, cottonseed, cottonseed
cake, diet (dairy cow), diet (poultry), diets
(mixed), feed, feed components, feed, mixed,
feed, mixed (primarily maize also maize, oats,
wheat), feed (bran), feed (broiler chicken),
feed (cattle), feed (chicken), feed (dairy), feed
(developing pig), feed (mill run, from
wheat), feed (miscellaneous), feed (pig), feed
(poultry), feed (poultry, pig), feed (starter
chicken), feedstuff, fish meal, forage grass,
grain, bruised, grain, mixed feed, grains (no
specification), hay, maize, maize ears, maize
flakes, maize germ, maize germ/bran, maize
gluten, maize kernels, maize meal, maize oil
cake, maize screenings, maize stalks (pith),
maize, "Baby", maize, hybrid, maize, shelled,
maize, unshelled, maize, white, maize grain,
artificially dried, maize grain, crib dried,
maize grain, ensiled, milk production mixed
feed, *Paspalum palidosum*, straw, peanut
hulls/skins, rice bran, rice germ, rice germ
cake, rye, silage, sorghum, soybeans,
soybeans, extracted, sunflower cake, tapioca,
triticale, wheat, wheat bran, wheat bran and
chana testa, wheat soya meal

Oilseed rape meal   may contain the following
mycotoxins:

ALTERNARIOL
incidence: 2/30, conc. range: ≤81 µg/kg,
Ø conc.: 68 µg/kg, country: UK[187]
see also cereals, mixture, sunflower, sunflower
seed meal, sunflower seeds, ensiling

ALTERNARIOL MONOMETHYL ETHER
incidence: 2/30, conc. range: ≤60 µg/kg,
Ø conc.: 55 µg/kg, country: UK[187]
see also sunflower seed meal, sunflower seeds,
ensiling

TENUAZONIC ACID
incidence: 9/30, conc. range: ≤970 µg/kg,
Ø conc.: 730 µg/kg, country: UK[187]
see also feed, feed (GP No. 1), sunflower seed
meal, sunflower seeds, ensiling

Palm kernel expeller cake   may contain the
following mycotoxins:

AFLATOXIN B$_1$
incidence: 98/98* (containing aflatoxins),
conc. range: 10–400 µg/kg, country: UK[36],
*imported?
see also alfalfa, *Ambadi* cake, animal feedstuffs
(dairy cake), bagasse, barley, bengalgram
husk, bird food, bird food, wild, biri testa,
blackgram, blackgram husk, bran, broiler
mixed feed, calf fattening mixed feed, calf
fattening mixed feed (containing 4–20 %
peanut products), *Carthamus* cake, castor
cake, cereals, cereal products, chick pea,
coconut cake, cocos, concentrate, mixed,
concentrates, cotton cake, cottonseed,
cottonseed (dehulled), cottonseed cake,
cottonseed extract, cottonseed meal,
cottonseed meal (ammoniated), cottonseed
meal (decorticated), cottonseed meats,
cottonseed products, crumbles, crumbles,
grower, cycad meal, dairy cattle feed, dairy
cattle feed (containing 2–5 % peanut
products), dairy cattle feed (containing
6–10 % peanut products), dairy cattle feed
(containing 6–12 % peanut products), dairy
cattle feed (containing more than 20 %
peanut products), diets, mixed, dog food, egg
production mixed feed, feed, feed and
ingredients, feed, compound, feed, layer,
feed, mixed, feed (beef), feed (broilers), feed

(calf), feed (cat), feed (cattle), feed (chicken), feed (dairy), feed (dog), feed (dug), feed (fish), feed (gluten), feed (horse), feed (miscellaneous), feed (pig), feed (poultry), feed (poultry, pig), feed (rabbit), feed (sheep), feed (50–60 % maize), fish meal, grain by-products, grains (no specification), greengram, hay/silage, horsegram, husk, *Jagni* cake, legume mixture, linseed, linseed cake, livol, *Mahua* cake, maize, maize germ, maize gluten, maize grits, maize husk, maize meal, maize oil cake, maize powder, maize screenings, maize, ground, maize, hybrid, maize, preharvest, maize, yellow, maize (dark grains), *Makhana* (*Euryale ferox* Salisb) puffs, manioc, milk production mixed feed, mung testa, murkool, mustard cake, neem cake, niger cake, oats, palm kernels, palm products, peanut cake, peanut cake (deoiled), peanut expeller, peanut hay, peanut meal, peanut, kernels, peanut, shells, peanuts, pellets, finisher, pig meal and pellets, pigeon pea, poultry feeds (peanut containing), rapeseed cake, redgram husk, rice, rice bran, rice bran (deoiled), rice chaff, rice crack, rice germ, rice germ cake, rice meal, rice straw, rice (damaged), rice (polish), safflower cake, sal seed cake, sesame, sesame cake, sorghum, soybean meal, soybeans, sunflower, sunflower cake, sunflower flour, tapioca, wheat, wheat bran, wheat bran and chana testa

Palm kernels   may contain the following mycotoxins:

Aflatoxin B_1
incidence: 6/7, conc. range: 21–50 µg/kg (5 sa), 51–100 µg/kg (1 sa), country: UK[267]
incidence: 21/27, conc. range: 10–660 µg/kg, Ø conc.: 91 µg/kg, country: investigated in Germany[298]
see also alfalfa, *Ambadi* cake, animal feedstuffs (dairy cake), bagasse, barley, bengalgram husk, bird food, bird food, wild, biri testa, blackgram, blackgram husk, bran, broiler mixed feed, calf fattening mixed feed, calf fattening mixed feed (containing 4–20 % peanut products), *Carthamus* cake, castor

cake, cereals, cereal products, chick pea, coconut cake, cocos, concentrate, mixed, concentrates, cotton cake, cottonseed, cottonseed (dehulled), cottonseed cake, cottonseed extract, cottonseed meal, cottonseed meal (ammoniated), cottonseed meal (decorticated), cottonseed meats, cottonseed products, crumbles, crumbles, grower, cycad meal, dairy cattle feed, dairy cattle feed (containing 2–5 % peanut products), dairy cattle feed (containing 6–10 % peanut products), dairy cattle feed (containing 6–12 % peanut products), dairy cattle feed (containing more than 20 % peanut products), diets, mixed, dog food, egg production mixed feed, feed, feed and ingredients, feed, compound, feed, layer, feed, mixed, feed (beef), feed (broilers), feed (calf), feed (cat), feed (cattle), feed (chicken), feed (dairy), feed (dog), feed (dug), feed (fish), feed (gluten), feed (horse), feed (miscellaneous), feed (pig), feed (poultry), feed (poultry, pig), feed (rabbit), feed (sheep), feed (50–60 % maize), fish meal, grain by-products, grains (no specification), greengram, hay/silage, horsegram, husk, *Jagni* cake, legume mixture, linseed, linseed cake, livol, *Mahua* cake, maize, maize germ, maize gluten, maize grits, maize husk, maize meal, maize oil cake, maize powder, maize screenings, maize, ground, maize, hybrid, maize, preharvest, maize, yellow, maize (dark grains), *Makhana* (*Euryale ferox* Salisb) puffs, manioc, milk production mixed feed, mung testa, murkool, mustard cake, neem cake, niger cake, oats, palm kernel expeller cake, palm products, peanut cake, peanut cake (deoiled), peanut expeller, peanut hay, peanut meal, peanut, kernels, peanut, shells, peanuts, pellets, finisher, pig meal and pellets, pigeon pea, poultry feeds (peanut containing), rapeseed cake, redgram husk, rice, rice bran, rice bran (deoiled), rice chaff, rice crack, rice germ, rice germ cake, rice meal, rice straw, rice (damaged), rice (polish), safflower cake, sal seed cake, sesame, sesame cake, sorghum, soybean meal, soybeans, sunflower, sunflower cake, sunflower flour, tapioca, wheat,

wheat bran, wheat bran and chana
testa

Palm products   may contain the following
mycotoxins:

AFLATOXIN B_1
incidence: 11/15*, conc. range: 1–11 µg/kg,
country: UK[94], *imported?
see also alfalfa, *Ambadi* cake, animal
feedstuffs (dairy cake), bagasse, barley,
bengalgram husk, bird food, bird food, wild,
biri testa, blackgram, blackgram husk, bran,
broiler mixed feed, calf fattening mixed feed,
calf fattening mixed feed (containing 4–20 %
peanut products), *Carthamus* cake, castor
cake, cereals, cereal products, chick pea,
coconut cake, cocos, concentrate, mixed,
concentrates, cotton cake, cottonseed,
cottonseed (dehulled), cottonseed cake,
cottonseed extract, cottonseed meal,
cottonseed meal (ammoniated), cottonseed
meal (decorticated), cottonseed meats,
cottonseed products, crumbles, crumbles,
grower, cycad meal, dairy cattle feed, dairy
cattle feed (containing 2–5 % peanut
products), dairy cattle feed (containing
6–10 % peanut products), dairy cattle feed
(containing 6–12 % peanut products), dairy
cattle feed (containing more than 20 %
peanut products), diets, mixed, dog food, egg
production mixed feed, feed, feed and
ingredients, feed, compound, feed, layer,
feed, mixed, feed (beef), feed (broilers), feed
(calf), feed (cat), feed (cattle), feed (chicken),
feed (dairy), feed (dog), feed (dug), feed
(fish), feed (gluten), feed (horse), feed
(miscellaneous), feed (pig), feed (poultry),
feed (poultry, pig), feed (rabbit), feed
(sheep), feed (50–60 % maize), fish meal,
grain by-products, grains (no specification),
greengram, hay/silage, horsegram, husk, *Jagni*
cake, legume mixture, linseed, linseed cake,
livol, *Mahua* cake, maize, maize germ, maize
gluten, maize grits, maize husk, maize meal,
maize oil cake, maize powder, maize
screenings, maize, ground, maize, hybrid,
maize, preharvest, maize, yellow, maize (dark
grains), *Makhana* (*Euryale ferox* Salisb) puffs,

manioc, milk production mixed feed, mung
testa, murkool, mustard cake, neem cake,
niger cake, oats, palm kernel expeller cake,
palm kernels, peanut cake, peanut cake
(deoiled), peanut expeller, peanut hay, peanut
meal, peanut, kernels, peanut, shells, peanuts,
pellets, finisher, pig meal and pellets, pigeon
pea, poultry feeds (peanut containing),
rapeseed cake, redgram husk, rice, rice bran,
rice bran (deoiled), rice chaff, rice crack, rice
germ, rice germ cake, rice meal, rice straw,
rice (damaged), rice (polish), safflower cake,
sal seed cake, sesame, sesame cake, sorghum,
soybean meal, soybeans, sunflower, sunflower
cake, sunflower flour, tapioca, wheat, wheat
bran, wheat bran and chana testa

AFLATOXINS (TOTAL)
incidence: 14/15*, conc. range: 1–22 µg/kg,
country: UK[94], *imported?
see also cottonseed, maize germ, maize
gluten, maize (dark grains), rice bran,
soybeans, sunflower

CITRININ
incidence: 1/15*, conc.: 7 µg/kg, country:
UK[94], *imported?
see also barley, barley, oats, barley-soybean
diet, cottonseed cake, feed, feed, mixed, feed
(cattle), feed (pig), fish meal, hay, maize,
maize, white, *Makhana* (*Euryale ferox* Salisb)
puffs, oats, peas and beans, rice bran, rice
germ, wheat, wheat and other grains
(moldy), wheat bran

OCHRATOXIN A
incidence: 2/15*, conc. range: 1–6 µg/kg,
country: UK[94], *imported?
see also alfalfa, barley, barley, oats, barley
(high moisture), barley-soybean diet, bird
food, domestic, bird food, wild, broilers feed,
cereal grains, citrus pulp, coconut, expeller,
corn cob mix silage, diet (dairy cow), diet
(poultry), diet (starter), dog food, eat, egg
production mixed feed, feed, feed wheat, oat
and barley, feed, commercial mix, feed,
mixed, feed, mixed (pelleted), feed (broilers),
feed (cat), feed (cattle), feed (cereals), feed
(pig), feed ec (pig), feed (poultry), feed ec
(poultry), feed (poultry, pig), feed ec (rabbit),

feed (trout), grain, mixed feed, grains, mixed, grains (heated), hay, horse bean, maize, maize feed, milo, maize gluten, maize meal, maize, white, *Makhana* (*Euryale ferox* Salisb) puffs, milk production mixed feed, millet, oat and barley (hammer-milled), oats, peanut cake, peas, peas and beans, pet food, pig feedstuffs, pig grower diet, pig meal, piglet diet, poultry feedstuffs, rice bran, rice germ, rice germ cake, rye, sorghum, soybean groats, sunflower, sunflower seeds, extracted, tapioca, triticale, *Vicia faba*, wheat, wheat and barley, wheat bran, wheat hay, wheat, oats

Paspalum palidosum, straw   may contain the following mycotoxins:

Zearalenone
incidence: 1/1, conc.: 422 μg/kg, country: India[129]
see also alfalfa, barley, barley, husked, barley, unhusked (naked), barley and feed, bone meal, bran, broilers feed, *Carthamus* cake, chick pea, concentrate, mixed, corn cob mix silage, cotton cake, cottonseed, cottonseed cake, diet (dairy cow), diet (poultry), diets (mixed), feed, feed components, feed, mixed, feed, mixed (primarily maize also maize, oats, wheat), feed (bran), feed (broiler chicken), feed (cattle), feed (chicken), feed (dairy), feed (developing pig), feed (mill run, from wheat), feed (miscellaneous), feed (pig), feed (poultry), feed (poultry, pig), feed (starter chicken), feedstuff, fish meal, forage grass, grain, bruised, grain, mixed feed, grains (no specification), hay, maize, maize ears, maize flakes, maize germ, maize germ/bran, maize gluten, maize kernels, maize meal, maize oil cake, maize screenings, maize stalks (pith), maize, "Baby", maize, hybrid, maize, shelled, maize, unshelled, maize, white, maize grain, artificially dried, maize grain, crib dried, maize grain, ensiled, milk production mixed feed, oats, peanut hulls/skins, rice bran, rice germ, rice germ cake, rye, silage, sorghum, soybeans, soybeans, extracted, sunflower cake, tapioca, triticale, wheat, wheat bran, wheat bran and chana testa, wheat soya meal

Peanut cake   may contain the following mycotoxins:

Aflatoxin B$_1$
incidence: 7/10, conc. range: 113.5–2250 μg/kg, country: India[85]
incidence: 1/1, conc.: 3182 μg/kg, country: India[129]
incidence: ?/10, conc. range: 80–120 μg/kg, country: India[183]
incidence: ?/10, conc. range: 40–80 μg/kg, country: India[183]
incidence: ?/10, conc. range: 40–60 μg/kg, country: India[183]
incidence: 291/301, conc. range: ≤25 μg/kg (27 sa), 26–50 μg/kg (32 sa), 51–100 μg/kg (44 sa), 101–200 μg/kg (49 sa), 201–500 μg/kg (67 sa), 501–1000 μg/kg (40 sa), 1001–1500 μg/kg (13 sa), 1501–2000 μg/kg (4 sa), 2001–3000 μg/kg (7 sa), 3001–4000 μg/kg (3 sa), 4001–5000 μg/kg (1 sa), >5000 μg/kg (4 sa), Ø conc.: 450 μg/kg, country: India[247]
incidence: 159/248, conc. range: tr–515.6 μg/kg, Ø conc.: 157.8 μg/kg, country: India[253]
incidence: 4/12, conc. range: 80–290 μg/kg, Ø conc.: 180 μg/kg, country: Taiwan[306]
incidence: 1/1*, conc.: 300 μg/kg, country: Denmark[316], *imported from Congo
incidence: 7/7, conc. range: 200–26,700 μg/kg, Ø conc.: 10,180 μg/kg, country: India[321]
incidence: 31/32*, conc. range: 19–455 μg/kg, Ø conc.: 137.4 μg/kg, country: Nigeria[330], *ncac
incidence: 13/26, conc. range: 16–2000 μg/kg, country: India[358]
incidence: 10/10, conc. range: tr–3000 μg/kg, country: India[383]
see also alfalfa, *Ambadi* cake, animal feedstuffs (dairy cake), bagasse, barley, bengalgram husk, bird food, bird food, wild, biri testa, blackgram, blackgram husk, bran, broiler mixed feed, calf fattening mixed feed, calf fattening mixed feed (containing 4–20% peanut products), *Carthamus* cake, castor cake, cereals, cereal products, chick pea, coconut cake, cocos, concentrate, mixed, concentrates, cotton cake, cottonseed,

cottonseed (dehulled), cottonseed cake, cottonseed extract, cottonseed meal, cottonseed meal (ammoniated), cottonseed meal (decorticated), cottonseed meats, cottonseed products, crumbles, crumbles, grower, cycad meal, dairy cattle feed, dairy cattle feed (containing 2–5 % peanut products), dairy cattle feed (containing 6–10 % peanut products), dairy cattle feed (containing 6–12 % peanut products), dairy cattle feed (containing more than 20 % peanut products), diets, mixed, dog food, egg production mixed feed, feed, feed and ingredients, feed, compound, feed, layer, feed, mixed, feed (beef), feed (broilers), feed (calf), feed (cat), feed (cattle), feed (chicken), feed (dairy), feed (dog), feed (dug), feed (fish), feed (gluten), feed (horse), feed (miscellaneous), feed (pig), feed (poultry), feed (poultry, pig), feed (rabbit), feed (sheep), feed (50–60 % maize), fish meal, grain by-products, grains (no specification), greengram, hay/silage, horsegram, husk, *Jagni* cake, legume mixture, linseed, linseed cake, livol, *Mahua* cake, maize, maize germ, maize gluten, maize grits, maize husk, maize meal, maize oil cake, maize powder, maize screenings, maize, ground, maize, hybrid, maize, preharvest, maize, yellow, maize (dark grains), *Makhana* (*Euryale ferox* Salisb) puffs, manioc, milk production mixed feed, mung testa, murkool, mustard cake, neem cake, niger cake, oats, palm kernel expeller cake, palm kernels, palm products, peanut cake (deoiled), peanut expeller, peanut hay, peanut meal, peanut, kernels, peanut, shells, peanuts, pellets, finisher, pig meal and pellets, pigeon pea, poultry feeds (peanut containing), rapeseed cake, redgram husk, rice, rice bran, rice bran (deoiled), rice chaff, rice crack, rice germ, rice germ cake, rice meal, rice straw, rice (damaged), rice (polish), safflower cake, sal seed cake, sesame, sesame cake, sorghum, soybean meal, soybeans, sunflower, sunflower cake, sunflower flour, tapioca, wheat, wheat bran, wheat bran and chana testa

Aflatoxin B$_2$
incidence: 1/1, conc.: 487 µg/kg, country: India[129]
incidence: ?/10, conc. range: tr–10 µg/kg, country: India[183]
incidence: ?/10, conc. range: tr–20 µg/kg, country: India[183]
incidence: ?/10, conc.: 40 µg/kg, country: India[183]
incidence: 142/301, conc. range: nc, Ø conc.: 80.4 µg/kg, country: India[247]
incidence: 1/1*, conc.: 40 µg/kg, country: Denmark[316], *imported from Congo
incidence: 2/26, conc. range: <1 µg/kg, country: India[358]
see also animal feedstuff (dairy cake), bird food, bird food, wild, biri testa, blackgram, blackgram husk, cottonseed, cottonseed cake, cottonseed extract, cottonseed meal, cottonseed meal (ammoniated), cottonseed meats, dog food, egg production mixed feed, feed, compound, feed (cat), feed (cattle), feed (dog), feed (pig), feed (poultry), feed (rabbit), feed (sheep), fish meal, horsegram, maize, maize gluten, maize husk, maize, ground, maize, preharvest, mung testa, mustard cake, niger cake, peanut cake (deoiled), peanut expeller, peanut hay, peanut meal, peanut, kernels, peanuts, redgram husk, rice bran, rice bran (deoiled), rice chaff, rice meal, rice (polish), sal seed cake, sesame cake, sorghum, soybean meal, soybeans, wheat, wheat bran

Aflatoxin G$_1$
incidence: ?/10, conc. range: tr µg/kg, country: India[183]
incidence: 1/1*, conc.: 60 µg/kg, country: Denmark[316], *imported from Congo
incidence: 9/26, conc. range: <1 µg/kg, country: India[358]
see also animal feedstuffs (dairy cake), bird food, bird food, wild, concentrate, mixed, cottonseed, cottonseed cake, feed (cat), feed (chicken), feed (dog), maize, maize, ground, maize, preharvest, meat meal, milk production mixed feed, murkool, peanut expeller, peanut hay, peanut meal, peanuts, rice bran, rice germ, sorghum, soybeans, wheat, wheat bran, wheat bran and chana testa

Aflatoxin
incidence: 291/301, conc. range:
8–6280 μg/kg, country: India[247]
incidence: ?/55, conc. range: ≤1875 μg/kg,
Ø conc.: 587 μg/kg, country: India[250]
see also blackgram husk, bread crumbs,
broiler finisher, broiler starter, cotton cake,
cottonseed, cottonseed cake, cottonseed
extract, cottonseed meal, feed, feed (cattle),
feed (cow), feed (dog), feed (horse), feed
(maize, gluten), feed (pig), feed (poultry),
feed (rabbit), feed (rat/mice), feed (sheep),
feeds, grain, fish meal, flour (wheat),
groundnut cake, grower's mash, horsegram,
layer's mash, maize, maize gluten, maize,
white, milo, peanut cake (deoiled), peanut,
kernels, peanut (oil cake), pearlmillet, pig
breeder's mash, pig finisher, pig starter, pod
with haulms, poultry breeder's mash, rabbit
pellets, redgram husk, rice, rice bran
(deoiled), rice, broken, rice (polish), sesame
cake, silk worm pupae, sorghum, soybean
cake, soybean meal, wheat, wheat bran

Aflatoxins
incidence: 10/27, conc. range: 10–29 μg/kg
(2 sa), 31–49 μg/kg (1 sa), 50–100 μg/kg
(3 sa), >100 μg/kg (4 sa), country: India[349]
see also animal feed (maize), animal feed
(mixed), barley, copra meal, cottonseed,
cottonseed fines, cottonseed meal, cottonseed
meats, feed, feed ingredients (miscellaneous),
feed, mixed, feed (excluding peanuts,
suspect), feed (goat), feed (pig), feed
(poultry), feedstuff, grain, grain, mixed feed,
maize, maize, shelled, millet, peanut meal,
peanut meal and by-products, peanuts,
protein concentrates, rice, rice bran,
sorghum, soybean meal, sunflower

Neosolaniol
incidence: ?/10, conc.: 10 μg/kg, country:
India[183]
see also feed, feed components, feed, mixed,
feed (poultry), wheat

Ochratoxin A
incidence: 1/27, conc. range: 50–100 μg/kg,
country: India[349]

see also alfalfa, barley, barley, oats, barley
(high moisture), barley-soybean diet, bird
food, domestic, bird food, wild, broilers feed,
cereal grains, citrus pulp, coconut, expeller,
corn cob mix silage, diet (dairy cow), diet
(poultry), diet (starter), dog food, eat, egg
production mixed feed, feed, feed wheat, oat
and barley, feed, commercial mix, feed,
mixed, feed, mixed (pelleted), feed (broilers),
feed (cat), feed (cattle), feed (cereals), feed
(pig), feed ec (pig), feed (poultry), feed ec
(poultry), feed (poultry, pig), feed ec (rabbit),
feed (trout), grain, mixed feed, grains, mixed,
grains (heated), hay, horse bean, maize,
maize feed, milo, maize gluten, maize meal,
maize, white, *Makhana* (*Euryale ferox* Salisb)
puffs, milk production mixed feed, millet, oat
and barley (hammer-milled), oats, palm
products, peas, peas and beans, pet food, pig
feedstuffs, pig grower diet, pig meal, piglet
diet, poultry feedstuffs, rice bran, rice germ,
rice germ cake, rye, sorghum, soybean groats,
sunflower, sunflower seeds, extracted,
tapioca, triticale, *Vicia faba*, wheat, wheat and
barley, wheat bran, wheat hay, wheat, oats
see also Groundnut cake

Peanut cake (deoiled)   may contain the
following mycotoxins:

Aflatoxin B$_1$
incidence: 76/79, conc. range: ≤25 μg/kg
(12 sa), 26–50 μg/kg (10 sa), 51–100 μg/kg
(17 sa), 101–200 μg/kg (15 sa), 201–500 μg/kg
(10 sa), 501–1000 μg/kg (7 sa),
1001–1500 μg/kg (5 sa), Ø conc.: 239.6 μg/kg,
country: India[247]
see also alfalfa, *Ambadi* cake, animal feedstuffs
(dairy cake), bagasse, barley, bengalgram
husk, bird food, bird food, wild, biri testa,
blackgram, blackgram husk, bran, broiler
mixed feed, calf fattening mixed feed, calf
fattening mixed feed (containing 4–20 %
peanut products), *Carthamus* cake, castor
cake, cereals, cereal products, chick pea,
coconut cake, cocos, concentrate, mixed,
concentrates, cotton cake, cottonseed,
cottonseed (dehulled), cottonseed cake,
cottonseed extract, cottonseed meal,

cottonseed meal (ammoniated), cottonseed meal (decorticated), cottonseed meats, cottonseed products, crumbles, crumbles, grower, cycad meal, dairy cattle feed, dairy cattle feed (containing 2–5 % peanut products), dairy cattle feed (containing 6–10 % peanut products), dairy cattle feed (containing 6–12 % peanut products), dairy cattle feed (containing more than 20 % peanut products), diets, mixed, dog food, egg production mixed feed, feed, feed and ingredients, feed, compound, feed, layer, feed, mixed, feed (beef), feed (broilers), feed (calf), feed (cat), feed (cattle), feed (chicken), feed (dairy), feed (dog), feed (dug), feed (fish), feed (gluten), feed (horse), feed (miscellaneous), feed (pig), feed (poultry), feed (poultry, pig), feed (rabbit), feed (sheep), feed (50–60 % maize), fish meal, grain by-products, grains (no specification), greengram, hay/silage, horsegram, husk, *Jagni* cake, legume mixture, linseed, linseed cake, livol, *Mahua* cake, maize, maize germ, maize gluten, maize grits, maize husk, maize meal, maize oil cake, maize powder, maize screenings, maize, ground, maize, hybrid, maize, preharvest, maize, yellow, maize (dark grains), *Makhana* (*Euryale ferox* Salisb) puffs, manioc, milk production mixed feed, mung testa, murkool, mustard cake, neem cake, niger cake, oats, palm kernel expeller cake, palm kernels, palm products, peanut cake, peanut expeller, peanut hay, peanut meal, peanut, kernels, peanut, shells, peanuts, pellets, finisher, pig meal and pellets, pigeon pea, poultry feeds (peanut containing), rapeseed cake, redgram husk, rice, rice bran, rice bran (deoiled), rice chaff, rice crack, rice germ, rice germ cake, rice meal, rice straw, rice (damaged), rice (polish), safflower cake, sal seed cake, sesame, sesame cake, sorghum, soybean meal, soybeans, sunflower, sunflower cake, sunflower flour, tapioca, wheat, wheat bran, wheat bran and chana testa

A F L A T O X I N B$_2$
incidence: 39/79, conc. range: nc, Ø conc.: 49.3 µg/kg, country: India[247]

see also animal feedstuff (dairy cake), bird food, bird food, wild, biri testa, blackgram, blackgram husk, cottonseed, cottonseed cake, cottonseed extract, cottonseed meal, cottonseed meal (ammoniated), cottonseed meats, dog food, egg production mixed feed, feed, compound, feed (cat), feed (cattle), feed (dog), feed (pig), feed (poultry), feed (rabbit), feed (sheep), fish meal, horsegram, maize, maize gluten, maize husk, maize, ground, maize, preharvest, mung testa, mustard cake, niger cake, peanut cake, peanut expeller, peanut hay, peanut meal, peanut, kernels, peanuts, redgram husk, rice bran, rice bran (deoiled), rice chaff, rice meal, rice (polish), sal seed cake, sesame cake, sorghum, soybean meal, soybeans, wheat, wheat bran

A F L A T O X I N
incidence: 76/79, conc. range: 9–1623 µg/kg, country: India[247]
see also blackgram husk, bread crumbs, broiler finisher, broiler starter, cotton cake, cottonseed, cottonseed cake, cottonseed extract, cottonseed meal, feed, feed (cattle), feed (cow), feed (dog), feed (horse), feed (maize, gluten), feed (pig), feed (poultry), feed (rabbit), feed (rat/mice), feed (sheep), feeds, grain, fish meal, flour (wheat), groundnut cake, grower's mash, horsegram, layer's mash, maize, maize gluten, maize, white, milo, peanut cake, peanut, kernels, peanut (oil cake), pearlmillet, pig breeder's mash, pig finisher, pig starter, pod with haulms, poultry breeder's mash, rabbit pellets, redgram husk, rice, rice bran (deoiled), rice, broken, rice (polish), sesame cake, silk worm pupae, sorghum, soybean cake, soybean meal, wheat, wheat bran

Peanut expeller   may contain the following mycotoxins:

A F L A T O X I N B$_1$
incidence: 23/28*, conc. range: 30–1385 µg/kg, Ø conc.: 347.74 µg/kg, country: Denmark[316], *imported from Argentina, Brazil, Ghana, Senegal, Indonesia, Kenya, Nigeria, Sudan, Uganda

see also alfalfa, *Ambadi* cake, animal feedstuffs (dairy cake), bagasse, barley, bengalgram husk, bird food, bird food, wild, biri testa, blackgram, blackgram husk, bran, broiler mixed feed, calf fattening mixed feed, calf fattening mixed feed (containing 4–20 % peanut products), *Carthamus* cake, castor cake, cereals, cereal products, chick pea, coconut cake, cocos, concentrate, mixed, concentrates, cotton cake, cottonseed, cottonseed (dehulled), cottonseed cake, cottonseed extract, cottonseed meal, cottonseed meal (ammoniated), cottonseed meal (decorticated), cottonseed meats, cottonseed products, crumbles, crumbles, grower, cycad meal, dairy cattle feed, dairy cattle feed (containing 2–5 % peanut products), dairy cattle feed (containing 6–10 % peanut products), dairy cattle feed (containing 6–12 % peanut products), dairy cattle feed (containing more than 20 % peanut products), diets, mixed, dog food, egg production mixed feed, feed, feed and ingredients, feed, compound, feed, layer, feed, mixed, feed (beef), feed (broilers), feed (calf), feed (cat), feed (cattle), feed (chicken), feed (dairy), feed (dog), feed (dug), feed (fish), feed (gluten), feed (horse), feed (miscellaneous), feed (pig), feed (poultry), feed (poultry, pig), feed (rabbit), feed (sheep), feed (50–60 % maize), fish meal, grain by-products, grains (no specification), greengram, hay/silage, horsegram, husk, *Jagni* cake, legume mixture, linseed, linseed cake, livol, *Mahua* cake, maize, maize germ, maize gluten, maize grits, maize husk, maize meal, maize oil cake, maize powder, maize screenings, maize, ground, maize, hybrid, maize, preharvest, maize, yellow, maize (dark grains), *Makhana* (*Euryale ferox* Salisb) puffs, manioc, milk production mixed feed, mung testa, murkool, mustard cake, neem cake, niger cake, oats, palm kernel expeller cake, palm kernels, palm products, peanut cake, peanut cake (deoiled), peanut hay, peanut meal, peanut, kernels, peanut, shells, peanuts, pellets, finisher, pig meal and pellets, pigeon pea, poultry feeds (peanut containing),

rapeseed cake, redgram husk, rice, rice bran, rice bran (deoiled), rice chaff, rice crack, rice germ, rice germ cake, rice meal, rice straw, rice (damaged), rice (polish), safflower cake, sal seed cake, sesame, sesame cake, sorghum, soybean meal, soybeans, sunflower, sunflower cake, sunflower flour, tapioca, wheat, wheat bran, wheat bran and chana testa

AFLATOXIN B$_2$
incidence: 17/28*, conc. range: 56–694 µg/kg, Ø conc.: 224.18 µg/kg, country: Denmark[316], *imported from Argentina, Brazil, Ghana, Senegal, Indonesia, Kenya, Nigeria, Sudan, Uganda
see also animal feedstuff (dairy cake), bird food, bird food, wild, biri testa, blackgram, blackgram husk, cottonseed, cottonseed cake, cottonseed extract, cottonseed meal, cottonseed meal (ammoniated), cottonseed meats, dog food, egg production mixed feed, feed, compound, feed (cat), feed (cattle), feed (dog), feed (pig), feed (poultry), feed (rabbit), feed (sheep), fish meal, horsegram, maize, maize gluten, maize husk, maize, ground, maize, preharvest, mung testa, mustard cake, niger cake, peanut cake, peanut cake (deoiled), peanut hay, peanut meal, peanut, kernels, peanuts, redgram husk, rice bran, rice bran (deoiled), rice chaff, rice meal, rice (polish), sal seed cake, sesame cake, sorghum, soybean meal, soybeans, wheat, wheat bran

AFLATOXIN G$_1$
incidence: 14/28*, conc. range: 28–694 µg/kg, Ø conc.: 187.21 µg/kg, country: Denmark[316], *imported from Argentina, Brazil, Ghana, Senegal, Indonesia, Kenya, Nigeria, Sudan, Uganda
see also animal feedstuffs (dairy cake), bird food, bird food, wild, concentrate, mixed, cottonseed, cottonseed cake, feed (cat), feed (chicken), feed (dog), maize, maize, ground, maize, preharvest, meat meal, milk production mixed feed, murkool, peanut cake, peanut hay, peanut meal, peanuts, rice bran, rice germ, sorghum, soybeans, wheat, wheat bran, wheat bran and chana testa

AFLATOXIN G_2
incidence: 6/28*, conc. range: 27–556 µg/kg,
Ø conc.: 171.17 µg/kg, country: Denmark[316],
*imported from Argentina, Brazil, Ghana,
Senegal, Indonesia, Kenya, Nigeria, Sudan,
Uganda
see also animal feedstuffs (dairy cake), bird
food, feed (cat), feed (dog), maize, maize,
preharvest, peanut hay, peanut meal, peanuts,
soybeans

Peanut hay may contain the following
mycotoxins:

AFLATOXIN B_1
incidence: 1/1, conc.: 200 µg/kg, country:
Australia[325]
see also alfalfa, *Ambadi* cake, animal feedstuffs
(dairy cake), bagasse, barley, bengalgram
husk, bird food, bird food, wild, biri testa,
blackgram, blackgram husk, bran, broiler
mixed feed, calf fattening mixed feed, calf
fattening mixed feed (containing 4–20 %
peanut products), *Carthamus* cake, castor
cake, cereals, cereal products, chick pea,
coconut cake, cocos, concentrate, mixed,
concentrates, cotton cake, cottonseed,
cottonseed (dehulled), cottonseed cake,
cottonseed extract, cottonseed meal,
cottonseed meal (ammoniated), cottonseed
meal (decorticated), cottonseed meats,
cottonseed products, crumbles, crumbles,
grower, cycad meal, dairy cattle feed, dairy
cattle feed (containing 2–5 % peanut
products), dairy cattle feed (containing
6–10 % peanut products), dairy cattle feed
(containing 6–12 % peanut products), dairy
cattle feed (containing more than 20 %
peanut products), diets, mixed, dog food, egg
production mixed feed, feed, feed and
ingredients, feed, compound, feed, layer,
feed, mixed, feed (beef), feed (broilers), feed
(calf), feed (cat), feed (cattle), feed (chicken),
feed (dairy), feed (dog), feed (dug), feed
(fish), feed (gluten), feed (horse), feed
(miscellaneous), feed (pig), feed (poultry),
feed (poultry, pig), feed (rabbit), feed
(sheep), feed (50–60 % maize), fish meal,
grain by-products, grains (no specification),

greengram, hay/silage, horsegram, husk, *Jagni*
cake, legume mixture, linseed, linseed cake,
livol, *Mahua* cake, maize, maize germ, maize
gluten, maize grits, maize husk, maize meal,
maize oil cake, maize powder, maize
screenings, maize, ground, maize, hybrid,
maize, preharvest, maize, yellow, maize (dark
grains), *Makhana* (*Euryale ferox* Salisb) puffs,
manioc, milk production mixed feed, mung
testa, murkool, mustard cake, neem cake,
niger cake, oats, palm kernel expeller cake,
palm kernels, palm products, peanut cake,
peanut cake (deoiled), peanut expeller,
peanut meal, peanut, kernels, peanut, shells,
peanuts, pellets, finisher, pig meal and pellets,
pigeon pea, poultry feeds (peanut
containing), rapeseed cake, redgram husk,
rice, rice bran, rice bran (deoiled), rice chaff,
rice crack, rice germ, rice germ cake, rice
meal, rice straw, rice (damaged), rice
(polish), safflower cake, sal seed cake, sesame,
sesame cake, sorghum, soybean meal,
soybeans, sunflower, sunflower cake,
sunflower flour, tapioca, wheat, wheat bran,
wheat bran and chana testa

AFLATOXIN B_2
incidence: 1/1, conc.: 4 µg/kg, country:
Australia[325]
see also animal feedstuff (dairy cake), bird
food, bird food, wild, biri testa, blackgram,
blackgram husk, cottonseed, cottonseed cake,
cottonseed extract, cottonseed meal,
cottonseed meal (ammoniated), cottonseed
meats, dog food, egg production mixed feed,
feed, compound, feed (cat), feed (cattle), feed
(dog), feed (pig), feed (poultry), feed
(rabbit), feed (sheep), fish meal, horsegram,
maize, maize gluten, maize husk, maize,
ground, maize, preharvest, mung testa,
mustard cake, niger cake, peanut cake,
peanut cake (deoiled), peanut expeller,
peanut meal, peanut, kernels, peanuts,
redgram husk, rice bran, rice bran (deoiled),
rice chaff, rice meal, rice (polish), sal seed
cake, sesame cake, sorghum, soybean meal,
soybeans, wheat, wheat bran

Aflatoxin G$_1$
incidence: 1/1, conc.: 200 µg/kg, country:
Australia[325]
see also animal feedstuffs (dairy cake), bird
food, bird food, wild, concentrate, mixed,
cottonseed, cottonseed cake, feed (cat), feed
(chicken), feed (dog), maize, maize, ground,
maize, preharvest, meat meal, milk
production mixed feed, murkool, peanut
cake, peanut expeller, peanut meal, peanuts,
rice bran, rice germ, sorghum, soybeans,
wheat, wheat bran, wheat bran and chana
testa

Aflatoxin G$_2$
incidence: 1/1, conc.: 20 µg/kg, country:
Australia[325]
see also animal feedstuffs (dairy cake), bird
food, feed (cat), feed (dog), maize, maize,
preharvest, peanut expeller, peanut meal,
peanuts, soybeans

Peanut hulls/skins may contain the following
mycotoxins:

Zearalenone
incidence: 2/12, conc. range: 70–150 µg/kg
(2 sa), country: USA[370]
see also alfalfa, barley, barley, husked, barley,
unhusked (naked), barley and feed, bone meal,
bran, broilers feed, *Carthamus* cake, chick pea,
concentrate, mixed, corn cob mix silage,
cotton cake, cottonseed, cottonseed cake, diet
(dairy cow), diet (poultry), diets (mixed), feed,
feed components, feed, mixed, feed, mixed
(primarily maize also maize, oats, wheat), feed
(bran), feed (broiler chicken), feed (cattle),
feed (chicken), feed (dairy), feed (developing
pig), feed (mill run, from wheat), feed
(miscellaneous), feed (pig), feed (poultry),
feed (poultry, pig), feed (starter chicken),
feedstuff, fish meal, forage grass, grain,
bruised, grain, mixed feed, grains (no
specification), hay, maize, maize ears, maize
flakes, maize germ, maize germ/bran, maize
gluten, maize kernels, maize meal, maize oil
cake, maize screenings, maize stalks (pith),
maize, "Baby", maize, hybrid, maize, shelled,
maize, unshelled, maize, white, maize grain,

artificially dried, maize grain, crib dried, maize
grain, ensiled, milk production mixed feed,
oats, *Paspalum palidosum*, straw, rice bran, rice
germ, rice germ cake, rye, silage, sorghum,
soybeans, soybeans, extracted, sunflower cake,
tapioca, triticale, wheat, wheat bran, wheat
bran and chana testa, wheat soya meal

Peanut meal Peanut meal for feed may
contain the following mycotoxins:

Aflatoxin B$_1$
incidence: 366/366*, conc. range:
10–750 µg/kg, Ø conc.: 71 µg/kg, country:
Poland[76], *imported
incidence: 1/2, conc.: 500 µg/kg, country:
Australia[121]
incidence: 3/4, conc. range: 112–6500 µg/kg,
country: USA[245]
incidence: 1/1, conc.: 11 µg/kg, country:
USA[272]
incidence: 13/16, conc. range: 60–300 µg/kg,
Ø conc.: 140 µg/kg, country: Ireland[279]
incidence: 16/18*, conc. range:
<20–1120 µg/kg, country: Indonesia, USA,
Brazil, India[305], *ncac
incidence: 20/22*, conc. range:
54–2520 µg/kg, Ø conc.: 701.75 µg/kg,
country: Denmark[316], *imported from
Argentina, Brazil, Nigeria, Senegal
incidence: 1/1* **, conc.: 269 µg/kg, country:
USA[378], *ncac, **de-oiled
incidence: 1/1*, conc.: 35.7 µg/kg, country:
USA[378], *ncac
incidence: ?/35, conc. range: <10–2400 µg/kg,
country: The Netherlands[379]
incidence: 2/2*, conc. range: 164–3527 µg/kg,
Ø conc.: 1845.5 µg/kg, country: UK[407],
*imported from Argentina and India
see also alfalfa, *Ambadi* cake, animal feedstuffs
(dairy cake), bagasse, barley, bengalgram
husk, bird food, bird food, wild, biri testa,
blackgram, blackgram husk, bran, broiler
mixed feed, calf fattening mixed feed, calf
fattening mixed feed (containing 4–20 %
peanut products), *Carthamus* cake, castor
cake, cereals, cereal products, chick pea,
coconut cake, cocos, concentrate, mixed,

concentrates, cotton cake, cottonseed, cottonseed (dehulled), cottonseed cake, cottonseed extract, cottonseed meal, cottonseed meal (ammoniated), cottonseed meal (decorticated), cottonseed meats, cottonseed products, crumbles, crumbles, grower, cycad meal, dairy cattle feed, dairy cattle feed (containing 2–5 % peanut products), dairy cattle feed (containing 6–10 % peanut products), dairy cattle feed (containing 6–12 % peanut products), dairy cattle feed (containing more than 20 % peanut products), diets, mixed, dog food, egg production mixed feed, feed, feed and ingredients, feed, compound, feed, layer, feed, mixed, feed (beef), feed (broilers), feed (calf), feed (cat), feed (cattle), feed (chicken), feed (dairy), feed (dog), feed (dug), feed (fish), feed (gluten), feed (horse), feed (miscellaneous), feed (pig), feed (poultry), feed (poultry, pig), feed (rabbit), feed (sheep), feed (50–60 % maize), fish meal, grain by-products, grains (no specification), greengram, hay/silage, horsegram, husk, *Jagni* cake, legume mixture, linseed, linseed cake, livol, *Mahua* cake, maize, maize germ, maize gluten, maize grits, maize husk, maize meal, maize oil cake, maize powder, maize screenings, maize, ground, maize, hybrid, maize, preharvest, maize, yellow, maize (dark grains), *Makhana* (*Euryale ferox* Salisb) puffs, manioc, milk production mixed feed, mung testa, murkool, mustard cake, neem cake, niger cake, oats, palm kernel expeller cake, palm kernels, palm products, peanut cake, peanut cake (deoiled), peanut expeller, peanut hay, peanut, kernels, peanut, shells, peanuts, pellets, finisher, pig meal and pellets, pigeon pea, poultry feeds (peanut containing), rapeseed cake, redgram husk, rice, rice bran, rice bran (deoiled), rice chaff, rice crack, rice germ, rice germ cake, rice meal, rice straw, rice (damaged), rice (polish), safflower cake, sal seed cake, sesame, sesame cake, sorghum, soybean meal, soybeans, sunflower, sunflower cake, sunflower flour, tapioca, wheat, wheat bran, wheat bran and chana testa

Aflatoxin B$_2$
incidence: 366/366*, conc. range: 2-200 µg/kg, Ø conc.: 16 µg/kg, country: Poland[76], *imported
incidence: 1/1, conc.: 3 µg/kg, country: USA[272]
incidence: 10/16, conc. range: 20–90 µg/kg, Ø conc.: 40 µg/kg, country: Ireland[279]
incidence: 16/18*, conc. range: <20–600 µg/kg, country: Indonesia, USA, Brazil, India[305], *ncac
incidence: 15/22*, conc. range: 30–556 µg/kg, Ø conc.: 301.73 µg/kg, country: Denmark[316], *imported from Argentina, Brazil, Nigeria, Senegal
see also animal feedstuff (dairy cake), bird food, bird food, wild, biri testa, blackgram, blackgram husk, cottonseed, cottonseed cake, cottonseed extract, cottonseed meal, cottonseed meal (ammoniated), cottonseed meats, dog food, egg production mixed feed, feed, compound, feed (cat), feed (cattle), feed (dog), feed (pig), feed (poultry), feed (rabbit), feed (sheep), fish meal, horsegram, maize, maize gluten, maize husk, maize, ground, maize, preharvest, mung testa, mustard cake, niger cake, peanut cake, peanut cake (deoiled), peanut expeller, peanut hay, peanut, kernels, peanuts, redgram husk, rice bran, rice bran (deoiled), rice chaff, rice meal, rice (polish), sal seed cake, sesame cake, sorghum, soybean meal, soybeans, wheat, wheat bran

Aflatoxin G$_1$
incidence: 1/1, conc.: 11 µg/kg, country: USA[272]
incidence: 7/16, conc. range: 40–150 µg/kg, Ø conc.: 70 µg/kg, country: Ireland[279]
incidence: 16/18*, conc. range: <20–450 µg/kg, country: Indonesia, USA, Brazil, India[305], *ncac
incidence: 16/22*, conc. range: 20–694 µg/kg, Ø conc.: 272.06 µg/kg, country: Denmark[316], *imported from Argentina, Brazil, Nigeria, Senegal
see also animal feedstuffs (dairy cake), bird food, bird food, wild, concentrate, mixed, cottonseed, cottonseed cake, feed (cat),

feed (chicken), feed (dog), maize, maize, ground, maize, preharvest, meat meal, milk production mixed feed, murkool, peanut cake, peanut expeller, peanut hay, peanuts, rice bran, rice germ, sorghum, soybeans, wheat, wheat bran, wheat bran and chana testa

Aflatoxin G_2
incidence: 1/1, conc.: 5 μg/kg, country: USA[272]
incidence: 3/16, conc. range: 30–50 μg/kg, Ø conc.: 40 μg/kg, country: Ireland[279]
incidence: 16/18*, conc. range: <20–90 μg/kg, country: Indonesia, USA, Brazil, India[305], *ncac
incidence: 11/22*, conc. range: 28–111 μg/kg, Ø conc.: 82.36 μg/kg, country: Denmark[316], *imported from Argentina, Brazil, Nigeria, Senegal
incidence: 2/2, conc. range: 27–601 μg/kg, Ø conc.: 314 μg/kg, country: UK[407], imported from Argentina and India
see also animal feedstuffs (dairy cake), bird food, feed (cat), feed (dog), maize, maize, preharvest, peanut expeller, peanut hay, peanuts, soybeans

Aflatoxins
incidence: 1/1, conc.: 72 μg/kg, country: USA[280]
see also animal feed (maize), animal feed (mixed), barley, copra meal, cottonseed, cottonseed fines, cottonseed meats, cottonseed meal, feed, feed ingredients (miscellaneous), feed, mixed, feed (excluding peanuts, suspect), feed (goat), feed (pig), feed (poultry), feedstuff, grain, grain, mixed feed, maize, maize, shelled, peanut cake, peanut meal and by-products, peanuts, protein concentrates, rice, rice bran, sorghum, soybean meal, sunflower

Peanut meal and by-products   may contain the following mycotoxins:

Aflatoxins
incidence: 45/82, conc. range: 5–>2000 μg/kg, country: Australia[21]
see also animal feed (maize), animal feed (mixed), barley, copra meal, cottonseed, cottonseed fines, cottonseed meal, cottonseed meats, feed, feed ingredients (miscellaneous), feed, mixed, feed (excluding peanuts, suspect), feed (goat), feed (pig), feed (poultry), feedstuff, grain, grain, mixed feed, maize, maize, shelled, peanut cake, peanut meal, peanuts, protein concentrates, rice, rice bran, sorghum, soybean meal, sunflower

Peanut, kernels   Peanut kernels for feed may contain the following mycotoxins:

Aflatoxin B_1
incidence: 3/20, Ø conc.: 30 μg/kg, country: Egypt[16]
see also alfalfa, *Ambadi* cake, animal feedstuffs (dairy cake), bagasse, barley, bengalgram husk, bird food, bird food, wild, biri testa, blackgram, blackgram husk, bran, broiler mixed feed, calf fattening mixed feed, calf fattening mixed feed (containing 4–20 % peanut products), *Carthamus* cake, castor cake, cereals, cereal products, chick pea, coconut cake, cocos, concentrate, mixed, concentrates, cotton cake, cottonseed, cottonseed (dehulled), cottonseed cake, cottonseed extract, cottonseed meal, cottonseed meal (ammoniated), cottonseed meal (decorticated), cottonseed meats, cottonseed products, crumbles, crumbles, grower, cycad meal, dairy cattle feed, dairy cattle feed (containing 2–5 % peanut products), dairy cattle feed (containing 6–10 % peanut products), dairy cattle feed (containing 6–12 % peanut products), dairy cattle feed (containing more than 20 % peanut products), diets, mixed, dog food, egg production mixed feed, feed, feed, and ingredients, feed, compound, feed, layer, feed, mixed, feed (beef), feed (broilers), feed (calf), feed (cat), feed (cattle), feed (chicken), feed (dairy), feed (dog), feed (dug), feed (fish), feed (gluten), feed (horse), feed (miscellaneous), feed (pig), feed (poultry), feed (poultry, pig), feed (rabbit), feed (sheep), feed (50–60 % maize), fish meal,

grain by-products, grains (no specification), greengram, hay/silage, horsegram, husk, *Jagni* cake, legume mixture, linseed, linseed cake, livol, *Mahua* cake, maize, maize germ, maize gluten, maize grits, maize husk, maize meal, maize oil cake, maize powder, maize screenings, maize, ground, maize, hybrid, maize, preharvest, maize, yellow, maize (dark grains), *Makhana* (*Euryale ferox* Salisb) puffs, manioc, milk production mixed feed, mung testa, murkool, mustard cake, neem cake, niger cake, oats, palm kernel expeller cake, palm kernels, palm products, peanut cake, peanut cake (deoiled), peanut expeller, peanut hay, peanut meal, peanut, kernels, peanut, shells, peanuts, pellets, finisher, pig meal and pellets, pigeon pea, poultry feeds (peanut containing), rapeseed cake, redgram husk, rice, rice bran, rice bran (deoiled), rice chaff, rice crack, rice germ, rice germ cake, rice meal, rice straw, rice (damaged), rice (polish), safflower cake, sal seed cake, sesame, sesame cake, sorghum, soybean meal, soybeans, sunflower, sunflower cake, sunflower flour, tapioca, wheat, wheat bran, wheat bran and chana testa

Aflatoxin B$_2$
incidence: 3/20, Ø conc.: 15 μg/kg, country: Egypt[16]
see also animal feedstuff (dairy cake), bird food, bird food, wild, biri testa, blackgram, blackgram husk, cottonseed, cottonseed cake, cottonseed extract, cottonseed meal, cottonseed meal (ammoniated), cottonseed meats, dog food, egg production mixed feed, feed, compound, feed (cat), feed (cattle), feed (dog), feed (pig), feed (poultry), feed (rabbit), feed (sheep), fish meal, horsegram, maize, maize gluten, maize husk, maize, ground, maize, preharvest, mung testa, mustard cake, niger cake, peanut cake, peanut cake (deoiled), peanut expeller, peanut hay, peanut meal, peanuts, redgram husk, rice bran, rice bran (deoiled), rice chaff, rice meal, rice (polish), sal seed cake, sesame cake, sorghum, soybean meal, soybeans, wheat, wheat bran

Aflatoxin
incidence: 24/77*, conc. range: 11–30 μg/kg (6 sa), 31–100 μg/kg (6 sa), ≤1776 μg/kg (12 sa), country: India[311], *ncac
incidence: 47/150*, conc. range: 6–10 μg/kg (22 sa), 14–20 μg/kg (12 sa), 25–30 μg/kg (5 sa), 40–50 μg/kg (3 sa), 60–125 μg/kg (5 sa), country: Israel[360], *ncac
see also blackgram husk, bread crumbs, broiler finisher, broiler starter, cotton cake, cottonseed, cottonseed cake, cottonseed extract, cottonseed meal, feed, feed (cattle), feed (cow), feed (dog), feed (horse), feed (maize, gluten), feed (pig), feed (poultry), feed (rabbit), feed (rat/mice), feed (sheep), feeds, grain, fish meal, flour (wheat), groundnut cake, grower's mash, horsegram, layer's mash, maize, maize gluten, maize, white, milo, peanut cake, peanut cake (deoiled), peanut (oil cake), pearlmillet, pig breeder's mash, pig finisher, pig starter, pod with haulms, poultry breeder's mash, rabbit pellets, redgram husk, rice, rice bran (deoiled), rice, broken, rice (polish), sesame cake, silk worm pupae, sorghum, soybean cake, soybean meal, wheat, wheat bran

Peanut, shells may contain the following mycotoxins:

Aflatoxin B$_1$
incidence: 4/20, Ø conc.: 400 μg/kg, country: Egypt[16]
see also alfalfa, *Ambadi* cake, animal feedstuffs (dairy cake), bagasse, barley, bengalgram husk, bird food, bird food, wild, biri testa, blackgram, blackgram husk, bran, broiler mixed feed, calf fattening mixed feed, calf fattening mixed feed (containing 4–20 % peanut products), *Carthamus* cake, castor cake, cereals, cereal products, chick pea, coconut cake, cocos, concentrate, mixed, concentrates, cotton cake, cottonseed, cottonseed (dehulled), cottonseed cake, cottonseed extract, cottonseed meal, cottonseed meal (ammoniated), cottonseed meal (decorticated), cottonseed meats, cottonseed products, crumbles, crumbles, grower, cycad meal, dairy cattle feed, dairy

cattle feed (containing 2–5 % peanut products), dairy cattle feed (containing 6–10 % peanut products), dairy cattle feed (containing 6–12 % peanut products), dairy cattle feed (containing more than 20 % peanut products), diets, mixed, dog food, egg production mixed feed, feed, feed and ingredients, feed, compound, feed, layer, feed, mixed, feed (beef), feed (broilers), feed (calf), feed (cat), feed (cattle), feed (chicken), feed (dairy), feed (dog), feed (dug), feed (fish), feed (gluten), feed (horse), feed (miscellaneous), feed (pig), feed (poultry), feed (poultry, pig), feed (rabbit), feed (sheep), feed (50–60 % maize), fish meal, grain by-products, grains (no specification), greengram, hay/silage, horsegram, husk, *Jagni* cake, legume mixture, linseed, linseed cake, livol, *Mahua* cake, maize, maize germ, maize gluten, maize grits, maize husk, maize meal, maize oil cake, maize powder, maize screenings, maize, ground, maize, hybrid, maize, preharvest, maize, yellow, maize (dark grains), *Makhana* (*Euryale ferox* Salisb) puffs, manioc, milk production mixed feed, mung testa, murkool, mustard cake, neem cake, niger cake, oats, palm kernel expeller cake, palm kernels, palm products, peanut cake, peanut cake (deoiled), peanut expeller, peanut hay, peanut meal, peanut, kernels, peanuts, pellets, finisher, pig meal and pellets, pigeon pea, poultry feeds (peanut containing), rapeseed cake, redgram husk, rice, rice bran, rice bran (deoiled), rice chaff, rice crack, rice germ, rice germ cake, rice meal, rice straw, rice (damaged), rice (polish), safflower cake, sal seed cake, sesame, sesame cake, sorghum, soybean meal, soybeans, sunflower, sunflower cake, sunflower flour, tapioca, wheat, wheat bran, wheat bran and chana testa

AFLATOXIN $B_1 + B_2 + G_1$
incidence: 1/2, conc. range: 21–50 µg/kg, (1 sa with a maximum of 48 µg/kg), country: Guatemala[375]

Peanut (oil cake)   may contain the following mycotoxins:

AFLATOXIN B
incidence: 4/12, conc. range: ≥25–>100 µg/kg, country: France[46]
see also barley, cocoa (oilcake), cottonseed cake, feed, feed (cattle), feed (maize, gluten), feed (pig), feed (poultry), flour (wheat), lucern (dried), maize, maize gluten, maize grains, oats, rice, broken, sorghum, soybean (oil cake), sunflower (oilcake), wheat

AFLATOXIN B + G
incidence: 19/25, conc. range: 10–4500 µg/kg, Ø conc.: 330 µg/kg, country: France[45]
incidence: 8/12, conc. range: ≤ 4000 µg/kg or more, country: Malaya[388]
see also feed (cattle), feed (pig), feed (poultry), feed (rabbit), feed (rat), mice feed

AFLATOXIN
incidence: 7/7, conc. range: <30 µg/kg (3 sa), >30 µg/kg (1 sa), >200 µg/kg (2 sa), >600 µg/kg (1 sa), country: India[380]
see also blackgram husk, bread crumbs, broiler finisher, broiler starter, cotton cake, cottonseed, cottonseed cake, cottonseed extract, cottonseed meal, feed, feed (cattle), feed (cow), feed (dog), feed (horse), feed (maize, gluten), feed (pig), feed (poultry), feed (rabbit), feed (rat/mice), feed (sheep), feeds, grain, fish meal, flour (wheat), groundnut cake, grower's mash, horsegram, layer's mash, maize, maize gluten, maize, white, peanut cake, peanut cake (deoiled), peanut, kernels, pearlmillet, pig breeder's mash, pig finisher, pig starter, pod with haulms, poultry breeder's mash, rabbit pellets, redgram husk, rice, rice bran (deoiled), rice, broken, rice (polish), sesame cake, sorghum, soybean cake, soybean meal, wheat, wheat bran

Peanuts   Peanuts for feed may contain the following mycotoxins:

AFLATOXIN B_1
incidence: 10/10*, conc. range: tr µg/kg (9 sa), 17 µg/kg (1 sa) country: UK[36], *imported?
incidence: 2/2*, conc. range: 43.5–432 µg/kg,

Ø conc.: 237.8 μg/kg, country: UK[122], *ncac
incidence: 1/1, conc.: 500 μg/kg, country:
Australia[181]
incidence: 22/37, conc. range: <20 μg/kg
(1 sa), 21–50 μg/kg (1 sa), 51–100 μg/kg
(3 sa), 101–1000 μg/kg (9 sa), >1000 μg/kg
(8 sa), country: UK[267]
incidence: 37/37, conc. range: 10–3300 μg/kg,
Ø conc.: 890 μg/kg, country: investigated in
Germany[298]
incidence: 1/1*, conc.: 56 μg/kg, country:
Denmark[316], *imported from Congo
incidence: 20/28*, conc. range: 1–20 μg/kg
(8 sa), 21–100 μg/kg (10 sa), 101–300 μg/kg
(2 sa), country: USA[370], *includes 15 peanut
hulls, 7 peanut skin, and 6 peanut meal sa
see also alfalfa, *Ambadi* cake, animal feedstuffs
(dairy cake), bagasse, barley, bengalgram
husk, bird food, bird food, wild, biri testa,
blackgram, blackgram husk, bran, broiler
mixed feed, calf fattening mixed feed, calf
fattening mixed feed (containing 4–20 %
peanut products), *Carthamus* cake, castor
cake, cereals, cereal products, chick pea,
coconut cake, cocos, concentrate, mixed,
concentrates, cotton cake, cottonseed,
cottonseed (dehulled), cottonseed cake,
cottonseed extract, cottonseed meal,
cottonseed meal (ammoniated), cottonseed
meal (decorticated), cottonseed meats,
cottonseed products, crumbles, crumbles,
grower, cycad meal, dairy cattle feed, dairy
cattle feed (containing 2–5 % peanut
products), dairy cattle feed (containing
6–10 % peanut products), dairy cattle feed
(containing 6–12 % peanut products), dairy
cattle feed (containing more than 20 %
peanut products), diets, mixed, dog food, egg
production mixed feed, feed, feed and
ingredients, feed, compound, feed, layer,
feed, mixed, feed (beef), feed (broilers), feed
(calf), feed (cat), feed (cattle), feed (chicken),
feed (dairy), feed (dog), feed (dug), feed
(fish), feed (gluten), feed (horse), feed
(miscellaneous), feed (pig), feed (poultry),
feed (poultry, pig), feed (rabbit), feed
(sheep), feed (50–60 % maize), fish meal,

grain by-products, grains (no specification),
greengram, hay/silage, horsegram, husk, *Jagni*
cake, legume mixture, linseed, linseed cake,
livol, *Mahua* cake, maize, maize germ, maize
gluten, maize grits, maize husk, maize meal,
maize oil cake, maize powder, maize
screenings, maize, ground, maize, hybrid,
maize, preharvest, maize, yellow, maize (dark
grains), *Makhana* (*Euryale ferox* Salisb) puffs,
manioc, milk production mixed feed, mung
testa, murkool, mustard cake, neem cake,
niger cake, oats, palm kernel expeller cake,
palm kernels, palm products, peanut cake,
peanut cake (deoiled), peanut expeller,
peanut hay, peanut meal, peanut, kernels,
peanut, shells, pellets, finisher, pig meal and
pellets, pigeon pea, poultry feeds (peanut
containing), rapeseed cake, redgram husk,
rice, rice bran, rice bran (deoiled), rice chaff,
rice crack, rice germ, rice germ cake, rice
meal, rice straw, rice (damaged), rice
(polish), safflower cake, sal seed cake, sesame,
sesame cake, sorghum, soybean meal,
soybeans, sunflower, sunflower cake,
sunflower flour, tapioca, wheat, wheat bran,
wheat bran and chana testa

AFLATOXIN B$_2$
incidence: 2/2*, conc. range: 10.5–74 μg/kg,
Ø conc.: 42.3 μg/kg, country: UK[122], *ncac
see also animal feedstuff (dairy cake), bird
food, bird food, wild, biri testa, blackgram,
blackgram husk, cottonseed, cottonseed cake,
cottonseed extract, cottonseed meal,
cottonseed meal (ammoniated), cottonseed
meats, dog food, egg production mixed feed,
feed, compound, feed (cat), feed (cattle), feed
(dog), feed (pig), feed (poultry), feed
(rabbit), feed (sheep), fish meal, horsegram,
maize, maize gluten, maize husk, maize,
ground, maize, preharvest, mung testa,
mustard cake, niger cake, peanut cake,
peanut cake (deoiled), peanut expeller,
peanut hay, peanut meal, peanut, kernels,
redgram husk, rice bran, rice bran (deoiled),
rice chaff, rice meal, rice (polish), sal seed
cake, sesame cake, sorghum, soybean meal,
soybeans, wheat, wheat bran

Aflatoxin G_1
incidence: 2/2*, conc. range:
93.5–115.5 µg/kg, Ø conc.: 104.5 µg/kg,
country: UK[122], *ncac
incidence: 1/1*, conc.: 56 µg/kg, country:
Denmark[316], *imported from Congo
see also animal feedstuffs (dairy cake), bird
food, bird food, wild, concentrate, mixed,
cottonseed, cottonseed cake, feed (cat), feed
(chicken), feed (dog), maize, maize, ground,
maize, preharvest, meat meal, milk
production mixed feed, murkool, peanut
cake, peanut expeller, peanut hay, peanut
meal, rice bran, rice germ, sorghum,
soybeans, wheat, wheat bran, wheat bran and
chana testa

Aflatoxin G_2
incidence: 2/2*, conc. range: 23–33 µg/kg,
Ø conc.: 28 µg/kg, country: UK[122], *ncac
see also animal feedstuffs (dairy cake), bird
food, feed (cat), feed (dog), maize, maize,
preharvest, peanut expeller, peanut hay,
peanut meal, soybeans

Aflatoxins
incidence: 50/50, conc. range: 3–22,000 µg/kg,
Ø conc.: 1685 µg/kg, country: USA[35]
incidence: 9/56*, conc. range: <2–57 µg/kg,
Ø conc.: 17 µg/kg, country: USA[329], *ncac
incidence: 18/79*, conc. range: <2–154 µg/kg,
Ø conc.: 25 µg/kg, country: USA[329], *ncac
incidence: 16/17*, conc. range: <2–311 µg/kg,
Ø conc.: 33 µg/kg, country: USA[329], *ncac
see also animal feed (maize), animal feed
(mixed), barley, copra meal, cottonseed,
cottonseed fines, cottonseed meal, cottonseed
meats, feed, feed ingredients (miscellaneous),
feed, mixed, feed (excluding peanuts,
suspect), feed (goat), feed (pig), feed
(poultry), feedstuff, grain, grain, mixed feed,
maize, maize, shelled, peanut cake, peanut
meal, peanut meal and by-products, protein
concentrates, rice, rice bran, sorghum,
soybean meal, sunflower

Cyclopiazonic Acid
incidence: 45/50, conc. range:
<50–2900 µg/kg, Ø conc.: 460 µg/kg,
country: USA[35]

see also chick mash, feed, groundnut cake,
maize, millet, little, rice bran, wheat

Diacetoxyscirpenol
incidence: 2/25*, conc. range:
410–2030 µg/kg, Ø conc.: 1220 µg/kg,
country: India[199], *ncac
see also alfalfa, barley, feed, feed components,
feed (dog), feed (fish), feed (mink), feed
(pig), feed (poultry), feed (reindeer), feeds,
grain, feedstuff, forage grass, grain, mixed
feed, grains (no specification), maize, maize
gluten, oat and barley (hammer-milled), oats,
soybeans, wheat

T-2 Toxin
incidence: 3/31*, conc. range:
170–38,890 µg/kg, Ø conc.: 25,270 µg/kg,
country: India[199], *ncac
incidence: 3/25*, conc. range:
630–1460 µg/kg, Ø conc.: 1043.3 µg/kg,
country: India[199], *ncac
see also alfalfa, barley, bran, diet (grower),
diet (poultry), feed, feed components, feed,
layer, feed, mixed, feed (dog), feed (fish),
feed (mink), feed (pig), feed (poultry),
feedstuff, forage grass, grain, mixed feed,
grains (no specification), hay, maize, maize
germ/bran, maize gluten, maize meal, maize
screenings, maize stalks (pith), oat and barley
(hammer-milled), oats, pig grower diet, piglet
diet, rye, sorghum, triticale, wheat

Pearlmillet Pearlmillet for feed may contain
the following mycotoxins:

Aflatoxin
incidence: ?/16, conc. range: ≤416 µg/kg,
country: India[250]
see also blackgram husk, bread crumbs,
broiler finisher, broiler starter, cotton cake,
cottonseed, cottonseed cake, cottonseed
extract, cottonseed meal, feed, feed (cattle),
feed (cow), feed (dog), feed (horse), feed
(maize, gluten), feed (pig), feed (poultry),
feed (rabbit), feed (rat/mice), feed (sheep),
feeds, grain, fish meal, flour (wheat),
groundnut cake, grower's mash, horsegram,
layer's mash, maize, maize gluten, maize,

white, milo, peanut cake, peanut cake
(deoiled), peanut, kernels, peanut (oil cake),
pig breeder's mash, pig finisher, pig starter,
pod with haulms, poultry breeder's mash,
rabbit pellets, redgram husk, rice, rice bran
(deoiled), rice, broken, rice (polish), sesame
cake, silk worm pupae, sorghum, soybean
cake, soybean meal, wheat, wheat bran

Peas   Peas for feed may contain the
following mycotoxins:

Ochratoxin A
incidence: 1/6, conc.: 4 μg/kg, country: The
Netherlands[103]
see also alfalfa, barley, barley, oats, barley
(high moisture), barley-soybean diet, bird
food, domestic, bird food, wild, broilers feed,
cereal grains, citrus pulp, coconut, expeller,
corn cob mix silage, diet (dairy cow), diet
(poultry), diet (starter), dog food, eat, egg
production mixed feed, feed, feed wheat, oat
and barley, feed, commercial mix, feed,
mixed, feed, mixed (pelleted), feed (broilers),
feed (cat), feed (cattle), feed (cereals), feed
(pig), feed ec (pig), feed (poultry), feed ec
(poultry), feed (poultry, pig), feed ec (rabbit),
feed (trout), grain, mixed feed, grains, mixed,
grains (heated), hay, horse bean, maize,
maize feed, milo, maize gluten, maize meal,
maize, white, *Makhana* (*Euryale ferox* Salisb)
puffs, milk production mixed feed, millet, oat
and barley (hammer-milled), oats, palm
products, peanut cake, peas and beans, pet
food, pig feedstuffs, pig grower diet, pig meal,
piglet diet, poultry feedstuffs, rice bran, rice
germ, rice germ cake, rye, sorghum, soybean
groats, sunflower, sunflower seeds, extracted,
tapioca, triticale, *Vicia faba*, wheat, wheat and
barley, wheat bran, wheat hay, wheat, oats

Peas and beans   Peas and beans for feed may
contain the following mycotoxins:

Citrinin
incidence: 1/15*, conc.: 9 μg/kg, country:
UK[94], *imported?
see also barley, barley, oats, barley-soybean
diet, cottonseed cake, feed, feed, mixed, feed

(cattle), feed (pig), fish meal, hay, maize,
maize, white, *Makhana* (*Euryale ferox* Salisb)
puffs, oats, palm products, rice bran, rice
germ, wheat, wheat and other grains
(moldy), wheat bran

Ochratoxin A
incidence: 7/15, conc. range: ≤33 μg/kg,
country: UK[12]
incidence: 1/15*, conc.: 33 μg/kg, country:
UK[94], *imported?
see also alfalfa, barley, barley, oats, barley
(high moisture), barley-soybean diet, bird
food, domestic, bird food, wild, broilers feed,
cereal grains, citrus pulp, coconut, expeller,
corn cob mix silage, diet (dairy cow), diet
(poultry), diet (starter), dog food, eat, egg
production mixed feed, feed, feed wheat, oat
and barley, feed, commercial mix, feed,
mixed, feed, mixed (pelleted), feed (broilers),
feed (cat), feed (cattle), feed (cereals), feed
(pig), feed ec (pig), feed (poultry), feed ec
(poultry), feed (poultry, pig), feed ec (rabbit),
feed (trout), grain, mixed feed, grains, mixed,
grains (heated), hay, horse bean, maize,
maize feed, milo, maize gluten, maize meal,
maize, white, *Makhana* (*Euryale ferox* Salisb)
puffs, milk production mixed feed, millet, oat
and barley (hammer-milled), oats, palm
products, peanut cake, peas, pet food, pig
feedstuffs, pig grower diet, pig meal, piglet
diet, poultry feedstuffs, rice bran, rice germ,
rice germ cake, rye, sorghum, soybean groats,
sunflower, sunflower seeds, extracted,
tapioca, triticale, *Vicia faba*, wheat, wheat and
barley, wheat bran, wheat hay, wheat, oats

Pet food   may contain the following
mycotoxins:

Ochratoxin A
incidence: 19/40, conc. range:
0.21–13.1 μg/kg, country: Austria, Poland[350]
see also alfalfa, barley, barley, oats, barley
(high moisture), barley-soybean diet, bird
food, domestic, bird food, wild, broilers feed,
cereal grains, citrus pulp, coconut, expeller,
corn cob mix silage, diet (dairy cow), diet
(poultry), diet (starter), dog food, eat, egg

production mixed feed, feed, feed wheat, oat and barley, feed, commercial mix, feed, mixed, feed, mixed (pelleted), feed (broilers), feed (cat), feed (cattle), feed (cereals), feed (pig), feed ec (pig), feed (poultry), feed ec (poultry), feed (poultry, pig), feed ec (rabbit), feed (trout), grain, mixed feed, grains, mixed, grains (heated), hay, horse bean, maize, maize feed, milo, maize gluten, maize meal, maize, white, *Makhana* (*Euryale ferox* Salisb) puffs, milk production mixed feed, millet, oat and barley (hammer-milled), oats, palm products, peanut cake, peas, peas and beans, pig feedstuffs, pig grower diet, pig meal, piglet diet, poultry feedstuffs, rice bran, rice germ, rice germ cake, rye, sorghum, soybean groats, sunflower, sunflower seeds, extracted, tapioca, triticale, *Vicia faba*, wheat, wheat and barley, wheat bran, wheat hay, wheat, oats

Pellets, finisher   may contain the following mycotoxins:

Aflatoxin B$_1$
incidence: 1/1, conc.: 20 µg/kg, country: Australia[121]
see also alfalfa, *Ambadi* cake, animal feedstuffs (dairy cake), bagasse, barley, bengalgram husk, bird food, bird food, wild, biri testa, blackgram, blackgram husk, bran, broiler mixed feed, calf fattening mixed feed, calf fattening mixed feed (containing 4–20 % peanut products), *Carthamus* cake, castor cake, cereals, cereal products, chick pea, coconut cake, cocos, concentrate, mixed, concentrates, cotton cake, cottonseed, cottonseed (dehulled), cottonseed cake, cottonseed extract, cottonseed meal, cottonseed meal (ammoniated), cottonseed meal (decorticated), cottonseed meats, cottonseed products, crumbles, crumbles, grower, cycad meal, dairy cattle feed, dairy cattle feed (containing 2–5 % peanut products), dairy cattle feed (containing 6–10 % peanut products), dairy cattle feed (containing 6–12 % peanut products), dairy cattle feed (containing more than 20 % peanut products), diets, mixed, dog food, egg production mixed feed, feed, feed and

ingredients, feed, compound, feed, layer, feed, mixed, feed (beef), feed (broilers), feed (calf), feed (cat), feed (cattle), feed (chicken), feed (dairy), feed (dog), feed (dug), feed (fish), feed (gluten), feed (horse), feed (miscellaneous), feed (pig), feed (poultry), feed (poultry, pig), feed (rabbit), feed (sheep), feed (50–60 % maize), fish meal, grain by-products, grains (no specification), greengram, hay/silage, horsegram, husk, *Jagni* cake, legume mixture, linseed, linseed cake, livol, *Mahua* cake, maize, maize germ, maize gluten, maize grits, maize husk, maize meal, maize oil cake, maize powder, maize screenings, maize, ground, maize, hybrid, maize, preharvest, maize, yellow, maize (dark grains), *Makhana* (*Euryale ferox* Salisb) puffs, manioc, milk production mixed feed, mung testa, murkool, mustard cake, neem cake, niger cake, oats, palm kernel expeller cake, palm kernels, palm products, peanut cake, peanut cake (deoiled), peanut expeller, peanut hay, peanut meal, peanut, kernels, peanut, shells, peanuts, pig meal and pellets, pigeon pea, poultry feeds (peanut containing), rapeseed cake, redgram husk, rice, rice bran, rice bran (deoiled), rice chaff, rice crack, rice germ, rice germ cake, rice meal, rice straw, rice (damaged), rice (polish), safflower cake, sal seed cake, sesame, sesame cake, sorghum, soybean meal, soybeans, sunflower, sunflower cake, sunflower flour, tapioca, wheat, wheat bran, wheat bran and chana testa

Pig breeder's mash   may contain the following mycotoxins:

Aflatoxin
incidence: 3/6, conc. range: 0–260 µg/kg, country: Nigeria[109]
see also blackgram husk, bread crumbs, broiler finisher, broiler starter, cotton cake, cottonseed, cottonseed cake, cottonseed extract, cottonseed meal, feed, feed (cattle), feed (cow), feed (dog), feed (horse), feed (maize, gluten), feed (pig), feed (poultry), feed (rabbit), feed (rat/mice), feed (sheep), feeds, grain, fish meal, flour (wheat),

groundnut cake, grower's mash, horsegram, layer's mash, maize, maize gluten, maize, white, milo, peanut cake, peanut cake (deoiled), peanut, kernels, peanut (oil cake), pearlmillet, pig finisher, pig starter, pod with haulms, poultry breeder's mash, rabbit pellets, redgram husk, rice, rice bran (deoiled), rice, broken, rice (polish), sesame cake, silk worm pupae, sorghum, soybean cake, soybean meal, wheat, wheat bran

Pig diet, fattening   may contain the following mycotoxins:

Rubratoxin B
incidence: 1/7345, conc.: nc, country: Hungary[209]
see also diet (grower), diet (poultry), diet (starter), pig grower diet, soybean groats, wheat

Pig Feed
see Feed (pig)

Pig feedstuffs   may contain the following mycotoxins:

Ochratoxin A
incidence: 27/39, conc. range: 1–>20 μg/kg, country: Czechoslovakia[110]
see also alfalfa, barley, barley, oats, barley (high moisture), barley-soybean diet, bird food, domestic, bird food, wild, broilers feed, cereal grains, citrus pulp, coconut, expeller, corn cob mix silage, diet (dairy cow), diet (poultry), diet (starter), dog food, eat, egg production mixed feed, feed, feed wheat, oat and barley, feed, commercial mix, feed, mixed, feed, mixed (pelleted), feed (broilers), feed (cat), feed (cattle), feed (cereals), feed (pig), feed ec (pig), feed (poultry), feed ec (poultry), feed (poultry, pig), feed ec (rabbit), feed (trout), grain, mixed feed, grains, mixed, grains (heated), hay, horse bean, maize, maize feed, milo, maize gluten, maize meal, maize, white, *Makhana* (*Euryale ferox* Salisb) puffs, milk production mixed feed, millet, oat and barley (hammer-milled), oats, palm products, peanut cake, peas, peas and beans,

pet food, pig grower diet, pig meal, piglet diet, poultry feedstuffs, rice bran, rice germ, rice germ cake, rye, sorghum, soybean groats, sunflower, sunflower seeds, extracted, tapioca, triticale, *Vicia faba*, wheat, wheat and barley, wheat bran, wheat hay, wheat, oats

Pig finisher   may contain the following mycotoxins:

Aflatoxin
incidence: 1/5, conc.: 260 μg/kg, country: Nigeria[109]
see also blackgram husk, bread crumbs, broiler finisher, broiler starter, cotton cake, cottonseed, cottonseed cake, cottonseed extract, cottonseed meal, feed, feed (cattle), feed (cow), feed (dog), feed (horse), feed (maize, gluten), feed (pig), feed (poultry), feed (rabbit), feed (rat/mice), feed (sheep), feeds, grain, fish meal, flour (wheat), groundnut cake, grower's mash, horsegram, layer's mash, maize, maize gluten, maize, white, milo, peanut cake, peanut cake (deoiled), peanut, kernels, peanut (oil cake), pearlmillet, pig breeder's mash, pig starter, pod with haulms, poultry breeder's mash, rabbit pellets, redgram husk, rice bran (deoiled), rice, broken, rice (polish), sesame cake, silk worm pupae, sorghum, soybean cake, soybean meal, wheat, wheat bran

Pig grower diet   may contain the following mycotoxins:

Ochratoxin A
incidence: 1/7345, conc.: nc, country: Hungary[209]
see also alfalfa, barley, barley, oats, barley (high moisture), barley-soybean diet, bird food, domestic, bird food, wild, broilers feed, cereal grains, citrus pulp, coconut, expeller, corn cob mix silage, diet (dairy cow), diet (poultry), diet (starter), dog food, eat, egg production mixed feed, feed, feed wheat, oat and barley, feed, commercial mix, feed, mixed, feed, mixed (pelleted), feed (broilers), feed (cat), feed (cattle), feed (cereals), feed (pig), feed ec (pig), feed (poultry), feed ec

(poultry), feed (poultry, pig), feed ec (rabbit), feed (trout), grain, mixed feed, grains, mixed, grains (heated), hay, horse bean, maize, maize feed, milo, maize gluten, maize meal, maize, white, *Makhana* (*Euryale ferox* Salisb) puffs, milk production mixed feed, millet, oat and barley (hammer-milled), oats, palm products, peanut cake, peas, peas and beans, pet food, pig feedstuffs, pig meal, piglet diet, poultry feedstuffs, rice bran, rice germ, rice germ cake, rye, sorghum, soybean groats, sunflower, sunflower seeds, extracted, tapioca, triticale, *Vicia faba*, wheat, wheat and barley, wheat bran, wheat hay, wheat, oats

RUBRATOXIN B
incidence: 1/7345, conc. range: nc, country: Hungary[209]
see also diet (grower), diet (poultry), diet (starter), pig diet, fattening, soybean groats, wheat

T-2 TOXIN
incidence: 1/7345, conc.: nc, country: Hungary[209]
see also alfalfa, barley, bran, diet (grower), diet (poultry), feed, feed components, feed, layer, feed, mixed, feed (dog), feed (fish), feed (mink), feed (pig), feed (poultry), feedstuff, forage grass, grain, mixed feed, grains (no specification), hay, maize, maize germ/bran, maize gluten, maize meal, maize screenings, maize stalks (pith), oat and barley (hammer-milled), oats, peanuts, piglet diet, rye, sorghum, triticale, wheat

Pig meal may contain the following mycotoxins:

OCHRATOXIN A
incidence: 12/96, conc. range: <25–250 µg/kg, country: UK[30]
see also alfalfa, barley, barley, oats, barley (high moisture), barley-soybean diet, bird food, domestic, bird food, wild, broilers feed, cereal grains, citrus pulp, coconut, expeller, corn cob mix silage, diet (dairy cow), diet (poultry), diet (starter), dog food, eat, egg production mixed feed, feed, feed wheat, oat

and barley, feed, commercial mix, feed, mixed, feed, mixed (pelleted), feed (broilers), feed (cat), feed (cattle), feed (cereals), feed (pig), feed ec (pig), feed (poultry), feed ec (poultry), feed (poultry, pig), feed ec (rabbit), feed (trout), grain, mixed feed, grains, mixed, grains (heated), hay, horse bean, maize, maize feed, milo, maize gluten, maize meal, maize, white, *Makhana* (*Euryale ferox* Salisb) puffs, milk production mixed feed, millet, oat and barley (hammer-milled), oats, palm products, peanut cake, peas, peas and beans, pet food, pig feedstuffs, pig grower diet, piglet diet, poultry feedstuffs, rice bran, rice germ, rice germ cake, rye, sorghum, soybean groats, sunflower, sunflower seeds, extracted, tapioca, triticale, *Vicia faba*, wheat, wheat and barley, wheat bran, wheat hay, wheat, oats

Pig meal and pellets may contain the following mycotoxins:

AFLATOXIN B$_1$
incidence: 13/291, conc. range: <10–375 µg/kg, country: UK[30]
see also alfalfa, *Ambadi* cake, animal feedstuffs (dairy cake), bagasse, barley, bengalgram husk, bird food, bird food, wild, biri testa, blackgram, blackgram husk, bran, broiler mixed feed, calf fattening mixed feed, calf fattening mixed feed (containing 4–20% peanut products), *Carthamus* cake, castor cake, cereals, cereal products, chick pea, coconut cake, cocos, concentrate, mixed, concentrates, cotton cake, cottonseed, cottonseed (dehulled), cottonseed cake, cottonseed extract, cottonseed meal, cottonseed meal (ammoniated), cottonseed meal (decorticated), cottonseed meats, cottonseed products, crumbles, crumbles, grower, cycad meal, dairy cattle feed, dairy cattle feed (containing 2–5% peanut products), dairy cattle feed (containing 6–10% peanut products), dairy cattle feed (containing 6–12% peanut products), dairy cattle feed (containing more than 20% peanut products), diets, mixed, dog food, egg production mixed feed, feed, feed and ingredients, feed, compound, feed, layer,

feed, mixed, feed (beef), feed (broilers), feed (calf), feed (cat), feed (cattle), feed (chicken), feed (dairy), feed (dog), feed (dug), feed (fish), feed (gluten), feed (horse), feed (miscellaneous), feed (pig), feed (poultry), feed (poultry, pig), feed (rabbit), feed (sheep), feed (50–60 % maize), fish meal, grain by-products, grains (no specification), greengram, hay/silage, horsegram, husk, *Jagni* cake, legume mixture, linseed, linseed cake, livol, *Mahua* cake, maize, maize germ, maize gluten, maize grits, maize husk, maize meal, maize oil cake, maize powder, maize screenings, maize, ground, maize, hybrid, maize, preharvest, maize, yellow, maize (dark grains), *Makhana* (*Euryale ferox* Salisb) puffs, manioc, milk production mixed feed, mung testa, murkool, mustard cake, neem cake, niger cake, oats, palm kernel expeller cake, palm kernels, palm products, peanut cake, peanut cake (deoiled), peanut expeller, peanut hay, peanut meal, peanut, kernels, peanut, shells, peanuts, pellets, finisher, pigeon pea, poultry feeds (peanut containing), rapeseed cake, redgram husk, rice, rice bran, rice bran (deoiled), rice chaff, rice crack, rice germ, rice germ cake, rice meal, rice straw, rice (damaged), rice (polish), safflower cake, sal seed cake, sesame, sesame cake, sorghum, soybean meal, soybeans, sunflower, sunflower cake, sunflower flour, tapioca, wheat, wheat bran, wheat bran and chana testa

Pig starter   may contain the following mycotoxins:

Aflatoxin
incidence: 3/5, conc. range: 0–260 μg/kg, country: Nigeria[109]
see also blackgram husk, bread crumbs, broiler finisher, broiler starter, cotton cake, cottonseed, cottonseed cake, cottonseed extract, cottonseed meal, feed, feed (cattle), feed (cow), feed (dog), feed (horse), feed (maize, gluten), feed (pig), feed (poultry), feed (rabbit), feed (rat/mice), feed (sheep), feeds, grain, fish meal, flour (wheat), groundnut cake, grower's mash, horsegram,

layer's mash, maize, maize gluten, maize, white, milo, peanut cake, peanut cake (deoiled), peanut, kernels, peanut (oil cake), pearlmillet, pig breeder's mash, pig finisher, pod with haulms, poultry breeder's mash, rabbit pellets, redgram husk, rice, rice bran (deoiled), rice, broken, rice (polish), sesame cake, silk worm pupae, sorghum, soybean cake, soybean meal, wheat, wheat bran

Pigeon pea   Pigeon pea for feed may contain the following mycotoxins:

Aflatoxin B_1
incidence: 3/10, conc. range: 10–60 μg/kg, country: India[183]
incidence: 5/10, conc. range: tr–10 μg/kg, country: India[183]
see also alfalfa, *Ambadi* cake, animal feedstuffs (dairy cake), bagasse, barley, bengalgram husk, bird food, bird food, wild, biri testa, blackgram, blackgram husk, bran, broiler mixed feed, calf fattening mixed feed, calf fattening mixed feed (containing 4–20 % peanut products), *Carthamus* cake, castor cake, cereals, cereal products, chick pea, coconut cake, cocos, concentrate, mixed, concentrates, cotton cake, cottonseed, cottonseed (dehulled), cottonseed cake, cottonseed extract, cottonseed meal, cottonseed meal (ammoniated), cottonseed meal (decorticated), cottonseed meats, cottonseed products, crumbles, crumbles, grower, cycad meal, dairy cattle feed, dairy cattle feed (containing 2–5 % peanut products), dairy cattle feed (containing 6–10 % peanut products), dairy cattle feed (containing 6–12 % peanut products), dairy cattle feed (containing more than 20 % peanut products), diets, mixed, dog food, egg production mixed feed, feed, feed and ingredients, feed, compound, feed, layer, feed, mixed, feed (beef), feed (broilers), feed (calf), feed (cat), feed (cattle), feed (chicken), feed (dairy), feed (dog), feed (dug), feed (fish), feed (gluten), feed (horse), feed (miscellaneous), feed (pig), feed (poultry), feed (poultry, pig), feed (rabbit), feed (sheep), feed (50–60 % maize), fish meal,

grain by-products, grains (no specification), greengram, hay/silage, horsegram, husk, *Jagni* cake, legume mixture, linseed, linseed cake, livol, *Mahua* cake, maize, maize germ, maize gluten, maize grits, maize husk, maize meal, maize oil cake, maize powder, maize screenings, maize, ground, maize, hybrid, maize, preharvest, maize, yellow, maize (dark grains), *Makhana* (*Euryale ferox* Salisb) puffs, manioc, milk production mixed feed, mung testa, murkool, mustard cake, neem cake, niger cake, oats, palm kernel expeller cake, palm kernels, palm products, peanut cake, peanut cake (deoiled), peanut expeller, peanut hay, peanut meal, peanut, kernels, peanut, shells, peanuts, pellets, finisher, pig meal and pellets, poultry feeds (peanut containing), rapeseed cake, redgram husk, rice, rice bran, rice bran (deoiled), rice chaff, rice crack, rice germ, rice germ cake, rice meal, rice straw, rice (damaged), rice (polish), safflower cake, sal seed cake, sesame, sesame cake, sorghum, soybean meal, soybeans, sunflower, sunflower cake, sunflower flour, tapioca, wheat, wheat bran, wheat bran and chana testa

Piglet diet   may contain the following mycotoxins:

Ochratoxin A
incidence: 2/7345, conc. range: nc, country: Hungary[209]
see also alfalfa, barley, barley, oats, barley (high moisture), barley-soybean diet, bird food, domestic, bird food, wild, broilers feed, cereal grains, citrus pulp, coconut, expeller, corn cob mix silage, diet (dairy cow), diet (poultry), diet (starter), dog food, eat, egg production mixed feed, feed, feed wheat, oat and barley, feed, commercial mix, feed, mixed, feed, mixed (pelleted), feed (broilers), feed (cat), feed (cattle), feed (cereals), feed (pig), feed ec (pig), feed (poultry), feed ec (poultry), feed (poultry, pig), feed ec (rabbit), feed (trout), grain, mixed feed, grains, mixed, grains (heated), hay, horse bean, maize, maize feed, milo, maize gluten, maize meal, maize, white, *Makhana* (*Euryale ferox* Salisb)

puffs, milk production mixed feed, millet, oat and barley (hammer-milled), oats, palm products, peanut cake, peas, peas and beans, pet food, pig feedstuffs, pig grower diet, pig meal, poultry feedstuffs, rice bran, rice germ, rice germ cake, rye, sorghum, soybean groats, sunflower, sunflower seeds, extracted, tapioca, triticale, *Vicia faba*, wheat, wheat and barley, wheat bran, wheat hay, wheat, oats

Patulin
incidence: 1/7345, conc.: nc, country: Hungary[209]
see also diet (grower), hay, maize meal, wheat

T-2 Toxin
incidence: 2/7345, conc. range: nc, country: Hungary[209]
see also alfalfa, barley, bran, diet (grower), diet (poultry), feed, feed components, feed, layer, feed, mixed, feed (dog), feed (fish), feed (mink), feed (pig), feed (poultry), feedstuff, forage grass, grain, mixed feed, grains (no specification), hay, maize, maize germ/bran, maize gluten, maize meal, maize screenings, maize stalks (pith), oat and barley (hammer-milled), oats, peanuts, pig grower diet, rye, sorghum, triticale, wheat

Pod with haulms   may contain the following mycotoxins:

Aflatoxin
incidence: 24/80, conc. range: 10–30 µg/kg (15 sa), 31–100 µg/kg (7 sa), >500 µg/kg (2 sa), country: India[311]
see also blackgram husk, bread crumbs, broiler finisher, broiler starter, cotton cake, cottonseed, cottonseed cake, cottonseed extract, cottonseed meal, feed, feed (cattle), feed (cow), feed (dog), feed (horse), feed (maize, gluten), feed (pig), feed (poultry), feed (rabbit), feed (rat/mice), feed (sheep), feeds, grain, fish meal, flour (wheat), groundnut cake, grower's mash, horsegram, layer's mash, maize, maize gluten, maize, white, milo, peanut cake, peanut cake (deoiled), peanut, kernels, peanut (oil cake), pearlmillet, pig breeder's mash, pig finisher,

pig starter, poultry breeder's mash, rabbit pellets, redgram husk, rice, rice bran (deoiled), rice, broken, rice (polish), sesame cake, silk worm pupae, sorghum, soybean cake, soybean meal, wheat, wheat bran

Poultry breeder's mash   may contain the following mycotoxins:

AFLATOXIN
incidence: 7/8, conc. range: 0–231 µg/kg, country: Nigeria[109]
see also blackgram husk, bread crumbs, broiler finisher, broiler starter, cotton cake, cottonseed, cottonseed cake, cottonseed extract, cottonseed meal, feed, feed (cattle), feed (cow), feed (dog), feed (horse), feed (maize, gluten), feed (pig), feed (poultry), feed (rabbit), feed (rat/mice), feed (sheep), feeds, grain, fish meal, flour (wheat), groundnut cake, grower's mash, horsegram, layer's mash, maize, maize gluten, maize, white, milo, peanut cake, peanut cake (deoiled), peanut, kernels, peanut (oil cake), pearlmillet, pig breeder's mash, pig finisher, pig starter, pod with haulms, rabbit pellets, redgram husk, rice, rice bran (deoiled), rice, broken, rice (polish), sesame cake, silk worm pupae, sorghum, soybean cake, soybean meal, wheat, wheat bran

POULTRY FEED
see Feed (poultry)

Poultry feedstuffs   may contain the following mycotoxins:

OCHRATOXIN A
incidence: 31/47, conc. range: 1–>20 µg/kg, country: Czechoslovakia[110]
see also alfalfa, barley, barley, oats, barley (high moisture), barley-soybean diet, bird food, domestic, bird food, wild, broilers feed, cereal grains, citrus pulp, coconut, expeller, corn cob mix silage, diet (dairy cow), diet (poultry), diet (starter), dog food, eat, egg production mixed feed, feed, feed wheat, oat and barley, feed, commercial mix, feed, mixed, feed, mixed (pelleted), feed (broilers), feed (cat), feed (cattle), feed (cereals), feed (pig), feed ec (pig), feed (poultry), feed ec (poultry), feed (poultry, pig), feed ec (rabbit), feed (trout), grain, mixed feed, grains, mixed, grains (heated), hay, horse bean, maize, maize feed, milo, maize gluten, maize meal, maize, white, *Makhana* (*Euryale ferox* Salisb) puffs, milk production mixed feed, millet, oat and barley (hammer-milled), oats, palm products, peanut cake, peas, peas and beans, pet food, pig feedstuffs, pig grower diet, pig meal, piglet diet, rice bran, rice germ, rice germ cake, rye, sorghum, soybean groats, sunflower, sunflower seeds, extracted, tapioca, triticale, *Vicia faba*, wheat, wheat and barley, wheat bran, wheat hay, wheat, oats

Poultry feeds (peanut containing)   may contain the following mycotoxins:

AFLATOXIN B$_1$
incidence: 2/11, conc.: 20 µg/kg, Ø conc.: 20 µg/kg, country: Ireland[279]
see also alfalfa, *Ambadi* cake, animal feedstuffs (dairy cake), bagasse, barley, bengalgram husk, bird food, bird food, wild, biri testa, blackgram, blackgram husk, bran, broiler mixed feed, calf fattening mixed feed, calf fattening mixed feed (containing 4–20 % peanut products), *Carthamus* cake, castor cake, cereals, cereal products, chick pea, coconut cake, cocos, concentrate, mixed, concentrates, cotton cake, cottonseed, cottonseed (dehulled), cottonseed cake, cottonseed extract, cottonseed meal, cottonseed meal (ammoniated), cottonseed meal (decorticated), cottonseed meats, cottonseed products, crumbles, crumbles, grower, cycad meal, dairy cattle feed, dairy cattle feed (containing 2–5 % peanut products), dairy cattle feed (containing 6–10 % peanut products), dairy cattle feed (containing 6–12 % peanut products), dairy cattle feed (containing more than 20 % peanut products), diets, mixed, dog food, egg production mixed feed, feed, feed and ingredients, feed, compound, feed, layer, feed, mixed, feed (beef), feed (broilers), feed (calf), feed (cat), feed (cattle), feed (chicken),

feed (dairy), feed (dog), feed (dug), feed (fish), feed (gluten), feed (horse), feed (miscellaneous), feed (pig), feed (poultry), feed (poultry, pig), feed (rabbit), feed (sheep), feed (50–60 % maize), fish meal, grain by-products, grains (no specification), greengram, hay/silage, horsegram, husk, *Jagni* cake, legume mixture, linseed, linseed cake, livol, *Mahua* cake, maize, maize germ, maize gluten, maize grits, maize husk, maize meal, maize oil cake, maize powder, maize screenings, maize, ground, maize, hybrid, maize, preharvest, maize, yellow, maize (dark grains), *Makhana* (*Euryale ferox* Salisb) puffs, manioc, milk production mixed feed, mung testa, murkool, mustard cake, neem cake, niger cake, oats, palm kernel expeller cake, palm kernels, palm products, peanut cake, peanut cake (deoiled), peanut expeller, peanut hay, peanut meal, peanut, kernels, peanut, shells, peanuts, pellets, finisher, pig meal and pellets, pigeon pea, rapeseed cake, redgram husk, rice, rice bran, rice bran (deoiled), rice chaff, rice crack, rice germ, rice germ cake, rice meal, rice straw, rice (damaged), rice (polish), safflower cake, sal seed cake, sesame, sesame cake, sorghum, soybean meal, soybeans, sunflower, sunflower cake, sunflower flour, tapioca, wheat, wheat bran, wheat bran and chana testa

Protein concentrates may contain the following mycotoxins:

Aflatoxins
incidence: 19/31, conc. range: 5–≤500 μg/kg, country: Poland[84]
see also animal feed (maize), animal feed (mixed), barley, copra meal, cottonseed, cottonseed fines, cottonseed meal, cottonseed meats, feed, feed ingredients (miscellaneous), feed, mixed, feed (excluding peanuts, suspect), feed (goat), feed (pig), feed (poultry), feedstuff, grain, grain, mixed feed, maize, maize, shelled, millet, peanut cake, peanut meal, peanut meal and by-products, peanuts, rice, rice bran, sorghum, soybean meal, sunflower

Rabbit pellets may contain the following mycotoxins:

Aflatoxin
incidence: 6/9, conc. range: 0–231 μg/kg, country: Nigeria[109]
see also blackgram husk, bread crumbs, broiler finisher, broiler starter, cotton cake, cottonseed, cottonseed cake, cottonseed extract, cottonseed meal, feed, feed (cattle), feed (cow), feed (dog), feed (horse), feed (maize, gluten), feed (pig), feed (poultry), feed (rabbit), feed (rat/mice), feed (sheep), feeds, grain, fish meal, flour (wheat), groundnut cake, grower's mash, horsegram, layer's mash, maize, maize gluten, maize, white, milo, peanut cake, peanut cake (deoiled), peanut, kernels, peanut (oil cake), pearlmillet, pig breeder's mash, pig finisher, pig starter, pod with haulms, poultry breeder's mash, redgram husk, rice, rice bran (deoiled), rice, broken, rice (polish), sesame cake, silk worm pupae, sorghum, soybean cake, soybean meal, wheat, wheat bran

Rapeseed cake may contain the following mycotoxins:

Aflatoxin B_1
incidence: 1/4, conc.: 22 μg/kg, country: India[247]
see also alfalfa, *Ambadi* cake, animal feedstuffs (dairy cake), barley, bengalgram husk, bird food, bird food, wild, biri testa, blackgram, blackgram husk, bran, broiler mixed feed, calf fattening mixed feed, calf fattening mixed feed (containing 4–20 % peanut products), *Carthamus* cake, castor cake, cereals, cereal products, chick pea, coconut cake, cocos, concentrate, mixed, concentrates, cotton cake, cottonseed, cottonseed (dehulled), cottonseed cake, cottonseed extract, cottonseed meal, cottonseed meal (ammoniated), cottonseed meal (decorticated), cottonseed meats, cottonseed products, crumbles, crumbles, grower, cycad meal, dairy cattle feed, dairy cattle feed (containing 2–5 % peanut products), dairy cattle feed (containing 6–10 % peanut

products), dairy cattle feed (containing 6–12 % peanut products), dairy cattle feed (containing more than 20 % peanut products), diets, mixed, dog food, egg production mixed feed, feed, feed and ingredients, feed, compound, feed, layer, feed, mixed, feed (beef), feed (broilers), feed (calf), feed (cat), feed (cattle), feed (chicken), feed (dairy), feed (dog), feed (dug), feed (fish), feed (gluten), feed (horse), feed (miscellaneous), feed (pig), feed (poultry), feed (poultry, pig), feed (rabbit), feed (sheep), feed (50–60 % maize), fish meal, grain by-products, grains (no specification), greengram, hay/silage, horsegram, husk, *Jagni* cake, legume mixture, linseed, linseed cake, livol, *Mahua* cake, maize, maize germ, maize gluten, maize grits, maize husk, maize meal, maize oil cake, maize powder, maize screenings, maize, ground, maize, hybrid, maize, preharvest, maize (dark grains), maize, yellow, *Makhana* (*Euryale ferox* Salisb) puffs, manioc, milk production mixed feed, mung testa, murkool, mustard cake, neem cake, niger cake, oats, palm kernel expeller cake, palm kernels, palm products, peanut cake, peanut cake (deoiled), peanut expeller, peanut hay, peanut meal, peanut, kernels, peanut, shells, peanuts, pellets, finisher, pig meal and pellets, pigeon pea, poultry feeds (peanut containing), redgram husk, rice, rice bran, rice bran (deoiled), rice chaff, rice crack, rice germ, rice germ cake, rice meal, rice straw, rice (damaged), rice (polish), safflower cake, sal seed cake, sesame, sesame cake, sorghum, soybean meal, soybeans, sunflower, sunflower cake, sunflower flour, tapioca, wheat, wheat bran, wheat bran and chana testa

Rat chow   may contain the following mycotoxins:

FUMONISIN B$_1$
incidence: 1/1, conc.: 220 µg/kg, country: USA[154]
see also barley, bird food, wild, dog food, feed, feed, complete ration, feed, general, feed, layer, feed, maize-based, feed, mixed,

feed, pelleted ration, feed, screenings, feed, sweet, feed (broilers), feed (cat), feed (chicken), feed (dog), feed (gluten), feed (horse), feed (maize), feed (pig), feed (poultry), feed (rat), feed (rodent), forage grass, maize, maize and maize screenings, maize bran, maize ears, maize fine fractions, maize flakes, maize germ, maize germ/bran, maize germ meal, maize gluten, maize grits, maize kernels, maize meal, maize powder, maize screenings, maize, "Baby", maize, ground, maize, preharvest, maize, sweet feed, maize/oats mix, silage, sorghum, soybeans, wheat

FUMONISIN B$_2$
incidence: 1/1, conc.: 20 µg/kg, country: USA[154]
see also barley, bird food, wild, dog food, feed, feed, maize-based, feed, mixed, feed (cat), feed (dog), feed (gluten), feed (horse), feed (maize), feed (poultry), feed (rodent), maize, maize bran, maize fine fractions, maize flakes, maize germ, maize germ/bran, maize germ meal, maize gluten, maize grits, maize kernels, maize meal, maize powder, maize screenings, maize, "Baby", maize, ground, maize, preharvest, wheat

Redgram husk   may contain the following mycotoxins:

AFLATOXIN B$_1$
incidence: 3/16, conc. range: ≤25 µg/kg (2 sa), 26–50 µg/kg (1 sa), Ø conc.: 22.7 µg/kg, country: India[247]
see also alfalfa, *Ambadi* cake, animal feedstuffs (dairy cake), barley, bengalgram husk, bird food, bird food, wild, biri testa, blackgram, blackgram husk, bran, broiler mixed feed, calf fattening mixed feed, calf fattening mixed feed (containing 4–20 % peanut products), *Carthamus* cake, castor cake, cereals, cereal products, chick pea, coconut cake, cocos, concentrate, mixed, concentrates, cotton cake, cottonseed, cottonseed (dehulled), cottonseed cake, cottonseed extract, cottonseed meal, cottonseed meal

(ammoniated), cottonseed meal (decorticated), cottonseed meats, cottonseed products, crumbles, crumbles, grower, cycad meal, dairy cattle feed, dairy cattle feed (containing 2–5 % peanut products), dairy cattle feed (containing 6–10 % peanut products), dairy cattle feed (containing 6–12 % peanut products), dairy cattle feed (containing more than 20 % peanut products), diets, mixed, dog food, egg production mixed feed, feed, feed and ingredients, feed, compound, feed, layer, feed, mixed, feed (beef), feed (broilers), feed (calf), feed (cat), feed (cattle), feed (chicken), feed (dairy), feed (dog), feed (dug), feed (fish), feed (gluten), feed (horse), feed (miscellaneous), feed (pig), feed (poultry), feed (poultry, pig), feed (rabbit), feed (sheep), feed (50–60 % maize), fish meal, grain by-products, grains (no specification), greengram, hay/silage, horsegram, husk, *Jagni* cake, legume mixture, linseed, linseed cake, livol, *Mahua* cake, maize, maize germ, maize gluten, maize grits, maize husk, maize meal, maize oil cake, maize powder, maize screenings, maize, ground, maize, hybrid, maize, preharvest, maize, yellow, maize (dark grains), *Makhana* (*Euryale ferox* Salisb) puffs, manioc, milk production mixed feed, mung testa, murkool, mustard cake, neem cake, niger cake, oats, palm kernel expeller cake, palm kernels, palm products, peanut cake, peanut cake (deoiled), peanut expeller, peanut hay, peanut meal, peanut, kernels, peanut, shells, peanuts, pellets, finisher, pig meal and pellets, pigeon pea, poultry feeds (peanut containing), rapeseed cake, rice, rice bran, rice bran (deoiled), rice chaff, rice crack, rice germ, rice germ cake, rice meal, rice straw, rice (damaged), rice (polish), safflower cake, sal seed cake, sesame, sesame cake, sorghum, soybean meal, soybeans, sunflower, sunflower cake, sunflower flour, tapioca, wheat, wheat bran, wheat bran and chana testa

AFLATOXIN B$_2$
incidence: 2/16, conc. range: nc, Ø conc.: 6.5 μg/kg, country: India[247]

see also animal feedstuffs (dairy cake), bird food, bird food, wild, biri testa, blackgram, blackgram husk, cottonseed, cottonseed cake, cottonseed extract, cottonseed meal, cottonseed meal (ammoniated), cottonseed meats, dog food, egg production mixed feed, feed, compound, feed (cat), feed (cattle), feed (dog), feed (pig), feed (poultry), feed (rabbit), feed (sheep), fish meal, horsegram, maize, maize gluten, maize husk, maize, ground, maize, preharvest, mung testa, mustard cake, niger cake, peanut cake, peanut cake (deoiled), peanut expeller, peanut hay, peanut meal, peanuts, peanut, kernels, rice bran, rice bran (deoiled), rice chaff, rice meal, rice (polish), sal seed cake, sesame cake, sorghum, soybean meal, soybeans, wheat, wheat bran

AFLATOXIN
incidence: 3/16, conc. range: 6–46 μg/kg, country: India[247]
see also blackgram husk, bread crumbs, broiler finisher, broiler starter, cotton cake, cottonseed, cottonseed cake, cottonseed extract, cottonseed meal, feed, feed (cattle), feed (cow), feed (dog), feed (horse), feed (maize, gluten), feed (pig), feed (poultry), feed (rabbit), feed (rat/mice), feed (sheep), feeds, grain, fish meal, flour (wheat), groundnut cake, grower's mash, horsegram, layer's mash, maize, maize gluten, maize, white, milo, peanut cake, peanut cake (deoiled), peanut, kernels, peanut (oil cake), pearlmillet, pig breeder's mash, pig finisher, pig starter, pod with haulms, poultry breeder's mash, rabbit pellets, rice, rice bran (deoiled), rice, broken, rice (polish), sesame cake, silk worm pupae, sorghum, soybean cake, soybean meal, wheat, wheat bran

Rice   Rice for feed may contain the following mycotoxins:

AFLATOXIN B$_1$
incidence: 1/4, conc.: 86 μg/kg, country: India[253]
see also alfalfa, *Ambadi* cake, animal feedstuffs (dairy cake), bagasse, barley, bengalgram

husk, bird food, bird food, wild, biri testa, blackgram, blackgram husk, bran, broiler mixed feed, calf fattening mixed feed, calf fattening mixed feed (containing 4–20% peanut products), *Carthamus* cake, castor cake, cereals, cereal products, chick pea, coconut cake, cocos, concentrate, mixed, concentrates, cotton cake, cottonseed, cottonseed (dehulled), cottonseed cake, cottonseed extract, cottonseed meal, cottonseed meal (ammoniated), cottonseed meal (decorticated), cottonseed meats, cottonseed products, crumbles, crumbles, grower, cycad meal, dairy cattle feed, dairy cattle feed (containing 2–5% peanut products), dairy cattle feed (containing 6–10% peanut products), dairy cattle feed (containing 6–12% peanut products), dairy cattle feed (containing more than 20% peanut products), diets, mixed, dog food, egg production mixed feed, feed, feed and ingredients, feed, compound, feed, layer, feed, mixed, feed (beef), feed (broilers), feed (calf), feed (cat), feed (cattle), feed (chicken), feed (dairy), feed (dog), feed (dug), feed (fish), feed (gluten), feed (horse), feed (miscellaneous), feed (pig), feed (poultry), feed (poultry, pig), feed (rabbit), feed (sheep), feed (50–60% maize), fish meal, grain by-products, grains (no specification), greengram, hay/silage, horsegram, husk, *Jagni* cake, legume mixture, linseed, linseed cake, livol, *Mahua* cake, maize, maize germ, maize gluten, maize grits, maize husk, maize meal, maize oil cake, maize powder, maize screenings, maize, ground, maize, hybrid, maize, preharvest, maize, yellow, maize (dark grains), *Makhana* (*Euryale ferox* Salisb) puffs, manioc, milk production mixed feed, mung testa, murkool, mustard cake, neem cake, niger cake, oats, palm kernel expeller cake, palm kernels, palm products, peanut cake, peanut cake (deoiled), peanut expeller, peanut hay, peanut meal, peanut, kernels, peanut, shells, peanuts, pellets, finisher, pig meal and pellets, pigeon pea, poultry feeds (peanut containing), rapeseed cake, redgram husk, rice bran, rice bran (deoiled), rice chaff, rice crack, rice germ, rice germ cake, rice meal, rice straw, rice (damaged), rice (polish), safflower cake, sal seed cake, sesame, sesame cake, sorghum, soybean meal, soybeans, sunflower, sunflower cake, sunflower flour, tapioca, wheat, wheat bran, wheat bran and chana testa
Redgram huskRice

AFLATOXIN
incidence: 7/7, conc. range: <30 μg/kg (3 sa), >30 μg/kg (1 sa), >200 μg/kg (2 sa), >600 μg/kg (1 sa), country: India[380]
see also blackgram husk, bread crumbs, broiler finisher, broiler starter, cotton cake, cottonseed, cottonseed cake, cottonseed extract, cottonseed meal, feed, feed (cattle), feed (cow), feed (dog), feed (horse), feed (maize, gluten), feed (pig), feed (poultry), feed (rabbit), feed (rat/mice), feed (sheep), feeds, grain, fish meal, flour (wheat), groundnut cake, grower's mash, horsegram, layer's mash, maize, maize gluten, maize, white, milo, peanut cake, peanut cake (deoiled), peanut, kernels, peanut (oil cake), pearlmillet, pig breeder's mash, pig finisher, pig starter, pod with haulms, poultry breeder's mash, rabbit pellets, redgram husk, rice bran (deoiled), rice, broken, rice (polish), sesame cake, silk worm pupae, sorghum, soybean cake, soybean meal, wheat, wheat bran

AFLATOXINS
incidence: 24/157*, conc. range: <2–52 μg/kg, Ø conc.: 11 μg/kg, country: USA[329], *ncac
incidence: 23/134*, conc. range: <2–282 μg/kg, Ø conc.: 52 μg/kg, country: USA[329], *ncac
see also animal feed (maize), animal feed (mixed), barley, copra meal, cottonseed, cottonseed fines, cottonseed meal, cottonseed meats, feed, feed ingredients (miscellaneous), feed, mixed, feed (excluding peanuts, suspect), feed (goat), feed (pig), feed (poultry), feedstuff, grain, grain, mixed feed, maize, maize, shelled, millet, peanut cake, peanut meal, peanut meal and by-products, peanuts, protein concentrates, rice bran, sorghum, soybean meal, sunflower

Rice bran   Rice bran for feed may contain the following mycotoxins:

AFLATOXIN B_1
incidence: 1/5, conc.: 30 µg/kg, country: Egypt[16]
incidence: 35/40*, conc. range: 1–13 µg/kg, country: UK[94], *imported?
incidence: 29/40, conc. range: 1–21 µg/kg, Ø conc.: 6.8 µg/kg, country: UK[96]
incidence: 9/9, conc. range: 36–71 µg/kg, country: Indonesia[426]
see also alfalfa, *Ambadi* cake, animal feedstuffs (dairy cake), bagasse, barley, bengalgram husk, bird food, bird food, wild, biri testa, blackgram, blackgram husk, bran, broiler mixed feed, calf fattening mixed feed, calf fattening mixed feed (containing 4–20 % peanut products), *Carthamus* cake, castor cake, cereals, cereal products, chick pea, coconut cake, cocos, concentrate, mixed, concentrates, cotton cake, cottonseed, cottonseed (dehulled), cottonseed cake, cottonseed extract, cottonseed meal, cottonseed meal (ammoniated), cottonseed meal (decorticated), cottonseed meats, cottonseed products, crumbles, crumbles, grower, cycad meal, dairy cattle feed, dairy cattle feed (containing 2–5 % peanut products), dairy cattle feed (containing 6–10 % peanut products), dairy cattle feed (containing 6–12 % peanut products), dairy cattle feed (containing more than 20 % peanut products), diets, mixed, dog food, egg production mixed feed, feed, feed and ingredients, feed, compound, feed, layer, feed, mixed, feed (beef), feed (broilers), feed (calf), feed (cat), feed (cattle), feed (chicken), feed (dairy), feed (dog), feed (dug), feed (fish), feed (gluten), feed (horse), feed (miscellaneous), feed (pig), feed (poultry), feed (poultry, pig), feed (rabbit), feed (sheep), feed (50–60 % maize), fish meal, grain by-products, grains (no specification), greengram, hay/silage, horsegram, husk, *Jagni* cake, legume mixture, linseed, linseed cake, livol, *Mahua* cake, maize, maize germ, maize gluten, maize grits, maize husk, maize meal, maize oil cake, maize powder, maize screenings, maize, ground, maize, hybrid, maize, preharvest, maize, yellow, maize (dark grains), *Makhana* (*Euryale ferox* Salisb) puffs, manioc, milk production mixed feed, mung testa, murkool, mustard cake, neem cake, niger cake, oats, palm kernel expeller cake, palm kernels, palm products, peanut cake, peanut cake (deoiled), peanut expeller, peanut hay, peanut meal, peanut, kernels, peanut, shells, peanuts, pellets, finisher, pig meal and pellets, pigeon pea, poultry feeds (peanut containing), rapeseed cake, redgram husk, rice, rice bran (deoiled), rice chaff, rice crack, rice germ, rice germ cake, rice meal, rice straw, rice (damaged), rice (polish), safflower cake, sal seed cake, sesame, sesame cake, sorghum, soybean meal, soybeans, sunflower, sunflower cake, sunflower flour, tapioca, wheat, wheat bran, wheat bran and chana testa

AFLATOXIN B_2
incidence: 9/40, conc. range: 1–2 µg/kg, Ø conc.: 1.2 µg/kg, country: UK[96]
see also animal feedstuffs (dairy cake), bird food, bird food, wild, biri testa, blackgram, blackgram husk, cottonseed, cottonseed cake, cottonseed extract, cottonseed meal, cottonseed meal (ammoniated), cottonseed meats, dog food, egg production mixed feed, feed, compound, feed (cat), feed (cattle), feed (dog), feed (pig), feed (poultry), feed (rabbit), feed (sheep), fish meal, horsegram, maize, maize gluten, maize husk, maize, ground, maize, preharvest, mung testa, mustard cake, niger cake, peanut cake, peanut cake (deoiled), peanut expeller, peanut hay, peanut meal, peanuts, peanut, kernels, redgram husk, rice bran (deoiled), rice chaff, rice meal, rice (polish), sal seed cake, sesame cake, sorghum, soybean meal, soybeans, wheat, wheat bran

AFLATOXIN G_1
incidence: 1/5, conc.: 20 µg/kg, country: Egypt[16]
incidence: 5/40, conc. range: 2–3 µg/kg, Ø conc.: 1.4 µg/kg, country: UK[96]
see also animal feedstuffs (dairy cake), bird food, bird food, wild, concentrate, mixed,

cottonseed, cottonseed cake, feed (cat), feed (chicken), feed (dog), maize, maize, ground, maize, preharvest, meat meal, milk production mixed feed, murkool, peanut cake, peanut expeller, peanut hay, peanut meal, peanuts, rice germ, sorghum, soybeans, wheat, wheat bran, wheat bran and chana testa

AFLATOXINS (TOTAL)

incidence: 40/40*, conc. range: 1–19 µg/kg, country: UK[94], *imported?
see also cottonseed, maize germ, maize gluten, maize (dark grains), palm products, soybeans, sunflower

AFLATOXINS

incidence: 3/14, conc. range: 10–100 µg/kg, country: India[349]
see also animal feed (maize), animal feed (mixed), barley, copra meal, cottonseed, cottonseed fines, cottonseed meal, cottonseed meats, feed, feed ingredients (miscellaneous), feed, mixed, feed (excluding peanuts, suspect), feed (goat), feed (pig), feed (poultry), feedstuff, grain, grain, mixed feed, maize, maize, shelled, millet, peanut cake, peanut meal, peanut meal and by-products, peanuts, protein concentrates, rice, sorghum, soybean meal, sunflower

CITRININ

incidence: 1/4, conc.: 20 µg/kg, country: Egypt[16]
see also barley, barley, oats, barley-soybean diet, cottonseed cake, feed, feed, mixed, feed (cattle), feed (pig), fish meal, hay, maize, maize, white, *Makhana* (*Euryale ferox* Salisb) puffs, oats, palm products, peas and beans, rice germ, wheat, wheat and other grains (moldy), wheat bran

CYCLOPIAZONIC ACID

incidence: 6*/40, conc. range: 100–220 µg/kg, Ø conc.: 151.7 µg/kg, country: UK[96], *mycotoxin suspected but identity of peak not unequivocally confirmed
see also chick mash, feed, groundnut cake, maize, millet, little, peanuts, wheat

DEOXYNIVALENOL

incidence: 1/4, conc.: 88 µg/kg, country: Egypt[16]
see also barley, barley, husked, barley, unhusked (naked), barley (pressed), bone meal, bran, broilers feed, calf fattening mixed feed, coconut, expeller, corn cob mix silage, cottonseed, cottonseed cake, dairy cattle feed, egg production mixed feed, feed, feed components, feed, commercial mix, feed, mixed, feed, mixed (primarily maize), feed (barley), feed (cattle), feed (chicken), feed (dog), feed (fish), feed (mill run, from wheat), feed (mink), feed (pig), feed (poultry), feed (reindeer), feeds, grain, feeds, industrial, feedstuff, feedstuffs (rapeseed, turnip, fish meal, concentrates), fish meal, grain, mixed feed, grains, mixed, grains (no specification), maize, maize ears, maize fibre, maize germ, maize germ/bran, maize germ meal, maize gluten, maize kernels, maize meal, maize powder, maize screenings, maize stalks (pith), maize, "Baby", maize, hybrid, maize, white, oats, rice germ cake, rye, silage, sorghum, soybeans, triticale, wheat, wheat and barley, wheat, red hard winter, wheat, soft white winter, wheat, spring, wheat, winter

MONILIFORMIN

incidence: 1*/40, conc.: 70 µg/kg, country: UK[96], *mycotoxin suspected but identity of peak not unequivocally confirmed
see also barley, feed, mixed, feed (poultry), maize, maize flakes, maize germ, maize germ/bran, maize gluten, maize meal, maize screenings, maize, "Baby", oats, triticale, wheat, wheat, summer, wheat, winter

OCHRATOXIN A

incidence: 7/40, conc. range: ≤ 6 µg/kg, country: UK[12]
incidence: 2/3, Ø conc.: 9 µg/kg, country: Egypt[16]
incidence: 3/40*, conc. range: 1–6 µg/kg, country: UK[94], *imported?
incidence: 38*/40, conc. range: 1–12 µg/kg, Ø conc.: 3.1 µg/kg, country: UK[96], *mycotoxin suspected but identity of peak not unequivocally confirmed

incidence: 1/14, conc. range: 10–29 μg/kg, country: India[349]
see also alfalfa, barley, barley, oats, barley (high moisture), barley-soybean diet, bird food, domestic, bird food, wild, broilers feed, cereal grains, citrus pulp, coconut, expeller, corn cob mix silage, diet (dairy cow), diet (poultry), diet (starter), dog food, eat, egg production mixed feed, feed, feed wheat, oat and barley, feed, commercial mix, feed, mixed, feed, mixed (pelleted), feed (broilers), feed (cat), feed (cattle), feed (cereals), feed (pig), feed ec (pig), feed (poultry), feed ec (poultry), feed (poultry, pig), feed ec (rabbit), feed (trout), grain, mixed feed, grains, mixed, grains (heated), hay, horse bean, maize, maize feed, milo, maize gluten, maize meal, maize, white, *Makhana* (*Euryale ferox* Salisb) puffs, milk production mixed feed, millet, oat and barley (hammer-milled), oats, palm products, peanut cake, peas, peas and beans, pet food, pig feedstuffs, pig grower diet, pig meal, piglet diet, poultry feedstuffs, rice germ, rice germ cake, rye, sorghum, soybean groats, sunflower, sunflower seeds, extracted, tapioca, triticale, *Vicia faba*, wheat, wheat and barley, wheat bran, wheat hay, wheat, oats

Zearalenone
incidence: 1/40*, conc.: 44 μg/kg, country: UK[94], *imported?
see also alfalfa, barley, barley, husked, barley, unhusked (naked), barley and feed, bone meal, bran, broilers feed, *Carthamus* cake, chick pea, concentrate, mixed, corn cob mix silage, cotton cake, cottonseed, cottonseed cake, diet (dairy cow), diet (poultry), diets (mixed), feed, feed components, feed, mixed, feed, mixed (primarily maize also maize, oats, wheat), feed (bran), feed (broiler chicken), feed (cattle), feed (chicken), feed (dairy), feed (developing pig), feed (mill run, from wheat), feed (miscellaneous), feed (pig), feed (poultry), feed (poultry, pig), feed (starter chicken), feedstuff, fish meal, forage grass, grain, bruised, grain, mixed feed, grains (no specification), hay, maize, maize ears, maize flakes, maize germ, maize germ/bran, maize gluten, maize kernels, maize meal, maize oil

cake, maize screenings, maize stalks (pith), maize, "Baby", maize, hybrid, maize, shelled, maize, unshelled, maize, white, maize grain, artificially dried, maize grain, crib dried, maize grain, ensiled, milk production mixed feed, oats, *Paspalum palidosum*, straw, peanut hulls/skins, rice germ, rice germ cake, rye, silage, sorghum, soybeans, soybeans, extracted, sunflower cake, tapioca, triticale, wheat, wheat bran, wheat bran and chana testa, wheat soya meal

Rice bran (deoiled)   may contain the following mycotoxins:

Aflatoxin B_1
incidence: 48/143, conc. range: ≤25 μg/kg (36 sa), 26–50 μg/kg (5 sa), 51–100 μg/kg (4 sa), 101–200 μg/kg (2 sa), 201–500 μg/kg (1 sa), Ø conc.: 31.1 μg/kg, country: India[247]
see also alfalfa, *Ambadi* cake, animal feedstuffs (dairy cake), bagasse, barley, bengalgram husk, bird food, bird food, wild, biri testa, blackgram, blackgram husk, bran, broiler mixed feed, calf fattening mixed feed, calf fattening mixed feed (containing 4–20% peanut products), *Carthamus* cake, castor cake, cereals, cereal products, chick pea, coconut cake, cocos, concentrate, mixed, concentrates, cotton cake, cottonseed, cottonseed (dehulled), cottonseed cake, cottonseed extract, cottonseed meal, cottonseed meal (ammoniated), cottonseed meal (decorticated), cottonseed meats, cottonseed products, crumbles, crumbles, grower, cycad meal, dairy cattle feed, dairy cattle feed (containing 2–5% peanut products), dairy cattle feed (containing 6–10% peanut products), dairy cattle feed (containing 6–12% peanut products), dairy cattle feed (containing more than 20% peanut products), diets, mixed, dog food, egg production mixed feed, feed, feed and ingredients, feed, compound, feed, layer, feed, mixed, feed (beef), feed (broilers), feed (calf), feed (cat), feed (cattle), feed (chicken), feed (dairy), feed (dog), feed (dug), feed (fish), feed (gluten), feed (horse), feed (miscellaneous), feed (pig), feed (poultry),

feed (poultry, pig), feed (rabbit), feed
(sheep), feed (50–60 % maize), fish meal,
grain by-products, grains (no specification),
greengram, hay/silage, horsegram, husk, *Jagni*
cake, legume mixture, linseed, linseed cake,
livol, *Mahua* cake, maize, maize germ, maize
gluten, maize grits, maize husk, maize meal,
maize oil cake, maize powder, maize
screenings, maize, ground, maize, hybrid,
maize, preharvest, maize, yellow, maize (dark
grains), *Makhana* (*Euryale ferox* Salisb) puffs,
manioc, milk production mixed feed, mung
testa, murkool, mustard cake, neem cake,
niger cake, oats, palm kernel expeller cake,
palm kernels, palm products, peanut cake,
peanut cake (deoiled), peanut expeller,
peanut hay, peanut meal, peanut, kernels,
peanut, shells, peanuts, pellets, finisher, pig
meal and pellets, pigeon pea, poultry feeds
(peanut containing), rapeseed cake, redgram
husk, rice, rice bran, rice chaff, rice crack,
rice germ, rice germ cake, rice meal, rice
straw, rice (damaged), rice (polish), safflower
cake, sal seed cake, sesame, sesame cake,
sorghum, soybean meal, soybeans, sunflower,
sunflower cake, sunflower flour, tapioca,
wheat, wheat bran, wheat bran and chana
testa

Aflatoxin B_2
incidence: 13/143, conc. range: nc, Ø conc.:
10.6 µg/kg, country: India[247]
see also animal feedstuffs (dairy cake), bird
food, bird food, wild, biri testa, blackgram,
blackgram husk, cottonseed, cottonseed cake,
cottonseed extract, cottonseed meal,
cottonseed meal (ammoniated), cottonseed
meats, dog food, egg production mixed feed,
feed, compound, feed (cat), feed (cattle), feed
(dog), feed (pig), feed (poultry), feed
(rabbit), feed (sheep), fish meal, horsegram,
maize, maize gluten, maize husk, maize,
ground, maize, preharvest, mung testa,
mustard cake, niger cake, peanut cake,
peanut cake (deoiled), peanut expeller,
peanut hay, peanut meal, peanuts, peanut,
kernels, redgram husk, rice bran, rice chaff,
rice meal, rice (polish), sal seed cake, sesame

cake, sorghum, soybean meal, soybeans,
wheat, wheat bran

Aflatoxin
incidence: 48/143, conc. range: 2–421 µg/kg,
country: India[247]
see also blackgram husk, bread crumbs,
broiler finisher, broiler starter, cotton cake,
cottonseed, cottonseed cake, cottonseed
extract, cottonseed meal, feed, feed (cattle),
feed (cow), feed (dog), feed (horse), feed
(maize, gluten), feed (pig), feed (poultry),
feed (rabbit), feed (rat/mice), feed (sheep),
feeds, grain, fish meal, flour (wheat),
groundnut cake, grower's mash, horsegram,
layer's mash, maize, maize gluten, maize,
white, milo, peanut cake, peanut cake
(deoiled), peanut, kernels, peanut (oil cake),
pearlmillet, pig breeder's mash, pig finisher,
pig starter, pod with haulms, poultry
breeder's mash, rabbit pellets, redgram husk,
rice, rice, broken, rice (polish), sesame cake,
silk worm pupae, sorghum, soybean cake,
soybean meal, wheat, wheat bran

Rice chaff   may contain the following
mycotoxins:

Aflatoxin B_1
incidence: 1/1, conc.: 16 µg/kg, country:
India[129]
see also alfalfa, *Ambadi* cake, animal feedstuffs
(dairy cake), bagasse, barley, bengalgram
husk, bird food, bird food, wild, biri testa,
blackgram, blackgram husk, bran, broiler
mixed feed, calf fattening mixed feed, calf
fattening mixed feed (containing 4–20 %
peanut products), *Carthamus* cake, castor
cake, cereals, cereal products, chick pea,
coconut cake, cocos, concentrate, mixed,
concentrates, cotton cake, cottonseed,
cottonseed (dehulled), cottonseed cake,
cottonseed extract, cottonseed meal,
cottonseed meal (ammoniated), cottonseed
meal (decorticated), cottonseed meats,
cottonseed products, crumbles, crumbles,
grower, cycad meal, dairy cattle feed, dairy
cattle feed (containing 2–5 % peanut
products), dairy cattle feed (containing

6–10 % peanut products), dairy cattle feed (containing 6–12 % peanut products), dairy cattle feed (containing more than 20 % peanut products), diets, mixed, dog food, egg production mixed feed, feed, feed and ingredients, feed, compound, feed, layer, feed, mixed, feed (beef), feed (broilers), feed (calf), feed (cat), feed (cattle), feed (chicken), feed (dairy), feed (dog), feed (dug), feed (fish), feed (gluten), feed (horse), feed (miscellaneous), feed (pig), feed (poultry), feed (poultry, pig), feed (rabbit), feed (sheep), feed (50–60 % maize), fish meal, grain by-products, grains (no specification), greengram, hay/silage, horsegram, husk, *Jagni* cake, legume mixture, linseed, linseed cake, livol, *Mahua* cake, maize, maize germ, maize gluten, maize grits, maize husk, maize meal, maize oil cake, maize powder, maize screenings, maize, ground, maize, hybrid, maize, preharvest, maize, yellow, maize (dark grains), *Makhana* (*Euryale ferox* Salisb) puffs, manioc, milk production mixed feed, mung testa, murkool, mustard cake, neem cake, niger cake, oats, palm kernel expeller cake, palm kernels, palm products, peanut cake, peanut cake (deoiled), peanut expeller, peanut hay, peanut meal, peanut, kernels, peanut, shells, peanuts, pellets, finisher, pig meal and pellets, pigeon pea, poultry feeds (peanut containing), rapeseed cake, redgram husk, rice, rice bran, rice bran (deoiled), rice crack, rice germ, rice germ cake, rice meal, rice straw, rice (damaged), rice (polish), safflower cake, sal seed cake, sesame, sesame cake, sorghum, soybean meal, soybeans, sunflower, sunflower cake, sunflower flour, tapioca, wheat, wheat bran, wheat bran and chana testa

AFLATOXIN B$_2$
incidence: 1/1, conc.: 3 µg/kg, country: India[129]
see also animal feedstuffs (dairy cake), bird food, bird food, wild, biri testa, blackgram, blackgram husk, cottonseed, cottonseed cake, cottonseed extract, cottonseed meal, cottonseed meal (ammoniated), cottonseed meats, dog food, egg production mixed feed, feed, compound, feed (cat), feed (cattle), feed

(dog), feed (pig), feed (poultry), feed (rabbit), feed (sheep), fish meal, horsegram, maize, maize gluten, maize husk, maize, ground, maize, preharvest, mung testa, mustard cake, niger cake, peanut cake, peanut cake (deoiled), peanut expeller, peanut hay, peanut meal, peanuts, peanut, kernels, redgram husk, rice bran, rice bran (deoiled), rice meal, rice (polish), sal seed cake, sesame cake, sorghum, soybean meal, soybeans, wheat, wheat bran

Rice crack   may contain the following mycotoxins:

AFLATOXIN B$_1$
incidence: 3/5, Ø conc.: 10 µg/kg, country: Egypt[16]
see also alfalfa, *Ambadi* cake, animal feedstuffs (dairy cake), bagasse, barley, bengalgram husk, bird food, bird food, wild, biri testa, blackgram, blackgram husk, bran, broiler mixed feed, calf fattening mixed feed, calf fattening mixed feed (containing 4–20 % peanut products), *Carthamus* cake, castor cake, cereals, cereal products, chick pea, coconut cake, cocos, concentrate, mixed, concentrates, cotton cake, cottonseed, cottonseed (dehulled), cottonseed cake, cottonseed extract, cottonseed meal, cottonseed meal (ammoniated), cottonseed meal (decorticated), cottonseed meats, cottonseed products, crumbles, crumbles, grower, cycad meal, dairy cattle feed, dairy cattle feed (containing 2–5 % peanut products), dairy cattle feed (containing 6–10 % peanut products), dairy cattle feed (containing 6–12 % peanut products), dairy cattle feed (containing more than 20 % peanut products), diets, mixed, dog food, egg production mixed feed, feed, feed and ingredients, feed, compound, feed, layer, feed, mixed, feed (beef), feed (broilers), feed (calf), feed (cat), feed (cattle), feed (chicken), feed (dairy), feed (dog), feed (dug), feed (fish), feed (gluten), feed (horse), feed (miscellaneous), feed (pig), feed (poultry), feed (poultry, pig), feed (rabbit), feed (sheep), feed (50–60 % maize), fish meal,

grain by-products, grains (no specification), greengram, hay/silage, horsegram, husk, *Jagni* cake, legume mixture, linseed, linseed cake, livol, *Mahua* cake, maize, maize germ, maize gluten, maize grits, maize husk, maize meal, maize oil cake, maize powder, maize screenings, maize, ground, maize, hybrid, maize, preharvest, maize, yellow, maize (dark grains), *Makhana* (*Euryale ferox* Salisb) puffs, manioc, milk production mixed feed, mung testa, murkool, mustard cake, neem cake, niger cake, oats, palm kernel expeller cake, palm kernels, palm products, peanut cake, peanut cake (deoiled), peanut expeller, peanut hay, peanut meal, peanut, kernels, peanut, shells, peanuts, pellets, finisher, pig meal and pellets, pigeon pea, poultry feeds (peanut containing), rapeseed cake, redgram husk, rice, rice bran, rice bran (deoiled), rice chaff, rice germ, rice germ cake, rice meal, rice straw, rice (damaged), rice (polish), safflower cake, sal seed cake, sesame, sesame cake, sorghum, soybean meal, soybeans, sunflower, sunflower cake, sunflower flour, tapioca, wheat, wheat bran, wheat bran and chana testa

Rice germ   may contain the following mycotoxins:

AFLATOXIN B$_1$
incidence: 3/5, Ø conc.: 10 μg/kg, country: Egypt[16]
see also alfalfa, *Ambadi* cake, animal feedstuffs (dairy cake), bagasse, barley, bengalgram husk, bird food, bird food, wild, biri testa, blackgram, blackgram husk, bran, broiler mixed feed, calf fattening mixed feed, calf fattening mixed feed (containing 4–20 % peanut products), *Carthamus* cake, castor cake, cereals, cereal products, chick pea, coconut cake, cocos, concentrate, mixed, concentrates, cotton cake, cottonseed, cottonseed (dehulled), cottonseed cake, cottonseed extract, cottonseed meal, cottonseed meal (ammoniated), cottonseed meal (decorticated), cottonseed meats, cottonseed products, crumbles, crumbles, grower, cycad meal, dairy cattle feed, dairy

cattle feed (containing 2–5 % peanut products), dairy cattle feed (containing 6–10 % peanut products), dairy cattle feed (containing 6–12 % peanut products), dairy cattle feed (containing more than 20 % peanut products), diets, mixed, dog food, egg production mixed feed, feed, feed and ingredients, feed, compound, feed, layer, feed, mixed, feed (beef), feed (broilers), feed (calf), feed (cat), feed (cattle), feed (chicken), feed (dairy), feed (dog), feed (dug), feed (fish), feed (gluten), feed (horse), feed (miscellaneous), feed (pig), feed (poultry), feed (poultry, pig), feed (rabbit), feed (sheep), feed (50–60 % maize), fish meal, grain by-products, grains (no specification), greengram, hay/silage, horsegram, husk, *Jagni* cake, legume mixture, linseed, linseed cake, livol, *Mahua* cake, maize, maize germ, maize gluten, maize grits, maize husk, maize meal, maize oil cake, maize powder, maize screenings, maize, ground, maize, hybrid, maize, preharvest, maize, yellow, maize (dark grains), *Makhana* (*Euryale ferox* Salisb) puffs, manioc, milk production mixed feed, mung testa, murkool, mustard cake, neem cake, niger cake, oats, palm kernel expeller cake, palm kernels, palm products, peanut cake, peanut cake (deoiled), peanut expeller, peanut hay, peanut meal, peanut, kernels, peanut, shells, peanuts, pellets, finisher, pig meal and pellets, pigeon pea, poultry feeds (peanut containing), rapeseed cake, redgram husk, rice, rice bran, rice bran (deoiled), rice chaff, rice crack, rice germ cake, rice meal, rice straw, rice (damaged), rice (polish), safflower cake, sal seed cake, sesame, sesame cake, sorghum, soybean meal, soybeans, sunflower, sunflower cake, sunflower flour, tapioca, wheat, wheat bran, wheat bran and chana testa

AFLATOXIN G$_1$
incidence: 3/5, Ø conc.: 10 μg/kg, country: Egypt[16]
see also animal feedstuffs (dairy cake), bird food, bird food, wild, concentrate, mixed, cottonseed, cottonseed cake, feed (cat), feed (chicken), feed (dog), maize, maize, ground,

maize, preharvest, meat meal, milk production mixed feed, murkool, peanut cake, peanut expeller, peanut hay, peanut meal, peanuts, rice bran, sorghum, soybeans, wheat, wheat bran, wheat bran and chana testa

CITRININ
incidence: 1/4, conc.: 4 µg/kg, country: Egypt[16]
see also barley, barley, oats, barley-soybean diet, cottonseed cake, feed, feed, mixed, feed (cattle), feed (pig), fish meal, hay, maize, maize, white, *Makhana* (*Euryale ferox* Salisb) puffs, oats, palm products, peas and beans, rice bran, wheat, wheat and other grains (moldy), wheat bran

OCHRATOXIN A
incidence: 1/3, Ø conc.: 577 µg/kg, country: Egypt[16]
see also alfalfa, barley, barley, oats, barley (high moisture), barley-soybean diet, bird food, domestic, bird food, wild, broilers feed, cereal grains, citrus pulp, coconut, expeller, corn cob mix silage, diet (dairy cow), diet (poultry), diet (starter), dog food, eat, egg production mixed feed, feed, feed wheat, oat and barley, feed, commercial mix, feed, mixed, feed, mixed (pelleted), feed (broilers), feed (cat), feed (cattle), feed (cereals), feed (pig), feed ec (pig), feed (poultry), feed ec (poultry), feed (poultry, pig), feed ec (rabbit), feed (trout), grain, mixed feed, grains, mixed, grains (heated), hay, horse bean, maize, maize feed, milo, maize gluten, maize meal, maize, white, *Makhana* (*Euryale ferox* Salisb) puffs, milk production mixed feed, millet, oat and barley (hammer-milled), oats, palm products, peanut cake, peas, peas and beans, pet food, pig feedstuffs, pig grower diet, pig meal, piglet diet, poultry feedstuffs, rice bran, rice germ cake, rye, sorghum, soybean groats, sunflower, sunflower seeds, extracted, tapioca, triticale, *Vicia faba*, wheat, wheat and barley, wheat bran, wheat hay, wheat, oats

ZEARALENONE
incidence: 2/4, conc. range: 24–38 µg/kg, Ø conc.: 31 µg/kg, country: Egypt[16]

see also alfalfa, barley, barley, husked, barley, unhusked (naked), barley and feed, bone meal, bran, broilers feed, *Carthamus* cake, chick pea, concentrate, mixed, corn cob mix silage, cotton cake, cottonseed, cottonseed cake, diet (dairy cow), diet (poultry), diets (mixed), feed, feed components, feed, mixed, feed, mixed (primarily maize also maize, oats, wheat), feed (bran), feed (broiler chicken), feed (cattle), feed (chicken), feed (dairy), feed (developing pig), feed (mill run, from wheat), feed (miscellaneous), feed (pig), feed (poultry), feed (poultry, pig), feed (starter chicken), feedstuff, fish meal, forage grass, grain, bruised, grain, mixed feed, grains (no specification), hay, maize, maize ears, maize flakes, maize germ, maize germ/bran, maize gluten, maize kernels, maize meal, maize oil cake, maize screenings, maize stalks (pith), maize, "Baby", maize, hybrid, maize, shelled, maize, unshelled, maize, white, maize grain, artificially dried, maize grain, crib dried, maize grain, ensiled, milk production mixed feed, oats, *Paspalum palidosum*, straw, peanut hulls/skins, rice bran, rye, silage, sorghum, soybeans, soybeans, extracted, sunflower cake, tapioca, triticale, wheat, wheat bran, wheat bran and chana testa, wheat soya meal

Rice germ cake   may contain the following mycotoxins:

AFLATOXIN B$_1$
incidence: 3/5, Ø conc.: 25 µg/kg, country: Egypt[16]
see also alfalfa, *Ambadi* cake, animal feedstuffs (dairy cake), bagasse, barley, bengalgram husk, bird food, bird food, wild, biri testa, blackgram, blackgram husk, bran, broiler mixed feed, calf fattening mixed feed, calf fattening mixed feed (containing 4–20 % peanut products), *Carthamus* cake, castor cake, cereals, cereal products, chick pea, coconut cake, cocos, concentrate, mixed, concentrates, cotton cake, cottonseed, cottonseed (dehulled), cottonseed cake, cottonseed extract, cottonseed meal, cottonseed meal (ammoniated), cottonseed meal (decorticated), cottonseed meats, cottonseed

products, crumbles, crumbles, grower, cycad meal, dairy cattle feed, dairy cattle feed (containing 2–5 % peanut products), dairy cattle feed (containing 6–10 % peanut products), dairy cattle feed (containing 6–12 % peanut products), dairy cattle feed (containing more than 20 % peanut products), diets, mixed, dog food, egg production mixed feed, feed, feed and ingredients, feed, compound, feed, layer, feed, mixed, feed (beef), feed (broilers), feed (calf), feed (cat), feed (cattle), feed (chicken), feed (dairy), feed (dog), feed (dug), feed (fish), feed (gluten), feed (horse), feed (miscellaneous), feed (pig), feed (poultry), feed (poultry, pig), feed (rabbit), feed (sheep), feed (50–60 % maize), fish meal, grain by-products, grains (no specification), greengram, hay/silage, horsegram, husk, *Jagni* cake, legume mixture, linseed, linseed cake, livol, *Mahua* cake, maize, maize germ, maize gluten, maize grits, maize husk, maize meal, maize oil cake, maize powder, maize screenings, maize, ground, maize, hybrid, maize, preharvest, maize, yellow, maize (dark grains), *Makhana* (*Euryale ferox* Salisb) puffs, manioc, milk production mixed feed, mung testa, murkool, mustard cake, neem cake, niger cake, oats, palm kernel expeller cake, palm kernels, palm products, peanut cake, peanut cake (deoiled), peanut expeller, peanut hay, peanut meal, peanut, kernels, peanut, shells, peanuts, pellets, finisher, pig meal and pellets, pigeon pea, poultry feeds (peanut containing), rapeseed cake, redgram husk, rice, rice bran, rice bran (deoiled), rice chaff, rice crack, rice germ, rice meal, rice straw, rice (damaged), rice (polish), safflower cake, sal seed cake, sesame, sesame cake, sorghum, soybean meal, soybeans, sunflower, sunflower cake, sunflower flour, tapioca, wheat, wheat bran, wheat bran and chana testa

DEOXYNIVALENOL
incidence: 2/4, conc. range: 200–264 µg/kg, Ø conc.: 232 µg/kg, country: Egypt[16]
see also barley, barley, husked, barley, unhusked (naked), barley (pressed), bone meal, bran, broilers feed, calf fattening mixed

feed, coconut, expeller, corn cob mix silage, cottonseed, cottonseed cake, dairy cattle feed, egg production mixed feed, feed, feed components, feed, commercial mix, feed, mixed, feed, mixed (primarily maize), feed (barley), feed (cattle), feed (chicken), feed (dog), feed (fish), feed (mill run, from wheat), feed (mink), feed (pig), feed (poultry), feed (reindeer), feeds, grain, feeds, industrial, feedstuff, feedstuffs (rapeseed, turnip, fish meal, concentrates), fish meal, grain, mixed feed, grains, mixed, grains (no specification), maize, maize ears, maize fibre, maize germ, maize germ/bran, maize germ meal, maize gluten, maize kernels, maize meal, maize powder, maize screenings, maize stalks (pith), maize, "Baby", maize, hybrid, maize, white, oats, rice bran, rye, silage, sorghum, soybeans, triticale, wheat, wheat and barley, wheat, red hard winter, wheat, soft white winter, wheat, spring, wheat, winter

OCHRATOXIN A
incidence: 1/3*, Ø conc.: 4 µg/kg, country: Egypt[16], *suspected but identity of peak not unequivocally confirmed
see also alfalfa, barley, barley, oats, barley (high moisture), barley-soybean diet, bird food, domestic, bird food, wild, broilers feed, cereal grains, citrus pulp, coconut, expeller, corn cob mix silage, diet (dairy cow), diet (poultry), diet (starter), dog food, eat, egg production mixed feed, feed wheat, oat and barley, feed, commercial mix, feed, mixed, feed, mixed (pelleted), feed (broilers), feed (cat), feed (cattle), feed (cereals), feed (pig), feed ec (pig), feed (poultry), feed ec (poultry), feed (poultry, pig), feed ec (rabbit), feed (trout), grain, mixed feed, grains, mixed, grains (heated), hay, horse bean, maize, maize feed, milo, maize gluten, maize meal, maize, white, *Makhana* (*Euryale ferox* Salisb) puffs, milk production mixed feed, millet, oat and barley (hammer-milled), oats, palm products, peanut cake, peas, peas and beans, pet food, pig feedstuffs, pig grower diet, pig meal, piglet diet, poultry feedstuffs, rice bran, rice germ, rye, sorghum, soybean groats,

sunflower, sunflower seeds, extracted, tapioca, triticale, *Vicia faba*, wheat, wheat and barley, wheat bran, wheat hay, wheat, oats

Rice meal   Rice meal for feed may contain the following mycotoxins:

AFLATOXIN B_1
incidence: 8/22, conc. range: 1–52.8 µg/kg, Ø conc.: 20.5 µg/kg, country: Colombia[125]
see also alfalfa, *Ambadi* cake, animal feedstuffs (dairy cake), bagasse, barley, bengalgram husk, bird food, bird food, wild, biri testa, blackgram, blackgram husk, bran, broiler mixed feed, calf fattening mixed feed, calf fattening mixed feed (containing 4–20 % peanut products), *Carthamus* cake, castor cake, cereals, cereal products, chick pea, coconut cake, cocos, concentrate, mixed, concentrates, cotton cake, cottonseed, cottonseed (dehulled), cottonseed cake, cottonseed extract, cottonseed meal, cottonseed meal (ammoniated), cottonseed meal (decorticated), cottonseed meats, cottonseed products, crumbles, crumbles, grower, cycad meal, dairy cattle feed, dairy cattle feed (containing 2–5 % peanut products), dairy cattle feed (containing 6–10 % peanut products), dairy cattle feed (containing 6–12 % peanut products), dairy cattle feed (containing more than 20 % peanut products), diets, mixed, dog food, egg production mixed feed, feed, feed and ingredients, feed, compound, feed, layer, feed, mixed, feed (beef), feed (broilers), feed (calf), feed (cat), feed (cattle), feed (chicken), feed (dairy), feed (dog), feed (dug), feed (fish), feed (gluten), feed (horse), feed (miscellaneous), feed (pig), feed (poultry), feed (poultry, pig), feed (rabbit), feed (sheep), feed (50–60 % maize), fish meal, grain by-products, grains (no specification), greengram, hay/silage, horsegram, husk, *Jagni* cake, legume mixture, linseed, linseed cake, livol, *Mahua* cake, maize, maize germ, maize gluten, maize grits, maize husk, maize meal, maize oil cake, maize powder, maize screenings, maize, ground, maize, hybrid, maize, preharvest, maize, yellow, maize (dark grains), *Makhana* (*Euryale ferox* Salisb) puffs, manioc, milk production mixed feed, mung testa, murkool, mustard cake, neem cake, niger cake, oats, palm kernel expeller cake, palm kernels, palm products, peanut cake, peanut cake (deoiled), peanut expeller, peanut hay, peanut meal, peanut, kernels, peanut, shells, peanuts, pellets, finisher, pig meal and pellets, pigeon pea, poultry feeds (peanut containing), rapeseed cake, redgram husk, rice, rice bran, rice bran (deoiled), rice chaff, rice crack, rice germ, rice germ cake, rice straw, rice (damaged), rice (polish), safflower cake, sal seed cake, sesame, sesame cake, sorghum, soybean meal, soybeans, sunflower, sunflower cake, sunflower flour, tapioca, wheat, wheat bran, wheat bran and chana testa

AFLATOXIN B_2
incidence: 4/22, conc. range: 1.1–3.3 µg/kg, Ø conc.: 2.4 µg/kg, country: Colombia[125]
see also animal feedstuffs (dairy cake), bird food, bird food, wild, biri testa, blackgram, blackgram husk, cottonseed, cottonseed cake, cottonseed extract, cottonseed meal, cottonseed meal (ammoniated), cottonseed meats, dog food, egg production mixed feed, feed, compound, feed (cat), feed (cattle), feed (dog), feed (pig), feed (poultry), feed (rabbit), feed (sheep), fish meal, horsegram, maize, maize gluten, maize husk, maize, ground, maize, preharvest, mung testa, mustard cake, niger cake, peanut cake, peanut cake (deoiled), peanut expeller, peanut hay, peanut meal, peanuts, peanut, kernels, redgram husk, rice bran, rice bran (deoiled), rice chaff, rice (polish), sal seed cake, sesame cake, sorghum, soybean meal, soybeans, wheat, wheat bran

Rice straw   may contain the following mycotoxins:

AFLATOXIN B_1
incidence: 3/35, conc. range: 3–4 µg/kg, country: India, Bangladesh[129]
see also alfalfa, *Ambadi* cake, animal feedstuffs (dairy cake), bagasse, barley, bengalgram

husk, bird food, bird food, wild, biri testa,
blackgram, blackgram husk, bran, broiler
mixed feed, calf fattening mixed feed, calf
fattening mixed feed (containing 4–20 %
peanut products), *Carthamus* cake, castor
cake, cereals, cereal products, chick pea,
coconut cake, cocos, concentrate, mixed,
concentrates, cotton cake, cottonseed,
cottonseed (dehulled), cottonseed cake,
cottonseed extract, cottonseed meal,
cottonseed meal (ammoniated), cottonseed
meal (decorticated), cottonseed meats,
cottonseed products, crumbles, crumbles,
grower, cycad meal, dairy cattle feed, dairy
cattle feed (containing 2–5 % peanut
products), dairy cattle feed (containing
6–10 % peanut products), dairy cattle feed
(containing 6–12 % peanut products), dairy
cattle feed (containing more than 20 %
peanut products), diets, mixed, dog food, egg
production mixed feed, feed, feed and
ingredients, feed, compound, feed, layer,
feed, mixed, feed (beef), feed (broilers), feed
(calf), feed (cat), feed (cattle), feed (chicken),
feed (dairy), feed (dog), feed (dug), feed
(fish), feed (gluten), feed (horse), feed
(miscellaneous), feed (pig), feed (poultry),
feed (poultry, pig), feed (rabbit), feed
(sheep), feed (50–60 % maize), fish meal,
grain by-products, grains (no specification),
greengram, hay/silage, horsegram, husk, *Jagni*
cake, legume mixture, linseed, linseed cake,
livol, *Mahua* cake, maize, maize germ, maize
gluten, maize grits, maize husk, maize meal,
maize oil cake, maize powder, maize
screenings, maize, ground, maize, hybrid,
maize, preharvest, maize, yellow, maize (dark
grains), *Makhana* (*Euryale ferox* Salisb) puffs,
manioc, milk production mixed feed, mung
testa, murkool, mustard cake, neem cake,
niger cake, oats, palm kernel expeller cake,
palm kernels, palm products, peanut cake,
peanut cake (deoiled), peanut expeller,
peanut hay, peanut meal, peanut, kernels,
peanut, shells, peanuts, pellets, finisher, pig
meal and pellets, pigeon pea, poultry feeds
(peanut containing), rapeseed cake, redgram
husk, rice, rice bran, rice bran (deoiled), rice

chaff, rice crack, rice germ, rice germ cake,
rice meal, rice (damaged), rice (polish),
safflower cake, sal seed cake, sesame, sesame
cake, sorghum, soybean meal, soybeans,
sunflower, sunflower cake, sunflower flour,
tapioca, wheat, wheat bran, wheat bran and
chana testa

Rice, broken   may contain the following
mycotoxins:

Aflatoxin B
incidence: ?/5, conc. range: 14–20 µg/kg,
Ø conc.: 17.7 µg/kg, country: Pakistan[162]
see also barley, cocoa (oilcake), cottonseed
cake, feed, feed (cattle), feed (maize, gluten),
feed (pig), feed (poultry), flour (wheat),
lucern (dried), maize, maize gluten, maize
grains, oats, peanut (oil cake), sorghum,
soybean (oil cake), sunflower (oilcake), wheat

Aflatoxin G
incidence: ?/5, conc. range: 5–25 µg/kg,
Ø conc.: 17.3 µg/kg, country: Pakistan[162]
see also cottonseed cake, feed (maize, gluten),
maize gluten, maize grains

Aflatoxin
incidence: 2/6, conc. range: ≤125 µg/kg,
country: India[250]
see also blackgram husk, bread crumbs,
broiler finisher, broiler starter, cotton cake,
cottonseed, cottonseed cake, cottonseed
extract, cottonseed meal, feed, feed (cattle),
feed (cow), feed (dog), feed (horse), feed
(maize, gluten), feed (pig), feed (poultry),
feed (rabbit), feed (rat/mice), feed (sheep),
feeds, grain, fish meal, flour (wheat),
groundnut cake, grower's mash, horsegram,
layer's mash, maize, maize gluten, maize,
white, milo, peanut cake, peanut cake
(deoiled), peanut, kernels, peanut (oil cake),
pearlmillet, pig breeder's mash, pig finisher,
pig starter, pod with haulms, poultry
breeder's mash, rabbit pellets, redgram husk,
rice, rice bran (deoiled), rice (polish), sesame
cake, silk worm pupae, sorghum, soybean
cake, soybean meal, wheat, wheat bran

Rice (damaged)   may contain the following mycotoxins:

Aflatoxin B₁
incidence: 2/5, conc. range: 7–107 µg/kg, Ø conc.: 57 µg/kg, country: India[247]
see also alfalfa, *Ambadi* cake, animal feedstuffs (dairy cake), bagasse, barley, bengalgram husk, bird food, bird food, wild, biri testa, blackgram, blackgram husk, bran, broiler mixed feed, calf fattening mixed feed, calf fattening mixed feed (containing 4–20% peanut products), *Carthamus* cake, castor cake, cereals, cereal products, chick pea, coconut cake, cocos, concentrate, mixed, concentrates, cotton cake, cottonseed, cottonseed (dehulled), cottonseed cake, cottonseed extract, cottonseed meal, cottonseed meal (ammoniated), cottonseed meal (decorticated), cottonseed meats, cottonseed products, crumbles, crumbles, grower, cycad meal, dairy cattle feed, dairy cattle feed (containing 2–5% peanut products), dairy cattle feed (containing 6–10% peanut products), dairy cattle feed (containing 6–12% peanut products), dairy cattle feed (containing more than 20% peanut products), diets, mixed, dog food, egg production mixed feed, feed, feed and ingredients, feed, compound, feed, layer, feed, mixed, feed (beef), feed (broilers), feed (calf), feed (cat), feed (cattle), feed (chicken), feed (dairy), feed (dog), feed (dug), feed (fish), feed (gluten), feed (horse), feed (miscellaneous), feed (pig), feed (poultry), feed (poultry, pig), feed (rabbit), feed (sheep), feed (50–60% maize), fish meal, grain by-products, grains (no specification), greengram, hay/silage, horsegram, husk, *Jagni* cake, legume mixture, linseed, linseed cake, livol, *Mahua* cake, maize, maize germ, maize gluten, maize grits, maize husk, maize meal, maize oil cake, maize powder, maize screenings, maize, ground, maize, hybrid, maize, preharvest, maize, yellow, maize (dark grains), *Makhana* (*Euryale ferox* Salisb) puffs, manioc, milk production mixed feed, mung testa, murkool, mustard cake, neem cake, niger cake, oats, palm kernel expeller cake, palm kernels, palm products, peanut cake, peanut cake (deoiled), peanut expeller, peanut hay, peanut meal, peanut, kernels, peanut, shells, peanuts, pellets, finisher, pig meal and pellets, pigeon pea, poultry feeds (peanut containing), rapeseed cake, redgram husk, rice, rice bran, rice bran (deoiled), rice chaff, rice crack, rice germ, rice germ cake, rice meal, rice straw, rice (polish), safflower cake, sal seed cake, sesame, sesame cake, sorghum, soybean meal, soybeans, sunflower, sunflower cake, sunflower flour, tapioca, wheat, wheat bran, wheat bran and chana testa

Rice (polish)   for feed may contain the following mycotoxins:

Aflatoxin B₁
incidence: 41/168, conc. range: ≤25 µg/kg (25 sa), 26–50 µg/kg (7 sa), 51–100 µg/kg (3 sa), 101–200 µg/kg (3 sa), 201–500 µg/kg (2 sa), 501–1000 µg/kg (1 sa), Ø conc.: 58.9 µg/kg, country: India[247]
incidence: 1/?*, conc.: 8 µg/kg, country: Japan[313], *from Egypt
see also alfalfa, *Ambadi* cake, animal feedstuffs (dairy cake), bagasse, barley, bengalgram husk, bird food, bird food, wild, biri testa, blackgram, blackgram husk, bran, broiler mixed feed, calf fattening mixed feed, calf fattening mixed feed (containing 4–20% peanut products), *Carthamus* cake, castor cake, cereals, cereal products, chick pea, coconut cake, cocos, concentrate, mixed, concentrates, cotton cake, cottonseed, cottonseed (dehulled), cottonseed cake, cottonseed extract, cottonseed meal, cottonseed meal (ammoniated), cottonseed meal (decorticated), cottonseed meats, cottonseed products, crumbles, crumbles, grower, cycad meal, dairy cattle feed, dairy cattle feed (containing 2–5% peanut products), dairy cattle feed (containing 6–10% peanut products), dairy cattle feed (containing 6–12% peanut products), dairy cattle feed (containing more than 20% peanut products), diets, mixed, dog food, egg production mixed feed, feed, feed and

ingredients, feed, compound, feed, layer, feed, mixed, feed (beef), feed (broilers), feed (calf), feed (cat), feed (cattle), feed (chicken), feed (dairy), feed (dog), feed (dug), feed (fish), feed (gluten), feed (horse), feed (miscellaneous), feed (pig), feed (poultry), feed (poultry, pig), feed (rabbit), feed (sheep), feed (50–60 % maize), fish meal, grain by-products, grains (no specification), greengram, hay/silage, horsegram, husk, *Jagni* cake, legume mixture, linseed, linseed cake, livol, *Mahua* cake, maize, maize germ, maize gluten, maize grits, maize husk, maize meal, maize oil cake, maize powder, maize screenings, maize, ground, maize, hybrid, maize, preharvest, maize, yellow, maize (dark grains), *Makhana* (*Euryale ferox* Salisb) puffs, manioc, milk production mixed feed, mung testa, murkool, mustard cake, neem cake, niger cake, oats, palm kernel expeller cake, palm kernels, palm products, peanut cake, peanut cake (deoiled), peanut expeller, peanut hay, peanut meal, peanut, kernels, peanut, shells, peanuts, pellets, finisher, pig meal and pellets, pigeon pea, poultry feeds (peanut containing), rapeseed cake, redgram husk, rice, rice bran, rice bran (deoiled), rice chaff, rice crack, rice germ, rice germ cake, rice meal, rice straw, rice (damaged), safflower cake, sal seed cake, sesame, sesame cake, sorghum, soybean meal, soybeans, sunflower, sunflower cake, sunflower flour, tapioca, wheat, wheat bran, wheat bran and chana testa

AFLATOXIN B$_2$
incidence: 9/168, conc. range: nc, Ø conc.: 44.5 μg/kg, country: India[247]
incidence: 1/?*, conc.: 2 μg/kg, country: Japan[313], *from Egypt
see also animal feedstuffs (dairy cake), bird food, bird food, wild, biri testa, blackgram, blackgram husk, cottonseed, cottonseed cake, cottonseed extract, cottonseed meal, cottonseed meal (ammoniated), cottonseed meats, dog food, egg production mixed feed, feed, compound, feed (cat), feed (cattle), feed (dog), feed (pig), feed (poultry), feed (rabbit), feed (sheep), fish meal, horsegram,

maize, maize gluten, maize husk, maize, ground, maize, preharvest, mung testa, mustard cake, niger cake, peanut cake, peanut cake (deoiled), peanut expeller, peanut hay, peanut meal, peanuts, peanut, kernels, redgram husk, rice bran, rice bran (deoiled), rice chaff, rice meal, sal seed cake, sesame cake, sorghum, soybean meal, soybeans, wheat, wheat bran

AFLATOXIN
incidence: 41/168, conc. range: 3–782 μg/kg, country: India[247]
incidence: ?/12, conc. range: ≤200 μg/kg, Ø conc.: 23 μg/kg, country: India[250]
see also blackgram husk, bread crumbs, broiler finisher, broiler starter, cotton cake, cottonseed, cottonseed cake, cottonseed extract, cottonseed meal, feed, feed (cattle), feed (cow), feed (dog), feed (horse), feed (maize, gluten), feed (pig), feed (poultry), feed (rabbit), feed (rat/mice), feed (sheep), feeds, grain, fish meal, flour (wheat), groundnut cake, grower's mash, horsegram, layer's mash, maize, maize gluten, maize, white, milo, peanut cake, peanut cake (deoiled), peanut, kernels, peanut (oil cake), pearlmillet, pig breeder's mash, pig finisher, pig starter, pod with haulms, poultry breeder's mash, rabbit pellets, redgram husk, rice, rice bran (deoiled), rice, broken, sesame cake, silk worm pupae, sorghum, soybean cake, soybean meal, wheat, wheat bran

Rye Rye for feed may contain the following mycotoxins:

DEOXYNIVALENOL
incidence: 31/31*, conc. range: <10–≤10,000 μg/kg, country: Germany[192], *ncac
incidence: 17/50* **, conc. range: ≤3090 μg/kg, Ø conc.: 490 μg/kg, country: Germany[355], *ncac, **cg
incidence: 2/19* **, conc. range: 120–130 μg/kg, Ø conc.: 125 μg/kg, country: Germany[355], *ncac, **og
incidence: 14/46*, conc. range: ≤102 μg/kg, Ø conc.: 54 μg/kg, country: Lithuania[399], *ncac

see also barley, barley, husked, barley, unhusked (naked), barley (pressed), bone meal, bran, broilers feed, calf fattening mixed feed, coconut, expeller, corn cob mix silage, cottonseed, cottonseed cake, dairy cattle feed, egg production mixed feed, feed, feed components, feed, commercial mix, feed, mixed, feed, mixed (primarily maize), feed (barley), feed (cattle), feed (chicken), feed (dog), feed (fish), feed (mill run, from wheat), feed (mink), feed (pig), feed (poultry), feed (reindeer), feeds, grain, feeds, industrial, feedstuff, feedstuffs (rapeseed, turnip, fish meal, concentrates), fish meal, grain, mixed feed, grains, mixed, grains (no specification), maize, maize ears, maize fibre, maize germ, maize germ/bran, maize germ meal, maize gluten, maize kernels, maize meal, maize powder, maize screenings, maize stalks (pith), maize, "Baby", maize, hybrid, maize, white, oats, rice bran, rice germ cake, silage, sorghum, soybeans, triticale, wheat, wheat and barley, wheat, red hard winter, wheat, soft white winter, wheat, spring, wheat, winter

Enniatin A$_1$
incidence: 1/1*, conc.: tr μg/kg, country: Finland[205], *ncac
see also barley, oats, wheat, wheat, summer, wheat, winter

Enniatin B
incidence: 1/1*, conc.: 47 μg/kg, country: Finland[205], *ncac
see also barley, oats, wheat, wheat, summer, wheat, winter

Enniatin B$_1$
incidence: 1/1*, conc.: tr μg/kg, country: Finland[205], *ncac
see also barley, oats, wheat, wheat, summer, wheat, winter

HT-2 Toxin
incidence: 12/46*, conc. range: ≤353 μg/kg, Ø conc.: 61 μg/kg, country: Lithuania[399], *ncac
see also barley, feed components, feed (dog), feed (fish), feed (pig), feed (poultry), feed (reindeer), grain, mixed feed, grains, mixed,

grains (no specification), maize, maize germ/bran, maize gluten, maize meal, maize screenings, oats, silage, wheat

Nivalenol
incidence: 1/46*, conc. range: ≤20 μg/kg, Ø conc.: 20 μg/kg, country: Lithuania[399], *ncac
see also barley, barley, husked, barley, unhusked (naked), barley (pressed), bran, feed, feed components, feed (cattle), feed (poultry), feed (reindeer), feeds, industrial, maize, maize ears, maize germ, maize germ/bran, maize gluten, maize kernels, maize meal, maize powder, maize screenings, maize, "Baby", oats, silage, triticale, wheat

Ochratoxin A
incidence: 4/37* **, conc. range: 4.7–8.8 μg/kg, Ø conc.: 6.8 μg/kg, country: Poland[339], *ncac, **cg
incidence: 5/46* **, conc. range: 2–35.3 μg/kg, Ø conc.: 14.5 μg/kg, country: Poland[339], *ncac, **eg
incidence: 3/52* **, conc. range: 0.8–2.6 μg/kg, Ø conc.: 1.4 μg/kg, country: Poland[341], *ncac, **cg
incidence: 18/48* **, conc. range: 0.2–10 μg/kg, Ø conc.: 3.2 μg/kg, country: Poland[341], *ncac, **eg
see also alfalfa, barley, barley, oats, barley (high moisture), barley-soybean diet, bird food, domestic, bird food, wild, broilers feed, cereal grains, citrus pulp, coconut, expeller, corn cob mix silage, diet (dairy cow), diet (poultry), diet (starter), dog food, eat, egg production mixed feed, feed, feed wheat, oat and barley, feed, commercial mix, feed, mixed, feed, mixed (pelleted), feed (broilers), feed (cat), feed (cattle), feed (cereals), feed (pig), feed ec (pig), feed (poultry), feed ec (poultry), feed (poultry, pig), feed ec (rabbit), feed (trout), grain, mixed feed, grains, mixed, grains (heated), hay, horse bean, maize, maize feed, milo, maize gluten, maize meal, maize, white, *Makhana* (*Euryale ferox* Salisb) puffs, milk production mixed feed, millet, oat and barley (hammer-milled), oats, palm products, peanut cake, peas, peas and beans,

pet food, pig feedstuffs, pig grower diet, pig meal, piglet diet, poultry feedstuffs, rice bran, rice germ, rice germ cake, sorghum, soybean groats, sunflower, sunflower seeds, extracted, tapioca, triticale, *Vicia faba*, wheat, wheat and barley, wheat bran, wheat hay, wheat, oats

T-2 TOXIN
incidence: 1/7345, conc.: nc, country: Hungary[209]
incidence: 1/46*, conc. range: ≤52 µg/kg, Ø conc.: 34 µg/kg, country: Lithuania[399], *ncac
see also alfalfa, barley, bran, diet (grower), diet (poultry), feed, feed components, feed, layer, feed, mixed, feed (dog), feed (fish), feed (mink), feed (pig), feed (poultry), feedstuff, forage grass, grain, mixed feed, grains (no specification), hay, maize, maize germ/bran, maize gluten, maize meal, maize screenings, maize stalks (pith), oat and barley (hammer-milled), oats, peanuts, pig grower diet, piglet diet, sorghum, triticale, wheat

ZEARALENONE
incidence: 2/6, conc. range: 7.2–9.3 µg/kg, Ø conc.: 8.2 µg/kg, country: Germany[107]
see also alfalfa, barley, barley, husked, barley, unhusked (naked), barley and feed, bone meal, bran, broilers feed, *Carthamus* cake, chick pea, concentrate, mixed, corn cob mix silage, cotton cake, cottonseed, cottonseed cake, diet (dairy cow), diet (poultry), diets (mixed), feed, feed components, feed, mixed, feed, mixed (primarily maize also maize, oats, wheat), feed (bran), feed (broiler chicken), feed (cattle), feed (chicken), feed (dairy), feed (developing pig), feed (mill run, from wheat), feed (miscellaneous), feed (pig), feed (poultry), feed (poultry, pig), feed (starter chicken), feedstuff, fish meal, forage grass, grain, bruised, grain, mixed feed, grains (no specification), hay, maize, maize ears, maize flakes, maize germ, maize germ/bran, maize gluten, maize kernels, maize meal, maize oil cake, maize screenings, maize stalks (pith), maize, "Baby", maize, hybrid, maize, shelled, maize, unshelled, maize, white, maize grain, artificially dried, maize grain, crib dried,

maize grain, ensiled, milk production mixed feed, oats, *Paspalum palidosum*, straw, peanut hulls/skins, rice bran, rice germ, rice germ cake, silage, sorghum, soybeans, soybeans, extracted, sunflower cake, tapioca, triticale, wheat, wheat bran, wheat bran and chana testa, wheat soya meal

Safflower cake   may contain the following mycotoxins:

AFLATOXIN B₁
incidence: 7/8, conc. range: tr–167.4 µg/kg, Ø conc.: 118.4 µg/kg, country: India[253]
see also alfalfa, *Ambadi* cake, animal feedstuffs (dairy cake), bagasse, barley, bengalgram husk, bird food, bird food, wild, biri testa, blackgram, blackgram husk, bran, broiler mixed feed, calf fattening mixed feed, calf fattening mixed feed (containing 4–20 % peanut products), *Carthamus* cake, castor cake, cereals, cereal products, chick pea, coconut cake, cocos, concentrate, mixed, concentrates, cotton cake, cottonseed, cottonseed (dehulled), cottonseed cake, cottonseed extract, cottonseed meal, cottonseed meal (ammoniated), cottonseed meal (decorticated), cottonseed meats, cottonseed products, crumbles, crumbles, grower, cycad meal, dairy cattle feed, dairy cattle feed (containing 2–5 % peanut products), dairy cattle feed (containing 6–10 % peanut products), dairy cattle feed (containing 6–12 % peanut products), dairy cattle feed (containing more than 20 % peanut products), diets, mixed, dog food, egg production mixed feed, feed, feed and ingredients, feed, compound, feed, layer, feed, mixed, feed (beef), feed (broilers), feed (calf), feed (cat), feed (cattle), feed (chicken), feed (dairy), feed (dog), feed (dug), feed (fish), feed (gluten), feed (horse), feed (miscellaneous), feed (pig), feed (poultry), feed (poultry, pig), feed (rabbit), feed (sheep), feed (50–60 % maize), fish meal, grain by-products, grains (no specification), greengram, hay/silage, horsegram, husk, *Jagni* cake, legume mixture, linseed, linseed cake, livol, *Mahua* cake, maize, maize germ, maize gluten, maize grits, maize husk, maize meal,

maize oil cake, maize powder, maize screenings, maize, ground, maize, hybrid, maize, preharvest, maize, yellow, maize (dark grains), *Makhana* (*Euryale ferox* Salisb) puffs, manioc, milk production mixed feed, mung testa, murkool, mustard cake, neem cake, niger cake, oats, palm kernel expeller cake, palm kernels, palm products, peanut cake, peanut cake (deoiled), peanut expeller, peanut hay, peanut meal, peanut, kernels, peanut, shells, peanuts, pellets, finisher, pig meal and pellets, pigeon pea, poultry feeds (peanut containing), rapeseed cake, redgram husk, rice, rice bran, rice bran (deoiled), rice chaff, rice crack, rice germ, rice germ cake, rice meal, rice straw, rice (damaged), rice (polish), sal seed cake, sesame, sesame cake, sorghum, soybean meal, soybeans, sunflower, sunflower cake, sunflower flour, tapioca, wheat, wheat bran, wheat bran and chana testa

Sal seed cake may contain the following mycotoxins:

AFLATOXIN B$_1$
incidence: 13/21, conc. range:
71.7–180.2 µg/kg, Ø conc.: 105.1 µg/kg,
country: India[253]
see also alfalfa, *Ambadi* cake, animal feedstuffs (dairy cake), bagasse, barley, bengalgram husk, bird food, bird food, wild, biri testa, blackgram, blackgram husk, bran, broiler mixed feed, calf fattening mixed feed, calf fattening mixed feed (containing 4–20 % peanut products), *Carthamus* cake, castor cake, cereals, cereal products, chick pea, coconut cake, cocos, concentrate, mixed, concentrates, cotton cake, cottonseed, cottonseed (dehulled), cottonseed cake, cottonseed extract, cottonseed meal, cottonseed meal (ammoniated), cottonseed meal (decorticated), cottonseed meats, cottonseed products, crumbles, crumbles, grower, cycad meal, dairy cattle feed, dairy cattle feed (containing 2–5 % peanut products), dairy cattle feed (containing 6–10 % peanut products), dairy cattle feed (containing 6–12 % peanut products), dairy cattle feed (containing more than 20 %

peanut products), diets, mixed, dog food, egg production mixed feed, feed, feed and ingredients, feed, compound, feed, layer, feed, mixed, feed (beef), feed (broilers), feed (calf), feed (cat), feed (cattle), feed (chicken), feed (dairy), feed (dog), feed (dug), feed (fish), feed (gluten), feed (horse), feed (miscellaneous), feed (pig), feed (poultry), feed (poultry, pig), feed (rabbit), feed (sheep), feed (50–60 % maize), fish meal, grain by-products, grains (no specification), greengram, hay/silage, horsegram, husk, *Jagni* cake, legume mixture, linseed, linseed cake, livol, *Mahua* cake, maize, maize germ, maize gluten, maize grits, maize husk, maize meal, maize oil cake, maize powder, maize screenings, maize, ground, maize, hybrid, maize, preharvest, maize, yellow, maize (dark grains), *Makhana* (*Euryale ferox* Salisb) puffs, manioc, milk production mixed feed, mung testa, murkool, mustard cake, neem cake, niger cake, oats, palm kernel expeller cake, palm kernels, palm products, peanut cake, peanut cake (deoiled), peanut expeller, peanut hay, peanut meal, peanut, kernels, peanut, shells, peanuts, pellets, finisher, pig meal and pellets, pigeon pea, poultry feeds (peanut containing), rapeseed cake, redgram husk, rice, rice bran, rice bran (deoiled), rice chaff, rice crack, rice germ, rice germ cake, rice meal, rice straw, rice (damaged), rice (polish), safflower cake, sesame, sesame cake, sorghum, soybean meal, soybeans, sunflower, sunflower cake, sunflower flour, tapioca, wheat, wheat bran, wheat bran and chana testa

AFLATOXIN B$_2$
incidence: 5/21, conc. range: 44.5–71.7 µg/kg, Ø conc.: 58.1 µg/kg, country: India[253]
see also animal feedstuffs (dairy cake), bird food, bird food, wild, biri testa, blackgram, blackgram husk, cottonseed, cottonseed cake, cottonseed extract, cottonseed meal, cottonseed meal (ammoniated), cottonseed meats, dog food, egg production mixed feed, feed, compound, feed (cat), feed (cattle), feed (dog), feed (pig), feed (poultry), feed (rabbit), feed (sheep), fish meal, horsegram,

maize, maize gluten, maize husk, maize, ground, maize, preharvest, mung testa, mustard cake, niger cake, peanut cake, peanut cake (deoiled), peanut expeller, peanut hay, peanut meal, peanuts, peanut, kernels, redgram husk, rice bran, rice bran (deoiled), rice chaff, rice meal, rice (polish), sesame cake, sorghum, soybean meal, soybeans, wheat, wheat bran

Sesame   Sesame for feed may contain the following mycotoxins:

AFLATOXIN B$_1$
incidence: 2/3, conc. range: 10 μg/kg,
Ø conc.range: 10 μg/kg, country: investigated in Germany[298]
see also alfalfa, *Ambadi* cake, animal feedstuffs (dairy cake), bagasse, barley, bengalgram husk, bird food, bird food, wild, biri testa, blackgram, blackgram husk, bran, broiler mixed feed, calf fattening mixed feed, calf fattening mixed feed (containing 4–20 % peanut products), *Carthamus* cake, castor cake, cereals, cereal products, chick pea, coconut cake, cocos, concentrate, mixed, concentrates, cotton cake, cottonseed, cottonseed (dehulled), cottonseed cake, cottonseed extract, cottonseed meal, cottonseed meal (ammoniated), cottonseed meal (decorticated), cottonseed meats, cottonseed products, crumbles, crumbles, grower, cycad meal, dairy cattle feed, dairy cattle feed (containing 2–5 % peanut products), dairy cattle feed (containing 6–10 % peanut products), dairy cattle feed (containing 6–12 % peanut products), dairy cattle feed (containing more than 20 % peanut products), diets, mixed, dog food, egg production mixed feed, feed, feed and ingredients, feed, compound, feed, layer, feed, mixed, feed (beef), feed (broilers), feed (calf), feed (cat), feed (cattle), feed (chicken), feed (dairy), feed (dog), feed (dug), feed (fish), feed (gluten), feed (horse), feed (miscellaneous), feed (pig), feed (poultry), feed (poultry, pig), feed (rabbit), feed (sheep), feed (50–60 % maize), fish meal, grain by-products, grains (no specification),

greengram, hay/silage, horsegram, husk, *Jagni* cake, legume mixture, linseed, linseed cake, livol, *Mahua* cake, maize, maize germ, maize gluten, maize grits, maize husk, maize meal, maize oil cake, maize powder, maize screenings, maize, ground, maize, hybrid, maize, preharvest, maize, yellow, maize (dark grains), *Makhana* (*Euryale ferox* Salisb) puffs, manioc, milk production mixed feed, mung testa, murkool, mustard cake, neem cake, niger cake, oats, palm kernel expeller cake, palm kernels, palm products, peanut cake, peanut cake (deoiled), peanut expeller, peanut hay, peanut meal, peanut, kernels, peanut, shells, peanuts, pellets, finisher, pig meal and pellets, pigeon pea, poultry feeds (peanut containing), rapeseed cake, redgram husk, rice, rice bran, rice bran (deoiled), rice chaff, rice crack, rice germ, rice germ cake, rice meal, rice straw, rice (damaged), rice (polish), safflower cake, sal seed cake, sesame cake, sorghum, soybean meal, soybeans, sunflower, sunflower cake, sunflower flour, tapioca, wheat, wheat bran, wheat bran and chana testa

Sesame cake   may contain the following mycotoxins:

AFLATOXIN B$_1$
incidence: 52/107, conc. range: ≤25 μg/kg (24 sa), 26–50 μg/kg (16 sa), 51–100 μg/kg (5 sa), 101–200 μg/kg (5 sa), 201–500 μg/kg (1 sa), 501–1000 μg/kg (1 sa), Ø conc.: 51.2 μg/kg, country: India[247]
incidence: 5/8, conc. range: 59.4–113.8 μg/kg, Ø conc.: 78.3 μg/kg, country: India[253]
see also alfalfa, *Ambadi* cake, animal feedstuffs (dairy cake), bagasse, barley, bengalgram husk, bird food, bird food, wild, biri testa, blackgram, blackgram husk, bran, broiler mixed feed, calf fattening mixed feed, calf fattening mixed feed (containing 4–20 % peanut products), *Carthamus cake*, castor cake, cereals, cereal products, chick pea, coconut cake, cocos, concentrate, mixed, concentrates, cotton cake, cottonseed, cottonseed (dehulled), cottonseed cake, cottonseed extract, cottonseed meal,

cottonseed meal (ammoniated), cottonseed meal (decorticated), cottonseed meats, cottonseed products, crumbles, crumbles, grower, cycad meal, dairy cattle feed, dairy cattle feed (containing 2–5 % peanut products), dairy cattle feed (containing 6–10 % peanut products), dairy cattle feed (containing 6–12 % peanut products), dairy cattle feed (containing more than 20 % peanut products), diets, mixed, dog food, egg production mixed feed, feed, feed and ingredients, feed, compound, feed, layer, feed, mixed, feed (beef), feed (broilers), feed (calf), feed (cat), feed (cattle), feed (chicken), feed (dairy), feed (dog), feed (dug), feed (fish), feed (gluten), feed (horse), feed (miscellaneous), feed (pig), feed (poultry), feed (poultry, pig), feed (rabbit), feed (sheep), feed (50–60 % maize), fish meal, grain by-products, grains (no specification), greengram, hay/silage, horsegram, husk, *Jagni* cake, legume mixture, linseed, linseed cake, livol, *Mahua* cake, maize, maize germ, maize gluten, maize grits, maize husk, maize meal, maize oil cake, maize powder, maize screenings, maize, ground, maize, hybrid, maize, preharvest, maize, yellow, maize (dark grains), *Makhana* (*Euryale ferox* Salisb) puffs, manioc, milk production mixed feed, mung testa, murkool, mustard cake, neem cake, niger cake, oats, palm kernel expeller cake, palm kernels, palm products, peanut cake, peanut cake (deoiled), peanut expeller, peanut hay, peanut meal, peanut, kernels, peanut, shells, peanuts, pellets, finisher, pig meal and pellets, pigeon pea, poultry feeds (peanut containing), rapeseed cake, redgram husk, rice, rice bran, rice bran (deoiled), rice chaff, rice crack, rice germ, rice germ cake, rice meal, rice straw, rice (damaged), rice (polish), safflower cake, sesame, sal seed cake, sorghum, soybean meal, soybeans, sunflower, sunflower cake, sunflower flour, tapioca, wheat, wheat bran, wheat bran and chana testa

Aflatoxin B$_2$
incidence: 8/107, conc. range: nc, Ø conc.: 16.2 µg/kg, country: India[247]

see also animal feedstuffs (dairy cake), bird food, bird food, wild, biri testa, blackgram, blackgram husk, cottonseed, cottonseed cake, cottonseed extract, cottonseed meal, cottonseed meal (ammoniated), cottonseed meats, dog food, egg production mixed feed, feed, compound, feed (cat), feed (cattle), feed (dog), feed (pig), feed (poultry), feed (rabbit), feed (sheep), fish meal, horsegram, maize, maize gluten, maize husk, maize, ground, maize, preharvest, mung testa, mustard cake, niger cake, peanut cake, peanut cake (deoiled), peanut expeller, peanut hay, peanut meal, peanuts, peanut, kernels, redgram husk, rice bran, rice bran (deoiled), rice chaff, rice meal, rice (polish), sal seed cake, sorghum, soybean meal, soybeans, wheat, wheat bran

Aflatoxin
incidence: 52/107, conc. range: 2–527 µg/kg, country: India[247]
see also blackgram husk, bread crumbs, broiler finisher, broiler starter, cotton cake, cottonseed, cottonseed cake, cottonseed extract, cottonseed meal, feed, feed (cattle), feed (cow), feed (dog), feed (horse), feed (maize, gluten), feed (pig), feed (poultry), feed (rabbit), feed (rat/mice), feed (sheep), feeds, grain, fish meal, flour (wheat), groundnut cake, grower's mash, horsegram, layer's mash, maize, maize gluten, maize, white, milo, peanut cake, peanut cake (deoiled), peanut, kernels, peanut (oil cake), pearlmillet, pig breeder's mash, pig finisher, pig starter, pod with haulms, poultry breeder's mash, rabbit pellets, redgram husk, rice, rice bran (deoiled), rice, broken, rice (polish), silk worm pupae, sorghum, soybean cake, soybean meal, wheat, wheat bran

Sheep Feed
see Feed (sheep)

Silage may contain the following mycotoxins:

15-Acetyldeoxynivalenol
incidence: 17/18, conc. range: 20–347 µg/kg, Ø conc.: 84 µg/kg, country: Germany[366]

see also barley, feed components, feed, commercial mix, feed (poultry), maize, maize ears, maize germ, maize germ/bran, maize gluten, maize kernels, maize meal, maize screenings, maize, "Baby", oats, wheat

DEOXYNIVALENOL
incidence: 18/18, conc. range: 323–3510 µg/kg, Ø conc.: 1426 µg/kg, country: Germany[366]
see also barley, barley, husked, barley, unhusked (naked), barley (pressed), bone meal, bran, broilers feed, calf fattening mixed feed, coconut, expeller, corn cob mix silage, cottonseed, cottonseed cake, dairy cattle feed, egg production mixed feed, feed, feed components, feed, commercial mix, feed, mixed, feed, mixed (primarily maize), feed (barley), feed (cattle), feed (chicken), feed (dog), feed (fish), feed (mill run, from wheat), feed (mink), feed (pig), feed (poultry), feed (reindeer), feeds, grain, feeds, industrial, feedstuff, feedstuffs (rapeseed, turnip, fish meal, concentrates), fish meal, grain, mixed feed, grains, mixed, grains (no specification), maize, maize ears, maize fibre, maize germ, maize germ/bran, maize germ meal, maize gluten, maize kernels, maize meal, maize powder, maize screenings, maize stalks (pith), maize, "Baby", maize, hybrid, maize, white, oats, rice bran, rice germ cake, rye, sorghum, soybeans, triticale, wheat, wheat and barley, wheat, red hard winter, wheat, soft white winter, wheat, spring, wheat, winter

DIACETOXYSCIRPENOL
incidence: 1/18, conc.: 64 µg/kg, country: Germany[366]
see also alfalfa, barley, feed, feed (dog), feed (fish), feed (mink), feed (pig), feed (reindeer), feeds, grain, feedstuff, grain, mixed feed, grains (no specification), maize, maize gluten, oat and barley (hammer-milled), oats, peanuts, soybeans, wheat

FUMONISIN B$_1$
incidence: 1/1, conc.: 13 µg/kg, country: USA[58]

see also barley, bird food, wild, dog food, feed, feed, complete ration, feed, general, feed, layer, feed, maize-based, feed, mixed, feed, pelleted ration, feed, screenings, feed, sweet, feed (broilers), feed (cat), feed (chicken), feed (dog), feed (gluten), feed (horse), feed (maize), feed (pig), feed (poultry), feed (rat), feed (rodent), forage grass, maize, maize and maize screenings, maize bran, maize ears, maize fine fractions, maize flakes, maize germ, maize germ/bran, maize germ meal, maize gluten, maize grits, maize kernels, maize meal, maize powder, maize screenings, maize, "Baby", maize, ground, maize, preharvest, maize, sweet feed, maize/oats mix, rat chow, sorghum, soybeans, wheat

HT-2 TOXIN
incidence: 16/18, conc. range: 5–47 µg/kg, Ø conc.: 21 µg/kg, country: Germany[366]
see also barley, feed components, feed (dog), feed (fish), feed (pig), feed (poultry), feed (reindeer), grain, mixed feed, grains, mixed, grains (no specification), maize, maize germ/bran, maize gluten, maize meal, maize screenings, oats, rye, wheat

MONOACETOXYSCIRPENOL
incidence: 8/18, conc. range: 14–51 µg/kg, Ø conc.: 32 µg/kg, country: Germany[366]
see also feed components, maize, maize gluten, maize meal, maize screenings, wheat

MYCOPHENOLIC ACID
incidence: 74/233, conc. range: 20–35,000 µg/kg, Ø conc.: 1400 µg/kg, country: Germany[393]

NIVALENOL
incidence: 15/18, conc. range: 113–2750 µg/kg, Ø conc.: 1049 µg/kg, country: Germany[366]
see also barley, barley, husked, barley, unhusked (naked), barley (pressed), bran, feed, feed components, feed (cattle), feed (poultry), feed (reindeer), feeds, industrial, maize, maize ears, maize germ, maize germ/bran, maize gluten, maize kernels, maize meal, maize powder, maize screenings, maize, "Baby", oats, rye, triticale, wheat

T-2 Tetraol
incidence: 1/18, conc.: 79 µg/kg, country:
Germany[366]
see also barley, feed components, maize,
wheat

Zearalenone
incidence: 18/18, conc. range: 2–1593 µg/kg,
Ø conc.: 116 µg/kg, country: Germany[366]
see also alfalfa, barley, barley, husked, barley,
unhusked (naked), barley and feed, bone
meal, bran, broilers feed, *Carthamus* cake,
chick pea, concentrate, mixed, corn cob mix
silage, cotton cake, cottonseed, cottonseed
cake, diet (dairy cow), diet (poultry), diets
(mixed), feed, feed components, feed, mixed,
feed, mixed (primarily maize also maize, oats,
wheat), feed (bran), feed (broiler chicken),
feed (cattle), feed (chicken), feed (dairy), feed
(developing pig), feed (mill run, from
wheat), feed (miscellaneous), feed (pig), feed
(poultry), feed (poultry, pig), feed (starter
chicken), feedstuff, fish meal, forage grass,
grain, bruised, grain, mixed feed, grains (no
specification), hay, maize, maize ears, maize
flakes, maize germ, maize germ/bran, maize
gluten, maize kernels, maize meal, maize oil
cake, maize screenings, maize stalks (pith),
maize, "Baby", maize, hybrid, maize, shelled,
maize, unshelled, maize, white, maize grain,
artificially dried, maize grain, crib dried,
maize grain, ensiled, milk production mixed
feed, oats, *Paspalum palidosum*, straw, peanut
hulls/skins, rice bran, rice germ, rice germ
cake, rye, sorghum, soybeans, soybeans,
extracted, sunflower cake, tapioca, triticale,
wheat, wheat bran, wheat bran and chana
testa, wheat soya meal

Silage (grass)   may contain the following
mycotoxins:

Roquefortine C
incidence: 9/12*, conc. range:
200–15,000 µg/kg, Ø conc.: 5633 µg/kg,
country: Germany[74], *molded
incidence: 5/12, conc. range: 100–300 µg/kg,
Ø conc.: 160 µg/kg, country: Germany[74]
see also silage (maize)

Silage (maize)   may contain the following
mycotoxins:

Roquefortine C
incidence: 12/12*, conc. range:
700–36,000 µg/kg, Ø conc.: 17,058 µg/kg,
country: Germany[74], *molded
incidence: 1/12, conc.: 200 µg/kg, country:
Germany[74]
see also silage (grass)

Silk worm pupae   may contain the following
mycotoxins:

Aflatoxin
incidence: 14/14, conc. range: <30 µg/kg
(4 sa), >30 µg/kg (6 sa), >200 µg/kg (2 sa),
>1750 µg/kg (2 sa), country: India[380]
see also blackgram husk, bread crumbs,
broiler finisher, broiler starter, cotton cake,
cottonseed, cottonseed cake, cottonseed
extract, cottonseed meal, feed, feed (cattle),
feed (cow), feed (dog), feed (horse), feed
(maize, gluten), feed (pig), feed (poultry),
feed (rabbit), feed (rat/mice), feed (sheep),
feeds, grain, fish meal, flour (wheat),
groundnut cake, grower's mash, horsegram,
layer's mash, maize, maize gluten, maize,
white, milo, peanut cake, peanut cake
(deoiled), peanut, kernels, peanut (oil cake),
pearlmillet, pig breeder's mash, pig finisher,
pig starter, pod with haulms, poultry
breeder's mash, rabbit pellets, redgram husk,
rice, rice bran (deoiled), rice, broken, rice
(polish), sesame cake, sorghum, soybean
cake, soybean meal, wheat, wheat bran

Sorghum   Sorghum for feed may contain the
following mycotoxins:

Aflatoxin B$_1$
incidence: 5/6, conc. range:
>100–2000 µg/kg, country: Cuba[106]
incidence: 1/4, conc.: 30 µg/kg, country:
Australia[121]
incidence: 11/45, conc. range: 1.4–42.6 µg/kg,
Ø conc.: 10.6 µg/kg, country: Colombia[125]
incidence: 4/7, conc. range: 3–6 µg/kg,
Ø conc.: 3.8 µg/kg, country: USA[134]

incidence: 4/17, conc. range: 1 µg/kg, country: USA[134]
incidence: 1/1, conc.: 30 µg/kg, country: Australia[181]
incidence: ?/10, conc. range: 10–20 µg/kg, country: India[183]
incidence: 12/30, conc. range: ≤25 µg/kg (7 sa), 26–50 µg/kg (3 sa), 51–100 µg/kg (1 sa), 101–200 µg/kg (1 sa), Ø conc.: 33.8 µg/kg, country: India[247]
incidence: 8/11, conc. range: 72.7–165.9 µg/kg, Ø conc.: 106 µg/kg, country: India[253]
incidence: 4/197*, conc. range: 6–45 µg/kg, Ø conc.: 17 µg/kg, country: USA[327], *ncac
incidence: 6/533, conc. range: 3–19 µg/kg, country: USA[384]
incidence: 9/44*, conc. range: 0.18–30.3 µg/kg, country: India[386], *ncac
incidence: 24/25*, conc. range: 2–830 µg/kg, country: India[386], *ncac, rain-affected
see also alfalfa, *Ambadi* cake, animal feedstuffs (dairy cake), bagasse, barley, bengalgram husk, bird food, bird food, wild, biri testa, blackgram, blackgram husk, bran, broiler mixed feed, calf fattening mixed feed, calf fattening mixed feed (containing 4–20 % peanut products), *Carthamus* cake, castor cake, cereals, cereal products, chick pea, coconut cake, cocos, concentrate, mixed, concentrates, cotton cake, cottonseed, cottonseed (dehulled), cottonseed cake, cottonseed extract, cottonseed meal, cottonseed meal (ammoniated), cottonseed meal (decorticated), cottonseed meats, cottonseed products, crumbles, crumbles, grower, cycad meal, dairy cattle feed, dairy cattle feed (containing 2–5 % peanut products), dairy cattle feed (containing 6–10 % peanut products), dairy cattle feed (containing 6–12 % peanut products), dairy cattle feed (containing more than 20 % peanut products), diets, mixed, dog food, egg production mixed feed, feed, feed and ingredients, feed, compound, feed, layer, feed, mixed, feed (beef), feed (broilers), feed (calf), feed (cat), feed (cattle), feed (chicken), feed (dairy), feed (dog), feed (dug), feed (fish), feed (gluten), feed (horse), feed (miscellaneous), feed (pig), feed (poultry), feed (poultry, pig), feed (rabbit), feed (sheep), feed (50–60 % maize), fish meal, grain by-products, grains (no specification), greengram, hay/silage, horsegram, husk, *Jagni* cake, legume mixture, linseed, linseed cake, livol, *Mahua* cake, maize, maize germ, maize gluten, maize grits, maize husk, maize meal, maize oil cake, maize powder, maize screenings, maize, ground, maize, hybrid, maize, preharvest, maize, yellow, maize (dark grains), *Makhana* (*Euryale ferox* Salisb) puffs, manioc, milk production mixed feed, mung testa, murkool, mustard cake, neem cake, niger cake, oats, palm kernel expeller cake, palm kernels, palm products, peanut cake, peanut cake (deoiled), peanut expeller, peanut hay, peanut meal, peanut, kernels, peanut, shells, peanuts, pellets, finisher, pig meal and pellets, pigeon pea, poultry feeds (peanut containing), rapeseed cake, redgram husk, rice, rice bran, rice bran (deoiled), rice chaff, rice crack, rice germ, rice germ cake, rice meal, rice straw, rice (damaged), rice (polish), safflower cake, sal seed cake, sesame, sesame cake, soybean meal, soybeans, sunflower, sunflower cake, sunflower flour, tapioca, wheat, wheat bran, wheat bran and chana testa

AFLATOXIN B$_2$
incidence: 2/45, conc. range: 1.1–3.7 µg/kg, Ø conc.: 2.4 µg/kg, country: Colombia[125]
incidence: 3/30, conc. range: nc µg/kg, Ø conc.: 15.4 µg/kg, country: India[247]
incidence: 4/11, conc. range: 72.7–99.3 µg/kg, Ø conc.: 86 µg/kg, country: India[253]
incidence: 1/197*, conc.: 9 µg/kg, country: USA[327], *ncac
see also animal feedstuffs (dairy cake), bird food, bird food, wild, biri testa, blackgram, blackgram husk, cottonseed, cottonseed cake, cottonseed extract, cottonseed meal, cottonseed meal (ammoniated), cottonseed meats, dog food, egg production mixed feed, feed, compound, feed (cat), feed (cattle), feed (dog), feed (pig), feed (poultry), feed (rabbit), feed (sheep), fish meal, horsegram, maize,

maize gluten, maize husk, maize, ground, maize, preharvest, mung testa, mustard cake, niger cake, peanut cake, peanut cake (deoiled), peanut expeller, peanut hay, peanut meal, peanuts, peanut, kernels, redgram husk, rice bran, rice bran (deoiled), rice chaff, rice meal, rice (polish), sal seed cake, sesame cake, soybean meal, soybeans, wheat, wheat bran

Aflatoxin $B_1 + B_2$
incidence: 5/5, conc. range: 1–4 µg/kg, Ø conc.: 3 µg/kg, country: USA[134]
see also cottonseed, feed (poultry), maize

Aflatoxin B
incidence: 9/12, conc. range: ≥25–>100 µg/kg, country: France[46]
see also barley, cocoa (oilcake), cottonseed cake, feed, feed (cattle), feed (maize, gluten), feed (pig), feed (poultry), flour (wheat), lucern (dried), maize, maize gluten, maize grains, oats, peanut (oil cake), rice, broken, soybean (oil cake), sunflower (oilcake), wheat

Aflatoxin G_1
incidence: 3/533, conc. range: 3–19 µg/kg, country: USA[384]
see also animal feedstuffs (dairy cake), bird food, bird food, wild, concentrate, mixed, cottonseed, cottonseed cake, feed (cat), feed (chicken), feed (dog), maize, maize, ground, maize, preharvest, meat meal, milk production mixed feed, murkool, peanut cake, peanut expeller, peanut hay, peanut meal, peanuts, rice bran, rice germ, soybeans, wheat, wheat bran, wheat bran and chana testa

Aflatoxin
incidence: 4/4, conc. range: 2–5 µg/kg, Ø conc.: 3 µg/kg, country: USA[134]
incidence: 12/30, conc. range: 6–145 µg/kg, country: India[247]
see also blackgram husk, bread crumbs, broiler finisher, broiler starter, cotton cake, cottonseed, cottonseed cake, cottonseed extract, cottonseed meal, feed, feed (cattle), feed (cow), feed (dog), feed (horse), feed (maize, gluten), feed (pig), feed (poultry), feed (rabbit), feed (rat/mice), feed (sheep),

feeds, grain, fish meal, flour (wheat), groundnut cake, grower's mash, horsegram, layer's mash, maize, maize gluten, maize, white, milo, peanut cake, peanut cake (deoiled), peanut, kernels, peanut (oil cake), pearlmillet, pig breeder's mash, pig finisher, pig starter, pod with haulms, poultry breeder's mash, rabbit pellets, redgram husk, rice, rice bran (deoiled), rice, broken, rice (polish), sesame cake, silk worm pupae, soybean cake, soybean meal, wheat, wheat bran

Aflatoxins
incidence: 4/50, conc. range: 5–500 µg/kg, country: Australia[21]
incidence: 2/12, conc. range: 12–60 µg/kg, Ø conc.: 36 µg/kg, country: USA[280]
incidence: 2/2*, conc. range: 87–1873 µg/kg, Ø conc.: 980 µg/kg, country: USA[280], *imported
incidence: 7/114*, conc. range: 3–20 µg/kg, Ø conc.: 10 µg/kg, country: USA[329], *ncac
incidence: 4/25*, conc. range: 4–9 µg/kg, Ø conc.: 6 µg/kg, country: USA[329], *ncac
incidence: 6/29, conc. range: 10–29 µg/kg (2 sa), 31–49 µg/kg (2 sa), >100 µg/kg (2 sa), country: India[349]
see also animal feed (maize), animal feed (mixed), barley, copra meal, cottonseed, cottonseed fines, cottonseed meal, cottonseed meats, feed, feed ingredients (miscellaneous), feed, mixed, feed (excluding peanuts, suspect), feed (goat), feed (pig), feed (poultry), feedstuff, grain, grain, mixed feed, maize, maize, shelled, millet, peanut cake, peanut meal, peanut meal and by-products, peanuts, protein concentrates, rice, rice bran, soybean meal, sunflower

Deoxynivalenol
incidence: 6/7, conc. range: 35–258 µg/kg, Ø conc.: 126.5 µg/kg, country: USA[134]
incidence: 5/5, conc. range: 6–103 µg/kg, Ø conc.: 34.8 µg/kg, country: USA[134]
incidence: 4/4, conc. range: 7–227 µg/kg, Ø conc.: 100 µg/kg, country: USA[134]
incidence: 4/4, conc. range: 21–1540 µg/kg, Ø conc.: 368.3 µg/kg, country: USA[134]

incidence: 5/17, conc. range: 20–558 µg/kg,
Ø conc.: 182.2 µg/kg, country: USA[134]
see also barley, barley, husked, barley,
unhusked (naked), barley (pressed), bone
meal, bran, broilers feed, calf fattening mixed
feed, coconut, expeller, corn cob mix silage,
cottonseed, cottonseed cake, dairy cattle feed,
egg production mixed feed, feed, feed
components, feed, commercial mix, feed,
mixed, feed, mixed (primarily maize), feed
(barley), feed (cattle), feed (chicken), feed
(dog), feed (fish), feed (mill run, from wheat),
feed (mink), feed (pig), feed (poultry), feed
(reindeer), feeds, grain, feeds, industrial,
feedstuff, feedstuffs (rapeseed, turnip, fish
meal, concentrates), fish meal, grain, mixed
feed, grains, mixed, grains (no specification),
maize, maize ears, maize fibre, maize germ,
maize germ/bran, maize germ meal, maize
gluten, maize kernels, maize meal, maize
powder, maize screenings, maize stalks (pith),
maize, "Baby", maize, hybrid, maize, white,
oats, rice bran, rice germ cake, rye, silage,
soybeans, triticale, wheat, wheat and barley,
wheat, red hard winter, wheat, soft white
winter, wheat, spring, wheat, winter

FUMONISIN B₁
incidence: 2/44*, conc. range: 150–500 µg/kg,
Ø conc.: 325 µg/kg, country: India[386], *ncac
incidence: 25/25*, conc. range:
70–7800 µg/kg, country: India[386], *ncac,
rain-affected
see also barley, bird food, wild, dog food,
feed, feed, complete ration, feed, general,
feed, layer, feed, maize-based, feed, mixed,
feed, pelleted ration, feed, screenings, feed,
sweet, feed (broilers), feed (cat), feed
(chicken), feed (dog), feed (gluten), feed
(horse), feed (maize), feed (pig), feed
(poultry), feed (rat), feed (rodent), forage
grass, maize, maize and maize screenings,
maize bran, maize ears, maize fine fractions,
maize flakes, maize germ, maize germ/bran,
maize germ meal, maize gluten, maize grits,
maize kernels, maize meal, maize powder,
maize screenings, maize, "Baby", maize,
ground, maize, preharvest, maize, sweet feed,

maize/oats mix, rat chow, silage, soybeans,
wheat

OCHRATOXIN A
incidence: 9/29, conc. range: 10–29 µg/kg
(2 sa), 30–49 µg/kg (1 sa), 50–100 µg/kg (1 sa),
100–≤400 µg/kg (5 sa), country: India[349]
see also alfalfa, barley, barley, oats, barley
(high moisture), barley-soybean diet, bird
food, domestic, bird food, wild, broilers feed,
cereal grains, citrus pulp, coconut, expeller,
corn cob mix silage, diet (dairy cow), diet
(poultry), diet (starter), dog food, eat, egg
production mixed feed, feed, feed wheat, oat
and barley, feed, commercial mix, feed,
mixed, feed, mixed (pelleted), feed (broilers),
feed (cat), feed (cattle), feed (cereals), feed
(pig), feed ec (pig), feed (poultry), feed ec
(poultry), feed (poultry, pig), feed ec (rabbit),
feed (trout), grain, mixed feed, grains, mixed,
grains (heated), hay, horse bean, maize,
maize feed, milo, maize gluten, maize meal,
maize, white, *Makhana* (*Euryale ferox* Salisb)
puffs, milk production mixed feed, millet, oat
and barley (hammer-milled), oats, palm
products, peanut cake, peas, peas and beans,
pet food, pig feedstuffs, pig grower diet, pig
meal, piglet diet, poultry feedstuffs, rice bran,
rice germ, rice germ cake, rye, soybean
groats, sunflower, sunflower seeds, extracted,
tapioca, triticale, *Vicia faba*, wheat, wheat and
barley, wheat bran, wheat hay, wheat, oats

T-2 TOXIN
incidence: 3/31*, conc. range:
1670–14,960 µg/kg, Ø conc.: 9987 µg/kg,
country: India[199], *ncac
incidence: 1/25*, conc. range: 15,000 µg/kg,
country: India[199], *ncac
see also alfalfa, barley, bran, diet (grower),
diet (poultry), feed, feed components, feed,
layer, feed, mixed, feed (dog), feed (fish),
feed (mink), feed (pig), feed (poultry),
feedstuff, forage grass, grain, mixed feed,
grains (no specification), hay, maize, maize
germ/bran, maize gluten, maize meal, maize
screenings, maize stalks (pith), oat and barley
(hammer-milled), oats, peanuts, pig grower
diet, piglet diet, rye, triticale, wheat

Zearalenone

incidence: 7/7, conc. range: 140–1643 µg/kg, Ø conc.: 530.1 µg/kg, country: USA[134]
incidence: 5/5, conc. range: 112–339 µg/kg, Ø conc.: 197.2 µg/kg, country: USA[134]
incidence: 4/4, conc. range: 26–95 µg/kg, Ø conc.: 49.5 µg/kg, country: USA[134]
incidence: 4/4, conc. range: 80–3099 µg/kg, Ø conc.: 579.3 µg/kg, country: USA[134]
incidence: 17/17, conc. range: 7–2024 µg/kg, Ø conc.: 443 µg/kg, country: USA[134]
incidence: 1/1, conc.: 12,000 µg/kg, country: USA[173]
incidence: ?/10, conc. range: 20–40 µg/kg, country: India[183]
incidence: 56/197*, conc. range: 200–6900 µg/kg, country: USA[327], *ncac
see also alfalfa, barley, barley, husked, barley, unhusked (naked), barley and feed, bone meal, bran, broilers feed, *Carthamus* cake, chick pea, concentrate, mixed, corn cob mix silage, cotton cake, cottonseed, cottonseed cake, diet (dairy cow), diet (poultry), diets (mixed), feed, feed components, feed, mixed, feed, mixed (primarily maize also maize, oats, wheat), feed (bran), feed (broiler chicken), feed (cattle), feed (chicken), feed (dairy), feed (developing pig), feed (mill run, from wheat), feed (miscellaneous), feed (pig), feed (poultry), feed (poultry, pig), feed (starter chicken), feedstuff, fish meal, forage grass, grain, bruised, grain, mixed feed, grains (no specification), hay, maize, maize ears, maize flakes, maize germ, maize germ/bran, maize gluten, maize kernels, maize meal, maize oil cake, maize screenings, maize stalks (pith), maize, "Baby", maize, hybrid, maize, shelled, maize, unshelled, maize, white, maize grain, artificially dried, maize grain, crib dried, maize grain, ensiled, milk production mixed feed, oats, *Paspalum palidosum*, straw, peanut hulls/skins, rice bran, rice germ, rice germ cake, rye, silage, soybeans, soybeans, extracted, sunflower cake, tapioca, triticale, wheat, wheat bran, wheat bran and chana testa, wheat soya meal

Soybean cake　may contain the following mycotoxins:

Aflatoxin

incidence: 1/8, conc.: 140 µg/kg, country: Nigeria[109]
incidence: 17/19*, conc. range: 11–30 µg/kg (11 sa), 31–100 µg/kg (6 sa), country: India[311], *ncac
see also blackgram husk, bread crumbs, broiler finisher, broiler starter, cotton cake, cottonseed, cottonseed cake, cottonseed extract, cottonseed meal, feed, feed (cattle), feed (cow), feed (dog), feed (horse), feed (maize, gluten), feed (pig), feed (poultry), feed (rabbit), feed (rat/mice), feed (sheep), feeds, grain, fish meal, flour (wheat), groundnut cake, grower's mash, horsegram, layer's mash, maize, maize gluten, maize, white, milo, peanut cake, peanut cake (deoiled), peanut, kernels, peanut (oil cake), pearlmillet, pig breeder's mash, pig finisher, pig starter, pod with haulms, poultry breeder's mash, rabbit pellets, redgram husk, rice, rice bran (deoiled), rice, broken, rice (polish), sesame cake, silk worm pupae, sorghum, soybean meal, wheat, wheat bran

Soybean groats　may contain the following mycotoxins:

Ochratoxin A

incidence: 1/7345, conc.: nc, country: Hungary[209]
see also alfalfa, barley, barley, oats, barley (high moisture), barley-soybean diet, bird food, domestic, bird food, wild, broilers feed, cereal grains, citrus pulp, coconut, expeller, corn cob mix silage, diet (dairy cow), diet (poultry), diet (starter), dog food, eat, egg production mixed feed, feed, feed wheat, oat and barley, feed, commercial mix, feed, mixed, feed, mixed (pelleted), feed (broilers), feed (cat), feed (cattle), feed (cereals), feed (pig), feed ec (pig), feed (poultry), feed ec (poultry), feed (poultry, pig), feed ec (rabbit), feed (trout), grain, mixed feed, grains, mixed, grains (heated), hay, horse bean, maize, maize feed, milo, maize gluten, maize meal, maize, white, *Makhana* (*Euryale ferox* Salisb) puffs, milk production mixed feed, millet, oat and barley (hammer-milled), oats, palm

products, peanut cake, peas, peas and beans, pet food, pig feedstuffs, pig grower diet, pig meal, piglet diet, poultry feedstuffs, rice bran, rice germ, rice germ cake, rye, sorghum, sunflower, sunflower seeds, extracted, tapioca, triticale, *Vicia faba*, wheat, wheat and barley, wheat bran, wheat hay, wheat, oats

Rubratoxin B
incidence: 4/7345, conc. range: nc, country: Hungary[209]
see also diet (grower), diet (poultry), diet (starter), pig diet (fattening), pig grower diet, wheat

Soybean meal   Soybean meal for feed may contain the following mycotoxins:

Aflatoxin B$_1$
incidence: 36/250, conc. range: ≤25 μg/kg (28 sa), 26–50 μg/kg (3 sa), 51–100 μg/kg (3 sa), 101–200 μg/kg (1 sa), 201–500 μg/kg (1 sa), Ø conc.: 26.7 μg/kg, country: India[247]
incidence: 3/118, conc. range: 5–25 μg/kg, Ø conc.: 18.3 μg/kg, country: Egypt[397]
incidence: 7/120, conc. range: 5–15 μg/kg, Ø conc.: 10.7 μg/kg, country: Egypt[397]
incidence: 2/75, conc. range: 5–25 μg/kg, Ø conc.: 15 μg/kg, country: Egypt[397]
see also alfalfa, *Ambadi* cake, animal feedstuffs (dairy cake), bagasse, barley, bengalgram husk, bird food, bird food, wild, biri testa, blackgram, blackgram husk, bran, broiler mixed feed, calf fattening mixed feed, calf fattening mixed feed (containing 4–20 % peanut products), *Carthamus* cake, castor cake, cereals, cereal products, chick pea, coconut cake, cocos, concentrate, mixed, concentrates, cotton cake, cottonseed, cottonseed (dehulled), cottonseed cake, cottonseed extract, cottonseed meal, cottonseed meal (ammoniated), cottonseed meal (decorticated), cottonseed meats, cottonseed products, crumbles, crumbles, grower, cycad meal, dairy cattle feed, dairy cattle feed (containing 2–5 % peanut products), dairy cattle feed (containing 6–10 % peanut products), dairy cattle feed (containing 6–12 % peanut products), dairy cattle feed (containing

more than 20 % peanut products), diets, mixed, dog food, egg production mixed feed, feed, feed and ingredients, feed, compound, feed, layer, feed, mixed, feed (beef), feed (broilers), feed (calf), feed (cat), feed (cattle), feed (chicken), feed (dairy), feed (dog), feed (dug), feed (fish), feed (gluten), feed (horse), feed (miscellaneous), feed (pig), feed (poultry), feed (poultry, pig), feed (rabbit), feed (sheep), feed (50–60 % maize), fish meal, grain by-products, grains (no specification), greengram, hay/silage, horsegram, husk, *Jagni* cake, legume mixture, linseed, linseed cake, livol, *Mahua* cake, maize, maize germ, maize gluten, maize grits, maize husk, maize meal, maize oil cake, maize powder, maize screenings, maize, ground, maize, hybrid, maize, preharvest, maize, yellow, maize (dark grains), *Makhana* (*Euryale ferox* Salisb) puffs, manioc, milk production mixed feed, mung testa, murkool, mustard cake, neem cake, niger cake, oats, palm kernel expeller cake, palm kernels, palm products, peanut cake, peanut cake (deoiled), peanut expeller, peanut hay, peanut meal, peanut, kernels, peanut, shells, peanuts, pellets, finisher, pig meal and pellets, pigeon pea, poultry feeds (peanut containing), rapeseed cake, redgram husk, rice, rice bran, rice bran (deoiled), rice chaff, rice crack, rice germ, rice germ cake, rice meal, rice straw, rice (damaged), rice (polish), safflower cake, sal seed cake, sesame, sesame cake, sorghum, soybeans, sunflower, sunflower cake, sunflower flour, tapioca, wheat, wheat bran, wheat bran and chana testa

Aflatoxin B$_2$
incidence: 11/107, conc. range: nc Ø conc.: 9.2 μg/kg, country: India[247]
see also animal feedstuffs (dairy cake), bird food, bird food, wild, biri testa, blackgram, blackgram husk, cottonseed, cottonseed cake, cottonseed extract, cottonseed meal, cottonseed meal (ammoniated), cottonseed meats, dog food, egg production mixed feed, feed, compound, feed (cat), feed (cattle), feed (dog), feed (pig), feed (poultry), feed (rabbit), feed (sheep), fish meal, horsegram,

maize, maize gluten, maize husk, maize, ground, maize, preharvest, mung testa, mustard cake, niger cake, peanut cake, peanut cake (deoiled), peanut expeller, peanut hay, peanut meal, peanuts, peanut, kernels, redgram husk, rice bran, rice bran (deoiled), rice chaff, rice meal, rice (polish), sal seed cake, sesame cake, sorghum, soybeans, wheat, wheat bran

AFLATOXIN
incidence: 36/250, conc. range: nd–256 µg/kg, country: India[247]
see also blackgram husk, bread crumbs, broiler finisher, broiler starter, cotton cake, cottonseed, cottonseed cake, cottonseed extract, cottonseed meal, feed, feed (cattle), feed (cow), feed (dog), feed (horse), feed (maize, gluten), feed (pig), feed (poultry), feed (rabbit), feed (rat/mice), feed (sheep), feeds, grain, fish meal, flour (wheat), groundnut cake, grower's mash, horsegram, layer's mash, maize, maize gluten, maize, white, milo, peanut cake, peanut cake (deoiled), peanut, kernels, peanut (oil cake), pearlmillet, pig breeder's mash, pig finisher, pig starter, pod with haulms, poultry breeder's mash, rabbit pellets, redgram husk, rice, rice bran (deoiled), rice, broken, rice (polish), sesame cake, silk worm pupae, sorghum, soybean cake, wheat, wheat bran

AFLATOXINS
incidence: 1/16, conc.: 5–50 µg/kg, country: Australia[21]
see also animal feed (maize), animal feed (mixed), barley, copra meal, cottonseed, cottonseed fines, cottonseed meal, cottonseed meats, feed, feed ingredients (miscellaneous), feed, mixed, feed (excluding peanuts, suspect), feed (goat), feed (pig), feed (poultry), feedstuff, grain, grain, mixed feed, maize, maize, shelled, millet, peanut cake, peanut meal, peanut meal and by-products, peanuts, protein concentrates, rice, rice bran, sorghum, sunflower

Soybean (oil cake)   may contain the following mycotoxins:

AFLATOXIN B
incidence: 24/51, conc. range: ≥25–>100 µg/kg, country: France[46]
see also barley, cocoa (oilcake), cottonseed cake, feed, feed (cattle), feed (maize, gluten), feed (pig), feed (poultry), flour (wheat), lucern (dried), maize, maize gluten, maize grains, oats, peanut (oil cake), rice, broken, sorghum, sunflower (oilcake), wheat

Soybeans   Soybeans for feed may contain the following mycotoxins:

AFLATOXIN B_1
incidence: 4/5, Ø conc.: 5 µg/kg, country: Egypt[16]
incidence: 1/20*, conc.: 4 µg/kg, country: UK[94], *imported?
incidence: 1/1*, conc.: 77.5 µg/kg, country: UK[122], *ncac
incidence: 2/866, conc. range: 7–10 µg/kg, Ø conc.: 8.5 µg/kg, country: USA[381]
see also alfalfa, *Ambadi* cake, animal feedstuffs (dairy cake), bagasse, barley, bengalgram husk, bird food, bird food, wild, biri testa, blackgram, blackgram husk, bran, broiler mixed feed, calf fattening mixed feed, calf fattening mixed feed (containing 4–20 % peanut products), *Carthamus* cake, castor cake, cereals, cereal products, chick pea, coconut cake, cocos, concentrate, mixed, concentrates, cotton cake, cottonseed, cottonseed (dehulled), cottonseed cake, cottonseed extract, cottonseed meal, cottonseed meal (ammoniated), cottonseed meal (decorticated), cottonseed meats, cottonseed products, crumbles, crumbles, grower, cycad meal, dairy cattle feed, dairy cattle feed (containing 2–5 % peanut products), dairy cattle feed (containing 6–10 % peanut products), dairy cattle feed (containing 6–12 % peanut products), dairy cattle feed (containing more than 20 % peanut products), diets, mixed, dog food, egg production mixed feed, feed, feed and ingredients, feed, compound, feed, layer, feed, mixed, feed (beef), feed (broilers), feed (calf), feed (cat), feed (cattle), feed (chicken), feed (dairy), feed (dog), feed (dug), feed

(fish), feed (gluten), feed (horse), feed (miscellaneous), feed (pig), feed (poultry), feed (poultry, pig), feed (rabbit), feed (sheep), feed (50–60 % maize), fish meal, grain by-products, grains (no specification), greengram, hay/silage, horsegram, husk, *Jagni* cake, legume mixture, linseed, linseed cake, livol, *Mahua* cake, maize, maize germ, maize gluten, maize grits, maize husk, maize meal, maize oil cake, maize powder, maize screenings, maize, ground, maize, hybrid, maize, preharvest, maize, yellow, maize (dark grains), *Makhana* (*Euryale ferox* Salisb) puffs, manioc, milk production mixed feed, mung testa, murkool, mustard cake, neem cake, niger cake, oats, palm kernel expeller cake, palm kernels, palm products, peanut cake, peanut cake (deoiled), peanut expeller, peanut hay, peanut meal, peanut, kernels, peanut, shells, peanuts, pellets, finisher, pig meal and pellets, pigeon pea, poultry feeds (peanut containing), rapeseed cake, redgram husk, rice, rice bran, rice bran (deoiled), rice chaff, rice crack, rice germ, rice germ cake, rice meal, rice straw, rice (damaged), rice (polish), safflower cake, sal seed cake, sesame, sesame cake, sorghum, soybean meal, sunflower, sunflower cake, sunflower flour, tapioca, wheat, wheat bran, wheat bran and chana testa

AFLATOXIN B₂
incidence: 1/1*, conc.: 16 µg/kg, country: UK[122], *ncac
see also animal feedstuffs (dairy cake), bird food, bird food, wild, biri testa, blackgram, blackgram husk, cottonseed, cottonseed cake, cottonseed extract, cottonseed meal, cottonseed meal (ammoniated), cottonseed meats, dog food, egg production mixed feed, feed, compound, feed (cat), feed (cattle), feed (dog), feed (pig), feed (poultry), feed (rabbit), feed (sheep), fish meal, horsegram, maize, maize gluten, maize husk, maize, ground, maize, preharvest, mung testa, mustard cake, niger cake, peanut cake, peanut cake (deoiled), peanut expeller, peanut hay, peanut meal, peanuts, peanut, kernels, redgram husk, rice bran, rice bran

(deoiled), rice chaff, rice meal, rice (polish), sal seed cake, sesame cake, sorghum, soybean meal, wheat, wheat bran

AFLATOXIN G₁
incidence: 1/1*, conc.: 8.1 µg/kg, country: UK[122], *ncac
incidence: 1/866, conc.: 4 µg/kg, country: USA[381]
see also animal feedstuffs (dairy cake), bird food, bird food, wild, concentrate, mixed, cottonseed, cottonseed cake, feed (cat), feed (chicken), feed (dog), maize, maize, ground, maize, preharvest, meat meal, milk production mixed feed, murkool, peanut cake, peanut expeller, peanut hay, peanut meal, peanuts, rice bran, rice germ, sorghum, wheat, wheat bran, wheat bran and chana testa

AFLATOXIN G₂
incidence: 1/1*, conc.: 2.8 µg/kg, country: UK[122], *ncac
see also animal feedstuffs (dairy cake), bird food, feed (cat), feed (dog), maize, maize, preharvest, peanut expeller, peanut hay, peanut meal, peanuts

AFLATOXINS (TOTAL)
incidence: 1/20*, conc.: 6 µg/kg, country: UK[94], *imported?
see also cottonseed, maize germ, maize gluten, maize (dark grains), palm products, rice bran, sunflower

DEOXYNIVALENOL
incidence: 8/24, conc. range: 50–490 µg/kg, Ø conc.: 161.3 µg/kg, country: USA[83]
incidence: 1/2, conc.: 580 µg/kg, country: The Netherlands[103]
see also barley, barley, husked, barley, unhusked (naked), barley (pressed), bone meal, bran, broilers feed, calf fattening mixed feed, coconut, expeller, corn cob mix silage, cottonseed, cottonseed cake, dairy cattle feed, egg production mixed feed, feed, feed components, feed, commercial mix, feed, mixed, feed, mixed (primarily maize), feed (barley), feed (cattle), feed (chicken), feed (dog), feed (fish), feed (mill run, from

wheat), feed (mink), feed (pig), feed
(poultry), feed (reindeer), feeds, grain, feeds,
industrial, feedstuff, feedstuffs (rapeseed,
turnip, fish meal, concentrates), fish meal,
grain, mixed feed, grains, mixed, grains (no
specification), maize, maize ears, maize fibre,
maize germ, maize germ/bran, maize germ
meal, maize gluten, maize kernels, maize
meal, maize powder, maize screenings, maize
stalks (pith), maize, "Baby", maize, hybrid,
maize, white, oats, rice bran, rice germ cake,
rye, silage, sorghum, triticale, wheat, wheat
and barley, wheat, red hard winter, wheat,
soft white winter, wheat, spring, wheat,
winter

DIACETOXYSCIRPENOL
incidence: 5/24, conc. range: 15–230 µg/kg,
Ø conc.: 77 µg/kg, country: USA[83]
see also alfalfa, barley, feed, feed components,
feed (dog), feed (fish), feed (mink), feed
(pig), feed (poultry), feed (reindeer), feeds,
grain, feedstuff, forage grass, grain, mixed
feed, grains (no specification), maize, maize
gluten, oat and barley (hammer-milled), oats,
peanuts, wheat

FUMONISIN B$_1$
incidence: 1/1, conc.: 8700 µg/kg, country:
Spain[155]
see also barley, bird food, wild, dog food,
feed, feed, complete ration, feed, general,
feed, layer, feed, maize-based, feed, mixed,
feed, pelleted ration, feed, screenings, feed,
sweet, feed (broilers), feed (cat), feed
(chicken), feed (dog), feed (gluten), feed
(horse), feed (maize), feed (pig), feed
(poultry), feed (rat), feed (rodent), forage
grass, maize, maize and maize screenings,
maize bran, maize ears, maize fine fractions,
maize flakes, maize germ, maize germ/bran,
maize germ meal, maize gluten, maize grits,
maize kernels, maize meal, maize powder,
maize screenings, maize, "Baby", maize,
ground, maize, preharvest, maize, sweet feed,
maize/oats mix, rat chow, silage, sorghum,
wheat

T-2 TOXIN EQUIVALENTS
incidence: 9/24, conc. range: 20–1070 µg/kg,
Ø conc.: 203.7 µg/kg, country: USA[83]

ZEARALENONE
incidence: 8/24, conc. range: 50–490 µg/kg,
Ø conc.: 161.3 µg/kg, country: USA[83]
incidence: 1/15*, conc. range: 160–500 µg/kg
(1 sa), country: USA[370], *includes 7 soybean
and 8 soybean meal sa
see also alfalfa, barley, barley, husked, barley,
unhusked (naked), barley and feed, bone
meal, bran, broilers feed, *Carthamus* cake,
chick pea, concentrate, mixed, corn cob mix
silage, cotton cake, cottonseed, cottonseed
cake, diet (dairy cow), diet (poultry), diets
(mixed), feed, feed components, feed, mixed,
feed, mixed (primarily maize also maize, oats,
wheat), feed (bran), feed (broiler chicken),
feed (cattle), feed (chicken), feed (dairy), feed
(developing pig), feed (mill run, from
wheat), feed (miscellaneous), feed (pig), feed
(poultry), feed (poultry, pig), feed (starter
chicken), feedstuff, fish meal, forage grass,
grain, bruised, grain, mixed feed, grains (no
specification), hay, maize, maize ears, maize
flakes, maize germ, maize germ/bran, maize
gluten, maize kernels, maize meal, maize oil
cake, maize screenings, maize stalks (pith),
maize, "Baby", maize, hybrid, maize, shelled,
maize, unshelled, maize, white, maize grain,
artificially dried, maize grain, crib dried,
maize grain, ensiled, milk production mixed
feed, oats, *Paspalum palidosum*, straw, peanut
hulls/skins, rice bran, rice germ, rice germ
cake, rye, silage, sorghum, soybeans,
extracted, sunflower cake, tapioca, triticale,
wheat, wheat bran, wheat bran and chana
testa, wheat soya meal

Soybeans, extracted   may contain the
following mycotoxins:

ZEARALENONE
incidence: 1/7*, conc.: 11 µg/kg, country: The
Netherlands[103], *imported?
see also alfalfa, barley, barley, husked, barley,
unhusked (naked), barley and feed, bone
meal, bran, broilers feed, *Carthamus* cake,

chick pea, concentrate, mixed, corn cob mix silage, cotton cake, cottonseed, cottonseed cake, diet (dairy cow), diet (poultry), diets (mixed), feed, feed components, feed, mixed, feed, mixed (primarily maize also maize, oats, wheat), feed (bran), feed (broiler chicken), feed (cattle), feed (chicken), feed (dairy), feed (developing pig), feed (mill run, from wheat), feed (miscellaneous), feed (pig), feed (poultry), feed (poultry, pig), feed (starter chicken), feedstuff, fish meal, forage grass, grain, bruised, grain, mixed feed, grains (no specification), hay, maize, maize ears, maize flakes, maize germ, maize germ/bran, maize gluten, maize kernels, maize meal, maize oil cake, maize screenings, maize stalks (pith), maize, "Baby", maize, hybrid, maize, shelled, maize, unshelled, maize, white, maize grain, artificially dried, maize grain, crib dried, maize grain, ensiled, milk production mixed feed, oats, *Paspalum palidosum*, straw, peanut hulls/skins, rice bran, rice germ, rice germ cake, rye, silage, sorghum, soybeans, sunflower cake, tapioca, triticale, wheat, wheat bran, wheat bran and chana testa, wheat soya meal

Sunflower Sunflower for feed may contain the following mycotoxins:

Aflatoxin B_1
incidence: 3/20*, conc. range: 1–15 µg/kg, country: UK[94], *imported?
see also alfalfa, *Ambadi* cake, animal feedstuffs (dairy cake), bagasse, barley, bengalgram husk, bird food, bird food, wild, biri testa, blackgram, blackgram husk, bran, broiler mixed feed, calf fattening mixed feed, calf fattening mixed feed (containing 4–20 % peanut products), *Carthamus* cake, castor cake, cereals, cereal products, chick pea, coconut cake, cocos, concentrate, mixed, concentrates, cotton cake, cottonseed, cottonseed (dehulled), cottonseed cake, cottonseed extract, cottonseed meal, cottonseed meal (ammoniated), cottonseed meal (decorticated), cottonseed meats, cottonseed products, crumbles, crumbles, grower, cycad meal, dairy cattle feed, dairy

cattle feed (containing 2–5 % peanut products), dairy cattle feed (containing 6–10 % peanut products), dairy cattle feed (containing 6–12 % peanut products), dairy cattle feed (containing more than 20 % peanut products), diets, mixed, dog food, egg production mixed feed, feed, feed and ingredients, feed, compound, feed, layer, feed, mixed, feed (beef), feed (broilers), feed (calf), feed (cat), feed (cattle), feed (chicken), feed (dairy), feed (dog), feed (dug), feed (fish), feed (gluten), feed (horse), feed (miscellaneous), feed (pig), feed (poultry), feed (poultry, pig), feed (rabbit), feed (sheep), feed (50–60 % maize), fish meal, grain by-products, grains (no specification), greengram, hay/silage, horsegram, husk, *Jagni* cake, legume mixture, linseed, linseed cake, livol, *Mahua* cake, maize, maize germ, maize gluten, maize grits, maize husk, maize meal, maize oil cake, maize powder, maize screenings, maize, ground, maize, hybrid, maize, preharvest, maize, yellow, maize (dark grains), *Makhana* (*Euryale ferox* Salisb) puffs, manioc, milk production mixed feed, mung testa, murkool, mustard cake, neem cake, niger cake, oats, palm kernel expeller cake, palm kernels, palm products, peanut cake, peanut cake (deoiled), peanut expeller, peanut hay, peanut meal, peanut, kernels, peanut, shells, peanuts, pellets, finisher, pig meal and pellets, pigeon pea, poultry feeds (peanut containing), rapeseed cake, redgram husk, rice, rice bran, rice bran (deoiled), rice chaff, rice crack, rice germ, rice germ cake, rice meal, rice straw, rice (damaged), rice (polish), safflower cake, sal seed cake, sesame, sesame cake, sorghum, soybean meal, soy-beans, sunflower cake, sunflower flour, tapioca, wheat, wheat bran, wheat bran and chana testa

Aflatoxins (Total)
incidence: 3/20*, conc. range: 1–17 µg/kg, country: UK[94], *imported?
see also cottonseed, maize germ, maize gluten, maize (dark grains), palm products, rice bran, soybeans

Aflatoxins
incidence: 5/10, conc. range: 30–49 µg/kg
(5 sa), country: India[349]
see also animal feed (maize), animal feed
(mixed), barley, copra meal, cottonseed,
cottonseed fines, cottonseed meal, cottonseed
meats, feed, feed, mixed, feed ingredients
(miscellaneous), feed (excluding peanuts,
suspect), feed (goat), feed (pig), feed
(poultry), feedstuff, grain, grain, mixed feed,
maize, maize, shelled, millet, peanut cake,
peanut meal, peanut meal and by-products,
peanuts, protein concentrates, rice, rice bran,
sorghum, soybean meal

Alternariol
incidence: 2/2*, conc. range: 357–1840 µg/kg,
Ø conc.: 1098.5 µg/kg, country: Italy[317], *ncac
see also cereals, mixture, oilseed rape meal,
sunflower seed meal, sunflower seeds, ensiling

Alternariol Methyl Ether
incidence: 1/2*, conc.: 129 µg/kg, country:
Italy[317], *ncac
see also barley, barley, oats, cereals, mixture,
oats, wheat, wheat, oats

Ochratoxin A
incidence: 1/10, conc. range: 30–49 µg/kg,
country: India[349]
see also alfalfa, barley, barley, oats, barley
(high moisture), barley-soybean diet, bird
food, domestic, bird food, wild, broilers feed,
cereal grains, citrus pulp, coconut, expeller,
corn cob mix silage, diet (dairy cow), diet
(poultry), diet (starter), dog food, eat, egg
production mixed feed, feed, feed wheat, oat
and barley, feed, commercial mix, feed,
mixed, feed, mixed (pelleted), feed (broilers),
feed (cat), feed (cattle), feed (cereals), feed
(pig), feed ec (pig), feed (poultry), feed ec
(poultry), feed (poultry, pig), feed ec (rabbit),
feed (trout), grain, mixed feed, grains, mixed,
grains (heated), hay, horse bean, maize,
maize feed, milo, maize gluten, maize meal,
maize, white, *Makhana* (*Euryale ferox* Salisb)
puffs, milk production mixed feed, millet, oat
and barley (hammer-milled), oats, palm
products, peanut cake, peas, peas and beans,
pet food, pig feedstuffs, pig grower diet, pig

meal, piglet diet, poultry feedstuffs, rice bran,
rice germ, rice germ cake, rye, sorghum,
soybean groats, sunflower seeds, extracted,
tapioca, triticale, *Vicia faba*, wheat, wheat and
barley, wheat bran, wheat hay, wheat, oats

Sunflower cake   may contain the following
mycotoxins:

Aflatoxin B$_1$
incidence: ?/10, conc. range: tr–20 µg/kg,
country: India[183]
incidence: 3/4, conc. range: 6.3–45 µg/kg,
Ø conc.: 31.5 µg/kg, country: India[247]
see also alfalfa, *Ambadi* cake, animal feedstuffs (dairy cake), bagasse, barley,
bengalgram husk, bird food, bird food, wild,
biri testa, blackgram, blackgram husk, bran,
broiler mixed feed, calf fattening mixed feed,
calf fattening mixed feed (containing 4–20 %
peanut products), *Carthamus* cake, castor
cake, cereals, cereal products, chick pea,
coconut cake, cocos, concentrate, mixed,
concentrates, cotton cake, cottonseed,
cottonseed (dehulled), cottonseed cake,
cottonseed extract, cottonseed meal,
cottonseed meal (ammoniated), cottonseed
meal (decorticated), cottonseed meats,
cottonseed products, crumbles, crumbles,
grower, cycad meal, dairy cattle feed, dairy
cattle feed (containing 2–5 % peanut
products), dairy cattle feed (containing
6–10 % peanut products), dairy cattle feed
(containing 6–12 % peanut products), dairy
cattle feed (containing more than 20 %
peanut products), diets, mixed, dog food, egg
production mixed feed, feed, feed and
ingredients, feed, compound, feed, layer,
feed, mixed, feed (beef), feed (broilers), feed
(calf), feed (cat), feed (cattle), feed (chicken),
feed (dairy), feed (dog), feed (dug), feed
(fish), feed (gluten), feed (horse), feed
(miscellaneous), feed (pig), feed (poultry),
feed (poultry, pig), feed (rabbit), feed
(sheep), feed (50–60 % maize), fish meal,
grain by-products, grains (no specification),
greengram, hay/silage, horsegram, husk, *Jagni*
cake, legume mixture, linseed, linseed cake,
livol, *Mahua* cake, maize, maize germ, maize

gluten, maize grits, maize husk, maize meal, maize oil cake, maize powder, maize screenings, maize, ground, maize, hybrid, maize, preharvest, maize, yellow, maize (dark grains), *Makhana* (*Euryale ferox* Salisb) puffs, manioc, milk production mixed feed, mung testa, murkool, mustard cake, neem cake, niger cake, oats, palm kernel expeller cake, palm kernels, palm products, peanut cake, peanut cake (deoiled), peanut expeller, peanut hay, peanut meal, peanut, kernels, peanut, shells, peanuts, pellets, finisher, pig meal and pellets, pigeon pea, poultry feeds (peanut containing), rapeseed cake, redgram husk, rice, rice bran, rice bran (deoiled), rice chaff, rice crack, rice germ, rice germ cake, rice meal, rice straw, rice (damaged), rice (polish), safflower cake, sal seed cake, sesame, sesame cake, sorghum, soybean meal, soybeans, sunflower, sunflower flour, tapioca, wheat, wheat bran, wheat bran and chana testa

ZEARALENONE
incidence: ?/10, conc. range: tr µg/kg, country: India[183]
incidence: 3/4, conc. range: 59.4–245.8 µg/kg, Ø conc.: 165.9 µg/kg, country: India[253]
see also alfalfa, barley, barley, husked, barley, unhusked (naked), barley and feed, bone meal, bran, broilers feed, *Carthamus* cake, chick pea, concentrate, mixed, corn cob mix silage, cotton cake, cottonseed, cottonseed cake, diet (dairy cow), diet (poultry), diet (mixed), feed, feed components, feed, mixed, feed, mixed (primarily maize also maize, oats, wheat), feed (bran), feed (broiler chicken), feed (cattle), feed (chicken), feed (dairy), feed (developing pig), feed (mill run, from wheat), feed (miscellaneous), feed (pig), feed (poultry), feed (poultry, pig), feed (starter chicken), feedstuff, fish meal, forage grass, grain, bruised, grain, mixed feed, grains (no specification), hay, maize, maize ears, maize flakes, maize germ, maize germ/bran, maize gluten, maize kernels, maize meal, maize oil cake, maize screenings, maize stalks (pith), maize, "Baby", maize, hybrid, maize, shelled, maize, unshelled, maize, white, maize grain,

artificially dried, maize grain, crib dried, maize grain, ensiled, milk production mixed feed, oats, *Paspalum palidosum*, straw, peanut hulls/skins, rice bran, rice germ, rice germ cake, rye, silage, sorghum, soybeans, soybeans, extracted, tapioca, triticale, wheat, wheat bran, wheat bran and chana testa, wheat soya meal

Sunflower flour   Sunflower flour for feed may contain the following mycotoxins:

AFLATOXIN B$_1$
incidence: 10/15, conc. range: >10–100 µg/kg, country: Cuba[106]
see also alfalfa, *Ambadi* cake, animal feedstuffs (dairy cake), bagasse, barley, bengalgram husk, bird food, bird food, wild, biri testa, blackgram, blackgram husk, bran, broiler mixed feed, calf fattening mixed feed, calf fattening mixed feed (containing 4–20 % peanut products), *Carthamus* cake, castor cake, cereals, cereal products, chick pea, coconut cake, cocos, concentrate, mixed, concentrates, cotton cake, cottonseed, cottonseed (dehulled), cottonseed cake, cottonseed extract, cottonseed meal, cottonseed meal (ammoniated), cottonseed meal (decorticated), cottonseed meats, cottonseed products, crumbles, crumbles, grower, cycad meal, dairy cattle feed, dairy cattle feed (containing 2–5 % peanut products), dairy cattle feed (containing 6–10 % peanut products), dairy cattle feed (containing 6–12 % peanut products), dairy cattle feed (containing more than 20 % peanut products), diets, mixed, dog food, egg production mixed feed, feed, feed and ingredients, feed, compound, feed, layer, feed, mixed, feed (beef), feed (broilers), feed (calf), feed (cat), feed (cattle), feed (chicken), feed (dairy), feed (dog), feed (dug), feed (fish), feed (gluten), feed (horse), feed (miscellaneous), feed (pig), feed (poultry), feed (poultry, pig), feed (rabbit), feed (sheep), feed (50–60 % maize), fish meal, grain by-products, grains (no specification), greengram, hay/silage, horsegram, husk, *Jagni* cake, legume mixture, linseed, linseed cake,

livol, *Mahua* cake, maize, maize germ, maize gluten, maize grits, maize husk, maize meal, maize oil cake, maize powder, maize screenings, maize, ground, maize, hybrid, maize, preharvest, maize, yellow, maize (dark grains), *Makhana* (*Euryale ferox* Salisb) puffs, manioc, milk production mixed feed, mung testa, murkool, mustard cake, neem cake, niger cake, oats, palm kernel expeller cake, palm kernels, palm products, peanut cake, peanut cake (deoiled), peanut expeller, peanut hay, peanut meal, peanut, kernels, peanut, shells, peanuts, pellets, finisher, pig meal and pellets, pigeon pea, poultry feeds (peanut containing), rapeseed cake, redgram husk, rice, rice bran, rice bran (deoiled), rice chaff, rice crack, rice germ, rice germ cake, rice meal, rice straw, rice (damaged), rice (polish), safflower cake, sal seed cake, sesame, sesame cake, sorghum, soybean meal, soybeans, sunflower, sunflower cake, tapioca, wheat, wheat bran, wheat bran and chana testa

Sunflower seed meal   may contain the following mycotoxins:

ALTERNARIOL
incidence: 18/22*, conc. range: 50–570 µg/kg, Ø conc.: 180 µg/kg, country: Argentina, India[187], *ec
see also cereals, mixture, oilseed rape meal, sunflower, sunflower seeds, ensiling

ALTERNARIOL MONOMETHYL ETHER
incidence: 16/22*, conc. range: 45–270 µg/kg, Ø conc.: 100 µg/kg, country: Argentina, India[187], *ec
see also oilseed rape meal, sunflower seeds, ensiling

TENUAZONIC ACID
incidence: 22/22*, conc. range: 510–5600 µg/kg, Ø conc.: 1900 µg/kg, country: Argentina, India[187], *ec
see also feed, feed (GP No. 1), oilseed rape meal, sunflower seeds, ensiling

Sunflower (oil cake)   may contain the following mycotoxins:

AFLATOXIN B
incidence: 11/15, conc. range: ≥25–>100 µg/kg, country: France[46]
see also barley, cocoa (oilcake), cottonseed cake, feed, feed (cattle), feed (maize, gluten), feed (pig), feed (poultry), flour (wheat), lucern (dried), maize, maize gluten, maize grains, oats, peanut (oil cake), rice, broken, sorghum, soybean (oil cake), wheat

Sunflower seeds, ensiling   may contain the following mycotoxins:

ALTERNARIOL
incidence: 20/20, conc. range: 250–980 µg/kg, Ø conc.: 661 µg/kg, country: Argentina[307]
incidence: 17/20, conc. range: 60–800 µg/kg, Ø conc.: 360 µg/kg, country: Argentina[307]
incidence: 4/20, conc. range: 800–1600 µg/kg, Ø conc.: 1070 µg/kg, country: Argentina[307]
see also cereals, mixture, oilseed rape meal, sunflower, sunflower seed meal

ALTERNARIOL MONOMETHYL ETHER
incidence: 1/20, conc.: 600 µg/kg, country: Argentina[307]
incidence: 1/20, conc.: 800 µg/kg, country: Argentina[307]
see also oilseed rape meal, sunflower seed meal

TENUAZONIC ACID
incidence: 16/20, conc. range: 3900–31,600 µg/kg, Ø conc.: 15,894 µg/kg, country: Argentina[307]
incidence: 13/20, conc. range: 3600–15,800 µg/kg, Ø conc.: 8000 µg/kg, country: Argentina[307]
incidence: 9/20, conc. range: 3120–6240 µg/kg, Ø conc.: 4500 µg/kg, country: Argentina[307]
see also feed, feed (GP No. 1), oilseed rape meal, sunflower seed meal

Sunflower seeds, extracted   may contain the following mycotoxins:

Ochratoxin A
incidence: 3/3, conc. range: 9–37 μg/kg,
Ø conc.: 27 μg/kg, country: The
Netherlands[103]
see also alfalfa, barley, barley, oats, barley (high
moisture), barley-soybean diet, bird food,
domestic, bird food, wild, broilers feed, cereal
grains, citrus pulp, coconut, expeller, corn cob
mix silage, diet (dairy cow), diet (poultry), diet
(starter), dog food, eat, egg production mixed
feed, feed wheat, oat and barley, feed, commercial mix, feed, mixed, feed, mixed (pelleted),
feed (broilers), feed (cat), feed (cattle), feed
(cereals), feed (pig), feedec(pig),feed(poultry),
feed ec (poultry), feed (poultry, pig), feed ec
(rabbit), feed (trout), grain, mixed feed, grains,
mixed, grains (heated), hay, horse bean, maize,
maize feed, milo, maize gluten, maize meal,
maize, white, *Makhana* (*Euryale ferox* Salisb)
puffs, milk production mixed feed, millet, oat
and barley (hammer-milled), oats, palm
products, peanut cake, peas, peas and beans,
pet food, pig feedstuffs, pig grower diet, pig
meal, piglet diet, poultry feedstuffs, rice bran,
rice germ, rice germ cake, rye, sorghum,
soybean groats, sunflower, tapioca, triticale,
Vicia faba, wheat, wheat and barley, wheat
bran, wheat hay, wheat, oats

Swine
see Pork

Tall fescue grass (*Festuga arundinacea* Screb)
may contain the following mycotoxins:

Ergopeptide Alkaloids
incidence: 9/11*, conc. range:
100–300 μg**/kg***, country: USA[200], *leaf
blades, **ergovaline, ***dry weight
incidence: 9/11*, conc. range:
300–2800 μg**/kg***, Ø conc.: 1167 μg/kg,
country: USA[200], *leaf sheaths, **ergovaline,
***dry weight

Ergot Alkaloids
incidence: 9/11*, conc. range:
400–1500 μg**/kg, Ø conc.: 956 μg/kg,
country: USA[200], *leaf blades, **ergonovine

incidence: 9/11*, conc. range:
700–13,800 μg**/kg, Ø conc.: 4133 μg/kg,
country: USA[200], *leaf sheaths, **ergonovine
incidence: 3/3*, conc. range:
3729.3–9440.8 μg/kg**, Ø conc.: 6469.2 μg/kg,
country: USA[400], *seed, **dry weight
incidence: 4/4*, conc. range:
907.9–1405.9 μg/kg**, Ø conc.: 1188.4 μg/kg,
country: USA[400], *culm, **dry weight
incidence: 4/4*, conc. range:
237.5–706.3 μg/kg**, Ø conc.: 495 μg/kg,
country: USA[400], *plant, **dry weight
see also feed, feedstuff, grain, grain diets

Ergovaline
incidence: 9/11, conc. range: 20–4620 μg/kg,
Ø conc.: 2100 μg/kg, country: Spain[396]
incidence: 8/8*, conc. range: 166–340 μg/kg,
Ø conc.: 231.5 μg/kg, country: USA[401],
*samples were taken in spring
incidence: 8/8*, conc. range: 314–839 μg/kg,
Ø conc.: 554.4 μg/kg, country: USA[401],
*samples were taken in fall

Tapioca   Tapioca for feed may contain the
following mycotoxins:

Aflatoxin B$_1$
incidence: 1/1, conc.: 270 μg/kg, country:
India[247]
see also alfalfa, *Ambadi* cake, animal feedstuffs
(dairy cake), bagasse, barley, bengalgram
husk, bird food, bird food, wild, biri testa,
blackgram, blackgram husk, bran, broiler
mixed feed, calf fattening mixed feed, calf
fattening mixed feed (containing 4–20 %
peanut products), *Carthamus* cake, castor
cake, cereals, cereal products, chick pea,
coconut cake, cocos, concentrate, mixed,
concentrates, cotton cake, cottonseed,
cottonseed (dehulled), cottonseed cake,
cottonseed extract, cottonseed meal,
cottonseed meal (ammoniated), cottonseed
meal (decorticated), cottonseed meats,
cottonseed products, crumbles, crumbles,
grower, cycad meal, dairy cattle feed, dairy
cattle feed (containing 2–5 % peanut
products), dairy cattle feed (containing
6–10 % peanut products), dairy cattle feed

(containing 6–12 % peanut products), dairy cattle feed (containing more than 20 % peanut products), diets, mixed, dog food, egg production mixed feed, feed, feed and ingredients, feed, compound, feed, layer, feed, mixed, feed (beef), feed (broilers), feed (calf), feed (cat), feed (cattle), feed (chicken), feed (dairy), feed (dog), feed (dug), feed (fish), feed (gluten), feed (horse), feed (miscellaneous), feed (pig), feed (poultry), feed (poultry, pig), feed (rabbit), feed (sheep), feed (50–60 % maize), fish meal, grain by-products, grains (no specification), greengram, hay/silage, horsegram, husk, *Jagni* cake, legume mixture, linseed, linseed cake, livol, *Mahua* cake, maize, maize germ, maize gluten, maize grits, maize husk, maize meal, maize oil cake, maize powder, maize screenings, maize, ground, maize, hybrid, maize, preharvest, maize, yellow, maize (dark grains), *Makhana* (*Euryale ferox* Salisb) puffs, manioc, milk production mixed feed, mung testa, murkool, mustard cake, neem cake, niger cake, oats, palm kernel expeller cake, palm kernels, palm products, peanut cake, peanut cake (deoiled), peanut expeller, peanut hay, peanut meal, peanut, kernels, peanut, shells, peanuts, pellets, finisher, pig meal and pellets, pigeon pea, poultry feeds (peanut containing), rapeseed cake, redgram husk, rice, rice bran, rice bran (deoiled), rice chaff, rice crack, rice germ, rice germ cake, rice meal, rice straw, rice (damaged), rice (polish), safflower cake, sal seed cake, sesame, sesame cake, sorghum, soybean meal, soybeans, sunflower, sunflower cake, sunflower flour, wheat, wheat bran, wheat bran and chana testa

OCHRATOXIN A
incidence: 3/6*, conc. range: 2–4 µg/kg, Ø conc.: 3 µg/kg, country: The Netherlands[103], *imported?
see also alfalfa, barley, barley, oats, barley (high moisture), barley-soybean diet, bird food, domestic, bird food, wild, broilers feed, cereal grains, citrus pulp, coconut, expeller, corn cob mix silage, diet (dairy cow), diet (poultry), diet (starter), dog food, eat, egg

production mixed feed, feed, feed wheat, oat and barley, feed, commercial mix, feed, mixed, feed, mixed (pelleted), feed (broilers), feed (cat), feed (cattle), feed (cereals), feed (pig), feed ec (pig), feed (poultry), feed ec (poultry), feed (poultry, pig), feed ec (rabbit), feed (trout), grain, mixed feed, grains, mixed, grains (heated), hay, horse bean, maize, maize feed, milo, maize gluten, maize meal, maize, white, *Makhana* (*Euryale ferox* Salisb) puffs, milk production mixed feed, millet, oat and barley (hammer-milled), oats, palm products, peanut cake, peas, peas and beans, pet food, pig feedstuffs, pig grower diet, pig meal, piglet diet, poultry feedstuffs, rice bran, rice germ, rice germ cake, rye, sorghum, soybean groats, sunflower, sunflower seeds, extracted, triticale, *Vicia faba*, wheat, wheat and barley, wheat bran, wheat hay, wheat, oats

ZEARALENONE
incidence: 6/6*, conc. range: 17–86 µg/kg, Ø conc.: 32 µg/kg, country: The Netherlands[103], *imported?
see also alfalfa, barley, barley, husked, barley, unhusked (naked), barley and feed, bone meal, bran, broilers feed, *Carthamus* cake, chick pea, concentrate, mixed, corn cob mix silage, cotton cake, cottonseed, cottonseed cake, diet (dairy cow), diet (poultry), diets (mixed), feed, feed components, feed, mixed, feed, mixed (primarily maize also maize, oats, wheat), feed (bran), feed (broiler chicken), feed (cattle), feed (chicken), feed (dairy), feed (developing pig), feed (mill run, from wheat), feed (miscellaneous), feed (pig), feed (poultry), feed (poultry, pig), feed (starter chicken), feedstuff, fish meal, grain, bruised, grain, mixed feed, grains (no specification), hay, maize, maize ears, maize flakes, maize germ, maize germ/bran, maize gluten, maize kernels, maize meal, maize oil cake, maize screenings, maize stalks (pith), maize, "Baby", maize, hybrid, maize, shelled, maize, unshelled, maize, white, maize grain, artificially dried, maize grain, crib dried, maize grain, ensiled, milk production mixed feed, oats, *Paspalum palidosum*, straw, peanut hulls/skins, rice bran, rice germ, rice germ

cake, rye, silage, sorghum, soybeans, soybeans, extracted, sunflower cake, triticale, wheat, wheat bran, wheat bran and chana testa, wheat soya meal

Triticale   Triticale for feed may contain the following mycotoxins:

DEOXYNIVALENOL
incidence: 26/32, conc. range: 50–580 µg/kg, Ø conc.: 256.5 µg/kg, country: Hungary[6]
incidence: 3/3, conc. range: 1100–11,200 µg/kg, Ø conc.: 7100 µg/kg, country: Australia[256]
incidence: ?/10*, conc. range: 1300–16,400 µg/kg, Ø conc.: 7990 µg/kg, country: Poland[290], *ncac
see also barley, husked, barley, unhusked (naked), barley (pressed), bone meal, bran, broilers feed, calf fattening mixed feed, coconut, expeller, corn cob mix silage, cottonseed, cottonseed cake, dairy cattle feed, egg production mixed feed, feed, feed components, feed, commercial mix, feed, mixed, feed, mixed (primarily maize), feed (barley), feed (cattle), feed (chicken), feed (dog), feed (fish), feed (mill run, from wheat), feed (mink), feed (pig), feed (poultry), feed (reindeer), feeds, grain, feeds, industrial, feedstuff, feedstuffs (rapeseed, turnip, fish meal, concentrates), fish meal, grain, mixed feed, grains, mixed, grains (no specification), maize, maize ears, maize fibre, maize germ, maize germ/bran, maize germ meal, maize gluten, maize kernels, maize meal, maize powder, maize screenings, maize stalks (pith), maize, "Baby", maize, hybrid, maize, white, oats, rice bran, rice germ cake, rye, silage, sorghum, soybeans, wheat, wheat and barley, wheat, red hard winter, wheat, soft white winter, wheat, spring, wheat, winter

MONILIFORMIN
incidence: ?/?* **, conc. range: 2400–15,700 µg/kg, Ø conc.: 6100 µg/kg, country: Poland[222], *ncac, **Fdk
incidence: ?/?* **, conc. range: 250 µg/kg, Ø conc.: 250 µg/kg, country: Poland[222], *ncac, **hlk

see also barley, feed, mixed, feed (poultry), maize, maize flakes, maize germ, maize germ/bran, maize gluten, maize meal, maize screenings, maize, "Baby", oats, rice bran, wheat, wheat, summer, wheat, winter

NIVALENOL
incidence: 1/32, conc.: 100 µg/kg, country: Hungary[6]
see also barley, barley, husked, barley, unhusked (naked), barley (pressed), bran, feed, feed components, feed (cattle), feed (poultry), feed (reindeer), feeds, industrial, maize, maize ears, maize germ, maize germ/bran, maize gluten, maize kernels, maize meal, maize powder, maize screenings, maize, "Baby", oats, silage, rye, wheat

OCHRATOXIN A
incidence: 1/32, conc.: 120 µg/kg, country: Hungary[6]
see also alfalfa, barley, barley, oats, barley (high moisture), barley-soybean diet, bird food, domestic, bird food, wild, broilers feed, cereal grains, citrus pulp, coconut, expeller, corn cob mix silage, diet (dairy cow), diet (poultry), diet (starter), dog food, eat, egg production mixed feed, feed, feed wheat, oat and barley, feed, commercial mix, feed, mixed, feed, mixed (pelleted), feed (broilers), feed (cat), feed (cattle), feed (cereals), feed (pig), feed ec (pig), feed (poultry), feed ec (poultry), feed (poultry, pig), feed ec (rabbit), feed (trout), grain, mixed feed, grains, mixed, grains (heated), hay, horse bean, maize, maize feed, milo, maize gluten, maize meal, maize, white, *Makhana* (*Euryale ferox* Salisb) puffs, milk production mixed feed, millet, oat and barley (hammer-milled), oats, palm products, peanut cake, peas, peas and beans, pet food, pig feedstuffs, pig grower diet, pig meal, piglet diet, poultry feedstuffs, rice bran, rice germ, rice germ cake, rye, sorghum, soybean groats, sunflower, sunflower seeds, extracted, tapioca, *Vicia faba*, wheat, wheat and barley, wheat bran, wheat hay, wheat, oats

T-2 TOXIN
incidence: 1/32, conc.: 160 µg/kg, country: Hungary[6]

see also alfalfa, barley, bran, diet (grower), diet (poultry), feed, feed components, feed, layer, feed, mixed, feed (dog), feed (fish), feed (mink), feed (pig), feed (poultry), feedstuff, forage grass, grain, mixed feed, grains (no specification), hay, maize, maize germ/bran, maize gluten, maize meal, maize screenings, maize stalks (pith), oat and barley (hammer-milled), oats, peanuts, pig grower diet, piglet diet, rye, sorghum, wheat

ZEARALENONE
incidence: 21/32, conc. range: 50–580 µg/kg, Ø conc.: 216.2 µg/kg, country: Hungary[6]
see also alfalfa, barley, barley, husked, barley, unhusked (naked), barley and feed, bone meal, bran, broilers feed, *Carthamus* cake, chick pea, concentrate, mixed, corn cob mix silage, cotton cake, cottonseed, cottonseed cake, diet (dairy cow), diet (poultry), diets (mixed), feed, feed components, feed, mixed, feed, mixed (primarily maize also maize, oats, wheat), feed (bran), feed (broiler chicken), feed (cattle), feed (chicken), feed (dairy), feed (developing pig), feed (mill run, from wheat), feed (miscellaneous), feed (pig), feed (poultry), feed (poultry, pig), feed (starter chicken), feedstuff, fish meal, grain, bruised, grain, mixed feed, grains (no specification), hay, maize, maize ears, maize flakes, maize germ, maize germ/bran, maize gluten, maize kernels, maize meal, maize oil cake, maize screenings, maize stalks (pith), maize, "Baby", maize, hybrid, maize, shelled, maize, unshelled, maize, white, maize grain, artificially dried, maize grain, crib dried, maize grain, ensiled, milk production mixed feed, oats, *Paspalum palidosum*, straw, peanut hulls/skins, rice bran, rice germ, rice germ cake, rye, silage, sorghum, soybeans, soybeans, extracted, sunflower cake, tapioca, wheat, wheat bran, wheat bran and chana testa, wheat soya meal

Vicia faba *Vicia faba* for feed may contain the following mycotoxins:

OCHRATOXIN A
incidence: 1/6, conc.: 4 µg/kg, country: The Netherlands[103]

see also alfalfa, barley, barley, oats, barley (high moisture), barley-soybean diet, bird food, domestic, bird food, wild, broilers feed, cereal grains, citrus pulp, coconut, expeller, corn cob mix silage, diet (dairy cow), diet (poultry), diet (starter), dog food, eat, egg production mixed feed, feed, feed wheat, oat and barley, feed, commercial mix, feed, mixed, feed, mixed (pelleted), feed (broilers), feed (cat), feed (cattle), feed (cereals), feed (pig), feed ec (pig), feed (poultry), feed ec (poultry), feed (poultry, pig), feed ec (rabbit), feed (trout), grain, mixed feed, grains, mixed, grains (heated), hay, horse bean, maize, maize feed, milo, maize gluten, maize meal, maize, white, *Makhana* (*Euryale ferox* Salisb) puffs, milk production mixed feed, millet, oat and barley (hammer-milled), oats, palm products, peanut cake, peas, peas and beans, pet food, pig feedstuffs, pig grower diet, pig meal, piglet diet, poultry feedstuffs, rice bran, rice germ, rice germ cake, rye, sorghum, soybean groats, sunflower, sunflower seeds, extracted, tapioca, triticale, wheat, wheat and barley, wheat bran, wheat hay, wheat, oats

Wheat Wheat for feed may contain the following mycotoxins:

3-ACETYLDEOXYNIVALENOL
incidence: 4/7*, conc. range: 35–106 µg/kg, Ø conc.: 57 µg/kg, country: Finland[9], *imported ?
incidence: 12/25, conc.: ≤250 µg/kg, Ø conc.: 93 µg/kg, country: Romania[54]
incidence: 50/84, conc. range: 3–18 µg/kg, Ø conc.: 6.7 µg/kg, country: Germany[63]
incidence: 24/78, conc. range: 3–12 µg/kg, Ø conc.: 7 µg/kg, country: Germany[63]
incidence: 20/80, conc. range: 2–25 µg/kg, Ø conc.: 11.9 µg/kg, country: Germany[63]
incidence: 17/80, conc. range: 13–120 µg/kg, Ø conc.: 49.5 µg/kg, country: Germany[63]
incidence: 13/78, conc. range: 10–44 µg/kg, Ø conc.: 20.2 µg/kg, country: Germany[63]
incidence: 28/45, conc. range: 10–1902 µg/kg, Ø conc.: 209.2 µg/kg, country: Germany[63]

incidence: 657/821* **, conc. range: 50–29,540 µg/kg, country: Poland[252], *wheat heads, **ncac
incidence: 1/97, conc.: 100–1000 µg/kg (1 sa), country: UK[268]
incidence: 9/13*, conc. range: 300–3000 µg/kg, Ø conc.: 1740 µg/kg, country: Poland[315], *Fdk
incidence: 67/276, conc. range: 2–1886 µg/kg, Ø conc.: 207 µg/kg, country: Germany[362]
see also barley, feed components, feed (poultry), feeds, grain, feeds, industrial, feedstuffs (rapeseed, turnip, fish meal, concentrates), maize, maize ears, maize gluten, maize kernels, maize meal, maize screenings, oats

15-Acetyldeoxynivalenol
incidence: 8/25, conc. range: ≤380 µg/kg, Ø conc.: 150 µg/kg, country: Romania[54]
incidence: 13/84, conc. range: 1–15 µg/kg, Ø conc.: 5 µg/kg, country: Germany[63]
incidence: 10/80, conc. range: 5–150 µg/kg, Ø conc.: 41.9 µg/kg, country: Germany[63]
incidence: 2/78, conc. range: 6–116 µg/kg, Ø conc.: 61 µg/kg, country: Germany[63]
incidence: 3/45, conc. range: 7–181 µg/kg, Ø conc.: 71 µg/kg, country: Germany[63]
incidence: 1/13*, conc.: 130 µg/kg, country: Japan[67], *ncac
incidence: 2/283, conc. range: ≤22 µg/kg, Ø conc.: 18 µg/kg, country: UK[310]
incidence: 7/169, conc. range: 3–54 µg/kg, Ø conc.: 15.6 µg/kg, country: Germany[362]
see also barley, feed components, feed, commercial mix, feed (poultry), maize, maize ears, maize germ, maize germ/bran, maize gluten, maize kernels, maize meal, maize screenings, maize, "Baby", oats, silage

Acetyldeoxynivalenol
incidence: 12/25*, conc. range: ≤38 µg/kg, Ø conc.: 16 µg/kg, country: Poland[392], *ncac

Aflatoxin B$_1$
incidence: 21/50*, conc. range: ≤100 µg/kg, (1 sa), 101–250 µg/kg (9 sa), 251–500 µg/kg (4 sa), 501–1000 µg/kg (5 sa), >1000 µg/kg (2 sa, with a maximum of 1248 µg/kg), country: India[28], *ncac

incidence: 7/13, conc. range: 10–700 µg/kg, Ø conc.: 201.4 µg/kg, country: Australia[121]
incidence: 5/5, conc. range: 20–700 µg/kg, Ø conc.: 182 µg/kg, country: Australia[181]
incidence: 16/244, conc. range: ≤25 µg/kg (14 sa), 51–100 µg/kg (2 sa, Ø conc.: 17.4 µg/kg) country: India[247]
incidence: 4/4, conc. range: 59.4–112.6 µg/kg, Ø conc.: 92.7 µg/kg, country: India[253]
incidence: 12/122, conc. range: 3–40 µg/kg, Ø conc.: 12 µg/kg, country: Australia[270]
incidence: 7/20, conc. range: 3–18 µg/kg, Ø conc.: 6.5 µg/kg, country: Australia[270]
incidence: 10/30, conc. range: 3–500 µg/kg, Ø conc.: 64 µg/kg, country: Australia[270]
incidence: 2/531, conc. range: 7 µg/kg, country: USA[384]
see also alfalfa, *Ambadi* cake, animal feedstuffs (dairy cake), bagasse, barley, bengalgram husk, bird food, bird food, wild, biri testa, blackgram, blackgram husk, bran, broiler mixed feed, calf fattening mixed feed, calf fattening mixed feed (containing 4–20 % peanut products), *Carthamus* cake, castor cake, cereals, cereal products, chick pea, coconut cake, cocos, concentrate, mixed, concentrates, cotton cake, cottonseed, cottonseed (dehulled), cottonseed cake, cottonseed extract, cottonseed meal, cottonseed meal (ammoniated), cottonseed meal (decorticated), cottonseed meats, cottonseed products, crumbles, crumbles, grower, cycad meal, dairy cattle feed, dairy cattle feed (containing 2–5 % peanut products), dairy cattle feed (containing 6–10 % peanut products), dairy cattle feed (containing 6–12 % peanut products), dairy cattle feed (containing more than 20 % peanut products), diets, mixed, dog food, egg production mixed feed, feed, feed and ingredients, feed, compound, feed, layer, feed, mixed, feed (beef), feed (broilers), feed (calf), feed (cat), feed (cattle), feed (chicken), feed (dairy), feed (dog), feed (dug), feed (fish), feed (gluten), feed (horse), feed (miscellaneous), feed (pig), feed (poultry), feed (poultry, pig), feed (rabbit), feed (sheep), feed (50–60 % maize), fish meal,

grain by-products, grains (no specification), greengram, hay/silage, horsegram, husk, *Jagni* cake, legume mixture, linseed, linseed cake, livol, *Mahua* cake, maize, maize germ, maize gluten, maize grits, maize husk, maize meal, maize oil cake, maize powder, maize screenings, maize, ground, maize, hybrid, maize, preharvest, maize, yellow, maize (dark grains), *Makhana* (*Euryale ferox* Salisb) puffs, manioc, milk production mixed feed, mung testa, murkool, mustard cake, neem cake, niger cake, oats, palm kernel expeller cake, palm kernels, palm products, peanut cake, peanut cake (deoiled), peanut expeller, peanut hay, peanut meal, peanut, kernels, peanut, shells, peanuts, pellets, finisher, pig meal and pellets, pigeon pea, poultry feeds (peanut containing), rapeseed cake, redgram husk, rice, rice bran, rice bran (deoiled), rice chaff, rice crack, rice germ, rice germ cake, rice meal, rice straw, rice (damaged), rice (polish), safflower cake, sal seed cake, sesame, sesame cake, sorghum, soybean meal, soybeans, sunflower, sunflower cake, sunflower flour, tapioca, wheat bran, wheat bran and chana testa

Aflatoxin B$_2$
incidence: 4/244, conc. range: nc, Ø conc.:
4.9 µg/kg, country: India[247]
incidence: 4/122, conc. range: 1–4 µg/kg,
country: Australia[270]
incidence: 9/30, conc. range: 1–150 µg/kg,
country: Australia[270]
see also animal feedstuff (dairy cake), bird food, bird food, wild, biri testa, blackgram, blackgram husk, cottonseed, cottonseed cake, cottonseed extract, cottonseed meal, cottonseed meal (ammoniated), cottonseed meats, dog food, egg production mixed feed, feed, compound, feed (cat), feed (cattle), feed (dog), feed (pig), feed (poultry), feed (rabbit), feed (sheep), fish meal, horsegram, maize, maize gluten, maize husk, maize, ground, maize, preharvest, mung testa, mustard cake, niger cake, peanut cake, peanut cake (deoiled), peanut expeller, peanut hay, peanut meal, peanuts, peanut, kernels, redgram husk, rice bran, rice bran

(deoiled), rice chaff, rice meal, rice (polish), sal seed cake, sesame cake, sorghum, soybean meal, soybeans, wheat bran

Aflatoxin B
incidence: 17/32, conc. range:
≥25–>100 µg/kg, country: France[46]
see also barley, cocoa (oil cake), cottonseed cake, feed, feed (cattle), feed (maize, gluten), feed (pig), feed (poultry), flour (wheat), lucern (dried), maize, maize gluten, maize grains, oats, peanut (oil cake), rice, broken, sorghum, soybean (oil cake), sunflower (oil cake)

Aflatoxin G$_1$
incidence: 2/531, conc.: 2 µg/kg, country:
USA[384]
see also animal feedstuffs (dairy cake), bird food, bird food, wild, concentrate, mixed, cottonseed, cottonseed cake, feed (cat), feed (chicken), feed (dog), maize, maize, ground, maize, preharvest, meat meal, milk production mixed feed, murkool, peanut cake, peanut expeller, peanut hay, peanut meal, peanuts, rice bran, rice germ, sorghum, soybeans, wheat bran, wheat bran and chana testa

Aflatoxin
incidence: 16/244, conc. range: 1–79 µg/kg, country: India[247]
incidence: 14/14, conc. range: <30 µg/kg (4 sa), >30 µg/kg (6 sa), >200 µg/kg (2 sa), >1750 µg/kg (2 sa), country: India[380]
see also blackgram husk, bread crumbs, broiler finisher, broiler starter, cotton cake, cottonseed, cottonseed cake, cottonseed extract, cottonseed meal, feed, feed (cattle), feed (cow), feed (dog), feed (horse), feed (maize, gluten), feed (pig), feed (poultry), feed (rabbit), feed (rat/mice), feed (sheep), feeds, grain, fish meal, flour (wheat), groundnut cake, grower's mash, horsegram, layer's mash, maize, maize gluten, maize, white, milo, peanut cake, peanut cake (deoiled), peanut, kernels, peanut (oil cake), pearlmillet, pig breeder's mash, pig finisher, pig starter, pod with haulms, poultry breeder's mash, rabbit pellets, redgram husk,

rice, rice bran (deoiled), rice, broken, rice (polish), sesame cake, silk worm pupae, sorghum, soybean cake, soybean meal, wheat bran

ALTERNARIOL METHYL ETHER
incidence: 3/17, conc. range: 4–8 μg/kg, country: Germany[294]
see also barley, barley, oats, cereals, mixture, oats, sunflower, wheat, oats

BEAUVERICIN
incidence: 13/13*, conc. range: 640–3500 μg/kg, country: Finland[234], *ncac
see also barley, maize, oats, wheat, summer, wheat, winter

CITRININ
incidence: 15/50*, conc. range: 1–10 μg/kg, country: UK[94], *imported?
incidence: 1/94, conc.: 4400 μg/kg, country: Australia[240]
see also barley, barley, oats, barley-soybean diet, cottonseed cake, feed, feed, mixed, feed (cattle), feed (pig), fish meal, hay, maize, maize, white, *Makhana* (*Euryale ferox* Salisb) puffs, oats, palm products, peas and beans, rice bran, rice germ, wheat and other grains (moldy), wheat bran

CYCLOPIAZONIC ACID
incidence: 1/1, conc. range: 20,000 μg/kg (estimated), country: India[33]
see also chick mash, feed, groundnut cake, maize, millet, little, peanuts, rice bran

DEOXYNIVALENOL
incidence: 47/47*, conc. range: 40–415 μg/kg, Ø conc.: 111 μg/kg, country: Germany[5], *ncac
incidence: 58/58*, conc. range: 67–1020 μg/kg, Ø conc.: 280 μg/kg, country: Germany[5], *ncac
incidence: 287/367, conc. range: 70–1560 μg/kg, Ø conc.: 298.9 μg/kg, country: Hungary[6]
incidence: 7/7*, conc. range: 68–700 μg/kg, Ø conc.: 170 μg/kg, country: Finland[9], *imported?
incidence: 40/53, conc. range: 50–3650 μg/kg, Ø conc.: 434.8 μg/kg, country: Canada[49]

incidence: 1/2, conc.: 500 μg/kg, country: Hungary[50]
incidence: 25/25, conc. range: ≤5600 μg/kg, Ø conc.: 1500 μg/kg, country: Romania[54]
incidence: 81/84, conc. range: 4–20,538 μg/kg, Ø conc.: 1691.6 μg/kg, country: Germany[63]
incidence: 54/78, conc. range: 3–1187 μg/kg, Ø conc.: 152.1 μg/kg, country: Germany[63]
incidence: 77/80, conc. range: 8–8969 μg/kg, Ø conc.: 595 μg/kg, country: Germany[63]
incidence: 77/80, conc. range: 4–4627 μg/kg, Ø conc.: 359.1 μg/kg, country: Germany[63]
incidence: 74/78, conc. range: 18–5412 μg/kg, Ø conc.: 334.8 μg/kg, country: Germany[63]
incidence: 43/45, conc. range: 19–6165 μg/kg, Ø conc.: 391.2 μg/kg, country: Germany[63]
incidence: 9/13*, conc. range: 30–1280 μg/kg, Ø conc.: 522.2 μg/kg, country: Japan[67], *ncac
incidence: 5/80, conc. range: 10–1300 μg/kg, country: Germany[68]
incidence: 1/3, conc.: 400 μg/kg, country: Brazil[80]
incidence: 3/15, conc. range: 560–590 μg/kg, Ø conc.: 576.7 μg/kg, country: Brazil[81]
incidence: 1/15, conc.: 470 μg/kg, country: Brazil[81]
incidence: 2/7, conc. range: 110–270 μg/kg, Ø conc.: 190 μg/kg, country: The Netherlands[103]
incidence: 5/9, conc. range: 6–173 μg/kg, Ø conc.: 23 μg/kg, country: Korea[108]
incidence: ?/81*, conc. range: 41–9330 μg/kg, Ø conc.: 1570 μg/kg, country: USA[126], *ncac, ncaps
incidence: 20/40*, conc. range: 8–1571 μg/kg, Ø conc.: 238.8 μg/kg, country: Norway[132], *probably feed
incidence: 13/13*, conc. range: 8–3193 μg/kg, Ø conc.: 419.9 μg/kg, country: Norway[132], *probably feed
incidence: 24/1291, conc. range: tr–1700 μg/kg, country: Australia[137]
incidence: 4/60, conc. range: tr–70 μg/kg, country: Australia[137]
incidence: 6/80, conc. range: 20–780 μg/kg, country: Australia[137]
incidence: 19/22* **, conc. range: 76–1654 μg/kg, Ø conc.: 452.8 μg/kg,

country: Germany[152], * and from different countries, **ncac
incidence: 94/94*, conc. range: <10–>10,000 µg/kg, country: Germany[192], *ncac
incidence: 8/14, conc. range: 110–1180 µg/kg, Ø conc.: 400 µg/kg, country: Sweden[197]
incidence: 23/29, conc. range: 60–360 µg/kg, Ø conc.: 190 µg/kg, country: Sweden[197]
incidence: 115/117, conc. range: 20–99 µg/kg (79 sa), 100–499 µg/kg (35 sa), 600 µg/kg (1 sa), country: UK[203]
incidence: 4/12, conc. range: 1500–10,800 µg/kg, country: New Zealand[211]
incidence: 3/3, conc. range: 3750–14,360 µg/kg, Ø conc.: 8700 µg/kg, country: South Africa[212]
incidence: 13/42*, conc. range: 7–309 µg/kg, Ø conc.: 94.7 µg/kg, country: Poland[214], *ncac
incidence: 9/15*, conc. range: 9–1285 µg/kg, Ø conc.: 335 µg/kg, country: Europe[214], *ncac
incidence: 501/737*, conc. range: 30–4300 µg/kg, country: Norway[224], *ncac
incidence: 2/3*, conc. range: 66–740 µg/kg, Ø conc.: 403 µg/kg, country: Japan[230], *ncac
incidence: 1/4*, conc.: 91 µg/kg, country: Japan[230], *ncac
incidence: 15/18*, conc. range: tr–4700 µg/kg, Ø conc.: 1000 µg/kg, country: Japan[231], *ncac
incidence: 2/2*, conc. range: 540–1100 µg/kg, Ø conc.: 829 µg/kg, country: Norway[235], *ncac
incidence: 2/2*, conc. range: <30 µg/kg, country: Norway[235], *ncac
incidence: 37–40%/61*, conc. range: 10–15,300 µg/kg, country: Poland[236], *ncac
incidence: 61–65%/68*, conc. range: 30–24,290 µg/kg, country: Poland[236], *ncac
incidence: 4/342, conc. range: 140–36,700 µg/kg, Ø conc.: 18,600 µg/kg, country: USA[237]
incidence: 2/94, conc. range: 20–200 µg/kg, Ø conc.: 110 µg/kg, country: Australia[240]
incidence: 1/1, conc.: 6900 µg/kg, country: Canada[249]
incidence: 821/821* **, conc. range: 210–30,400 µg/kg, country: Poland[252], *wheat heads, **ncac

incidence: 7/12, conc. range: ≤6700 µg/kg, Ø conc.: 1800 µg/kg, country: Australia[256]
incidence: 39/78, conc. range: nd–102 µg/kg, Ø conc.: 21 µg/kg, country: Poland[264]
incidence: 82/151*, conc. range: 5–1620 µg/kg, Ø conc.: 162 µg/kg, country: Japan[265], *ncac
incidence: 8/97, conc. range: <100 µg/kg (6 sa), 100–1000 µg/kg (1 sa), >1000 µg/kg (1 sa), country: UK[268]
incidence: ?/33*, conc. range: 2400–35,000 µg/kg, country: Poland[290], *ncac
incidence: 226/283, conc. range: ≤5175 µg/kg, Ø conc.: 100 µg/kg, country: UK[310]
incidence: 11/13*, conc. range: 15,900–39,600 µg/kg, Ø conc.: 28,190 µg/kg, country: Poland[315], *Fdk
incidence: 10/13*, conc. range: 400–3600 µg/kg, Ø conc.: 1700 µg/kg, country: Poland[315], *hlk
incidence: 79/99, conc. range: ≤9160 µg/kg, Ø conc.: 1500 µg/kg, country: Canada[346]
incidence: 104/150***, conc. range: ≤11,660 µg/kg, Ø conc.: 1540 µg/kg, country: Germany[355], *ncac, **cg
incidence: 25/46***, conc. range: ≤4220 µg/kg, Ø conc.: 760 µg/kg, country: Germany[355], *ncac, **og
incidence: 236/276, conc. range: 4–15,869 µg/kg, Ø conc.: 320.7 µg/kg, country: Germany[362]
incidence: 24/25*, conc. range: ≤371 µg/kg, Ø conc.: 104 µg/kg, country: Poland[392], *ncac
incidence: 53/84*, conc. range: ≤202 µg/kg, Ø conc.: 64 µg/kg, country: Lithuania[399], *ncac
incidence: 28/46*, conc. range: tr–580 µg/kg, Ø conc.: 30 µg/kg, country: France[404], *ncac
incidence: 62/69*, conc. range: tr–650 µg/kg, Ø conc.: 80 µg/kg, country: France[404], *ncac
incidence: 87/99, conc. range: <50–4300 µg/kg, Ø conc.: 1152 µg/kg, country: Hungary[411]
see also barley, barley, husked, barley, unhusked (naked), barley (pressed), bone meal, bran, broilers feed, calf fattening mixed feed, coconut, expeller, corn cob mix silage, cottonseed, cottonseed cake, dairy cattle feed,

egg production mixed feed, feed, feed
components, feed, commercial mix, feed,
mixed, feed, mixed (primarily maize), feed
(barley), feed (cattle), feed (chicken), feed
(dog), feed (fish), feed (mill run, from
wheat), feed (mink), feed (pig), feed
(poultry), feed (reindeer), feeds, grain, feeds,
industrial, feedstuff, feedstuffs (rapeseed,
turnip, fish meal, concentrates), fish meal,
grain, mixed feed, grains, mixed, grains (no
specification), maize, maize ears, maize fibre,
maize germ, maize germ/bran, maize germ
meal, maize gluten, maize kernels, maize
meal, maize powder, maize screenings, maize
stalks (pith), maize, "Baby", maize, hybrid,
maize, white, oats, rice bran, rice germ cake,
rye, silage, sorghum, soybeans, triticale,
wheat and barley, wheat, red hard winter,
wheat, soft white winter, wheat, spring,
wheat, winter

DIACETOXYSCIRPENOL
incidence: 5/59, conc. range: 300–2000 μg/kg,
country: Germany[68]
incidence: 2/4, conc.: 300 μg/kg, country:
Argentina[80]
incidence: 1/15, conc.: 600 μg/kg, country:
Brazil[81]
incidence: 1/97, conc.: <100 μg/kg (1 sa),
country: UK[268]
see also alfalfa, barley, feed, feed components,
feed (dog), feed (fish), feed (mink), feed
(pig), feed (poultry), feed (reindeer), feeds,
grain, feedstuff, forage grass, grain, mixed
feed, grains (no specification), maize, maize
gluten, oat and barley (hammer-milled), oats,
peanuts, soybeans

ENNIATIN A₁
incidence: 10/13*, conc. range: tr–6900 μg/kg,
country: Finland[234], *ncac
see also barley, oats, rye, wheat, summer,
wheat, winter

ENNIATIN B
incidence: 8/13*, conc. range: tr–1900 μg/kg,
country: Finland[234], *ncac
see also barley, oats, rye, wheat, summer,
wheat, winter

ENNIATIN B₁
incidence: 12/13*, conc. range: tr–4800 μg/kg,
country: Finland[234], *ncac
see also barley, oats, rye, wheat, summer,
wheat, winter

FUMONISIN B₁
incidence: 8/17, conc. range: 200–8800 μg/kg,
Ø conc.: 4800 μg/kg, country: Spain[155]
see also barley, bird food, wild, dog food,
feed, feed, complete ration, feed, general,
feed, layer, feed, maize-based, feed, mixed,
feed, pelleted ration, feed, screenings, feed,
sweet, feed (broilers), feed (cat), feed
(chicken), feed (dog), feed (gluten), feed
(horse), feed (maize), feed (pig), feed
(poultry), feed (rat), feed (rodent), forage
grass, maize, maize and maize screenings,
maize bran, maize ears, maize fine fractions,
maize flakes, maize germ, maize germ/bran,
maize germ meal, maize gluten, maize grits,
maize kernels, maize meal, maize powder,
maize screenings, maize, "Baby", maize,
ground, maize, preharvest, maize, sweet feed,
maize/oats mix, rat chow, silage, sorghum,
soybeans

FUMONISIN B₂
incidence: 1/17, conc.: 200 μg/kg, country:
Spain[155]
see also barley, bird food, wild, dog food,
feed, feed, maize-based, feed, mixed, feed
(cat), feed (dog), feed (gluten), feed (horse),
feed (maize), feed (poultry), feed (rodent),
maize, maize bran, maize fine fractions,
maize flakes, maize germ, maize germ/bran,
maize germ meal, maize gluten, maize grits,
maize kernels, maize meal, maize powder,
maize screenings, maize, "Baby", maize,
ground, maize, preharvest, rat chow

FUSARENONE-X
incidence: 3/13*, conc. range: 20 μg/kg,
Ø conc.: 20 μg/kg, country: Japan[67], *ncac
see also barley, feed components, feed
(poultry), maize, maize ears, maize
germ/bran, maize gluten, maize kernels,
maize meal, maize screenings, oats

HT-2 Toxin
incidence: 1/2, conc.: 200 µg/kg, country:
Hungary[50]
incidence: 6/84, conc. range: 2–20 µg/kg,
Ø conc.: 8.8 µg/kg, country: Germany[63]
incidence: 6/78, conc. range: 12–22 µg/kg,
Ø conc.: 17.2 µg/kg, country: Germany[63]
incidence: 1/80, conc.: 17 µg/kg, country:
Germany[63]
incidence: 7/78, conc. range: 8–150 µg/kg,
Ø conc.: 50.8 µg/kg, country: Germany[63]
incidence: 1/80, conc.: 100 µg/kg, country:
Germany[68]
incidence: 3/97, conc. range: <100 µg/kg
(3 sa), country: UK[268]
incidence: 82/283, conc. range: ≤193 µg/kg,
Ø conc.: 25 µg/kg, country: UK[310]
incidence: 1/?, conc.: 1000 µg/kg, country:
Canada[346]
incidence: 7/276, conc. range: 5–59 µg/kg,
Ø conc.: 25 µg/kg, country: Germany[362]
incidence: 20/84*, conc. range: ≤110 µg/kg,
Ø conc.: 34 µg/kg, country: Lithuania[399], *ncac
see also barley, feed components, feed (dog),
feed (fish), feed (pig), feed (poultry), feed
(reindeer), grain, mixed feed, grains, mixed,
grains (no specification), maize, maize
germ/bran, maize gluten, maize meal, maize
screenings, oats, rye, silage

Moniliformin
incidence: 8/13*, conc. range: tr–87 µg/kg,
Ø conc.: <40 µg/kg, country: Norway[204],
*ncac
incidence: 33/35*, conc. range: tr–420 µg/kg,
Ø conc.: 92 µg/kg, country: Norway[204], *ncac
incidence: 35/35*, conc. range: tr–950 µg/kg,
Ø conc.: 210 µg/kg, country: Norway[204],
*ncac
incidence: ?/?* **, conc. range:
7200–25,200 µg/kg, Ø conc.: 15,900 µg/kg,
country: Poland[222], *ncac, **Fdk
incidence: ?/?* **, conc. range: 250–700 µg/kg,
Ø conc.: 420 µg/kg, country: Poland[222],
*ncac, **hlk
incidence: 60–65 %/61*, conc. range:
10–720 µg/kg, country: Poland[236], *ncac
incidence: 89–95 %/68*, conc. range:
10–1720 µg/kg, country: Poland[236], *ncac

incidence: 48/78, conc. range: nd–495 µg/kg,
Ø conc.: 182 µg/kg, country: Poland[264]
incidence: 7/25*, conc. range: ≤198 µg/kg,
Ø conc.: 63 µg/kg, country: Poland[392], *ncac
see also barley, feed, mixed, feed (poultry),
maize, maize flakes, maize germ, maize
germ/bran, maize gluten, maize meal, maize
screenings, maize, "Baby", oats, rice bran,
triticale, wheat, summer, wheat, winter

Monoacetoxyscirpenol
incidence: 2/97, conc. range: <100 µg/kg
(2 sa), country: UK[268]
see also feed components, maize, maize
gluten, maize meal, maize screenings, silage

Neosolaniol
incidence: 27/35*, conc. range:
144.6–853.7 µg/kg, Ø conc.: 476.2 µg/kg,
country: Iran[238], *ncac
see also feed, feed components, feed, mixed,
feed (poultry), peanut cake

Nivalenol
incidence: 33/367, conc. range: 50–590 µg/kg,
Ø conc.: 177.3 µg/kg, country: Hungary[6]
incidence: 21/84, conc. range: 3–32 µg/kg,
Ø conc.: 9 µg/kg, country: Germany[63]
incidence: 33/78, conc. range: 3–58 µg/kg,
Ø conc.: 19.5 µg/kg, country: Germany[63]
incidence: 31/80, conc. range: 4–218 µg/kg,
Ø conc.: 42.8 µg/kg, country: Germany[63]
incidence: 47/80, conc. range: 1–188 µg/kg,
Ø conc.: 22.1 µg/kg, country: Germany[63]
incidence: 50/78, conc. range: 3–219 µg/kg,
Ø conc.: 33.3 µg/kg, country: Germany[63]
incidence: 15/45, conc. range: 5–47 µg/kg,
Ø conc.: 15.7 µg/kg, country: Germany[63]
incidence: 10/42, conc. range: ≥50–140 µg/kg,
Ø conc.: 85 µg/kg, country: Sweden[65]
incidence: 1/42, conc.: 60 µg/kg, country:
Sweden[65]
incidence: 1/96, conc.: 75 µg/kg, country:
Sweden[65]
incidence: 2/13*, conc. range: 40–1220 µg/kg,
Ø conc.: 630 µg/kg, country: Japan[67], *ncac
incidence: 2/15, conc. range: 200–400 µg/kg,
Ø conc.: 300 µg/kg, country: Brazil[81]
incidence: 1/15, conc.: 160 µg/kg, country:
Brazil[81]

incidence: 9/9, conc. range: 82–3169 µg/kg,
Ø conc.: 535 µg/kg, country: Korea[108]
incidence: 40/40*, conc. range: 15–887 µg/kg,
Ø conc.: 59.7 µg/kg, country: Norway[132],
*probably feed
incidence: 13/13*, conc. range: 30–150 µg/kg,
Ø conc.: 57.5 µg/kg, country: Norway[132],
*probably feed
incidence: 4/12, conc. range: 20–770 µg/kg,
country: New Zealand[211]
incidence: 37/42*, conc. range: 3–350 µg/kg,
Ø conc.: 47.9 µg/kg, country: Poland[214], *ncac
incidence: 9/15*, conc. range: 2–60 µg/kg,
Ø conc.: 23 µg/kg, country: Europe[214], *ncac
incidence: 3/737*, conc. range: 50–54 µg/kg,
Ø conc.: 54 µg/kg, country: Norway[224], *ncac
incidence: 2/3*, conc. range: 260–1630 µg/kg,
Ø conc.: 945 µg/kg, country: Japan[230], *ncac
incidence: 15/18*, conc. range: tr–7800 µg/kg,
Ø conc.: 1500 µg/kg, country: Japan[231], *ncac
incidence: 60–65 %/61*, conc. range:
10–14,200 µg/kg, country: Poland[236], *ncac
incidence: 89–95 %/68*, conc. range:
20–320 µg/kg, country: Poland[236], *ncac
incidence: 34/35*, conc. range:
49.1–1119.1 µg/kg, Ø conc.: 577.6 µg/kg,
country: Iran[238], *ncac
incidence: 23/78, conc. range: nd–99 µg/kg,
Ø conc.: 34 µg/kg, country: Poland[264]
incidence: 100/151*, conc. range:
5–4390 µg/kg, Ø conc.: 136 µg/kg, country:
Japan[265], *ncac
incidence: 21/97, conc. range: <100 µg/kg
(14 sa), 100–1000 µg/kg (5 sa), >1000 µg/kg
(2 sa), country: UK[268]
incidence: 226/283, conc. range: ≤428 µg/kg,
Ø conc.: 42 µg/kg, country: UK[310]
incidence: 104/276, conc. range: 2–145 µg/kg,
Ø conc.: 36.3 µg/kg, country: Germany[362]
incidence: 19/25*, conc. range: ≤453 µg/kg,
Ø conc.: 97 µg/kg, country: Poland[392], *ncac
see also barley, barley, husked, barley,
unhusked (naked), barley (pressed), bran,
feed, feed components, feed (cattle), feed
(poultry), feed (reindeer), feeds, industrial,
maize, maize ears, maize germ, maize
germ/bran, maize gluten, maize kernels,
maize meal, maize powder, maize screenings,
maize, "Baby", oats, rye, silage, triticale

OCHRATOXIN A

incidence: 2/14*, conc. range: 0.6–0.7 µg/kg,
Ø conc.: 0.7 µg/kg, country: Germany[5], *ncac
incidence: 7/29*, conc. range: 0.6–0.8 µg/kg,
Ø conc.: 0.6 µg/kg, country: Germany[5], *ncac
incidence: 9/367, conc. range: 90–320 µg/kg,
Ø conc.: 207.8 µg/kg, country: Hungary[6]
incidence: 8/64*, conc. range:
0.1–137.3 µg/kg, Ø conc.: 17.9 µg/kg,
country: Germany[13], *ncac
incidence: 1/3, conc.: 10 µg/kg, country:
Egypt[16]
incidence: 6/95, conc. range: 30–6000 µg/kg,
country: Canada[17]
incidence: 11/64, conc. range: ≤9.2 µg/kg,
Ø conc.: 1.65 µg/kg, country: UK[43]
incidence: 1/25, conc.: 37 µg/kg, country:
Romania[54]
incidence: 1/4, conc.: 40 µg/kg, country:
Brazil[80]
incidence: 15/125, conc. range: 5–100 µg/kg,
country: Poland[84]
incidence: 10/50*, conc. range: 1–17 µg/kg,
country: UK[94], *imported?
incidence: 22/117, conc. range: 0.3–9.9 µg/kg
(17 sa), 10–99.9 µg/kg (4 sa), 231 µg/kg (1 sa),
country: UK[203]
incidence: 2/7345, conc. range: nc, country:
Hungary[209]
incidence: 2/94, conc. range: 900–1100 µg/kg,
Ø conc.: 1000 µg/kg, country: Canada[240]
incidence: 1/1, conc.: 13.7 µg/kg, country:
UK[282]
incidence: 18/37***, conc. range:
0.6–1024 µg/kg, Ø conc.: 267 µg/kg, country:
Poland[339], *ncac, **cg
incidence: 8/34***, conc. range: 0.8–1.6 µg/kg,
Ø conc.: 1.2 µg/kg, country: Poland[339], *ncac,
**eg
incidence: 3/39***, conc. range: 0.5–1.2 µg/kg,
Ø conc.: 0.8 µg/kg, country: Poland[341], *ncac,
**eg
incidence: 1/?, conc.: 17 µg/kg, country:
Canada[346]
incidence: 74/92*, conc. range:
0.02–160 µg/kg, country: Croatia[371], *ncac
see also alfalfa, barley, barley (high moisture),
barley, oats, barley-soybean diet, bird food,

domestic, bird food, wild, broilers feed, cereal grains, citrus pulp, coconut, expeller, corn cob mix silage, diet (dairy cow), diet (poultry), diet (starter), dog food, eat, egg production mixed feed, feed, feed wheat, oat and barley, feed, commercial mix, feed, mixed, feed, mixed (pelleted), feed (broilers), feed (cat), feed (cattle), feed (cereals), feed (pig), feed ec (pig), feed (poultry), feed ec (poultry), feed (poultry, pig), feed ec (rabbit), feed (trout), grain, mixed feed, grains, mixed, grains (heated), hay, horse bean, maize, maize feed, milo, maize gluten, maize meal, maize, white, *Makhana* (*Euryale ferox* Salisb) puffs, milk production mixed feed, millet, oat and barley (hammer-milled), oats, palm products, peanut cake, peas, peas and beans, pet food, pig feedstuffs, pig grower diet, pig meal, piglet diet, poultry feedstuffs, rice bran, rice germ, rice germ cake, rye, sorghum, soybean groats, sunflower, sunflower seeds, extracted, tapioca, triticale, wheat and barley, wheat bran, wheat hay, wheat, oats

PATULIN
incidence: 1/7345, conc.: nc, country: Hungary[209]
see also diet (grower), hay, maize meal, piglet diet

RUBRATOXIN B
incidence: 1/7345, conc.: nc, country: Hungary[209]
see also diet (grower), diet (poultry), diet (starter), pig diet (fattening), pig grower diet, soybean groats

STERIGMATOCYSTIN
incidence: 1/1, conc.: 300 µg/kg, country: Canada[92]
incidence: 1/50*, conc.: 18 µg/kg, country: UK[94], *imported?
see also feed, mixed (crumbled), feed (excluding peanut meal, suspect)

T-2 TETRAOL
incidence: 1/4, conc.: 1680 µg/kg, country: Argentina[80]
see also barley, feed components, maize, silage

T-2 TOXIN
incidence: 24/367, conc. range: 80–370 µg/kg, Ø conc.: 193.8 µg/kg, country: Hungary[6]
incidence: 2/2, conc. range: 200–1900 µg/kg, Ø conc.: 1000.5 µg/kg, country: Hungary[50]
incidence: 6/25, conc. range: ≤63 µg/kg, Ø conc.: 26 µg/kg, country: Romania[54]
incidence: 22/84, conc. range: 3–249 µg/kg, Ø conc.: 82.5 µg/kg, country: Germany[63]
incidence: 5/78, conc. range: 10–12 µg/kg, Ø conc.: 11 µg/kg, country: Germany[63]
incidence: 9/80, conc. range: 10–136 µg/kg, Ø conc.: 51.7 µg/kg, country: Germany[63]
incidence: 3/80, conc. range: 4–16 µg/kg, Ø conc.: 10 µg/kg, country: Germany[63]
incidence: 18/45, conc. range: 3–94 µg/kg, Ø conc.: 20.1 µg/kg, country: Germany[63]
incidence: 1/59, conc.: 300 µg/kg, country: Germany[68]
incidence: 4/81, conc. range: 200–500 µg/kg, country: Germany[68]
incidence: 2/4, conc. range: 350–360 µg/kg, Ø conc.: 355 µg/kg, country: Argentina[80]
incidence: 1/15, conc.: 400 µg/kg, country: Brazil[81]
incidence: 1/15, conc.: 800 µg/kg, country: Brazil[81]
incidence: 6/283, conc. range: ≤21 µg/kg, Ø conc.: 16 µg/kg, country: UK[310]
incidence: 16/276, conc. range: 4–235 µg/kg, Ø conc.: 31 µg/kg, country: Germany[362]
incidence: 8/84*, conc. range: ≤58 µg/kg, Ø conc.: 33 µg/kg, country: Lithuania[399], *ncac
see also alfalfa, barley, bran, diet (grower), diet (poultry), feed, feed components, feed, layer, feed, mixed, feed (dog), feed (fish), feed (mink), feed (pig), feed (poultry), feedstuff, forage grass, grain, mixed feed, grains (no specification), hay, maize, maize germ/bran, maize gluten, maize meal, maize screenings, maize stalks (pith), oat and barley (hammer-milled), oats, peanuts, pig grower diet, piglet diet, rye, sorghum, triticale

T-2 TRIOL
incidence: 11/80, conc. range: 100–700 µg/kg, country: Germany[68]

incidence: 2/283, conc. range: ≤15 µg/kg,
Ø conc.: 14 µg/kg, country: UK[310]
see also barley, feed components, grain,
mixed feed, grains (no specification),
maize, oats

α-Zearalenol
incidence: 1/45, conc.: 8 µg/kg, country:
Germany[63]

β-Zearalenol
incidence: 1/84, conc.: 12 µg/kg, country:
Germany[63]

Zearalenone
incidence: 215/367, conc. range:
50–890 µg/kg, Ø conc.: 210.3 µg/kg, country:
Hungary[6]
incidence: 1/2, conc.: 200 µg/kg, country:
Hungary[50]
incidence: 25/25, conc. range: ≤170 µg/kg,
Ø conc.: 23 µg/kg, country: Romania[54]
incidence: 67/84, conc. range: 1–8036 µg/kg,
Ø conc.: 178 µg/kg, country: Germany[63]
incidence: 11/78, conc. range: 1–6 µg/kg,
Ø conc.: 3.2 µg/kg, country: Germany[63]
incidence: 9/80, conc. range: 1–15 µg/kg,
Ø conc.: 5.1 µg/kg, country: Germany[63]
incidence: 10/80, conc. range: 1–109 µg/kg,
Ø conc.: 20.3 µg/kg, country: Germany[63]
incidence: 15/78, conc. range: 1–20 µg/kg,
Ø conc.: 4.3 µg/kg, country: Germany[63]
incidence: 28/45, conc. range: 2–52 µg/kg,
Ø conc.: 11.1 µg/kg, country: Germany[63]
incidence: 6/13*, conc. range: 2–25 µg/kg,
Ø conc.: 12.5 µg/kg, country: Japan[67], *ncac
incidence: 6/59, conc. range: 10–200 µg/kg,
country: Germany[68]
incidence: 7/81, conc. range: 20–2000 µg/kg,
country: Germany[68]
incidence: 1/15, conc.: 400 µg/kg, country:
Brazil[81]
incidence: 2/15, conc. range: 130–210 µg/kg,
Ø conc.: 170 µg/kg, country: Brazil[81]
incidence: 1/7, conc.: 14 µg/kg, country: The
Netherlands[103]
incidence: 7/21, conc. range: 4.7–63.6 µg/kg,
Ø conc.: 26.9 µg/kg, country: Germany[107]
incidence: 5/9, conc. range: 3–1254 µg/kg,
Ø conc.: 141 µg/kg, country: Korea[108]

incidence: 1/40*, conc.: 23 µg/kg, country:
Norway[132], *probably feed
incidence: 2/13*, conc. range: 2–4 µg/kg,
Ø conc.: 3 µg/kg, country: Norway[132],
*probably feed
incidence: 1/1291, conc.: 40 µg/kg, country:
Australia[137]
incidence: 3/80, conc. range: 10–100 µg/kg,
Ø conc.: 46.7 µg/kg, country: Australia[137]
incidence: ?/10, conc. range: 10–40 µg/kg,
country: India[183]
incidence: 1/7345, conc.: nc, country:
Hungary[209]
incidence: 12/12, conc. range: 40–350 µg/kg,
country: New Zealand[211]
incidence: 1/42*, conc.: 76 µg/kg, country:
Poland[214], *ncac
incidence: 35/35*, conc. range:
1266.3–5487.5 µg/kg, Ø conc.: 3464.3 µg/kg,
country: Iran[238], *ncac
incidence: 1/97, conc.: <100 µg/kg (1 sa),
country: UK[268]
incidence: 19/42*, conc. range:
360–11,050 µg/kg, country: USA[273], *ncac
incidence: 19/102*, conc. range:
364–11,054 µg/kg, Ø conc.: 2721.6 µg/kg,
country: USA[300], *ncac
incidence: 17/283, conc. range: ≤188 µg/kg,
Ø conc.: 35 µg/kg, country: UK[310]
incidence: 5/13*, conc. range:
100–1800 µg/kg, Ø conc.: 760 µg/kg, country:
Poland[315], *Fdk
incidence: 16/?, conc. range: ≤110 µg/kg,
country: Canada[346]
incidence: 10/135* **, conc. range:
≤250 µg/kg, Ø conc.: 74 µg/kg, country:
Germany[355], *ncac, **cg
incidence: 2/46* **, conc. range: ≤55 µg/kg,
Ø conc.: 47 µg/kg, country: Germany[355],
*ncac, **cg
incidence: 55/276, conc. range: 1–123 µg/kg,
Ø conc.: 8.1 µg/kg, country: Germany[362]
see also alfalfa, barley, barley, husked, barley,
unhusked (naked), barley and feed, bone
meal, bran, broilers feed, *Carthamus* cake,
chick pea, concentrate, mixed, corn cob mix
silage, cotton cake, cottonseed, cottonseed
cake, diet (dairy cow), diet (poultry), diets

(mixed), feed, feed components, feed, mixed, feed, mixed (primarily maize also maize, oats, wheat), feed (bran), feed (broiler chicken), feed (cattle), feed (chicken), feed (dairy), feed (developing pig), feed (mill run, from wheat), feed (miscellaneous), feed (pig), feed (poultry), feed (poultry, pig), feed (starter chicken), feedstuff, fish meal, forage grass, grain, bruised, grain, mixed feed, grains (no specification), hay, maize, maize ears, maize flakes, maize germ, maize germ/bran, maize gluten, maize kernels, maize meal, maize oil cake, maize screenings, maize stalks (pith), maize, "Baby", maize, hybrid, maize, shelled, maize, unshelled, maize, white, maize grain, artificially dried, maize grain, crib dried, maize grain, ensiled, milk production mixed feed, oats, *Paspalum palidosum*, straw, peanut hulls/skins, rice bran, rice germ, rice germ cake, rye, silage, sorghum, soybeans, soybeans, extracted, sunflower cake, tapioca, triticale, wheat bran, wheat bran and chana testa, wheat soya meal

Wheat and barley   Wheat and barley for feed may contain the following mycotoxins:

DEOXYNIVALENOL
incidence: 1/94, conc.: 95 µg/kg, country: USA[240]
see also barley, barley, husked, barley, unhusked (naked), barley (pressed), bone meal, bran, broilers feed, calf fattening mixed feed, coconut, expeller, corn cob mix silage, cottonseed, cottonseed cake, dairy cattle feed, egg production mixed feed, feed, feed components, feed, commercial mix, feed, mixed, feed, mixed (primarily maize), feed (barley), feed (cattle), feed (chicken), feed (dog), feed (fish), feed (mill run, from wheat), feed (mink), feed (pig), feed (poultry), feed (reindeer), feeds, grain, feeds, industrial, feedstuff, feedstuffs (rapeseed, turnip, fish meal, concentrates), fish meal, grain, mixed feed, grains, mixed, grains (no specification), maize, maize ears, maize fibre, maize germ, maize germ/bran, maize germ meal, maize gluten, maize kernels, maize meal, maize powder, maize screenings, maize

stalks (pith), maize, "Baby", maize, hybrid, maize, white, oats, rice bran, rice germ cake, rye, silage, sorghum, soybeans, triticale, wheat, wheat, red hard winter, wheat, soft white winter, wheat, spring, wheat, winter

OCHRATOXIN A
incidence: 23/95, conc. range: ≤102 µg/kg, country: UK[12]
see also alfalfa, barley, barley, oats, barley (high moisture), barley-soybean diet, bird food, domestic, bird food, wild, broilers feed, cereal grains, citrus pulp, coconut, expeller, corn cob mix silage, diet (dairy cow), diet (poultry), diet (starter), dog food, eat, egg production mixed feed, feed, feed wheat, oat and barley, feed, commercial mix, feed, mixed, feed, mixed (pelleted), feed (broilers), feed (cat), feed (cattle), feed (cereals), feed (pig), feed ec (pig), feed (poultry), feed ec (poultry), feed (poultry, pig), feed ec (rabbit), feed (trout), grain, mixed feed, grains, mixed, grains (heated), hay, horse bean, maize, maize feed, milo, maize gluten, maize meal, maize, white, *Makhana* (*Euryale ferox* Salisb) puffs, milk production mixed feed, millet, oat and barley (hammer-milled), oats, palm products, peanut cake, peas, peas and beans, pet food, pig feedstuffs, pig grower diet, pig meal, piglet diet, poultry feedstuffs, rice bran, rice germ, rice germ cake, rye, sorghum, soybean groats, sunflower, sunflower seeds, extracted, tapioca, triticale, wheat, wheat bran, wheat hay, wheat, oats

Wheat and other grains (moldy)   may contain the following mycotoxins:

CITRININ
incidence: 13/18, conc. range: 70–80,000 µg/kg, country: Canada[92]
see also barley, barley, oats, barley-soybean diet, cottonseed cake, feed, feed, mixed, feed (cattle), feed (pig), fish meal, hay, maize, maize, white, *Makhana* (*Euryale ferox* Salisb) puffs, oats, palm products, peas and beans, rice bran, rice germ, wheat, wheat bran

Wheat bran   Wheat bran for feed may contain the following mycotoxins:

Aflatoxin B_1
incidence: 3/5, Ø conc.: 50 µg/kg, country:
Egypt[16]
incidence: 1/1, conc.: 3 µg/kg, country:
India[129]
incidence: 20/108, conc. range: ≤25 µg/kg
(10 sa), 26–50 µg/kg (7 sa), 51–100 µg/kg
(3 sa), Ø conc.: 26.1 µg/kg, country: India[247]
see also alfalfa, *Ambadi* cake, animal feedstuffs
(dairy cake), bagasse, barley, bengalgram
husk, bird food, bird food, wild, biri testa,
blackgram, blackgram husk, bran, broiler
mixed feed, calf fattening mixed feed, calf
fattening mixed feed (containing 4–20 %
peanut products), *Carthamus* cake, castor
cake, cereals, cereal products, chick pea,
coconut cake, cocos, concentrate, mixed,
concentrates, cotton cake, cottonseed,
cottonseed (dehulled), cottonseed cake,
cottonseed extract, cottonseed meal,
cottonseed meal (ammoniated), cottonseed
meal (decorticated), cottonseed meats,
cottonseed products, crumbles, crumbles,
grower, cycad meal, dairy cattle feed, dairy
cattle feed (containing 2–5 % peanut
products), dairy cattle feed (containing
6–10 % peanut products), dairy cattle feed
(containing 6–12 % peanut products), dairy
cattle feed (containing more than 20 %
peanut products), diets, mixed, dog food, egg
production mixed feed, feed, feed and
ingredients, feed, compound, feed, layer,
feed, mixed, feed (beef), feed (broilers), feed
(calf), feed (cat), feed (cattle), feed (chicken),
feed (dairy), feed (dog), feed (dug), feed
(fish), feed (gluten), feed (horse), feed
(miscellaneous), feed (pig), feed (poultry),
feed (poultry, pig), feed (rabbit), feed
(sheep), feed (50–60 % maize), fish meal,
grain by-products, grains (no specification),
greengram, hay/silage, horsegram, husk, *Jagni*
cake, legume mixture, linseed, linseed cake,
livol, *Mahua* cake, maize, maize germ, maize
gluten, maize grits, maize husk, maize meal,
maize oil cake, maize powder, maize
screenings, maize, ground, maize, hybrid,
maize, preharvest, maize, yellow, maize (dark
grains), *Makhana* (*Euryale ferox* Salisb) puffs,

manioc, milk production mixed feed, mung
testa, murkool, mustard cake, neem cake,
niger cake, oats, palm kernel expeller cake,
palm kernels, palm products, peanut cake,
peanut cake (deoiled), peanut expeller,
peanut hay, peanut meal, peanut, kernels,
peanut, shells, peanuts, pellets, finisher, pig
meal and pellets, pigeon pea, poultry feeds
(peanut containing), rapeseed cake, redgram
husk, rice, rice bran, rice bran (deoiled), rice
chaff, rice crack, rice germ, rice germ cake,
rice meal, rice straw, rice (damaged), rice
(polish), safflower cake, sal seed cake, sesame,
sesame cake, sorghum, soybean meal,
soybeans, sunflower, sunflower cake,
sunflower flour, tapioca, wheat, wheat bran
and chana testa

Aflatoxin B_2
incidence: 1/108, conc.: 3 µg/kg, country:
India[247]
see also animal feedstuff (dairy cake), bird
food, bird food, wild, biri testa, blackgram,
blackgram husk, cottonseed, cottonseed cake,
cottonseed extract, cottonseed meal,
cottonseed meal (ammoniated), cottonseed
meats, dog food, egg production mixed feed,
feed, compound, feed (cat), feed (cattle), feed
(dog), feed (pig), feed (poultry), feed
(rabbit), feed (sheep), fish meal, horsegram,
maize, maize gluten, maize husk, maize,
ground, maize, preharvest, mung testa,
mustard cake, niger cake, peanut cake,
peanut cake (deoiled), peanut expeller,
peanut hay, peanut meal, peanuts, peanut,
kernels, redgram husk, rice bran, rice bran
(deoiled), rice chaff, rice meal, rice (polish),
sal seed cake, sesame cake, sorghum, soybean
meal, soybeans, wheat

Aflatoxin G_1
incidence: 1/1, conc.: 3 µg/kg, country:
India[129]
see also animal feedstuffs (dairy cake), bird
food, bird food, wild, concentrate, mixed,
cottonseed, cottonseed cake, feed (cat), feed
(chicken), feed (dog), maize, maize, ground,
maize, preharvest, meat meal, milk
production mixed feed, murkool, peanut

cake, peanut expeller, peanut hay, peanut meal, peanuts, rice bran, rice germ, sorghum, soybeans, wheat, wheat bran and chana testa

AFLATOXIN
incidence: 20/108, conc. range: 5–67 µg/kg, country: India[247]
see also blackgram husk, bread crumbs, broiler finisher, broiler starter, cotton cake, cottonseed, cottonseed cake, cottonseed extract, cottonseed meal, feed, feed (cattle), feed (cow), feed (dog), feed (horse), feed (maize, gluten), feed (pig), feed (poultry), feed (rabbit), feed (rat/mice), feed (sheep), feeds, grain, fish meal, flour (wheat), groundnut cake, grower's mash, horsegram, layer's mash, maize, maize gluten, maize, white, milo, peanut cake, peanut cake (deoiled), peanut, kernels, peanut (oil cake), pearlmillet, pig breeder's mash, pig finisher, pig starter, pod with haulms, poultry breeder's mash, rabbit pellets, redgram husk, rice, rice bran (deoiled), rice, broken, rice (polish), sesame cake, silk worm pupae, sorghum, soybean cake, soybean meal, wheat

CITRININ
incidence: 17/24, conc. range: <0.5–230 µg/kg, country: Bulgaria[8]
see also barley, barley, oats, barley-soybean diet, cottonseed cake, feed, feed, mixed, feed (cattle), feed (pig), fish meal, hay, maize, maize, white, *Makhana* (*Euryale ferox* Salisb) puffs, oats, palm products, peas and beans, rice bran, rice germ, wheat, wheat and other grains (moldy)

OCHRATOXIN A
incidence: 17/24, conc. range: 0.5–2.8 µg/kg, Ø conc.: 1.37 µg/kg, country: Bulgaria[8]
incidence: 1/3, Ø conc.: 6 µg/kg, country: Egypt[16]
incidence: 3/4, conc. range: 2–6 µg/kg, Ø conc.: 4 µg/kg, country: The Netherlands[103]
see also alfalfa, barley, barley, oats, barley (high moisture), barley-soybean diet, bird food, domestic, bird food, wild, broilers feed, cereal grains, citrus pulp, coconut, expeller, corn cob mix silage, diet (dairy cow), diet (poultry), diet (starter), dog food, eat, egg

production mixed feed, feed, feed wheat, oat and barley, feed, commercial mix, feed, mixed, feed, mixed (pelleted), feed (broilers), feed (cat), feed (cattle), feed (cereals), feed (pig), feed ec (pig), feed (poultry), feed ec (poultry), feed (poultry, pig), feed ec (rabbit), feed (trout), grain, mixed feed, grains, mixed, grains (heated), hay, horse bean, maize, maize feed, milo, maize gluten, maize meal, maize, white, *Makhana* (*Euryale ferox* Salisb) puffs, milk production mixed feed, millet, oat and barley (hammer-milled), oats, palm products, peanut cake, peas, peas and beans, pet food, pig feedstuffs, pig grower diet, pig meal, piglet diet, poultry feedstuffs, rice bran, rice germ, rice germ cake, rye, sorghum, soybean groats, sunflower, sunflower seeds, extracted, tapioca, triticale, *Vicia faba*, wheat, wheat and barley, wheat hay, wheat, oats

ZEARALENONE
incidence: 2/4, conc. range: 43–67 µg/kg, Ø conc.: 55 µg/kg, country: Egypt[16]
incidence: 1/4, conc.: 22 µg/kg, country: The Netherlands[103]
see also alfalfa, barley, barley, husked, barley, unhusked (naked), barley and feed, bone meal, bran, broilers feed, *Carthamus* cake, chick pea, concentrate, mixed, corn cob mix silage, cotton cake, cottonseed, cottonseed cake, diet (dairy cow), diet (poultry), diets (mixed), feed, feed components, feed, mixed, feed, mixed (primarily maize also maize, oats, wheat), feed (bran), feed (broiler chicken), feed (cattle), feed (chicken), feed (dairy), feed (developing pig), feed (mill run, from wheat), feed (miscellaneous), feed (pig), feed (poultry), feed (poultry, pig), feed (starter chicken), feedstuff, fish meal, forage grass, grain, bruised, grain, mixed feed, grains (no specification), hay, maize, maize ears, maize flakes, maize germ, maize germ/bran, maize gluten, maize kernels, maize meal, maize oil cake, maize screenings, maize stalks (pith), maize, "Baby", maize, hybrid, maize, shelled, maize, unshelled, maize, white, maize grain, artificially dried, maize grain, crib dried, maize grain, ensiled, milk production mixed feed, oats, *Paspalum palidosum*, straw, peanut

hulls/skins, rice bran, rice germ, rice germ cake, rye, silage, sorghum, soybeans, soybeans, extracted, sunflower cake, tapioca, triticale, wheat, wheat bran and chana testa, wheat soya meal

Wheat bran and chana testa   may contain the following mycotoxins:

Aflatoxin B$_1$
incidence: 2/3, conc. range: 3–4 µg/kg, Ø conc.: 3.5 µg/kg, country: India[129]
see also alfalfa, *Ambadi* cake, animal feedstuffs (dairy cake), bagasse, barley, bengalgram husk, bird food, bird food, wild, biri testa, blackgram, blackgram husk, bran, broiler mixed feed, calf fattening mixed feed, calf fattening mixed feed (containing 4–20 % peanut products), *Carthamus* cake, castor cake, cereals, cereal products, chick pea, coconut cake, cocos, concentrate, mixed, concentrates, cotton cake, cottonseed, cottonseed (dehulled), cottonseed cake, cottonseed extract, cottonseed meal, cottonseed meal (ammoniated), cottonseed meal (decorticated), cottonseed meats, cottonseed products, crumbles, crumbles, grower, cycad meal, dairy cattle feed, dairy cattle feed (containing 2–5 % peanut products), dairy cattle feed (containing 6–10 % peanut products), dairy cattle feed (containing 6–12 % peanut products), dairy cattle feed (containing more than 20 % peanut products), diets, mixed, dog food, egg production mixed feed, feed, feed and ingredients, feed, compound, feed, layer, feed, mixed, feed (beef), feed (broilers), feed (calf), feed (cat), feed (cattle), feed (chicken), feed (dairy), feed (dog), feed (dug), feed (fish), feed (gluten), feed (horse), feed (miscellaneous), feed (pig), feed (poultry), feed (poultry, pig), feed (rabbit), feed (sheep), feed (50–60 % maize), fish meal, grain by-products, grains (no specification), greengram, hay/silage, horsegram, husk, *Jagni* cake, legume mixture, linseed, linseed cake, livol, *Mahua* cake, maize, maize germ, maize gluten, maize grits, maize husk, maize meal, maize oil cake, maize powder, maize

screenings, maize, ground, maize, hybrid, maize, preharvest, maize, yellow, maize (dark grains), *Makhana* (*Euryale ferox* Salisb) puffs, manioc, milk production mixed feed, mung testa, murkool, mustard cake, neem cake, niger cake, oats, palm kernel expeller cake, palm kernels, palm products, peanut cake, peanut cake (deoiled), peanut expeller, peanut hay, peanut meal, peanut, kernels, peanut, shells, peanuts, pellets, finisher, pig meal and pellets, pigeon pea, poultry feeds (peanut containing), rapeseed cake, redgram husk, rice, rice bran, rice bran (deoiled), rice chaff, rice crack, rice germ, rice germ cake, rice meal, rice straw, rice (damaged), rice (polish), safflower cake, sal seed cake, sesame, sesame cake, sorghum, soybean meal, soybeans, sunflower, sunflower cake, sunflower flour, tapioca, wheat, wheat bran

Aflatoxin G$_1$
incidence: 2/3, conc. range: 3 µg/kg, Ø conc.: 3 µg/kg, country: India[129]
see also animal feedstuffs (dairy cake), bird food, bird food, wild, concentrate, mixed, cottonseed, cottonseed cake, feed (cat), feed (chicken), feed (dog), maize, maize, ground, maize, preharvest, meat meal, milk production mixed feed, murkool, peanut cake, peanut expeller, peanut hay, peanut meal, peanuts, rice bran, rice germ, sorghum, soybeans, wheat, wheat bran

Zearalenone
incidence: 1/3, conc.: 316 µg/kg, country: India[129]
see also alfalfa, barley, barley, husked, barley, unhusked (naked), barley and feed, bone meal, bran, broilers feed, *Carthamus* cake, chick pea, concentrate, mixed, corn cob mix silage, cotton cake, cottonseed, cottonseed cake, diet (dairy cow), diet (poultry), diets (mixed), feed, feed components, feed, mixed, feed, mixed (primarily maize also maize, oats, wheat), feed (bran), feed (broiler chicken), feed (cattle), feed (chicken), feed (dairy), feed (developing pig), feed (mill run, from wheat), feed (miscellaneous), feed (pig), feed (poultry), feed (poultry, pig), feed (starter

chicken), feedstuff, fish meal, forage grass, grain, bruised, grain, mixed feed, grains (no specification), hay, maize, maize ears, maize flakes, maize germ, maize germ/bran, maize gluten, maize kernels, maize meal, maize oil cake, maize screenings, maize stalks (pith), maize, "Baby", maize, hybrid, maize, shelled, maize, unshelled, maize, white, maize grain, artificially dried, maize grain, crib dried, maize grain, ensiled, milk production mixed feed, oats, *Paspalum palidosum*, straw, peanut hulls/skins, rice bran, rice germ, rice germ cake, rye, silage, sorghum, soybeans, soybeans, extracted, sunflower cake, tapioca, triticale, wheat, wheat bran, wheat soya meal

Wheat Flour
see Flour (wheat)

Wheat hay   may contain the following mycotoxins:

Ochratoxin A
incidence: 1/95, conc. 30 μg/kg, country: Canada[17]
see also alfalfa, barley, barley, oats, barley (high moisture), barley-soybean diet, bird food, domestic, bird food, wild, broilers feed, cereal grains, citrus pulp, coconut, expeller, corn cob mix silage, diet (dairy cow), diet (poultry), diet (starter), dog food, eat, egg production mixed feed, feed, feed wheat, oat and barley, feed, commercial mix, feed, mixed, feed, mixed (pelleted), feed (broilers), feed (cat), feed (cattle), feed (cereals), feed (pig), feed ec (pig), feed (poultry), feed ec (poultry), feed (poultry, pig), feed ec (rabbit), feed (trout), grain, mixed feed, grains, mixed, grains (heated), hay, horse bean, maize, maize feed, milo, maize gluten, maize meal, maize, white, *Makhana* (*Euryale ferox* Salisb) puffs, milk production mixed feed, millet, oat and barley (hammer-milled), oats, palm products, peanut cake, peas, peas and beans, pet food, pig feedstuffs, pig grower diet, pig meal, piglet diet, poultry feedstuffs, rice bran, rice germ, rice germ cake, rye, sorghum, soybean groats, sunflower, sunflower seeds, extracted, tapioca, triticale, *Vicia faba*, wheat, wheat and barley, wheat bran, wheat, oats

Wheat soya meal   may contain the following mycotoxins:

Zearalenone
incidence: 2/4, conc. range: 20–36 μg/kg, Ø conc.: 28 μg/kg, country: Egypt[16]
see also alfalfa, barley, barley, husked, barley, unhusked (naked), barley and feed, bone meal, bran, broilers feed, *Carthamus* cake, chick pea, concentrate, mixed, corn cob mix silage, cotton cake, cottonseed, cottonseed cake, diet (dairy cow), diet (poultry), diets (mixed), feed, feed components, feed, mixed, feed, mixed (primarily maize also maize, oats, wheat), feed (bran), feed (broiler chicken), feed (cattle), feed (chicken), feed (dairy), feed (developing pig), feed (mill run, from wheat), feed (miscellaneous), feed (pig), feed (poultry), feed (poultry, pig), feed (starter chicken), feedstuff, fish meal, forage grass, grain, bruised, grain, mixed feed, grains (no specification), hay, maize, maize ears, maize flakes, maize germ, maize germ/bran, maize gluten, maize kernels, maize meal, maize oil cake, maize screenings, maize stalks (pith), maize, "Baby", maize, hybrid, maize, shelled, maize, unshelled, maize, white, maize grain, artificially dried, maize grain, crib dried, maize grain, ensiled, milk production mixed feed, oats, *Paspalum palidosum*, straw, peanut hulls/skins, rice bran, rice germ, rice germ cake, rye, silage, sorghum, soybeans, soybeans, extracted, sunflower cake, tapioca, triticale, wheat, wheat bran, wheat bran and chana testa

Wheat, oats   Wheat and oats for feed may contain the following mycotoxins:

Alternariol Methyl Ether
incidence: 1/2, conc.: 14 μg/kg, country: Germany[294]
see also barley, barley, oats, cereals, mixture, oats, sunflower, wheat

Ochratoxin A
incidence: 18/32, conc. range: 30–27,000 μg/kg, country: Canada[18]
see also alfalfa, barley, barley, oats, barley (high moisture), barley-soybean diet, bird

food, domestic, bird food, wild, broilers feed, cereal grains, citrus pulp, coconut, expeller, corn cob mix silage, diet (dairy cow), diet (poultry), diet (starter), dog food, eat, egg production mixed feed, feed, feed wheat, oat and barley, feed, commercial mix, feed, mixed, feed, mixed (pelleted), feed (broilers), feed (cat), feed (cattle), feed (cereals), feed (pig), feed ec (pig), feed (poultry), feed ec (poultry), feed (poultry, pig), feed ec (rabbit), feed (trout), grain, mixed feed, grains, mixed, grains (heated), hay, horse bean, maize, maize feed, milo, maize gluten, maize meal, maize, white, *Makhana* (*Euryale ferox* Salisb) puffs, milk production mixed feed, millet, oat and barley (hammer-milled), oats, palm products, peanut cake, peas, peas and beans, pet food, pig feedstuffs, pig grower diet, pig meal, piglet diet, poultry feedstuffs, rice bran, rice germ, rice germ cake, rye, sorghum, soybean groats, sunflower, sunflower seeds, extracted, tapioca, triticale, *Vicia faba*, wheat, wheat and barley, wheat bran, wheat hay

Wheat, red hard winter   for feed may contain the following mycotoxins:

Deoxynivalenol
incidence: 157/157*, conc. range: ≤100 µg/kg (12 sa), >100–2000 µg/kg (95 sa), >2000–4000 µg/kg (35 sa), >4000–8000 µg/kg (15 sa), Ø conc.: 1700 µg/kg, country: USA[201], *ncac
see also barley, barley, husked, barley, unhusked (naked), barley (pressed), bone meal, bran, broilers feed, calf fattening mixed feed, coconut, expeller, corn cob mix silage, cottonseed, cottonseed cake, dairy cattle feed, egg production mixed feed, feed, feed components, feed, commercial mix, feed, mixed, feed, mixed (primarily maize), feed (barley), feed (cattle), feed (chicken), feed (dog), feed (fish), feed (mill run, from wheat), feed (mink), feed (pig), feed (poultry), feed (reindeer), feeds, grain, feeds, industrial, feedstuff, feedstuffs (rapeseed, turnip, fish meal, concentrates), fish meal, grain, mixed feed, grains, mixed, grains (no specification), maize, maize ears, maize fibre,

maize germ, maize germ/bran, maize germ meal, maize gluten, maize kernels, maize meal, maize powder, maize screenings, maize stalks (pith), maize, "Baby", maize, hybrid, maize, white, oats, rice bran, rice germ cake, rye, silage, sorghum, soybeans, triticale, wheat, wheat and barley, wheat, soft white winter, wheat, spring, wheat, winter

Wheat, soft white winter   for feed may contain the following mycotoxins:

Deoxynivalenol
incidence: 8/8, conc. range: 90–450 µg/kg, Ø conc.: 218.8 µg/kg, country: Canada[198]
see also barley, barley, husked, barley, unhusked (naked), barley (pressed), bone meal, bran, broilers feed, calf fattening mixed feed, coconut, expeller, corn cob mix silage, cottonseed, cottonseed cake, dairy cattle feed, egg production mixed feed, feed, feed components, feed, commercial mix, feed, mixed, feed, mixed (primarily maize), feed (barley), feed (cattle), feed (chicken), feed (dog), feed (fish), feed (mill run, from wheat), feed (mink), feed (pig), feed (poultry), feed (reindeer), feeds, grain, feeds, industrial, feedstuff, feedstuffs (rapeseed, turnip, fish meal, concentrates), fish meal, grain, mixed feed, grains, mixed, grains (no specification), maize, maize ears, maize fibre, maize germ, maize germ/bran, maize germ meal, maize gluten, maize kernels, maize meal, maize powder, maize screenings, maize stalks (pith), maize, "Baby", maize, hybrid, maize, white, oats, rice bran, rice germ cake, rye, silage, sorghum, soybeans, triticale, wheat, wheat and barley, wheat, red hard winter, wheat, spring, wheat, winter

Wheat, spring   for feed may contain the following mycotoxins:

Deoxynivalenol
incidence: 120/206, conc. range: 900–7600 µg/kg, Ø conc.: 900 µg/kg, country: USA[370]
see also barley, barley, husked, barley, unhusked (naked), barley (pressed), bone

meal, bran, broilers feed, calf fattening mixed feed, coconut, expeller, corn cob mix silage, cottonseed, cottonseed cake, dairy cattle feed, egg production mixed feed, feed, feed components, feed, commercial mix, feed, mixed, feed, mixed (primarily maize), feed (barley), feed (cattle), feed (chicken), feed (dog), feed (fish), feed (mill run, from wheat), feed (mink), feed (pig), feed (poultry), feed (reindeer), feeds, grain, feeds, industrial, feedstuff, feedstuffs (rapeseed, turnip, fish meal, concentrates), fish meal, grain, mixed feed, grains, mixed, grains (no specification), maize, maize ears, maize fibre, maize germ, maize germ/bran, maize germ meal, maize gluten, maize kernels, maize meal, maize powder, maize screenings, maize stalks (pith), maize, "Baby", maize, hybrid, maize, white, oats, rice bran, rice germ cake, rye, silage, sorghum, soybeans, triticale, wheat, wheat and barley, wheat, red hard winter, wheat, soft white winter, wheat, winter

Wheat, summer for feed may contain the following mycotoxins:

BEAUVERICIN
incidence: 4/4*, conc. range: tr µg/kg, country: Finland[205], *ncac
incidence: 7/7*, conc. range: tr µg/kg, country: Finland[205], *ncac
see also barley, maize, oats, wheat, wheat, winter

ENNIATIN A
incidence: 4/4*, conc. range: 3–490 µg/kg, Ø conc.: 128.5 µg/kg, country: Finland[205], *ncac
incidence: 7/7*, conc. range: tr–5 µg/kg, country: Finland[205], *ncac
see also barley

ENNIATIN A₁
incidence: 4/4*, conc. range: tr–940 µg/kg, country: Finland[205], *ncac
incidence: 7/7*, conc. range: tr–15 µg/kg, country: Finland[205], *ncac
see also barley, oats, rye, wheat, wheat, winter

ENNIATIN B
incidence: 4/4*, conc. range: 120–18,300 µg/kg, Ø conc.: 5372.5 µg/kg, country: Finland[205], *ncac
incidence: 7/7*, conc. range: 31–160 µg/kg, Ø conc.: 94.29 µg/kg, country: Finland[205], *ncac
see also barley, oats, rye, wheat, wheat, winter

ENNIATIN B₁
incidence: 4/4*, conc. range: 27–5100 µg/kg, Ø conc.: 1449.3 µg/kg, country: Finland[205], *ncac
incidence: 7/7*, conc. range: tr–67 µg/kg, country: Finland[205], *ncac
see also barley, oats, rye, wheat, wheat, winter

MONILIFORMIN
incidence: 3/4*, conc. range: 58–810 µg/kg, Ø conc.: 372.7 µg/kg, country: Finland[205], *ncac
incidence: 6/7*, conc. range: 30–96 µg/kg, Ø conc.: 71.5 µg/kg, country: Finland[205], *ncac
incidence: 3/4*, conc. range: 58–810 µg/kg, Ø conc.: 373 µg/kg, country: Finland[312], *ncac
see also barley, feed, mixed, feed (poultry), maize, maize flakes, maize germ, maize germ/bran, maize gluten, maize meal, maize screenings, maize, "Baby", oats, rice bran, triticale, wheat, wheat, winter

Wheat, winter for feed may contain the following mycotoxins:

BEAUVERICIN
incidence: 2/3*, conc. range: tr µg/kg, country: Finland[205], *ncac
see also barley, maize, oats, wheat, wheat, summer

DEOXYNIVALENOL
incidence: 54/57*, conc. range: 200–9000 µg/kg, Ø conc.: 3600 µg/kg, country: USA[226], *ncac
incidence: 15/28, conc. range: 320–8530 µg/kg, country: Canada[233]
incidence: 201/207, conc. range: 400–40,000 µg/kg, Ø conc.: 2400 µg/kg, country: USA[370]

see also barley, barley, husked, barley, unhusked (naked), barley (pressed), bone meal, bran, broilers feed, calf fattening mixed feed, coconut, expeller, corn cob mix silage, cottonseed, cottonseed cake, dairy cattle feed, egg production mixed feed, feed, feed components, feed, commercial mix, feed, mixed, feed, mixed (primarily maize), feed (barley), feed (cattle), feed (chicken), feed (dog), feed (fish), feed (mill run, from wheat), feed (mink), feed (pig), feed (poultry), feed (reindeer), feeds, grain, feeds, industrial, feedstuff, feedstuffs (rapeseed, turnip, fish meal, concentrates), fish meal, grain, mixed feed, grains, mixed, grains (no specification), maize, maize ears, maize fibre, maize germ, maize germ/bran, maize germ meal, maize gluten, maize kernels, maize meal, maize powder, maize screenings, maize stalks (pith), maize, "Baby", maize, hybrid, maize, white, oats, rice bran, rice germ cake, rye, silage, sorghum, soybeans, triticale, wheat, wheat and barley, wheat, red hard winter, wheat, soft white winter, wheat, spring

ENNIATIN A$_1$
incidence: 1/3*, conc.: 6 µg/kg, country: Finland[205], *ncac
see also barley, oats, rye, wheat, wheat, summer

ENNIATIN B
incidence: 3/3*, conc. range: tr–170 µg/kg, country: Finland[205], *ncac
see also barley, oats, rye, wheat, wheat, summer

ENNIATIN B$_1$
incidence: 3/3*, conc. range: tr–39 µg/kg, country: Finland[205], *ncac
see also barley, oats, rye, wheat, wheat, summer

MONILIFORMIN
incidence: 1/3*, conc.: tr µg/kg, country: Finland[205], *ncac
incidence: 1/3*, conc.: <20 µg/kg, country: Finland[312], *ncac
see also barley, feed, mixed, feed (poultry), maize, maize flakes, maize germ, maize germ/bran, maize gluten, maize meal, maize screenings, maize, "Baby", oats, rice bran, triticale, wheat, wheat, summer

Literature

1. Dalcero, A., Magnoli, C., Hallak, C., Chiachiera, S.M., Palacio, G., Rosa, C.A.R., 2002: Detection of ochratoxin A in animal feed and capacity to produce this mycotoxin by *Aspergillus* section *Nigri* in Argentina. Food Additives and Contaminants 19, 1065–1072

2. Perkowski, J., Jelen, H., Kiecana, I., Golinski, P., 1997: Natural contamination of spring barley with group A trichothecenes mycotoxins in southeastern Poland. Food Additives and Contaminants 14, 321–325

3. Julian, A.M., Wareing, P.W., Phillips, S.I., Medlock, V.P.M., MacDonald, M.V., del Rio, L.E., 1995: Fungal contamination and selected mycotoxins in pre- and post-harvested maize in Honduras. Mycopathologia 129, 5–16

4. Maia, P.P., Bastos de Siqueira, M.E., 2002: Occurrence of aflatoxins B_1, B_2, G_1 and G_2 in some Brazilian pet foods. Food Additives and Contaminants 19, 1180–1183

5. Birzele, B., Prange, A., Krämer, J., 2000: Deoxynivalenol and ochratoxin A in German wheat and changes of level in relation to storage parameters. Food Additives and Contaminants 17, 1027–1035

6. Rafai, P., Bata, A., Jakab, L., Ványi, A., 2000: Evaluation of mycotoxin-contaminated cereals for their use in animal feeds in Hungary. Food Additives and Contaminants 17, 799–808

7. Pineiro, M.S., Silva, G.E., Scott, P.M., Lawrence, G.A., Stack, M.E., 1997: Fumonisin level in Uruguayan corn products. Journal of the Association of Official Analytical Chemists International 80, 825–828

8. Vrabcheva, T., Usleber, E., Dietrich, R., Märtlbauer, E., 2000: Co-occurrence of ochratoxin A and citrinin in cereals from Bulgarian villages with a history of Balkan endemic nephropathy. Journal of Agricultural and Food Chemistry 48, 2483–2488

9. Hietaniemi, V., Kumpulainen, J., 1991: Contents of *Fusarium* toxins in Finnish and imported grains and feeds. Food Additives and Contaminants 8, 171–182

10. Karppanen, E., Rizzo, A., Berg, S., Lindfors, E., Aho, R., 1985: *Fusarium* mycotoxins as a problem in Finnish feeds and cereals. Journal of Agricultural Science in Finland 57, 195–206

11. Lindfors, E., Berg, S., Rizzo, A., 1988: Determination of trichothecene mycotoxins as their trimethylsilyl and heptarobutyryl derivatives in feeds and grains. Proceedings of the Japanese Association of Mycotoxicology, Supplement 1, 57–58

12. Scudamore, K.A., 1996: Ochratoxin A in animal feed-effects of processing. Food Additives and Contaminants 13, 39–42

13. Bauer, J., Gareis, M., 1987: Ochratoxin A in der Nahrungskette. Journal of Veterinary Medicine Series B 34, 613–627 [German]

14. Solfrizzo, M., Avantaggiato, G., Visconti, A., 1998: Use of various clean-up procedures for the analysis of ochratoxin A in cereals. Journal of Chromatography A 815, 67–73

15. MAFF, 1999: Survey of ochratoxin A in grain traded by central depots 1997–1998. Food Surveillance Information Sheet No. 171

16. Abdelhamid, A.M., 1990: Occurrence of some mycotoxins (aflatoxin, ochratoxin, citrinin, zearalenone and vomitoxin) in various Egyptian feeds. Archive of Animal Nutrition 40, 647–664

17. Prior, M.G., 1976: Mycotoxin determinations on animal feedstuffs and tissues in western Canada. Canadian Journal of Comparative Medicine 40, 75–79

18. Scott, P.M., van Walbeck, W., Kennedy, B., Anyeti, D., 1972: Mycotoxins (ochratoxin A, citrinin and sterigmatocystin) and toxigenic fungi in grains and other agricultural products. Journal of Agricultural and Food Chemistry 20, 1103–1109

19. Prior, M.G., 1981: Mycotoxins in animal feedstuffs and tissues in western Canada. Canadian Journal of Comparative Medicine 45, 116–119

20. Abramson, D., Mills, J.T., Boycott, B.R., 1983: Mycotoxins and mycoflora in animal feedstuffs in western Canada. Canadian Journal of Comparative Medicine 47, 23–26

21. Connole, M.D., Blaney, B.J., McEwan, T., 1981: Mycotoxins in animal feeds and toxic fungi in Queensland, 1971–1980. Australian Veterinary Journal 57, 314–318

22. Shephard, G.S., Marasas, W.F.O., Leggott, N.L., Yazdanpanah, H., Rahimian, H., Safavi, N., 2000: Natural occurrence of fumonisins in corn from Iran. Journal of Agricultural and Food Chemistry 48, 1860–1864

23. Krogh, P., Hald, B., Englund, P., Rutquist, L., Swahn, O., 1974: Contamination of Swedish cereals with ochratoxin A. Acta Pathological et Microbiologica Scandinavica Section B 82, 301–302

24. Krogh, P., Hald, B., Pedersen, E.J., 1973: Occurrence of ochratoxin A and citrinin in cereals associated with mycotoxic porcine nephropathy. Acta Pathologica et Microbiologica Scandinavica Section B 81, 689–695

25. Pettersson, H., Kiessling, K.H., 1992: Mycotoxins in Swedish grains and mixed feeds. Journal of Environmental Pathology, Toxicology and Oncology 11, 105–107

26. Golinski, P., Garbarkiewicz-Seczesna, J., Chelkowski, J., Hult, K., Kostecki, M., 1991: Possible sources of ochratoxin A in human blood in Poland. In: Castegnaro, M., Plestina, R., Dirheimer, G., Chernozemsky, I.N., Bartsch, H., (eds.), Mycotoxins, Endemic Nephropathy and Urinary Tract Tumours, pp. 153–158. International Agency for Research on Cancer, Lyon

27. Juszkiewicz, T., Piskorska-Pliszcynska, J., 1977: Occurrence of mycotoxins in mixed feeds and concentrates. Medycyna Weterynaryjna 33, 193–196

28. Sinha, A.K., Sinha, K.K., 1990: Insect pests, *Aspergillus flavus* and aflatoxin contamination in stored wheat: a survey at North Bihar (India). Journal of stored Product Research 26, 223–226

29. Böhm, J., Leibetseder, J., 1987: Ochratoxin A – Vorkommen in österreichischen Schlachtschweinen. Zeitschrift für Tierphysiologie, Tierernährung und Futtermittelkunde 58, 41–42 [German]

30. Buckle, A.E., 1983: The occurrence of mycotoxins in cereals and animal feedstuffs. Veterinary Research Communications 7, 171–186

31. Bacha, H., Hadidane, R., Creppy, E.E., Regnault, C., Ellouze, F., Dirheimer, G., 1988: Monitoring and identification of fungal toxins in food products, animal feed and cereals in Tunisia. Journal of stored Product Research 24, 199–206

32. Chalam, R.V., Stahr, H.M., 1979: Thin layer determination of citrinin. Journal of the Association of Official Analytical Chemists International 62, 570–572

33. Balachandran, C., Parthasarathy, K.R., 1996: Occurrence of cyclopiazonic acid in feeds and feedstuffs in Tamil Nadu, India. Mycopathologia 133, 159–162

34. Stoltz, D.R., Widiastuti, R., Maryam, R., Tri Akoso, B., Unruh, A., Unruh, B., 1988: Suspected cyclopiazonic acid mycotoxicosis of quail in Indonesia. Toxicon 26, 39–40

35. Urano, T., Trucksess, W.M., Beaver, R.W., Wilson, D.M., Dorner, J.W., Dowell, F.E., 1992: Co–occurrence of cyclopiazonic acid and aflatoxins in corn and peanuts. Journal of the Association of Official Analytical Chemists International 75, 838–841

36. Anonymous, 1987: Mycotoxins. The eighteenth report of the steering group on food surveillance. The working party on naturally occurring toxicants in food: sub-group on mycotoxins. Food Surveillance Paper No. 18. London HSMO

37. Jarvis, B., 1982: The occurrence of mycotoxins in UK foods. Food Technology in Australia 34, 508–515

38. Hesseltine, C.W., Rogers, R.F., Shotwell, O.L., 1981: Aflatoxin and mold flora in North Carolina in 1977 corn crop. Mycologia 73, 216–228

39. Cohen, H., Lapointe, M., 1981: High pressure liquid chromatographic determination and fluorescence detection of aflatoxins in corn and dairy feeds. Journal of the Association of Official Analytical Chemists 64, 1372–1376

40. Couvillion, C.E., Jackson, J.R., Ingram, R.P., Bennett, L.W., McCoy, C.P., 1991: Potential natural exposure of Mississippi sandhill cranes to aflatoxin B_1. Journal of Wildlife Diseases 27, 650–656

41. Müller, H.-M., Reimann, J., Schumacher, U., Schwadorf, K., 1998: Natural occurrence of *Fusarium* toxins in oats harvested during five years in an area of southwest Germany. Food Additives and Contaminants 15, 801–806

42. Scudamore, K.A., Hetmanski, M.T., Nawaz, S., Naylor, J., Rainbird, S., 1997: Determination of mycotoxins in pet foods sold for domestic pets and wild birds using linked-column immunoassay clean-up and HPLC. Food Additives and Contaminants 4, 175–186

43. Scudamore, K.A., Patel, S., Breeze, V., 1999: Surveillance of stored grain from the harvest in the United Kingdom for ochratoxin A. Food Additives and Contaminants 16, 281–290

44. Dutton, M.F., Kinsey, A., 1995: Occurrence of mycotoxins in cereals and animal feedstuffs in Natal, South Africa 1994. Mycopathologia 131, 31–36

45. Jacquet, J., Boutibonnes, P., Teherani, A., 1970: Sur la présence des flavatoxine dans les aliments des animaux et dans les aliments d'origine animale destinés à l'homme. Bulletin de l'Academie Véterinare XLIII, 35–43 [French]

46. Lafont, P., Lafont, J., 1970: Contamination de produit céréaliers et d'aliments du bétail par l'aflatoxine. Food and Cosmetics Toxicology 8, 403–408 [French]

47. Maragos, C.M., Greer, J.I., 1997: Analysis of aflatoxin B_1 in corn using capillary electrophoresis with laser-induced fluorescence detection. Journal of Agricultural and Food Chemistry 45, 4337–4341

48. Abbas, H.K., Mirocha, C.J., Meronuck, R.A., Pokorny, J.D., Gould, S.L., Kommedahl, T., 1988: Mycotoxins and *Fusarium* spp. associated with infected ears of corn in Minnesota. Applied and Environmental Microbiology 54, 1930–1933

49. Abramson, D., Clear, R.M., Nowicki, T.W., 1987: *Fusarium* species and trichothecene mycotoxins in suspect samples of 1985 Manitoba wheat. Canadian Journal of Plant Science 67, 611–619

50. Bata, A., Ványi, A., Lásztity, R., 1983: Simultaneous detection of some fusariotoxins by gas-liquid chromatography. Journal of the Association of Official Analytical Chemists 66, 577–581

51. Bottalico, A., Lerario, P., Visconti, A., 1981: Occurrence of trichothecenes and zearalenone in preharvest *Fusarium*-infected ears of maize from some Austrian localities. Phytopathologia Mediterranea 20, 1–6

52. Bottalico, A., Logrieco, A., Ritieni, A., Moretti, A., Randazzo, G., Corda, P., 1995: Beauvericin and fumonisin B_1 in preharvest *Fusarium moniliforme* maize ear rot in Sardena. Food Additives and Contaminants 12, 599–607

53. Botallico, A., Logrieco, A., Visconti, A., 1989: *Fusarium* species and their mycotoxins in infected corn in Italy. Mycopathologia 107, 85–92

54. Curtui, V., Usleber, E., Dietrich, R., Lepschy, J., Märtlbauer, E., 1998: A survey on the occurrence of mycotoxins in wheat and maize from western Romania. Mycopathologia 143, 97–100

55. Gilbert, J., Shepherd, M.S., Startin, R., 1983: A survey of the occurrence of the trichothecene mycotoxin deoxynivalenol (vomitoxin) in UK grown barley and in imported maize by combined gas chromatography-mass spectrometry. Journal of the Science of Food and Agriculture 34, 86–92

56. Wilson, T.M., Ross, P.F., Rice, L.G., Osweiler, G.D., Nelson, H.A., Owens, D.L., Plattner, E.D., Reggiardo, C., Noon, T.H., Pickrell, J.W., 1990: Fumonisin B_1 levels associated with an epizootic of equine leukoencephalomalacia. Journal of Veterinary Diagnostic Investigation 2, 213–216

57. Thiel, P.G., Shephard, G.S., Sydenham, E.W., Marasas, W.F.O., Nelson, P.E., Wilson, T.M., 1991: Levels of fumonisin B_1 and B_2 associated with confirmed cases of equine leukoencephalomalacia. Journal of Agricultural and Food Chemistry 39, 109–111

58. Ross, P.F., Rice, L.G., Plattner, L.G., Osweiler, G.D., Wilson, T.M., Owens, D.L., Nelson, H.A., Richard, J.L., 1991: Concentrations of fumonisin B_1 in feeds associated with animal health problems. Mycopathologia 114, 129–135

59. Plattner, R.D., Norred, W.P., Bacon, C.W., Voss, K.W., Peterson, R., Shackelford, D.D., Weisleder, D., 1990: A method of detection of fumonisins in corn samples associated with field cases of equine leucoencephalomalacia. Mycologia 82, 698–702

60. Minervini, F., Bottalico, C., Pestka, J.J., Visconti, A., 1992: On the occurrence of fumonisins in feed in Italy. Atti della Società Italiana delle Sience Veterinarie 46, 1365–1368

61. Jemmali, M., Ueno, Y., Ishii, K., Frayssinet, C., Etienne, M., 1978: Natural occurrence of trichothecenes (nivalenol, deoxynivalenol, T_2) and zearalenone in corn. Experientia 34, 1333–1334

62. Karppanen, E., Rizzo, A., Berg, S., Lindfors, E., Aho, R., 1985: *Fusarium* mycotoxins as a problem in Finnish feeds and cereals. Journal of Agricultural Science in Finland 57, 195–206

63. Müller, H.-M., Reimann, J., Schuhmacher, U., Schwadorf, K., 1997: *Fusarium* toxins in wheat harvested during six years in an area of southwest Germany. Natural Toxins 5, 24–30

64. Okoye, Z.S.C., 1993: *Fusarium* mycotoxins nivalenol and 4-acetyl-nivanlenol (fusarenon-X) in mouldy maize harvested from farms in Jos district, Nigeria. Food Additives and Contaminants 10, 375–379

65. Pettersson, H., Hedman, R., Engström, B., Elwinger, K., Fossum, O., 1995: Nivalenol in Swedish cereals – occurrence, production and toxicity towards chickens. Food Additives and Contaminants 12, 373–376

66. Smith, T.K., Sousadias, M.G., 1993: Fusaric acid content of swine feedstuffs. Journal of Agricultural and Food Chemistry 41, 2296–2298

67. Sugiura, Y., Fukasaku, K., Tanaka, T., Matsui, Y., Ueno, Y., 1993: *Fusarium poae* and *Fusarium crookwellense*, fungi responsible for the natural occurrence of nivalenol in Hokkaido. Applied and Environmental Microbiology 59, 3334–3338

68. Thalmann, A., Matzenauer, S., Gruber-Schley, S., 1985: Untersuchungen über das Vorkommen von Fusarientoxinen in Getreide. Berichte über Landwirtschaft 63, 257–272 [German]

69. Thiel, P.G., Gelderblom, W.C.A., Marasas, W.F.O., Nelson, P.E., Wilson, T.W., 1986: Natural occurrence of moniliformin and fusarin C in corn screenings known to be hepatocarcinogenic in rats. Journal of Agricultural and Food Chemistry 34, 775–780

70. Ueno, Y., Lee, U.-S., Tanaka, T., Hasegawa, A., Matsuki, Y., 1986: Examination of Chinese and U.S.S.R. cereals for the *Fusarium* mycotoxins, nivalenol, deoxynivalenol and zearalenone. Toxicon 6, 618–621

71. Vesonder, R.F., 1983: VI.2. Natural occurrence in North America. Developments in Food Science 4, 210–217

72. Yuwai, K.E., Rao, K.S., Singh, K., Tanaka, T., Ueno, Y., 1994: Occurrence of nivalenol, deoxynivalenol, and zearalenone in imported cereals in Papua, New Guinea. Natural Toxins 2, 19–21

73. DFG, 1990: Ochratoxin A. Vorkommen und toxikologische Bedeutung. Mitteilung XII der Senatskomission zur Prüfung von Lebensmittelzusatz- und -inhaltsstoffen. VCH Verlagsgesellschaft mbH, Weinheim, Federal Republic of Germany [German]

74. Auerbach, H., Oldenburg, E., Weissbach, F., 1998: Incidence of *Penicillium roqueforti* and roquefortine C in silages. Journal of the Science of Food and Agriculture 76, 565–572

75. Balzer, I., Bogdanovic, C., Muzic, S., 1977: Natural contamination of corn (*Zea mais*) with mycotoxins in Yugoslavia. Annales de la Nutrition et de l'Alimentation 31, 425–430

76. Strelecki, E.L., Cader Strezelecka, B., 1988: A survey of aflatoxin levels in peanut meal imported into Poland for animal feedingstuffs. Food Additives and Contaminants 5, 597–599

77. Collet, J.-C., Regnier, J., 1977: Contamination par mycotoxines de mais conservés en cribs et visiblement altérés. Annales de la Nutrition et de l'Alimentation 31, 447–457 [French]

78. Dalcero, A., Magnoli, C., Luna, M., Ancasi, G., Reynoso, M.M., Chiacchiera, S., Miazzo, R., Palacio, G., 1998: Mycoflora and naturally occurring mycotoxins in poultry feeds in Argentina. Mycopathologia 141, 37–43

79. Dutton, M.F., Westlake, K., 1985: Occurrence of mycotoxins in cereals and animal feedstuffs in Natal, South Africa. Journal of the Association of Official Analytical Chemists 68, 839–842

80. Furlong, E.B., Valente Soares, L.M., Lasca, C.C., Kohara, E.Y., 1995: Mycotoxins and fungi in wheat stored in elevators in the state of Rio Grande do Sul, Brazil. Food Additives and Contaminants 12, 683–688

81. Furlong, E.D., Valente Soares, L.M., Lasca, C.C., Kohara, E.Y., 1995: Mycotoxins and fungi in wheat harvested during 1990 in test plots in the state of Sao Paulo. Mycopathologia 131, 185–190

82. Hald, B., Christensen, D.H., Krogh, P., 1983: Natural occurrence of the mycotoxin viomellein in barley and the associated quinone-producing penicillia. Applied and Environmental Microbiology 46, 1311–1317

83. Jacobsen, B.J., Harlin, K.S., Swanson, S.P., Lambert, R.J., Beasley, V.R., Sinclair, J.B., Wei, L.S., 1995: Occurrence of fungi and mycotoxins associated with field mold damaged soybeans in the midwest. Plant Disease 79, 86–88

84. Juskiewicz, T., Piskorska-Pliszcynska, J., 1984: Occurrence of mycotoxins in animal feeds. Bulletin Veterinair Institut Pulawy 27, 72–78

85. Pal, R., Varma, B.K., Srivastava, D.D., 1979: Aflatoxin in groundnut oil, groundnut cake and hydrogenated oil in Hapur (Uttar Pradesh, India) market. Journal of Food Science and Technology 16, 169–170

86. Romer, T., 1984: Mycotoxins in corn and corn milling products. Cereal Foods World 29, 459–462

87. Szebiotko, K., Chelkowski, J., Dopierala, G., Godlewska, B., Radomyska, W., 1981: Mycotoxins in cereal grain. Part I. Ochratoxin, citrinin, sterigmatocystin, penicillic acid and toxigenic fungi in cereal grain. Die Nahrung 25, 415–421

88. Mirocha, C.J., Christensen, C.M., 1974: Oestrogenic mycotoxins synthesized by *Fusarium*. In: Purchase, I. F.H., (ed.), Mycotoxins, pp. 142–144. Elsevier, Amsterdam

89. Bullerman, L.B., 1996: Occurrence of *Fusarium* and fumonisins on food grains and in foods. In: Jackson, L.S., De Vries, J., Bullerman, L.B., (eds.), Fumonisins in Food, pp. 19–38. Plenum Press, New York and London

90. Wilson, D.M., Tabor, W.H., Trucksess, M.W., 1976: Screening method for the detection of aflatoxin, ochratoxin, zearalenone, penicillic acid, and citrinin. Journal of the Association of Official Analytical Chemists 59, 125–127

91. Pineiro, M., Dawson, R., Costarrica, L.M., 1996: Monitoring program for mycotoxin contamination in Uruguayan food and feeds. Natural Toxins 4, 242–245

92. Scott, P.M., van Walbeck, W., Kennedy, B., Anyeti, D., 1972: Mycotoxins (Ochratoxin A, citrinin, and sterigmatocystin) and toxigenic fungi in grains and other agricultural products. Journal of Agricultural and Food Chemistry 20, 1103–1109

93. Scudamore, K.A., Atkin, P.M., Buckle, A.E., 1986: Natural occurrence of the naphthoquinone mycotoxins, xanthomegnin, viomellein and vioxanthin in cereals and animal feedstuffs. Journal of stored Product Research 22, 81–84

94. Scudamore, K.A., Hetmanski, M.T., Chan, H.K., Collins, S., 1997: Occurrence of mycotoxins in raw ingredients used for animal feeding stuffs in the United Kingdom in 1992. Food Additives and Contaminants 14, 157–173

95. Scudamore, K.A., Nawaz, S., Hetmanski, M.T., 1998: Mycotoxins in ingredients of animal feeding stuffs: II. determinations of mycotoxins in maize and maize products. Food Additives and Contaminants 15, 30–55

96. Scudamore, K.A., Nawaz, S., Hetmanski, M.T., Rainbird, S.C., 1998: Mycotoxins in ingredients of animal feeding stuffs: III. Determination of mycotoxins in rice bran. Food Additives and Contaminants 15, 185–194

97. Scudamore, K.A., Patel, S., 2000: Survey for aflatoxins, ochratoxin A, zearalenone and fumonisins in maize imported into the United Kingdom. Food Additives and Contaminants 17, 407–416

98. Shetty, P.H., Bhat, R.V., 1997: Natural occurrence of fumonisin B_1 and its co-occurrence with aflatoxin B_1 in Indian sorghum, maize, and poultry feeds. Journal of Agricultural and Food Chemistry 45, 2170–2173

99. Churchwell, M.I., Cooper, W.M., Howard, P.C., Doerge, D.R., 1997: Determination of fumonisins in rodent feed using HPLC with electrospray mass spectrometic detection. Journal of Agricultural and Food Chemistry 45, 2573–2578

100. Shephard, G.S., Sydenham, E.W., Thiel, P.G., Gelderblom, W.C.A., 1990: Quantitative determination of fumonisins B_1 and B_2 by high-performance liquid chromatography with fluorescence detection. Journal of Liquid Chromatography 13, 2077–2087

101. Sydenham, E.W., Shephard, G.S., Thiel, P.G., 1992: Liquid chromatographic determination of fumonisins B_1, B_2, and B_3 in foods and feeds. Journal of the Association of Official Analytical Chemists International 75, 313–318

102. Siame, B.A., Mpuchane, S.F., Gashe, B.A., Allotey, J., Teffera, G., 1998: Occurrence of aflatoxines, fumonisin B_1, and zearalenone in foods and feeds in Botswana. Journal of Food Protection 61, 1670–1673

103. Veldman, A., Borggreve, G.J., Mulders, E.J., van de Lagemaat, D., 1992: Occurrence of the mycotoxins ochratoxin A, zearalenone and deoxynivalenol in feed components. Food Additives and Contaminants 9, 647–655

104. Wang, D.S., Liang, Y.-X., Chau, N.T., Dien, L.D., Tanaka, T., Ueno, Y., 1995: Natural occurrence of *Fusarium* toxins and aflatoxin B_1 in corn for feed in North Vietnam. Natural Toxins 3, 445–449

105. Widiastuti, R., Maryam, R., Blaney, B.J., Stoltz, S., Stoltz, D.R., 1988: Cyclopiazonic acid in combination with aflatoxins, zearalenone and ochratoxin A in Indonesian corn. Mycopathologia 104, 153–156

106. Escobar, A., Sánchez Regueiro, Q., 2002: Determination of aflatoxin B_1 in food and feedstuffs in Cuba (1990 through 1996) using an immunoenzymatic reagent kit (Aflacen). Journal of Food Protection 65, 219–221

107. Ranfft, K., Gerstl, R., Mayer, G., 1990: Bestimmung und Vorkommen von Zearalenon in Getreide und Mischfuttermitteln. Zeitschrift für Lebensmittel-Untersuchung und -Forschung 191, 449–453 [German]

108. Lee, U.-S., Jang, H.-S., Tanaka, T., Hasegawa, A., Oh, Y.-J., Chos, C.-M., Sugiura, Y., Ueno, Y., 1986: Further survey on the *Fusarium* mycotoxins in Korean cereals. Food Additives and Contaminants 3, 253–261

109. Atawodi, S.E., Atiku, A.A., Lamorde, A.G., 1994: Aflatoxin contamination of Nigerian foods and feedingstuffs. Food and Chemical Toxicology 32, 61–63

110. Fukal, L., 1990: A survey of cereals, cereal products, feedstuffs and porcine kidneys for ochratoxin A by radioimmunoassay. Food Additives and Contaminants 7, 253–258

111. Ono, E.Y.S., Ono, A.M., Funos, F.Y., Medina, A.E., Oliveiras, T.C.R.M., Kawamura, O., Ueno, Y., Hirooka, E.Y., 2001: Evaluation of fumonisin-aflatoxin co-occurrence in Brazilian corn hybrids by ELISA. Food Additives and Contaminants 18, 719–729

112. Eskola, M., Parikka, P., Rizzo, A., 2001: Trichothecenes, ochratoxin A and contamination and *Fusarium* infection in Finnish cereal samples in 1998. Food Additives and Contaminants 18, 707–718

113. Munkvold, G., Stahr, H.M., Logrieco, A., Moretti, A., Ritieni, A., 1998: Occurrence of fusaproliferin and beauvericin in *Fusarium*-contaminated livestock feed in Iowa. Applied and Environmental Microbiology 64, 3923–3926

114. Siame, B.A., Lovelace, C.E.A., 1989: Natural occurrence of zearalenone and trichothecene toxins in maize-based animal feeds in Zambia. Journal of the Science of Food and Agriculture 49, 25–35

115. Böhm, J., Noonpugdee, Ch., Abdelhamid, A.M., Leibetseder, J., Schuh, M., 1985: On the occurrence of the estrogenic mycotoxin zearalenone (F-2 toxin) in Austrian feed samples analysed over a period of 5 years (1979–1984) related with fertility problems in cattle and swine. Bodenkultur 36, 333–338

116. Blaney, B.J., Moore, C.J., Tyler, A.L., 1984: Mycotoxins and fungal damage in maize harvested during 1982 in Far North Queensland. Australian Journal of Agricultural Research 35, 463–471

117. Yoshizawa, T., Yamashita, A., Chokethaworn, N., 1996: Occurrence of fumonisins and aflatoxins in corn from Thailand. Food Additives and Contaminants 13, 163–168

118. Widiastuti, R., Maryam, R., Blaney, B.J., Stoltz, S., Stoltz, D.R., 1988: Corn as a source of mycotoxins in Indonesian poultry feeds and the effectiveness of visual examination methods for detecting contamination. Mycopathologia 102, 45–49

119. Bagneris, R.W., Gaul, J.A., Ware, G.M., 1986: Liquid chromatographic determination of zearalenone and zearalenol in animal feeds and grains, using fluorescence detection. Journal of the Association of Official Analytical Chemists 69, 894–898

120. Terhune, S.J., Nguyen, N.V., Baxter, J.A., Pryde, D.H., Qureshi, S.A., 1984: Improved gas chromatographic method for quantitation of deoxynivalenol in wheat, corn, and feed. Journal of the Association of Official Analytical Chemists 67, 1102–1104

121. Bryden, W.L., Lloyd, A.B., Cumming, R.B., 1980: Aflatoxin contamination of Australian feeds and suspected cases of mycotoxicoses. Australian Veterinary Journal 56, 176–180

122. Hetmanski, M.T., Scudamore, K.A., 1989: A simple quantitative HPLC method for the determination of aflatoxins in cereals and animal feedstuffs using gel permeation chromatography clean-up. Food Additives and Contaminants 6, 35–48

123. Müller, H.-M., Reimann, J., Schumacher, U., Schwadorf, K., 1997: Natural occurrence of *Fusarium* toxins in barley harvested during five years in an area of southwest Germany. Mycopathologia 137, 185–192

124. Bauer, J., Wermter, R., Gedek, B., 1980: Zur Kontamination von Futtermitteln mit toxinbildenden Fusarienstämmen und deren Toxinen. Wiener tierärztliche Monatsschrift 67, 282–288 [German]

125. Céspedes, A., Diaz, G.J., 1997: Analysis of aflatoxins in poultry and pig feeds and feedstuffs used in Colombia. Journal of the Association of Official Analytical Chemists International 80, 1215–1219

126. Fernandez, C., Stack, M., Musser, S.M., 1994: Determination of deoxynivalenol in 1991 U.S. winter and spring wheat by high-performance thin-layer chromatography. Journal of the Association of Official Analytical Chemists International 77, 628–630

127. Ramakrishna, Y., Bhat, R.V., Vasanthi, S., 1990: Natural occurrence of mycotoxins in staple foods in India. Journal of Agricultural and Food Chemistry 38, 1857–1859

128. Sanchis, V., Sala, N., Palomes, A., Santamarina, P., Burdaspal, P.A., 1986: Occurrence of aflatoxin and aflatoxigenic molds in foods and feeds in Spain. Journal of Food Protection 49, 445–448

129. Phillips, S.I., Wareing, P.W., Dutta, A., Panigrahi, S., Medlok, V., 1996: The mycoflora and incidence of aflatoxin, zearalenone and sterigmatocystin in dairy feed and forage from Eastern India and Bangladesh. Mycopathologia 133, 15–21

130. Müller, H.-M., Schwadorf, K., 1993: Natural occurrence of *Fusarium* toxins in barley grown in a southwestern area of Germany. Bulletin of Environmental Contamination and Toxicology 51, 532–537

131. Hussein, H.M., Franich, R.A., Baxter, M., Andrew, I.G., 1989: Naturally occurring *Fusarium* toxins in New Zealand maize. Food Additives and Contaminants 6, 49–58

132. Sundheim, L., Nagayama, S., Kawamura, O., Tanaka, T., Brodal, G., Ueno, O., 1988: Trichothecenes and zearalenone in Norwegian barley and wheat. Norwegian Journal of Agricultural Sciences 2, 49–59

133. Abramson, D., Mills, J.T., Boycott, B.R., 1983: Mycotoxins and mycoflora in animal feedstuffs in western Canada. Canadian Journal of Comparative Medicine 47, 23–26

134. Hagler, W.M., Bowman, D.T., Jr., Babadoost, M., Haney, C.A., Swanson, S.P., 1987: Aflatoxin, zearalenone, and deoxynivalenol in North Carolina grain sorghum. Crop Science 27, 1273–1278

135. Funnell, H.S., 1979: Mycotoxins in animal feedstuffs in Ontario. Canadian Journal of Comparative Medicine 43, 243–246

136. Ram, B.P., Hart, P., Shotwell, O.L., Pestka, J., 1986: Enzyme-linked immunosorbent assay of aflatoxin B_1 in naturally contaminated corn and cottonseed. Journal of the Association of Official Analytical Chemists 69, 904–907

137. Blaney, B.J., Moore, C.J., Tyler, A.L., 1987: The mycotoxins – 4-deoxynivalenol, zearalenone and aflatoxin – in weather-damaged wheat harvested 1983–1985 in south-eastern Queensland. Australian Journal of Agricultural Research 38, 993–1000

138. Caramelli, H., Visconti, A., Minervini, F., Doko, M.B., Cantini Cortellezzi, G., Dondo, A., Guarda, F., 1993: Leucoencefalomalacia nell 'equino da fumonisine: prima segnalazione in Italia. lppologia 4

139. Park, D.L., Rua, S.M., Jr., Mirocha, C.J., Abd-Alla, E.-S.A.M., 1992: Mutagenic potentials of fumonisin contaminated corn following ammonia decontamination procedure. Mycopathologia 117, 105–108

140. Ross, P.F., Rice, L.G., Reagor, J.C., Osweiler, G.D., Wilson, T.M., Nelson, H.A., Owens, D.R, Plattner, R.D., Harlin, K.A., Richard, J.L., Colvin, B.M., Banton, M.I., 1991: Fumonisin B_1 concentrations in feeds from 45 confirmed equine leukoencephalomalacia cases. Journal of Veterinary Diagnostic Investigation 3, 238–241

141. Ueno, Y., Aoyama, S., Sugiura, Y., Wang, D.-S., Lee, U.-S., Hirooka, E.Y., Hara, S., Karki, T., Chen, G., Yu, S.-Z., 1993: A limited survey of fumonisins in corn and corn-based products in Asian countries. Mycotoxin Research 9, 27–34

142. Osweiler, G.D., Ross, P.F., Wilson, T.M., Nelson, P.E., Witte, S.T., Carson, T.L., Rice, L.G., Nelson, H.A., 1992: Characterization of an epizootic of pulmonary edema in swine associated with fumonisin in corn screenings. Journal of Veterinary Diagnostic Investigation 4, 53–59

143. Sydenham, E.W., Marasas, W.F.O., Shepard, G.S., Thiel, P.G., Hirooka, E.Y., 1992: Fumonisin concentrations in Brazilian feeds associated with field outbreaks of confirmed and suspected animal mycotoxicoses. Journal of Agricultural Food Chemistry 40, 994–997

144. Bane, D.P., Neumann, E.J., Hall, W.F., Harlin, K.F., Slife, L.N., 1992: Relationship between fumonisin contamination of feed and mystery swine disease. Mycopathologia 117, 121–124

145. Ung-Soo, L., Myond-Yur, L., Kwan-Sop, S., Yun-Sik, M., Chae-Min, C., Ueno, Y., 1994: Production of fumonisin B_1 and B_2 by *Fusarium moniliforme* isolated from Korean corn kernels for feed. Mycotoxin Research 10, 67–72

146. Viljoen, J.H., Marasas, W.F.O., Thiel, P.G., 1992: Proceedings of the 10th South African Maize Breeding Symposium, Du Plessis, J.D., Van Rensberg, J.B.J., McLaren, N.W., Flett, B.C., (eds.), Department of Agriculture, Potchefsroom, South Africa, pp. 26–37

147. Shepard, G.S., Thiel, P.G., Stockenström, S., Sydenham, E.W., 1996: Worldwide survey of fumonisin contamination of corn and corn-based products. Journal of the Association of Official Analytical Chemists International 79, 671–687

148. Pittet, A., Parisod, V., Schellenberg, M., 1992: Occurrence of fumonisins B_1 and B_2 in corn-based products from the Swiss market. Journal of Agricultural and Food Chemistry 40, 1352–1354

149. Wang, D.-S., Sugiura, Y., Ueno, Y., Buddhanont, P., Suttajit, M., 1993: Thailand Journal of Toxicology 9, 42–46

150. Scudamore, K.A., Chan, H.K., 1993: Proceedings UK Workshop on Occurrence and Significance of Mycotoxins, In: Scudamore, K.A., (ed.), Central Science Laboratory, MAFF, Slough, UK, pp. 42–48

151. Murphy, P., Rice, L., Ross, P., 1993: Fumonisin B_1, B_2 and B_3 content of Iowa, Wisconsin and Illinois corn and corn screenings. Journal of Agricultural and Food Chemistry 41, 263–266

152. Schothorst, R.C., Jekel, A.A., 2001: Determination of trichothecenes in wheat by capillary gas chromatography with flame ionisation detection. Food Chemistry 73, 111–117

153. Stack, M.E., Eppley, R.M., 1993: Liquid chromatographic determination of fumonisins B_1 and B_2 in corn and corn products. Journal of the Association of Official Analytical Chemists International 75, 834–837

154. Hopmans, E.C., Murphy, P.A., 1993: Detection of fumonisins B_1, B_2 and B_3 and hydrolyzed fumonisin B_1 in corn-containing foods. Journal of Agricultural and Food Chemistry 41, 1655–1658

155. Castellá, G., Bragulat, M.R., Cabanes, F.J., 1999: Surveillance of fumonisins in maize-based feeds and cereals from Spain. Journal of Agricultural and Food Chemistry 47, 4707–4710

156. Mirocha, C.J., Mackintosh, C.G., Mirza, U.A., Xie, W., Xu, Y., Chen, J., 1992: Occurrence of fumonisin in forage grass in New Zealand. Applied and Environmental Microbiology 58, 3196–3198

157. Preis, R.A., Vargas, E.A., 2000: A method for determining fumonisin B_1 in corn using immunoaffinity column clean-up and thin layer chromatography/densitometry. Food Additives and Contaminants 17, 463–468

158. Rice, L.G., Ross, P.F., 1994: Methods for detection and quantitation of fumonisins in corn, cereal products and animal excreta. Journal of Food Protection 57, 536–540

159. Sanchis V., Abadias, M., Oncins, L., Sala, N., Vinas, I., Canela, R., 1995: Fumonisins B_1 and B_2 and toxigenic *Fusarium* strains in feeds from the Spanish market. International Journal of Food Microbiology 27, 37–44

160. Seo, J.-A., Lee, Y.-W., 1999: Natural occurrence of the C series of fumonisins in moldy corn. Applied and Environmental Microbiology 65, 1331–1334

161. Thakur, R.A., Smith, J.S., 1996: Determination of fumonisins B_1 and B_2 and their major hydrolysis products in corn, feed, and meat, using HPLC. Journal of Agricultural Food Chemistry 44, 1047–1052

162. Afzal, M., Cheema, R.A., Chaudhary, R.A., 1979: Incidence of aflatoxins and aflatoxin producing fungi in animal feedstuffs. Mycopathologia 69, 149–151

163. Stoloff, L., 1976: Occurrence of mycotoxins in foods and feeds. In: Rodricks, J., (ed.), Mycotoxins and other fungal Related Food Problems, 149, 23–50, American Chemical Society, Washington, DC, Advances in Chemistry Series

164. Stoloff, L., 1976: Incidence, distribution and disposition of products containing aflatoxins. Proceedings of the American Phytopathological Society 3, 156–172

165. Cardeilhac, P.T., Nair, K.P.C., 1974: Hazards presented by mycotoxins and toxigenic-mold spores. Toxicology and Applied Pharmacology 30, 299–308

166. Smith, R.B., Jr., Griffin, J.M., Hamilton, P.B., 1976: Survey of aflatoxicosis in farm animals. Applied and Environmental Microbiology 31, 385–388

167. Hald, B., Krogh, P., 1970: Occurrence of aflatoxin in imported cottonseed products. Nordisk Veterinaermedicin 22, 39–47

168. Shreeve, B.J., Patterson, D.S.P., Roberts, B.A., 1975: Investigation of suspected cases of mycotoxicosis in farm animals in Britain. Veterinary Record 97, 275–278

169. Kiermeier F., Weiss, G., Behringer, G., Miller, M., Ranfft, K., 1977: Vorkommen und Gehalt an Aflatoxin M_1 in Molkerei-Anlieferungsmilch. Zeitschrift für Lebensmittel-Untersuchung und -Forschung 163, 171–174 [German]

170. Strzelecki, E.L., Gasiorowska, U.W., 1974: Aflatoxin B_1 in feedstuffs. Zentralblatt für Veterinärmedizin B 21, 395–400

171. Hamilton, P.B., Huff, W.E., Harris, J.R., Wyatt, R.D., 1977: Outbreaks of ochratoxicosis in poultry. Abstr. 77th Annual Meeting, American Society for Microbiology, New Orleans, Louisiana, May, 8–13, p. 248

172. Scott, P.M., van Walbeck, W., Harwig, J., Fennell, D.I., 1970: Occurrence of a mycotoxin, ochratoxin A, in wheat and isolation of ochratoxin A and citrinin producing strains of *Penicillium viridicatum*. Canadian Journal of Plant Science 50, 583–585

173. Mirocha, C.J., Schauerhammer, B., Pathre, S.V., 1974: Isolation, detection and quantitation of zearalenone in maize and barley. Journal of the Association of Official Analytical Chemists 57, 1104–1110

174. Mirocha, C.J., Pathre, S.V., Schauerhammer, B., Christensen, C.M., 1976: Natural occurrence of *Fusarium* toxins in feedstuffs. Applied and Environmental Microbiology 32, 553–556

175. Korpinen, E.-L., 1972: Natural occurrence of F-2 and F-2-producing *Fusarium* strains associated with field cases of bovine and swine infertility. Abstr. I.U.P.A.C. Symposium, Control of Mycotoxins, Kungälv, Sweden, Aug. 21–22, p. 21

176. Roine, K., Korpinen, E.-L., Kallela, K., 1971: Mycotoxicosis as a probable cause of infertility in dairy cows. A case report. Nordisk Veterinaermedicin 23, 628–633

177. Mirocha, C.J., Harrison, J., Nichols, A.A., McClintok, M., 1968: Detection of a fungal estrogen (F-2) in hay associated with infertility in dairy cattle. Applied Microbiology 16, 797–798

178. Miller, J.K., Hacking, A., Harrison, J., Gross, V.J., 1973: Stillbirth, neonatal mortality and small litters in pigs associated with the ingestion of *Fusarium* toxin by pregnant sows. Veterinary Record 93, 555–559

179. Hsu, W.E., Smalley, E.B., Strong, F.M., Ribelin, W.E., 1972: Identification of T-2 toxin in moldy corn associated with a lethal toxicosis in dairy cattle. Applied Microbiology 24, 684–690

180. Puls, R., Greenway, J.A., 1976: Fusariotoxicosis from barley in British Columbia. II. Analysis and toxicity of suspected barley. Canadian Journal of Comparative Medicine 40, 16–19

181. Bryden, W.L., Rajion, M.A., Lloyd, A.B., Cumming, R.B., 1975: Survey of Australian feedstuffs for toxigenic strains of *Aspergillus flavus* and for aflatoxin. Australian Veterinary Journal 51, 491–493

182. Lew, H., 1993: Die aktuelle Mykotoxinsituation in der heimischen Landwirtschaft. Veröffentlichung Bundesanstalt für Agrarbiologie Linz/Donau 21, 5–26 [German]

183. Nahdi, S., Nusrath, M., 1985: Aflatoxins and other mycotoxins in mixed feeds and its constituents. Indian Journal of Botany 8, 16–24

184. Dietrich, R., Märtelbauer, E., Terplan, G., 1989: Erhebungen über die Mykotoxinbelastungen von Getreide und zerealienhaltigen Lebensmitteln in Bayern. Vortrag 11. Mykotoxin-Workshop 22–24. Mai, Berlin [German]

185. Bauer, J., Gareis, M., Detzler, W., Gedek, B., Heinritzi, K., Kabilka, G., 1987: Zur Entgiftung von Mykotoxinen in Futtermitteln. Tierärztliche Umschau 42, 613–627 [German]

186. Gruber, S., Thalmann, A., 1982: Untersuchungen über das Vorkommen von Mykotoxinen (außer Aflatoxin) in Futtermitteln. BML-Forschungsbericht 78, H S 12 Oktober [German]

187. Nawaz, S., Scudamore, K.A., Rainbird, S.C., 1997: Mycotoxins in ingredients of animal feeding stuff: I. Determination of *Alternaria* mycotoxins in oilseed rape meal and sunflower seed meal. Food Additives and Contaminants 14, 249–262

188. Pitt, J.I., Hocking, A.D., Jackson, K.l., Mullins, J.D., Webley, D.J., 1998: The occurrence of *Alternaria* species and related mycotoxins in international wheat. Journal of Food Mycology 1, 103–113

189. Norred, W.P., Voss, K.A., Bacon, C.W., Riley, R.T., 1991: Effectiveness of ammonia treatment in detoxification of fumonisin-contaminated corn. Food and Chemical Toxicology 29, 815–819

190. Sydenham, E.W., van der Westhuizen, L., Stockenström, S., Shepard, G.S., Thiel, P.G., 1994: Fumonisin-contaminated maise: physical treatment for the partial decontamination of bulk shipments. Food Additives and Contaminants 11, 25–32

191. Haschek, W.M., Motelin, G., Ness, D.K., Harlin, K.S., Hall, W., Vesonder, R.F., Peterson, R.E., Beasley, V.R., 1992: Characterization of fumonisin toxicity in orally and intravenously dosed swine. Mycopathologia 117, 83–96

192. Lepschy, J., 1991: Mykotoxine, die vernachlässigten Gifte. Pflanzenschutz-Praxis 4, 26–28 [German]

193. Stoloff, L., 1976: Occurrence of mycotoxins in foods and feeds. In: Rodricks J. (ed.), Mycotoxins and other Fungal Related Food Problems. American Chemical Society, Washington, DC, Advances in Chemistry Series, 149, 23–50

194. Iyer, L., Jackson, K., Leclerc, J., 1994: Commercial measurment of phomopsin A in lupin seeds and stubble. In: Dracup, M., Palta, J., (eds.), Proceedings First Australian Lupin Technical Symposium. Department of Agriculture, Western Australia

195. Wood, G.E., 1992: Mycotoxins in foods and feeds in the United States. Journal of Animal Science 70, 3941–3949

196. Noonpudgee, Ch., Böhm, J., Abdelhamid, A.M., Leibetseder, J., Schuh, M., 1986: Über das Vorkommen von Desoxynivalenol (Vomitoxin, DON) in Futtermitteln für österreichische Nutztierbestände im Zeitraum von 1979 bis 1985. Bodenkultur 37, 87–94 [German]

197. Pettersson, H., Kiessling, K.-H., Sandholm, K., 1986: Occurrence of the trichothecene mycotoxin deoxynivalenol (Vomitoxin) in Swedish-grown cereals. Swedish Journal of Agricultural Research 16, 179–182

198. Teich, A.H., Shugar, L., Smid, A., 1987: Soft white winter wheat cultivar field-resistance to scab and deoxynivalenol accumulation. Cereal Research Communications 15, 109–114

199. Bhavanishankar, T.N., Shanta, T., 1987: Natural occurrence of *Fusarium* toxins in peanut, sorghum and maize from Mysore (India). Journal of the Science of Food and Agriculture 40, 327–332

200. Lyons, P.C., Plattner, R.D., Bacon, C.W., 1986: Occurrence of peptide and clavine ergot alkaloids in tall fescue grass. Science 232, 487–489

201. Shotwell, O.L., Bennett, G.A., Stubblefield, R.D., Shannon, G.M., Kwolek, W.F., Plattner, R.D., 1985: Deoxynivalenol in hard red winter wheat: relationship between toxin levels and factors that could be used in grading. Journal of the Association of Official Analytical Chemists 68, 954–957

202. Usleber, E., Abramson, D., Gessler, R., Smith, D.M., Clear, R.M., Märtlbauer, E., 1996: Natural contamination of Manitoba barley by 3,15-diacetyldeoxynivalenol and its detection by immunochromatography. Applied and Environmental Microbiology 62, 3858–3860

203. MacDonald, S., Prikett, T.J., Wildey, K.B., Chan, D., 2004: Survey of ochratoxin A and deoxynivalenol in stored grains from the 1999 harvest in the UK. Food Additives and Contaminants 21, 172–181

204. Uhlig, S., Torp, M., Jarp, J., Parich, A., Gutleb, A.C., Krska, R., 2004: Moniliformin in Norwegian grain. Food Additives and Contaminants 21, 598–606

205. Jestoi, M., Rokka, M., Yli-Mattila, T., Parikkas, P., Rizzo, A., Peltonen, K., 2004: Presence and concentrations of the *Fusarium*-related mycotoxins beauvericin, enniatins and moniliformin in Finnish grain samples. Food Additives and Contaminants 21, 794–802

206. Pietri, A., Bertuzzi, T., Pallaroni, L., Piva, G., 2004: Occurrence of mycotoxins and ergosterol in maize harvested over 5 years in Northern Italy. Food Additives and Contaminants 21, 479–487

207. Bankole, S.A., Mabekoje, O.O., 2004: Occurrence of aflatoxins and fumonisins in preharvest maize from south-western Nigeria. Food Additives and Contaminants 21, 251–255

208. Bean, G.A., Schillinger, J.A., Klarman, W.L., 1972: Occurrence of aflatoxins and aflatoxin-producing strains of *Aspergillus* spp. in soybeans. Applied Microbiology 24, 437–439

209. Sandor, G., 1984: Occurrence of mycotoxins in feeds, animal organs and secretions. Acta Veterinaria Hungarica 32, 57–69

210. Vedanayagam, S., Indulkar, A.S., Raghavendar Rao, S., 1971: Aflatoxins & *Aspergillus flavus* Link in Indian cottonseed. Indian Journal of Experimental Biology 9, 410–411

211. Agnew, M.P., Poole, P.R., Lauren, D.L., Ledgard, S.F., 1986: Presence of zearalenone and trichothecene mycotoxins in *Fusarium*-infected New Zealand grown wheat. New Zealand Veterinary Journal 34, 176–177

212. Sydenham, E.W., Thiel, P.G., Marasas, W.F.O., Nieuwenhuis, J.J., 1989: Occurrence of deoxynivalenol and nivalenol in *Fusarium graminearum* infected undergrade wheat in South Africa. Journal of Agricultural and Food Chemistry 37, 921–926

213. Marpegan, M.R., Perfumo, C.J., Godoy, H.M., Sala de Miguel, M., Diaz, E., Risso, M.A., 1988: Feed refusal of pigs caused by *Fusarium* mycotoxins in Argentina. Journal of Veterinary Medicine Series A 35, 610–616

214. Ueno, Y., Lee, U.-S., Tanaka, T., Hasagawa, A., Strzelecki, E., 1985: Natural co-occurrence of nivalenol and deoxynivalenol in Polish cereals. Microbiologie – Aliments – Nutrition 3, 321–326

215. Mirocha, C.J., Schauerhammer, B., 1973: Natural occurrence of zearalenone (F-2) in maize. 2nd International Congress of Plant Pathology, September 5–12, Minneapolis, Paper No. 0359 [Abstract]

216. Eppley, R.M., Stoloff, L., Trucksess, M.W., Chang, C.W., 1974: Survey of corn for *Fusarium* toxins. Journal of the Association of Official Analytical Chemists 57, 632–635

217. Shotwell, O.L., Hesseltine, C.W., Goulden, M.L., Vandegraft, E.E., 1970: Survey of corn for aflatoxin, zearalenone, and ochratoxin. Cereal Chemistry 47, 700–707

218. Shotwell, O.L., Hesseltine, C.W., Vandegraft, E.E., Goulden, M.L., 1971: Survey of corn from different regions for aflatoxin, ochratoxin and zearalenone. Cereal Science Today 16, 266–273

219. Nesheim, S., 1971: Ochratoxins: Occurrence, production, analysis, and toxicity. Proceedings 85th Annual Meeting, Journal of the Association of Official Analytical Chemists October 11–14, p. 23 [Abstract]

220. Abouzied, M.M., Askegard, S.D., Bird, C.B., Miller, B.M., 1995: Development of a rapid quantitative ELISA for determination of the mycotoxin fumonisin in food and feed. Journal of Clinical Ligand Assay 18, 145–149

221. Tseng, T.-C., Yuan, G.-F., Tseng, J., Hsiao, I.-W., Mirocha, C.J., 1985: Natural occurrence of *Fusarium* mycotoxins in grains and feeds in Taiwan. Botanical Bulletin of Academia Sinica 26, 83–95

222. Lew, H., Chelkowski, J., Wakulinski, W., Edinger, W., 1993: Moniliformin in wheat and *triticale* grain. Mycotoxin Research 9, 66–71

223. Leibetseder, J., 1982: Mykotoxine in der Schweineproduktion. Kraftfutter 60, 428–432 [German]

224. Langseth, W., Elen, O., 1996: Differences between barley, oats and wheat in the occurrence of deoxynivalenol and other trichothecenes in Norwegian grain. Journal of Phytopathology 144, 113–118

225. Bauer, J., Binder, S., 1993: Fumonisine in Futtermitteln: Vorkommen und Bedeutung einer neuen Gruppe von Fusarientoxinen. Tierärztliche Umschau 48, 718–727 [German]

226. Eppley, R.M., Trucksess, M.W., Nesheim, S., Thorpe, C.W., Wood, G.E., Pohland, A.E., 1984: Deoxynivalenol in winter wheat: thin layer chromatographic method and survey. Journal of the Association of Official Analytical Chemists 67, 43–45

227. Park, K.-J., Lee, Y.-W., 1990: Natural occurrence of *Fusarium* mycotoxins in Korean barley samples harvested in 1987 and 1989. Proceedings of the Japanese Association of Mycotoxicology 31, 37–41

228. Marasas, W.F.O., Kriek, N.P.J., Rensburg van, S.J., Steyn, M., Schalkwyk van, G.C., 1977: Occurrence of zearalenone and deoxynivalenol, mycotoxins produced by *Fusarium graminearum* Schwabe, in maize in Southern Africa. South African Journal of Science 73, 346–349

229. Jeswal, P., 1990: Mycotoxin contamination in cattle feeds from Bihar. National Academy Science Letters 13, 431–433

230. Tanaka, T., Hasegawa, A., Matsuki, Y., Ishii, K., Ueno, Y., 1985: Improved methodology for the simultaneous detection of the trichothecene mycotoxins deoxynivalenol and nivalenol in cereals. Food Additives and Contaminants 2, 125–137

231. Kamimura, H., Nishijima, M., Yasuda, K., Saito, K., Ibe, A., Nagayama, T., Ushiyama, H., Naoi, Y., 1981: Simultaneous detection of several *Fusarium* mycotoxins in cereals, grains, and foodstuffs. Journal of the Association of Official Analytical Chemists 64, 1067–1073

232. Mulders, E.J., Impelen-Peek van, H.A.M., 1986: Gas chromatographic determination of deoxynivalenol in cereals. Zeitschrift für Lebensmittel-Untersuchung und -Forschung 183, 406–409

233. Trenholm, H.L., Cochrane, W.P., Cohen, H., Elliot, J.I., Farnworth, E.R., Friend, D.W., Hamilton, R.M.G., Standish, J.F., Thompson, B.K., 1983: Survey of vomitoxin contamination of 1980 Ontario white winter wheat crop: results of survey and feeding trials. Journal of the Association of Official Analytical Chemists 66, 92–97

234. Logrieco, A., Rizzo, A., Ferracane, R., Ritieni, A., 2002: Occurrence of beauvericin and enniatins in wheat affected by *Fusarium avenaceum* head blight. Applied and Environmental Microbiology 68, 82–85

235. Langseth, W., Kosiak, B., Clasen, P.-E., Torp, M., Gareis, M., 1997: Toxicity and occurrence of *Fusarium* species and mycotoxins in late harvested and overwintered grain from Norway, 1993. Journal of Phytopathology 145, 409–416

236. Tomczak, M., Wisniewska, H., Stepien, L., Kostecki, M., Chelkowski, J., Golinski, P., 2002: Deoxynivalenol, nivalenol and moniliformin in wheat samples with head blight (scab) symptoms in Poland (1998–2000). European Journal of Plant Pathology 108, 625–630

237. Côté, L.M., Reynolds, J.D., Vesonder, R.F., Buck, W.B., Swanson, S.P., Coffey, R.T., Brown, D.C., 1984: Survey of vomitoxin-contaminated feed grains in midwestern United States, and associated health problems in swine. Journal of American Veterinary Medical Association 184, 189–192

238. Yazdanpanah, H., Khoshnood Mansour-Khani, M.J., Shafaati, A., Rahimian, H., Rasekh, H.R., Gilani, K., Moradkhani, M., 1997: Evaluation of natural occurrence of *Fusarium* mycotoxins in wheat fields of northern Iran. Cereal Research Communications 25, 337–341

239. Ellend, N., Binder, J., Krska, R., Horvath, E.M., 1997: Contamination of Austrian corn with *Fusarium* toxins in autumn 1996. Cereal Research Communications 25, 359–360

240. Abramson, D., Mills, J.T., Marquardt, R.R., Frohlich, A.A., 1997: Mycotoxins in fungal contaminated samples of animal feed from western Canada, 1982–1994. Canadian Journal of Veterinary Research 61, 49–52

241. Sinha, K.K., 1991: Contamination of freshly harvested maize kernels with *Fusarium* mycotoxins in Bihar State, India. Mycotoxin Research 7A, 178–183

242. Muñoz, L., Cardelle, M., Pereiro, M., Riguera, R., 1990: Occurrence of corn mycotoxins in Galicia (Northwest Spain). Journal of Agricultural and Food Chemistry 38, 1004–1006

243. Dalcero, A., Magnoli, C., Chiacchiera, S., Palacios, G., Reynoso, M., 1997: Mycoflora and incidence of aflatoxin B_1, zearalenone and deoxynivalenol in poultry feeds in Argentina. Mycopathologia 137, 179–184

244. Pettersson, H., Göransson, B., Kiessling, K.-H., Tideman, K., 1978: Aflatoxin in a Swedish grain sample. Nordisk Veterinaermedicin 30, 482–485

245. Chen, S.-C., Friedman, L., 1966: Aflatoxin determination in seed meal. Journal of the Association of Official Analytical Chemists 49, 28–33

246. Marsh, P.B., Simpson, M.E., Feretti, R.J., Campbell, T.C., Donoso, J., 1969: Relation of aflatoxins in cotton seeds at harvest to fluorescence in the fiber. Journal of Agricultural and Food Chemistry 17, 462–467

247. Dhavan, A.S., Choudary, M.R., 1995: Incidence of aflatoxins in animal feedstuffs: a decade's scenario in India. Journal of the Association of Official Analytical Chemists International 78, 693–698

248. Meister, U., Symmank, H., Dahlke, H., 1996: Untersuchung und Bewertung der Fumonisinkontamination von einheimischem und importiertem Getreide. Zeitschrift für Lebensmittel-Untersuchung und -Forschung 203, 528–533 [German]

249. Chavez, E.R., 1984: Vomitoxin-contaminated wheat in pig diets: pregnant and lactating gilts and weaners. Canadian Journal of Animal Science 64, 717–723

250. Sudhakara Reddy, P., Reddy, C.V., Ravindra Reddy, V., Rao, P.V., 1984: Occurrence of aflatoxin in some feed ingredients in three geogaphical regions of Andrah Pradesh. Indian Journal of Animal Science 54, 235–238

251. Vesonder, R.F., Ciegler, A., Rohwedder, W.K., Eppley, R., 1979: Reexamination of 1972 midwest corn for vomiting. Toxicon 17, 658–660

252. Visconti, A., Chelkowski, J., Bottalico, A., 1986: Deoxynivalenol and 3-acetyldeoxynivalenol – mycotoxins associated with wheat head fusariosis in Poland. Mycotoxin Research 2, 59–64

253. Patel, P.M., Netke, S.P., Gupta, B.S., Dabadghao, A.K., 1981: Survey of oilcakes and some feeds for the presence of aflatoxin. Indian Journal of Animal Science 51, 402–407

254. Caldwell, R.W., Tuite, J., 1974: Zearalenone in freshly harvested corn. Phytopathology 64, 752–753

255. Vesonder, R.F., Ciegler, A., Rogers, R.F., Burbridge, K.A., Bothast, R.J., Jensen, A.H., 1978: Survey of 1977 crop year preharvest corn for vomitoxin. Applied and Environmental Microbiology 36, 885–888

256. Tobin, N.F., 1988: Presence of deoxynivalenol in Australian wheat and triticale – New South Wales Northern Rivers region, 1983. Australian Journal of Experimental Agriculture 28, 107–110

257. Holcomb, M., Thompson, H.C., Jr., Hankins, L.J., 1993: Analysis of fumonisin B_1 in rodent feed by gradient elution HPLC using precolumn derivatization with FMOC and fluorescence detection. Journal of Agricultural and Food Chemistry 41, 764–767

258. Rheeder, J.P., Sydenham, E.W., Marasas, W.F.O., Thiel, P.G., Shepard, G.S., Schlechter, M., Stockenström, S., Cronje, D.W., Viljoen, J.H., 1995: Fungal infestation and mycotoxin contamination of South African commercial maize harvested in 1989 and 1990. South African Journal of Science 91, 127–131

259. Kostecki, M., Szczesna, J., Chelkowski, J., Wisniewska, H., 1995: Beauvericin and moniliformin production by Polish isolates of *Fusarium subglutinans* and natural cooccurrence of both mycotoxins in maize samples. Microbiologie – Aliments – Nutrition 13, 67–70

260. Pathre, S.V., Mirocha, C.J., 1979: Trichothecenes: natural occurrence and potential hazard. The Journal of the American Oil Chemists Society 86, 820–823

261. Hökby, E., Hult, K., Gatenbeck, S., Rutquist, L., 1979: Ochratoxin A and citrinin in 1976 crop of barley stored on farms in Sweden. Acta Agriculturæ Scandinavica 29, 174–178

262. Galtier, P., Jemmali, M., Larrieu, G., 1977: Enquête sur la présence éventuelle d'aflatoxine et d'ochratoxine A dans des maïs récoltés en France en 1973 et 1974. Annales de la Nutrition et de l'Alimentation 31, 381–389 [French]

263. Aucock, H.W., Marasas, W.F.O., Meyer, C.J., Chalmers, P., 1980: Field outbreaks of hyperestrogenism (Vulvo-vaginitis) in pigs consuming maize infected by *Fusarium graminearum* and contaminated with zearalenone. Journal of the South African Veterinary Association 51, 163–166

264. Golinski, P., Perkowski, J., Kostecki, M., Grabarkiewicz-Szczęsna, J., Chelkowski, J., 1996: *Fusarium* species and *Fusarium* toxins in wheat in Poland – a comparison with neighbour countries. Sydowia 48, 12–22

265. Yoshizawa, T., 1997: Geographic difference in trichothecene occurrence in Japanese wheat and barley. Bulletin of the Institute for Comprehensive Agricultural Science 5, 23–30

266. Fazekas, B., Tóthné, H.E., 1995: Incidence of fumonisin-B_1 mycotoxin in maize cultivated in Hungary. Magyar Állatorvosok Lapja 50, 515–518

267. Buckle, A.E., 1986: Moulds and mycotoxins in cereal and other components of compound animal feed. International Biodeterioration, Supplement. 22, 55–60

268. Scudamore, K.A., 1993: Occurrence and significance of mycotoxins in cereals grown and stored in the United Kingdom. Aspects of Applied Biology 36, 361–371

269. Logrieco, A., Moretti, A., Ritieni, A., Chelkowski, J., Altomare, C., Bottalico, A., Randazzo, G., 1993: Natural occurrence of beauvericin in preharvest *Fusarium subglutinans* infected corn ears in Poland. Journal of Agricultural Food Chemistry 41, 2149–2152

270. Blaney, B.L., 1986: Mycotoxins in water-damaged and mouldy wheat from temporary bulk stores in Queensland. Australian Journal of Agricultural Research 37, 561–565

271. Sabino, M., Prado, G., Inomata, E.I., de Oliveira Predoso, M., Garcia, R.V., 1989: Natural occurrence of aflatoxins and zearalenone in maize in Brazil. Part II. Food Additives and Contaminants 6, 327–331

272. Romer, T., 1975: Screening method for the detection of aflatoxins in mixed feeds and other agricultural commodities with subsequent confirmation and quantitative measurement of aflatoxins in positive samples. Journal of the Association of Official Analytical Chemists 58, 500–506

273. Shotwell, O.L., Goulden, M.L., Bennett, G.A., Plattner, R.D., Hesseltine, C.W., 1977: Survey of 1975 wheat and soybeans for aflatoxin, zearalenone, and ochratoxin. Journal of the Association of Official Analytical Chemists 60, 778–783

274. Pons, W.A., Jr., Franz, A.O., Jr., 1977: High performance liquid chromatography of aflatoxins in cottonseed products. Journal of the Association of Official Analytical Chemists 60, 89–95

275. Logrieco, A., Moretti, A., Ritieni, A., Bottalico, A., Corda, P., 1995: Occurrence and toxigenicity of *Fusarium proliferatum* from preharvest maize ear rot, and associated mycotoxins, in Italy. Plant Disease 79, 727–730

276. Bergsjø, B., Langseth, W., Nafstad, I., Høgset Jansen, J., Larsen, H.J.S., 1993: The effects of naturally deoxynivalenol-contaminated oats on the clinical condition, blood parameters, performance and carcass composition of growing pigs. Veterinary Research Communication 17, 283–294

277. Lillehoj, E.B., Fennell, D.I., Kwolek, W.F., 1977: Aflatoxin and *Aspergillus flavus* occurrence in 1975 corn at harvest from a limited region of Iowa. Cereal Chemistry 54, 366–372

278. Shotwell, O.L., Kwolek, W.F., Goulden, M.L., Jackson, L.K., 1975: Aflatoxin occurrence in some white corn under loan, 1971. I. Incidence and level. Cereal Chemistry 52, 373–380

279. Wheeler, B., 1969: An examination of cereals and feeds for aflatoxins. Irish Journal of Agricultural Research 8, 172–174

280. Wood, G.E., 1989: Aflatoxins in domestic and imported foods and feeds. Journal of the Association of Official Analytical Chemists 72, 543–548

281. Lappalainen, S., Nikulin, M., Berg, S., Parikka, P., Hintikka, E.-L., Pasanen, A.-L., 1996: *Fusarium* toxins and fungi associated with handling of grain on eight Finnish farms. Atmospheric Environment 30, 3059–3065

282. Sharman, M., MacDonald, S., Gilbert, J., 1992: Automated liquid chromatographic determination of ochratoxin A in cereals and animal products using immunoaffinity column clean-up. Journal of Chromatography 603, 285–289

283. Tuinstra, L.G., Haasnoot, W., 1983: Rapid determination of aflatoxin B_1 in Dutch feeding stuffs by high-performance liquid chromatography and post-column derivatization. Journal of Chromatography 282, 457–462

284. Loosmore, R.M., Allcroft, R., Tutton, E.A., Carnaghan, R.B.A., 1964: The presence of aflatoxin in a sample of cotton-seed cake. Veterinary Record 76, 64–65

285. Girgis, A.N., El-Sherif, S., Rofael, N., Nesheim, S., 1977: Aflatoxins in Egyptian foodstuffs. Journal of the Association of Official Analytical Chemists 60, 746–747

286. Lillehoj, E.B., Kwolek, W.F., Shannon, G.M., Shotwell, O.L., Hesseltine, C.W., 1975: Aflatoxin occurrence in 1973 corn at harvest. I. A limited survey in the southeastern U.S. Cereal Chemistry 52, 603–611

287. Price, R.L., Paulson, J.H., Lough, O.G., Gingg, C., Kurtz, A.G., 1985: Aflatoxin conversion by dairy cattle consuming naturally-contaminated whole cottonseed. Journal of Food Protection 48, 11–15

288. Scudamore, K., Hetmanski, M., 1992: Analysis of mycotoxins in cereals and animal feed using a multi-toxin gel permeation chromatography clean-up method. Mycotoxin Research 8, 37–47

289. Fazekas, B., Bajmócy, E., Glávits, R., Fenyvesi, A., Janyi, J., 1998: Fumonisin B_1 contamination of maize and experimental acute fumonisin toxicosis in pigs. Journal of Veterinary Medicine Series B 45, 171–181

290. Perkowski, J., Chelkowski, J., Błaźczak, P., Snijders, C.H.A., Wakulinski, W., 1991: A study of the correlations between the amount of deoxynivalenol in grain of wheat and triticale and percentage of *Fusarium* damaged kernels. Mycotoxin Research 7, 102–114

291. Langseth, W., Ellingsen, Y., Nyomen, U., Økland, E.M., 1989: High-performance liquid chromatographic determination of zearalenone and ochratoxin in cereals and feed. Journal of Chromatography 478, 269–274

292. Shotwell, O.L., Hesseltine, C.W., Goulden, M.L., 1969: Ochratoxin A: occurrence as natural contaminant of a corn sample. Applied Microbiology 17, 765–766

293. Ware, G.M., Thorpe, C.W., 1977: Analysis for zearalenone in corn by high pressure liquid chromatography with a fluorescence detector. Abstracts 91st Annual Meeting, Association of Official Analytical Chemists, Washington, DC, 47 (Abstract no. 148)

294. Gruber-Schley, S., Thalmann, A., 1988: Zum Vorkommen von *Alternaria* spp. und deren Toxine in Getreide und mögliche Zusammenhänge mit Leistungsminderungen landwirtschaftlicher Nutztiere. Landwirtschaftliche Forschung 41, 11–29 [German]

295. Madsen, A., Hald, B., Lillehoj, E., Mortensen, H.P., 1982: Feeding experiments with ochratoxin A contaminated barley for bacon pigs. 2. Naturally contaminated barley given for 6 weeks from 20 kg compared with normal barley supplemented with crystalline ochratoxin A and/or citrinin. Acta Agriculturæ Scandinavica 32, 369–372

296. Rambo, G.W., Tuite, J., Caldwell, R.W., 1974: *Apergillus flavus* and aflatoxin in preharvest corn from Indiana in 1971 and 1972. Cereal Chemistry 51, 848–853

297. Hald, B., Krogh, P., 1975: Detection of ochratoxin A in barley, using silicia gel minicolumns. Journal of the Association of Official Analytical Chemists 58, 156–158

298. Seibold, R., Ruch, W., 1977: Der Aflatoxingehalt im Milchpulver. Kraftfutter 60, 182–185 [German]

299. Shotwell, O.L., Goulden, M.L., Hesseltine, C.W., 1976: Aflatoxin M_1. Occurrence in stored and freshly harvested corn. Journal of Agricultural and Food Chemistry 24, 683–684

300. Hesseltine, C.W., Rogers, R.F., Shotwell, O., 1978: Fungi, especially *Gibberella zeae*, and zearalenone occurrence in wheat. Mycologia 70, 14–18

301. Binkerd, K.A., Scott, D.H., Everson, R.J., Sullivan, J.M., Robinson, F.R., 1993: Fumonisin contamination of the 1991 Indiana corn crop and its effects on horses. Journal of Veterinary Diagnostic Investigation 5, 653–655

302. Schmitt, S.G., Hurburgh, C.R., Jr., 1989: Distribution and measurement of aflatoxin in 1983 Iowa corn. Cereal Chemistry 66, 165–168

303. McMillian, W.W., Wilson, D.M., Widstrom, N.W., Gueldner, R.C., 1980: Incidence and level of aflatoxin in preharvest corn in south Georgia in 1978. Cereal Chemistry 57, 83–84

304. Whitten, M.E., 1969: Aflatoxins in cottonseed hulls. The Journal of the American Oil Chemists Society 47, 5–6

305. Manabe, M., Tsuruta, O., Sugimoto, T., Minamisawa, M., Matsuura, S., 1971: Aflatoxins in imported peanut meals. Journal of the Food Hygienics Society of Japan 12, 364–369

306. Tung, T.C., Ling, K.H., 1968: Study on aflatoxins of foodstuffs in Taiwan. Journal of Vitaminology 14, 48–52

307. Dacero, A.M., Combina, M., Etcheverry, M., Varsavsky, E., Rodriguez, M.I., 1997: Evaluation of *Alternaria* and its mycotoxins during ensiling of sunflower seeds. Natural Toxins 5, 20–23

308. Danielsen, S., Jensen, D.F., 1998: Relationships between germination, fumonisin content, and *Fusarium verticillioides* infection in selected maize samples from different regions of Costa Rica. Plant Pathology 47, 609–647

309. Hirooka, E.Y., Yamaguchi, M.M., Aoyama, S., Sugiura, Y., Ueno, Y., 1996: The natural occurrence of fumonisins in Brazilian corn kernels. Food Additives and Contaminants 13, 173–183

310. Edwards, S.G., 2003: *Fusarium* mycotoxins in UK wheat. Aspects of Applied Biology 68, 35–42

311. Waliyar, F., Reddy, S.V., Subramaniam, K., Reddy, T.Y., Rama Devi, K., Craufurd, P.Q., Wheeler, T.R., 2003: Importance of mycotoxins in food and feed in India. Aspects of Applied Biology 68, 147–154

312. Jestoi, M., Rooka, M., Rizzo, A., Peltonen, K., Parikka, P., Yli-Mattila, T., 2003: Moniliformin in Finnish grains: analysis with LC-MS/MS. Aspects of Applied Biology 68, 211–216

313. Norizuki, H., Minamisawa, M., Yammamoto, K., Sugimoto, T., 1987: Natural occurrence of aflatoxins in Egyptian polished rice. Proceedings of the Japanese Association of Mycotoxicology 26, 50–51

314. Kordić, B., Panin, M., Istoni, A., Nedeljković, V., 1986: Frequency and conditions of occurrence of mycotoxins in cereal grains and prevention of their production. Acta Veterinaria (Beograd) 36, 307–312

315. Perkowski, J., Wakulinski, W., Chelkowski, J., 1990: Natural occurrence of deoxynivalenol, 3-acetyldeoxynivalenol and zearalenone in wheat in 1988. Microbiologie – Aliments – Nutrition 8, 241–247

316. Krogh, P., Hald, B., 1969: Occurrence of aflatoxin in imported groundnut products. Nordisk Veterinaermedicin 21, 398–407

317. Logrieco, A., Bottalico, A., Visconti, A., Vurro, M., 1988: Natural occurrence of *Alternaria*-mycotoxins in some plant products. Microbiologie – Aliments – Nutrition 6, 13–17

318. Jurjevic, Z., Solfrizzo, M., Cvjetkovic, B., de Girolamo, A., Visconti, A., 2002: Occurrence of beauvericin in corn from Croatia. Food Technology and Biotechnology 40, 91–94

319. Ross, P.F., Ledet, A.E., Owens, D.L., Rice, L.G., Nelson, H.A., Osweiler, G.D., Wilson, T.M., 1993: Experimental equine leukoencephalomalacia, toxic hepatosis, and encephalopathy caused by corn naturally contaminated with fumonisins. Journal of Veterinary Diagnostic Investigation 5, 69–74

320. Dickens, J.W., Whitaker, T.B., 1981: Bright greenish-yellow fluorescence and aflatoxin in recently harvested yellow corn marketed in North Carolina. The Journal of the American Oil Chemists Society 59, 973A–975A

321. Balasubramanian, T., 1985: Incidence of aflatoxin B_1 in animal feed. Indian veterinary Journal 62, 982–988

322. Nepote, M.C., Piontelli, E., Saubois, A., 1997: Occurrence of *Aspergillus flavus* strains and aflatoxins in corn from Santa Fe, Argentina. Archivos Latinoamericanos de Nutricion 47, 262–264

323. Clear, R.M., Patrick, S.K., Gaba, D., 2000: Prevalence of fungi and fusariotoxins on barley seed from western Canada, 1995 to 1997. Canadian Journal of Plant Pathology 22, 44–50

324. Lee, L.S., Koltun, S.P., Buco, S., 1983: Aflatoxin distribution in fines and meats from decorticated cottonseed. The Journal of the American Oil Chemists Society 60, 1548–1549

325. McKenzie, R.A., Blaney, B.J., Connole, M.D., Fitzpatrick, L.A., 1981: Acute aflatoxicoses in calves fed peanut hay. Australian Veterinary Journal 57, 284–286

326. Wicklow, D.T., Stubblefield, R.D., Horn, B.W., Shotwell, O.D., 1988: Citreoviridin levels in *Eupenicillium ochrosalmoneum*-infested maize kernels at harvest. Applied and Environmental Microbiology 54, 1096–1098

327. Shotwell, O.L., Bennett, G.A., Goulden, M.L., Plattner, R.D., Hesseltine, C.W., 1980: Survey for zearalenone, aflatoxin, and ochratoxin in U.S. grain sorghum from 1975 and 1976 crops. Journal of the Association of Official Analytical Chemists 63, 922–926

328. Hunt, W.H., Semper, R.C., Liebe, E.B., 1976: Incidence of aflatoxin in corn and a field method for aflatoxin analysis. Cereal Chemistry 53, 227–232

329. Schroeder, H.W., Boller, R.A., 1973: Aflatoxin production of species and strains of the *Aspergillus flavus* group isolated from field crops. Applied Microbiology 25, 885–889

330. Akano, D.A., Atanda, O.O., 1989: The present level of aflatoxin in Nigerian groundnut cake ('kulikuli'). Letters in Applied Microbiology 10, 187–189

331. Blaney, B.J., McKenzie, R.A., Walters, J.R., Taylor, L.F., Bewg, W.S., Ryley, M.J., Maryam, R., 2000: Sorghum ergot (*Claviceps africana*) associated with agalactia and feed refusal in pigs and dairy cattle. Australian Veterinary Journal 78, 102–107

332. Reich, K, 1998: Feldstudie zum Vorkommen von Ochratoxin A und Zearalenon in Futtermitteln und im Blut von Zucht- und Mastschweinen mit besonderer Berücksichtigung der Futterherkunft und -lagerung. Freie Universität Berlin, Fachbereich Veterinärmedizin, Dissertation, [German]

333. Pettersson, H., Holmberg, T., Larsson, K., Kaspersson, A., 1989: Aflatoxins in acid treated grain in Sweden and occurrence of aflatoxin M_1 in milk. Journal of the Science of Food and Agriculture 48, 411–420

334. Yu, W., Yu, F.-Y., Undersander, D.J., Chu, F., 1999: Immunoassays of selected mycotoxins in hay, silage and mixed feed. Food and Agricultural Immunology 11, 307–319

335. Orsi, R.B., Correa, B., Possi, C.R., Schammass, E.A., Nogueira, J.R., Dias, S.M.C., Malozzi, M.A.B., 2000: Mycoflora and occurrence of fumonisins in freshly harvested and stored hybrid maize. Journal of stored Product Research 36, 75–87

336. Ono, E.Y.S., Sasaki, E.Y., Hashimoto, E.H., Hara, L.N., Correa, B., Itano, E.N., Sugiura, T., Ueno, Y., Hirooka, E.Y., 2002: Post-harvest storage of corn: effect of beginning moisture content on mycoflora and fumonisin contamination. Food Additives and Contaminants 19, 1081–1090

337. Tseng, T.-C., Liu, C.-Y., 2001: Occurrence of fumonisin B_1 in maize imported into Taiwan. International Journal of Food Microbiology 65, 23–26

338. Shephard, G.S., Marasas, W.F.O., Yazdanpanah, H., Rahimian, H., Safavi, N., Zarghi, A., Shafaati, A., Rasekh, H.R., 2002: Fumonisin B_1 in maize harvested in Iran during 1999. Food Additives and Contaminants 19, 676–679

339. Czerwiecki, L., Czajkowska, D., Witkowska-Gwiazdowska, A., 2002: On ochratoxin A and fungal flora in Polish cereals from conventional and ecological farms. Part 2: Occurrence of ochratoxin A and fungi in cereals in 1998. Food Additives and Contaminants 19, 1051–1057

340. Gunsen, U., Yarolu, T., 2002: Aflatoxin in dog and horse feeds in Turkey. Veterinary and Human Toxicology 44, 113–114

341. Czerwiecki, L., Czajkowska, D., Witkowska-Gwiazdowska, A., 2002: On ochratoxin A and fungal flora in Polish cereals from conventional and ecological farms. Part 1: Occurrence of ochratoxin A and fungi in cereals in 1997. Food Additives and Contaminants 19, 470–477

342. Perkowski, J., Basinski, T., 2002: Natural contamination of oat with group A trichothecene mycotoxins in Poland. Food Additives and Contaminants 19, 478–482

343. Krause, H.P., 1985: Untersuchungen zum Vorkommen der Mykotoxine Aflatoxin B_1, B_2, G_1, G_2 und Zearalenon im Schweinefutter (Fertigfuttermittel und betriebseigene Mischungen) im Raum Süd-Schleswig-Holstein. Freie Universität Berlin, Fachbereich Veterinärmedizin, Dissertation, [German]

344. Škrinjar, M., Stubblefield, R.D., Stojanović, E., Dimič, G., 1995: Occurrence of *Fusarium* species and zearalenone in dairy cattle feeds in Vojvodina. Acta Veterinaria Hungarica 43, 259–267

345. Chelkowski, J., Lew, H., Pettersson, H., 1994: *Fusarium poae* (Peck) Wollenw. – Occurrence in maize ears, nivalenol production and mycotoxin accumulation in cobs. Mycotoxin Research 10, 116–120

346. Campell, H., Choo, T.M., Vigier, B., Underhill, L., 2002: Comparison of mycotoxin profiles among cereal samples from eastern Canada. Canadian Journal of Botany 80, 526–532

347. Velutti, A., Marin, S., Sanchis, V., Ramos, A.J., 2001: Note. Occurrence of fumonisin B_1 in Spanish corn-based foods for animal and human consumption. Food Science and Technology International 7, 433–437

348. Visconti, A., Chelkowski, J., Solfrizzo, M., Bottalico, A., 1990: Mycotoxins in corn ears naturally infected with *Fusarium graminearum* and *F. crookwellense*. Canadian Journal of Plant Pathology 12, 187–189

349. Thirumala-Devi, K., Mayo, M.A., Reddy, G., Reddy, D.V.R., 2002: Occurrence of aflatoxins and ochratoxin A in Indian poultry feed. Journal of Food Protection 65, 1338–1340

350. Razzazi, E., Böhm, J., Grajewski, J., Szcepaniak, K., Kübber-Heiss, A.J., Iben, C.H., 2001: Residues of ochratoxin A in pet foods, canine and feline kidneys. Journal of Animal Physiology and Animal Nutrition 85, 212–216

351. Almeida, A.P., Fonseca, H., Fancelli, A.L., Direito, G.M., Ortega, E.M., Corrêa, B., 2002: Mycoflora and fumonisin contamination in Brazilian corn from sowing to harvest. Journal of Agricultural and Food Chemistry 50, 3877–3882

352. Abbas, H.K., Williams, W.P., Windham, G.L., Pringle H.C., III, Xie, W., Shier, W.T., 2002: Aflatoxin and fumonisin contamination of commercial corn (*Zea mays*) hybrids in Mississippi. Journal of Agricultural and Food Chemistry 50, 5246–5254

353. Kollarczik, B., Kaaden, O.-R., 1995: Mykotoxikologische Untersuchung von Futtermitteln. Kraftfutter 6, 264–265 [German]

354. Warner, R., Ram, B.P., Hart, P., Pestka, J.J., 1986: Screening for zearalenone in corn by competitive direct enzyme-linked immunosorbent assay. Journal of Agricultural and Food Chemistry 34, 714–717

355. Döll, S., Valenta, H., Dänicke, S., Flachowsky, G., 2002: *Fusarium* mycotoxins in conventionally and organically grown grain from Thuringia/Germany. Landbauforschung Völkenrode 52, 91–96

356. Sinha, B.K., Ranjan, K.S., Pandey, T.N., 2002: Mycotoxins, mycotoxigenic fungi and the biochemical changes in *Makhana* (*Euryale ferox* Salisb) puffs of North Bihar. Journal of Food Science and Technology 39, 38–41

357. Salem, D.A., 2002: Natural occurrence of aflatoxins in feedstuffs and milk of dairy farms in Assiut Province, Egypt. Wiener tierärztliche Monatsschrift 89, 86–91

358. Phutela, R.P., Kalra, M.S., 1979: A note on the incidence of aflatoxins in peanut and cottonseed cakes of Ludhiana. Indian Journal of Agricultural Research 13, 255–256

359. Juskiewicz, T., Piskorska-Pliszczyńska, J., 1977: Mycotoxins in grain for animal feeds. Annales de la Nutrition et de l'Alimentation 31, 489–493

360. Joffe, A.Z., 1970: The presence of aflatoxin in kernels from five years groundnut crops and of *Aspergillus flavus* isolates from kernels and soils. Plant and Soil 33, 91–96

361. Mazen, M.B., El-Kady, I.A., Saber, S.M., 1990: Survey of the mycoflora and mycotoxins of cotton seeds and cotton seed products in Egypt. Mycopathologia 110, 133–138

362. Müller, H.-M., Reimann, J., Schumacher, U., Schwadorf, K., 2001: Further survey of the occurrence of *Fusarium* toxins in wheat grown in southwest Germany. Archives of Animal Nutrition 54, 173–182

363. Lew, H., Adler, A., Edinger, W., Brodacz, W., Kiendler, E., Hinterholzer, J., 2001: Fusarien und ihre Toxine bei Mais in Österreich. Die Bodenkultur 52, 199–207 [German]

364. Park, J.W., Kim, E.K., Shon, D.H., Kim, Y.B., 2002: Occurrence of zearalenone in Korean barley and corn foods. Food Additives and Contaminants 19, 158–162

365. Camargos, S.M., Valente Soares, L.M., Sawazaki, E., Bolonhezi, D., Castro, J.L., Bortolleto, N., 2000: Fumonisins in corn cultivars in the state of Sao Paulo. Brazilian Journal of Microbiology 31, 226–229

366. Schollenberger, M., Müller, H.-M., Drochner, W., 2005: *Fusarium* toxin content of maize and maize products purchased in the years 2000 and 2001 in Germany. Mycotoxin Research 21, 26–28

367. Macri, A., Schollenberger, M., Dancea, Z., Drochner, W., 2003: Zearalenone and ergosterol contents in corn samples of Transylvania, Romania. Mycotoxin Research 19, 190–193

368. Vesonder, R., Haliburton, J., Stubblefield, R., Gilmore, W., Peterson, S., 1991: *Aspergillus flavus* and aflatoxins B_1, B_2 and M_1 in corn associated with equine death. Archives of Environmental Contamination and Toxicology 20, 151–153

369. Bottalico, A., Corda, P., Logrieco, A., Moretti, A., 1993: *Fusarium* ear rot and associated fumonisin B_1 in pre-harvest maize ears in Sardina in 1992. Petria 3, 13

370. Price, W.D., Lovell, R.A., McChesney, D.G., 1993: Naturally occurring toxins in feedstuffs: center for veterinary medicine perspective. Journal of Animal Science 71, 2556–2562

371. Puntarić, D., Bošnir, J., Šmit, Z., Škes, I., Baklaić, Z., 2001: Ochratoxin A in corn and wheat: geographical association with endemic nephropathy. Croatian Medical Journal 42, 175–180

372. Engels, R., Krämer, J., 1996: Incidence of Fusaria and occurrence of selected *Fusarium* mycotoxins on *Lolium* spp. in Germany. Mycotoxin Research 12, 31–40

373. Škringar, M., Stubblefield, R.D., Vujičić, I.F., 1992: Ochratoxigenic moulds and ochratoxin A in forages and grain feeds. Acta Veterinaria Hungarica 40, 185–190

374. Drábek, J., Piskač, A., 1979: The occurrence of aflatoxin in some components and complete feed mixture for pigs. Veterinári Medicína 10, 597–602 [Czech]

375. de Campos, M., Olszyna-Marzys, A.E., 1979: Aflatoxin contamination in grains and grain products during dry season in Guatemala. Bulletin of Environmental Contamination and Toxicology 22, 350–356

376. Kiermeier F., Weiss G., Behringer G., Miller M., Ranfft K., 1977: Vorkommen und Gehalt an Aflatoxin M_1 in Molkerei-Anlieferungsmilch. Zeitschrift für Lebensmittel-Untersuchung und – Forschung 163, 171–174 [German]

377. Margolles, E., Escobar, A., Rivero, R., 1991: Determination de aflatoxina B_1 en productos agricolas utilizando el juego de reactivos Davih – afla B_1. Revista de Salud Animal 13, 279–281 [Spanish]

378. El-Nakib, O., Pestka, J.J., Chu, F.S., 1981: Determination of aflatoxin B_1 in corn, wheat, and peanut butter by enzyme-linked immunosorbent assay and solid phase radioimmunoassay. Journal of the Association of Official Analytical Chemists 64, 1077–1082

379. Schuddeboom, L.J., 1983: Development of legislation concerning mycotoxins in dairy products in The Netherlands. Microbiologie – Aliments – Nutrition 1, 179–185

380. Jagadish Kumar, B., Veeranarayana Gowda, D.K., 1979: Some observations on aflatoxin levels in livestock feed. Indian veterinary Journal 56, 494–496

381. Shotwell, O.L., Hesseltine, C.W., Burmeister, H.R., Kwolek, W.F., Shannon, G.M., Hall, H.H., 1969: Survey of cereal grains and soybeans for the presence of aflatoxin. II. Corn and soybeans. Cereal Chemistry 46, 454–463

382. Ramirez, M.L., Pascale M., Chulze, S., Reynoso, M.M., March, G., Visconti, A., 1996: Natural occurrence of fumonisins and their correlation to *Fusarium* contamination in commercial corn hybrids growth in Argentina. Mycopathologia 135, 29–34

383. Yadagiri, B., Tulpule, P.G., 1974: Aflatoxin in buffalo milk. Indian Journal of Dairy Science 27, 293–297

384. Shotwell, O.L., Hesseltine, C.W., Burmeister, H.R., Kwolek, W.F., Shannon, G.M., Hall, H.H., 1969: Survey of cereal grains and soybeans for the presence of aflatoxin. I. Wheat grain, sorghum, and oats. Cereal Chemistry 46, 446–454

385. Shotwell, O.L., Bennett, G.A., Goulden, M.L., Shannon, G.M., Stubblefield, R.D., Hesseltine, C.W., 1980: Survey of 1977 midwest corn at harvest for aflatoxin. Cereal Foods World 25, 12–13

386. Vasanthi, S., Bhat, R.V., 1998: Mycotoxins in foods – Occurrence, health & economic significance & food control measures. Indian Journal of Medical Research 108, 212–224

387. Chu, F.S., Fan, T.S., Zhang, G.-Z., Xu, Y.-C., Faust, S., McMahon, P.L., 1987: Improved enzyme-linked immunosorbent assay for aflatoxin B_1 in agricultural commodities. Journal of the Association of Official Analytical Chemists 70, 854–857

388. Lim, H.K., Yeap, G.S., 1966: The occurrence of aflatoxin in Malayan imported oil cakes and groundnut kernels. The Malayan Agricultural Journal 45, 232–244

389. Al-Julaifi, M., Al-Falih, A.M., 2001: Detection of trichothecenes in animal feeds and foodstuffs during the years 1997 to 2000 in Saudi Arabia. Journal of Food Protection 64, 1603–1606

390. Ono, E.Y.S., Sugiura, Y., Homechin, M., Kamogae, M., Vizzoni, É., Ueno, Y., Hirooka, E., 1999: Effect of climatic conditions on natural mycoflora and fumonisins in freshly harvested corn of the State of Paraná, Brazil. Mycopathologia 147, 139–148

391. Jindal, N., Mahipal, S.K., Rottinghaus, G.E., 1999: Occurrence of fumonisin B_1 in maize and poultry feeds in Haryana, India. Mycopathologia 148, 37–40

392. Grabarkiewicz-Szcęsna, J., Kostecki, M., Goliński, P., Kiecana, I., 2001: Fusariotoxins in kernels of winter wheat cultivars field samples collected during 1993 in Poland. Nahrung/Food 45, 28–30

393. Schneweis, I., Meyer, K., Hörmansdorfer, S., Bauer, J., 2000: Mycophenolic acid in silage. Applied and Environmental Microbiology 66, 3639–3641

394. Machinski, M., Jr., Soares, L.M.V., Sawazaki, E., Bolonhezi, D., Castro, J.L., Bortello, N., 2001: Aflatoxins, ochratoxin A and zearalenone in Brazilian corn cultivars. Journal of the Science of Food and Agriculture 81, 1001–1007

395. Sharma, M., Márquez, C., 2001: Determination of aflatoxins in domestic pet foods (dog and cat) using immunoaffinity column and HPLC. Animal Feed Science and Technology 93, 109–114

396. Vázquez, B.R., Ciudad, A.G., Zabalgogeazcoa, I., Criado, B.G., 2001: Ergovaline levels in cultivars of *Festuca arundinacea*. Animal Feed Science and Technology 93, 169–176

397. El-Tahan, F.H., El-Tahan, M.H., Shebl, M.A., 2000: Occurrence of aflatoxins in cereal grains from four Egyptian governorates. Nahrung 44, 279–280

398. Silva, C.M.G., Vargas, E.A., 2001: A survey of zearalenone in corn using Romer Mycosep™ 224 column and high performance liquid chromatography. Food Additives and Contaminants 18, 39–45

399. Keblys, M., Flåøyen, A., Langseth, W., 2000: The occurrence of type A and B trichothecenes in Lithuanian cereals. Acta Agriculturæ Scandinavica, Section B, Soil and Plant Science 50, 155–160

400. Shelby, R.A., 1997: Analysis of ergot alkaloids in endophyte-infected tall fescue by liquid chromatography/electrospray ionization mass spectrometry. Journal of Agricultural and Food Chemistry 45, 4674–4679

401. Rottinghaus, G.E., Garner, G.B., Cornell, C.N., Ellis, J.L., 1991: HPLC method for quantitating ergovaline in endophyte-infested tall fescue: seasonal variation of ergovaline levels in stems with leaf sheaths, leaf blades, and seed heads. Journal of Agricultural and Food Chemistry 39, 112–115

402. Kpodo, K., Thrane, U., Hald, B., 2000: Fusaria and fumonisins in maize from Ghana and their co-occurrence with aflatoxins. International Journal of Food Microbiology 61, 147–157

403. Vargas, E.A., Preis, R.A., Castro, L., Silva, C.M.G., 2001: Co-occurrence of aflatoxins B_1, B_2, G_1, G_2, zearalenone and fumonisin B_1 in Brazilian corn. Food Additives and Contaminants 18, 981–986

404. Bakan, B., Cahagnier, B., Melcion, D., 1998: Natural occurrence of *Fusarium* toxins in domestic wheat and corn harvested in 1996 and 1997 and production of mycotoxin by *Fusarium* isolates from these samples. Revue de Médicine Véterinaire 149, 697

405. Schweitzer, S.H., Quist, C.F., Grimes, G.L., Forster, D.L., 2001: Aflatoxin levels in corn avaiable as wild turkey feed in Georgia. Journal of Wildlife Diseases 37, 657–659

406. Henke, S.E., Gallardo, V.C., Martinez, B., Bailey, R., 2001: Survey of aflatoxin concentrations in wild bird seed purchased in Texas. Journal of Wildlife Diseases 37, 831–835

407. Neal, G.E., Judah, D.J., Carthew, P., Verma, A., Latour, I., Weir, L., Coker, R.D., Nagler, M.J., Hoogenboom, L.A.P., 2001: Differences detected *in vivo* between samples of aflatoxin-contaminated peanut meal, following decontamination by two ammonia-based processes. Food Additives and Contaminants 18, 137–149

408. Visconti, A., Pascale, M., 1998: Determination of zearalenone in corn by means of immunoaffinity clean-up and high-performance liquid chromatography with fluorescence detection. Journal of Chromatography A 815, 133–140

409. Corrêra, B., Galhardo, M., Costa, E.O., Sabino, M., 1997: Distribution of molds and aflatoxins in dairy cattle feeds and raw milk. Revista de Microbiologia 28, 279–283

410. Campbell, H., Choo, T.M., Vigier, B., Underhill, L., 2000: Mycotoxins in barley and oat samples from eastern Canada. Canadian Journal of Plant Science 80, 977–980

411. Fazekas, B., Hajdu, E.T., Tahnyi, J., 2000: Natural deoxynivalenol (DON) contamination of wheat samples grown in 1998 as determined by high-performance liquid chromatography. Acta Veterinaria Hungarica 48, 151–160

412. Chaudhry, R.A., Afzal, H., Afzal, M., 1981: Incidence of aflatoxins in cottonseed cake and corn gluten feed at various storage conditions. Pakistan Veterinary Journal 1, 111–113

413. Porter, J.K., Bacon, C.W., Wray, E.M., Hagler, W.M., Jr., 1995: Fusaric acid in *Fusarium moniliforme* cultures, corn, and feeds toxic to livestock and the neurochemical effects in the brain and pineal gland of rats. Natural Toxins 3, 91–100

414. Ono, E.Y.S., Fungaro, M.H.P., Sofia, S.H., Figueira, E.L.Z., Gerage, A.C., Ichinoe, M., Sugiura, Y., Ueno, Y., Hirooka, E.Y., 2004: Trends of fumonisin contamination and animal intoxication through monitoring 1991 to 1997 corn crop in the State of Paraná, Brazil. Mycopathologia 158, 451–455

415. Hegazy, S.M., Azzam, A., Gabal, M.A., 1991: Interaction of naturally occurring aflatoxins in poultry feed and immunization against fowl cholera. Poultry Science 70, 2425–2428

416. Ngoko, Z., Marasas, W.F.O., Rheeder, J.P., Shepard, G.S., Wingfield, M.J., Cardwell, K.F., 2001: Fungal infection and mycotoxin contamination of maize in the humid forest and the western highlands of Cameroon. Phytoparasitica 29, 352–360

417. Domijan, A.-M., Peraica, M., Jurjevic, Ž., Ivić, D., Cvjetkovic, B., 2005: Fumonisin B_1, fumonisin B_2, zearalenone and ochratoxin A contamination of maize in Croatia. Food Additives and Contaminants 22, 677–680

418. Olsson, J., Börjesson, T., Lundstedt, T., Schnürer, J., 2002: Detection and quantification of ochratoxin A and deoxynivalenol in barley grains by GC-MS and electronic nose. International Journal of Food Microbiology 72, 203–214

419. Hennigen, M.R., Dick, T., 1995: Incidence and abundance of mycotoxins in maize in Rio Grande do Sul, Brazil. Food Additives and Contaminants 12, 677–681

420. Labuda, R., Parich, A., Berthiller, F., Tančinová, D., 2005: Incidence of trichothecenes and zearalenone in poultry feed mixtures from Slovakia. International Journal of Food Microbiology 105, 19–25

421. Beg, M.U., Al-Mutairi, M., Beg, K.R., Al-Mazeedi, H.M., Ali, L.N., Saeed, T., 2006: Mycotoxins in poultry feed in Kuwait. Archives of Environmental Contamination and Toxicology 50, 594–602

422. Gonzales, H.H.L., Martinez, E.J., Pacin, A.M., Resnik, S.L., Sydenham, E.W., 1999: Natural co-occurrence of fumonisins, deoxynivalenol, zearalenone and aflatoxins in field trial corn in Argentina. Food Additives and Contaminants 16, 565–569

423. Benkerroum, S., Tantaoui-Elaraki, A., 2001: Study of toxigenic moulds and mycotoxins in poultry feeds. Revue Médicine Véterinaire 152, 335–342

424. Vengušt, A., Tavčar, G., Žust, J., 1998: Contamination of feedingstuffs with some mycotoxins in Slovenia. Revue Médicine Véterinaire 149, 707

425. Baldissera, M.A., Santurio, J.M., Silva, J.B., Almeida, C.A.A., 1992: Aflatoxinas, ocratoxina A e zearalenona em grãos e rações para consumo animal no sul do Brasil. Ciência e Tecnologia dos Alimentos 12, 77–82 [Spanish]

426. Purwoko, H.M., Hald, B., Wolstrup, J., 1991: Aflatoxin content and number of fungi in poultry feedstuffs from Indonesia. Letters in Applied Microbiology 12, 212–215

427. Fischer, J.R., Jain, A.V., Shipes, D.A., Osborne, J.S., 1995: Aflatoxin contamination of corn used as bait for deer in the southeastern United States. Journal of Wildlife Diseases 31, 570–572

428. Dawlatana, M., Coker, R.D., Wild, C.P., Hassan, M.S., Blunden, G., 2002: The occurrence of mycotoxins in key commodities in Bangladesh surveillance results from 1993 to 1995. Journal of Natural Toxins 11, 379–386

429. Kichou, F., Walser, M.M., 1993: The natural occurrence of aflatoxin B_1 in Moroccan poultry feeds. Veterinary and Human Toxicology 35, 105–108

430. Labuda, R., Parich, A., Vekiru, E., Tančinova, D., 2005: Incidence of fumonisins, moniliformin and *Fusarium* species in poultry feed mixtures from Slovakia. Annales of Agricultural and Environmental Medicine 12, 81–86

431. Rumbeiha, W.K., Oehme, F.W., 1997: Fumonisin exposure to Kansans through consumption of corn-based market foods. Veterinary and Human Toxicology 39, 220–225

432. Stratton, G.W., Robinson, A.R., Smith, H.C., Kittilsen, L., Barbour, M., 1993: Levels of five mycotoxins in grains harvested in Atlantic Canada as measured by high performance liquid chromatography. Archives of Environmental Contamination and Toxicology 24, 399–409

Printed in the United States of America